Do
ORD
24 FEB 2000
DO 0244158 6
AF747227
BOURNEMOUTH UNIVERSITY LIBRARY
WITHDRAWN

# Handbook of
# Beach and Shoreface Morphodynamics

*To Julia, Renee, Kathleen, Kitty, Lorn, Susanne and Donna*
*The girls who often had to wait*
*While we played at the beach*

# Handbook of Beach and Shoreface Morphodynamics

*Edited by*

Andrew D. Short

*Coastal Studies Unit, School of Geosciences, University of Sydney, Australia*

JOHN WILEY & SONS, LTD

Chichester · New York · Weinheim · Brisbane · Singapore · Toronto

Copyright © 1999 by John Wiley & Sons Ltd,
Baffins Lane, Chichester,
West Sussex PO19 1UD, England

*National* 01243 779777
*International* ( + 44) 1243 779777
e-mail (for orders and customer service enquiries): cs-books@wiley.co.uk
Visit our Home Page on http://www.wiley.co.uk
or http://www.wiley.com

All rights reserved. No part of this publication may be reproduced, stored in a retrieval system, or transmitted, in any form or by any means, electronic, mechanical, photocopying, recording, scanning or otherwise, except under the terms of the Copyright, Designs and Patents Act 1988 or under the terms of a licence issued by the Copyright Licensing Agency, 90 Tottenham Court Road, London, UK W1P 9HE, without the permission in writing of John Wiley and Sons Ltd., Baffins Lane, Chichester, West Sussex, UK PO19 1UD.

*Other Wiley Editorial Offices*

John Wiley & Sons, Inc., 605 Third Avenue,
New York, NY 10158-0012, USA

WILEY-VCH Verlag GmbH, Pappelallee 3,
D-69469 Weinheim, Germany

Jacaranda Wiley Ltd, 33 Park Road, Milton,
Queensland 4064, Australia

John Wiley & Sons (Asia) Pte Ltd, 2 Clementi Loop #02-01,
Jin Xing Distripark, Singapore 129809

John Wiley & Sons (Canada) Ltd, 22 Worcester Road,
Rexdale, Ontario M9W 1L1, Canada

C2M
BOURNEMOUTH UNIVERSITY
LIBRARY
02441586
551.457 SHO

*Library of Congress Cataloging-in-Publication Data*

Handbook of beach and shoreface morphodynamics/edited by Andrew D. Short.
p. cm.
Includes bibliographical references and index.
ISBN 0-471-96570-7 (alk. paper)
1. Beaches. 2. Coast changes. 3. Beaches – Australia. 4. Coast changes – Australia. I. Short, Andrew D.
GB454.B3H35 1999
551.45′7 – dc21 99-15118
CIP

*British Library Cataloguing in Publication Data*

A catalogue record for this book is available from the British Library

ISBN 0 471 96570 7

Typeset in 9/11pt Times by Vision Typesetting, Manchester
Printed and bound in Great Britain by Antony Rowe Ltd, Chippenham
This book is printed on acid-free paper responsibly manufactured from sustainable forestry, in which at least two trees are planted for each one used for paper production.

# Contents

# Contributors


**Dr Troels Aagaard** — Institute of Geography, University of Copenhagen, Copenhagen DK-1350, Denmark
Chapter 4

**Dr Peter Cowell** — Coastal Studies Unit, School of Geosciences, University of Sydney, Sydney NSW 2006, Australia.
Chapter 3

**Mr David Hanslow** — Coastal Section, Department Land and Water Conservation, Newcastle NSW 2309, Australia.
Chapter 3

**Dr Patrick Hesp** — Department of Geography, Massey University, Palmerston North, New Zealand.
Chapters 6, 11, 12, 14

**Dr Michael Hughes** — School of Geosciences, University of Sydney, Sydney NSW 2006, Australia.
Chapter 5

**Dr Gerhard Masselink** — Department of Geography, University of Loughborough, Loughborough, Leics LE11 3TU, United Kingdom.
Chapters 4, 8, 9

**Justin Meleo** — Coastal Studies Unit, School of Geosciences, University of Sydney, Sydney, NSW 2006, Australia.
Chapter 3

**Associate Professor Andrew Short** — Coastal Studies Unit, School of Geosciences, University of Sydney, Sydney NSW 2006, Australia.
Chapters 1, 2, 7, 9, 10, 11, 12 13, 14

**Dr Ian Turner** — Water Research Laboratory, University of New South Wales, King St, Manly Vale, NSW 2093, Australia.
Chapters 5, 8

# Preface

Beaches are a part of the natural environment that most people know of and many make special trips to visit. They are a preferred site for much tourism, recreation and residential and vacation development. As such they attract considerable attention and pressure for development. Traditionally this attention has not been matched by knowledge, resulting in a proliferation of coastal problems. The scientific study of beach systems is therefore vital not only to the wellbeing of beaches, but also to their development and to people attracted to them.

What attracts people to beaches is a combination of phenomena. Beaches provide views and settings, seascapes and waves, surf and swash, all dynamic and ever-changing scenes. They provide the shore for walks and fishing, the dry beach for sitting and playing, and the surf zone for those who wish to venture into the area of swash, wave bores, breaking waves and currents.

The beach also provides one of the few environments where people can recreate in an egalitarian and natural way, as people freely choose their sandy or cobble patch to sit, and then begin to disrobe in full view down to bare feet, swimming costumes, g-strings or nothing. And no matter how heavily the beach is trampled during the day, the swash will overnight plane down the swash zone, with winds and storms periodically cleaning up the dry beach.

Most people who visit beaches know or care little about the dynamic processes that interact to form, maintain and continually change this narrow strip of the world's surface. Many sit and wonder at the never-ending procession of breaking waves. Some play in the surf fighting the power released by the breaking waves, while a very few master the waves riding them with their bodies, body boards, surf boards, wave skis and wind surfers, with a fearless handful venturing out to ride the biggest waves.

In addition to all the above, there are also another few who look at the beach and ask why? Why are there waves and where do they come from? Why is there beach sand, or cobbles or boulders and what is their origin? How do these interact to produce breaking waves, surf and swash, and how do these processes interact with the sediment to produce bars, troughs and channels? Why does the beach erode under some conditions and accrete during others? And what is the relation of the beach to the seabed beyond the breakers and the dunes behind?

In his travels the editor has come across many beach 'locals' who freely offer their opinion on why beaches change. Most of their explanations, while interesting, are usually ill-informed, often fanciful and usually wrong. Beaches seem to attract an unusually high proportion of people willing to venture forceful opinions on a topic they have observed, but know little about.

The authors of this book belong to the very small category of people who not only ask why, but have conducted scientific observation and experiments to provide answers based on fact, not fantasy. Their quest to understand some of these processes and the knowledge they have produced, form the basis of this book.

This book had it origin in Australian beach research undertaken by staff and graduate students of the Coastal Studies Unit (CSU) between 1976 and 1998, in the courses taught by the staff, and in the international literature on beaches. The book, while designed for an international audience, does have a definite geographical and Australian flavour, as all the authors have at one time been members of the CSU which is part of the Department of Geography (more recently the School of Geosciences) at the University of Sydney, and much of their own research has been undertaken on the Australian coast.

I am a coastal geographer and have been a member of CSU since 1977, teaching undergraduate coastal courses and supervising graduate students' research in coastal topics. All the other contributors were undergraduate and/or graduate students at CSU. Peter Cowell completed his doctorate on breaking waves under Don Wright in 1986, and has since been a staff member. Patrick Hesp took his doctorate under Don Wright on foredune morphodynamics in 1982. Michael Hughes completed his doctorate in 1989 on swash dynamics under the supervision of Peter Cowell. David Hanslow completed an MSc on shoreface dynamics under Peter Cowell in 1990. Troels Aagaard completed parts of his PhD research at CSU in 1988 while working in collaboration with Andy Short. Ian Turner undertook a PhD on the impact of groundwater on macro-tidal beaches completing in 1994; while Gerd Masselink examined the morphodynamics of macro-tidal beaches completing in 1995, both studying under the supervision of Andy Short. Justin Meleo is a postgradate student at CSU under the supervision of Peter Cowell.

Of the nine contributors, Short, Cowell, Hughes and Meleo are presently with the CSU, Hesp is at Massey University, Aagaard at the University of Copenhagen, Hanslow with the Coastal Section of the Department of Land and Water Conservation in Newcastle, Turner is at the University of New South Wales and Masselink with the University of Loughborough.

Perhaps equally importantly, five of the nine contributors grew up on the Australian coast, most live in beachfront suburbs, four are active surfers, two ex-lifeguards, while all feel at home in the surf. Their knowledge of the beach and surf comes not only from formal study and experiments, but also from countless hours of surfing, of experiencing the power and turbulence of breaking waves, of being dragged out to sea by rip currents and pounded by shorebreaks, but always admiring the strength and beauty of the beach and surf.

This book represents the combined efforts of nine people who between them know and have experienced everything a beach can offer. While this book will not answer every question about beaches, it goes a long way to presenting everything we and our international colleagues know about beaches at the end of the 20th century.

A. D. Short
Narrabeen Beach
March 1999

# Acknowledgments

This book was very much a joint effort by all the contributors from day one. Each contributor/s prepared all their own text and drafted the figures. All assisted in reviewing the chapters and some in the editing. Most of the illustrations were computer drafted by Peter Johnson of the Cartography section of the School of Geosciences, led by John Roberts, and assisted by Florian Heyder, Dirk Waldik, Marc Seigert, Regina Penzenstadler and Claudia Zipfel. In addition Katherine McLeod, of Surf Life Saving Australia, computer drafted all the figures for Chapter 13. Staff of the University of Sydney's Photography department prepared most of the photos.

Chapter 3 contributes toward the European Commission MAST-III program (contract MAS3-CT95-0002) with support from the Australian Department of Industry Science and Technology and the National Greenhouse Advisory Committee.

As this entire book was written/edited/prepared at my home office I thank my family for putting up with its intrusion into their home life, and for accompanying me on many of my coastal field expeditions.

At John Wiley the book was initiated by Amanda Hewes, nurtured by Emma Bottomley and Sally Wilkinson and finally edited and produced by Mandy Collison. Many thanks to all for their patience and assistance.

Andrew Short

# List of Notations

| | |
|---|---|
| $a$ | wave amplitude (m) |
| $a'$ | pore space factor ( – ) |
| $A$ | wave orbital amplitude (m) |
| $A$ | area of roller in vertical projection ($m^2$) |
| $A_{rip}$ | cross-sectional area of rip channel ($m^2$) |
| $A_{tide}$ | tidal amplitude |
| $b$ | bar height (m) |
| $B*$ | dimensionless bar parameter ( – ) |
| $B_0$ | dimensionless wave shape factor ( – ) |
| $c$ | sediment concentration (g/l) |
| $c_f$ | drag coefficient ( – ) |
| $c_0$ | sediment concentration at the bed ($g\,l^{-1}$, – ) |
| $C$ | wave celerity (m/s) |
| $C_e$ | aeolian sand transport coefficent |
| $C_g$ | wave group velocity ($m\,s^{-1}$) |
| $C_{uc(f)}$ | co-spectrum between $u$ and $c$ |
| $d$ | still water depth (m) |
| $d_0$ | wave orbital diameter (m) |
| $D$ | sediment grain size (m) |
| $D_{50}$ | median sediment size (m) |
| $D*$ | energy dissipation ($J\,m^{-2}\,s^{-1}$) |
| $D*$ | dimensionless particle size ( – ) |
| $e$ | efficiency factor |
| $e_b$ | bedload efficiency factor ( – ) |
| $e_s$ | suspended load efficiency factor ( – ) |
| $e_{\text{up}}$ | uprush efficency factor |
| $e_{\text{back}}$ | backwash efficency factor |
| $E$ | wave energy density ($J\,m^{-2}$) |
| $f$ | wave frequency ($s^{-1}$) |
| $f_w$ | wave friction factor (–) |
| $g$ | gravitational constant ($m\,s^{-2}$) |
| $h$ | mean water depth (m) |
| $h_{tr}$ | mean water depth below wave trough level (m) |
| $h_s$ | water depth in swash zone |
| $H$ | wave height (m) |
| $\bar{H}$ | mean wave height (m) |
| $H_{1/10}$ | mean wave height of the highest 1/10 of the waves (m) |
| $H_{1/3}$ | mean wave height of the highest 1/3 of the waves (m) |
| $H_b$ | breaker wave height |
| $H_o$ | deepwater wave height |
| $H_r$ | residual wave height at the shoreline (m) |
| $H_{rms}$ | root mean square wave height (m) |
| $H_s$ | significant wave height (m); $\approx H_{1/3}$ |
| $i$ | immersed weight sediment transport rate ($kg\,s^{-1}$) |
| $i_{up}$ | uprush transport per unit width of shoreline |
| $i_{back}$ | backwash transport per unit width of shoreline |
| $I_l$ | immersed weight longshore sediment transport rate ($kg\,s^{-1}$) |
| $J_0$ | zero-order Bessel function |
| $k$ | wave number ($m^{-1}$) |
| k | von Karman constant |
| $k_d$ | grain (bed) roughness (m) |
| $k_e$ | edge wave wave number ($m^{-1}$) |
| $k_m$ | (bed) roughness due to sediment movement (m) |
| $k_r$ | form drag (bed) roughness (m) |
| $k_s$ | hydraulic (bed) roughness (m) |
| $k_{up}$ | uprush fluid power |
| $k_y$ | longshore wave number ($m^{-1}$) |
| $k_{\text{back}}$ | backwash fluid power |
| $K$ | bar parameter ( – ) |
| $K$ | empirical constant ( – ) |
| $K$ | hydraulic conductivity |
| $Kd$ | aquifer transmissivity |
| $K'$ | threshold of aeolian sediment movement |
| $L$ | wave length (m) |
| $L_e$ | edge wave wave length (m) |
| $L_n$ | Laguerre polynomial of order $n$ |
| $l_s$ | mixing length scale (m) |
| MSR | mean spring tide range |
| $m$ | mass of fluid element |
| $n$ | edge wave node number ( – ), or the ratio of wave group velocity to wave celerity ( – ) |
| $N$ | number of observations ( – ) |
| $q$ | sediment transport rate ($m^3\,s^{-1}$) |
| $q_a$ | aeolian sand transport volume ($kg\,m^{-1}\,s^{-1}$) |
| $q_b$ | bedload sediment transport rate ($m^3\,s^{-1}$) |
| $q_s$ | suspended load sediment transport rate ($m^3\,s^{-1}$) |
| $Q$ | discharge ($m^3\,s^{-1}$) |

| | |
|---|---|
| $Q_{drift}$ | onshore discharge due to Stokes drift ($m^3 s^{-1}$) |
| $Q_l$ | longshore volumetric sediment transport rate ($m^3 s^{-1}$) |
| $Q_{roller}$ | onshore discharge due to roller ($m^3 s^{-1}$) |
| $p$ | total pressure ($N m^{-2}$) |
| $p_0$ | hydrostatic pressure ($N m^{-2}$) |
| $P$ | profile parameter ( – ) |
| $RTR$ | relative tide range |
| $R_{2\%}$ | wave runup height exceeded by 2% data |
| $s$ | relative sediment density; ≈ 2.65 for quartz particles in water ( – ) |
| $S$ | radiation stress ($J m^{-2}$) |
| $S_b$ | beachface specific yield |
| $S_{xx}$ | onshore component of shoreward-directed radiation stress ($J m^{-2}$) |
| $S_{xy}$ | onshore component of longshore-directed radiation stress ($J m^{-2}$) |
| $S_{yy}$ | longshore component of longshore-directed radiation stress ($J m^{-2}$) |
| $t$ | time (s) |
| $T$ | wave period (s) |
| $T_i$ | incident wave period |
| $T_{1/10}$ | mean wave period of the highest 1/10 of the waves (s) |
| $T_{1/3}$ | mean wave period of the highest 1/3 of the waves (s) |
| $T_e$ | edge wave period (s) |
| $T_p$ | peak wave period (s) |
| $T_z$ | zero-crossing wave period (s) |
| $TR$ | tide range |
| $T_{tide}$ | tidal period |
| $\tan\beta$ | beach gradient |
| $u$ | rate of groundwater flow |
| $u, u'$ | cross-shore (horizontal) orbital velocity ($m s^{-1}$) |
| $u*$ | shear velocity at the bed ($m s^{-1}$) |
| $u_m$ | maximum wave orbital velocity ($m s^{-1}$) |
| $u_{*,w}$ | wave-induced shear velocity at the bed ($m s^{-1}$) |
| $U, \bar{u}$ | cross-shore mean current velocity ($m s^{-1}$) |
| $U_{rip}$ | mean rip current velocity ($m s^{-1}$) |
| $Ur$ | Ursell Number ( – ) |
| $U_t$ | wind speed at height $K'$ ($m s^{-1}$) |
| $u_z$ | wind speed at height z ($m s^{-1}$) |
| $v, v'$ | longshore orbital velocity ($m s^{-1}$) |
| $V, v$ | longshore mean current velocity ($m s^{-1}$) |
| $V_{ep}$ | rate of vertical exit point fall |
| $v_{bar}$ | bar migration rate ($m s^{-1}$) |
| $w_s$ | sediment fall velocity ($m s^{-1}$) |
| $X_s$ | position relative to initial shoreline position |
| $x$ | cross-shore coordinate (m) |
| $x_{bar}$ | distance of bar crest to shoreline (m) |
| $x_{IT}$ | intertidal zone ($TR \tan\beta$) |
| $x_s$ | offshore distance of sloping part of beach before horizontal bottom (m) |
| $y$ | longshore coordinate (m) |
| $Z_s$ | maximum swash height |
| $Z_{is}$ | infragravity swash height |
| $z$ | vertical coordinate (m) |
| $z_0$ | zero intercept of vertical velocity profile (m) |
| $z_o$ | roughness length |
| $z_{oa}$ | apparent roughness parameter height where saltation occurs |
| $Z_{step}$ | step height |

*subscripts*

| | |
|---|---|
| b | breaker conditions |
| o | deep water wave conditions |

*symbols*

| | |
|---|---|
| $\alpha$ | angle of wave incidence |
| $\beta$ | beach/nearshore gradient (°) |
| $\beta_{eqm}$ | beach/nearshore equilibrium gradient (°) |
| $\Sigma$ | seepage face parameter |
| $\gamma$ | breaker index |
| $<\gamma>$ | random wave equivalent of the breaker index |
| $\delta$ | boundary layer thickness (m) |
| $\varepsilon$ | surf scaling parameter |
| $\varepsilon_s$ | eddy diffusion coefficient ($m^2 s^{-1}$) |
| $\eta$ | water surface elevation |
| $\bar{\eta}_s$ | wave set-up at the shoreline (m) |
| $\eta_r$ | ripple height (m) |
| s | wave set-up at the shoreline (m) |
| $\eta_r$ | ripple height (m) |
| $\theta$ | Shield's parameter |
| $\theta'$ | skin friction Shield's parameter |
| $\kappa$ | von Karman's constant; ≈ 0.4 |
| $\lambda$ | longshore wave length of rhythmic morphology (m) |
| $\lambda_c$ | cusp wavelength |
| $\lambda_r$ | ripple wavelength (m) |
| $\lambda_{rip}$ | rip spacing (m) |
| $\tau$ | eddy viscosity ($m^2 s^{-1}$) |
| $\xi$ | Iribarren number (surf similarity parameter) |
| $\rho$ | density of sea water ($kg m^{-3}$) |
| $\sigma$ | standard deviation |
| $\tau_{b,w}$ | bed shear stress due to waves ($N m^{-2}$) |
| $\tau_b$ | bed shear stress ($N m^{-2}$) |
| $\phi$ | angle of sediment repose (°) |
| $\chi$ | non-dimensional cross-shore distance (—) |
| $\Omega$ | dimensionless fall velocity (—) |
| $\omega$ | wave radian frequency (rad) |

# PART A

# BEACH SYSTEMS: DEFINITION AND GLOBAL PERSPECTIVE



The first two chapters set the basic approach of the book, which is both geographic and comprehensive, particularly in terms of physical beach processes. Chapter 1 introduces the beach system and the study of beaches, while Chapter 2 considers the geography of beaches through the influences that contribute to the global variation in beach systems.

# 1

# Beaches

*Andrew D. Short*

## 1.1 INTRODUCTION

Beaches occur on all sedimentary shorelines exposed to waves. They are amongst the most dynamic physical systems on the earth's surface, and occur on coasts wherever there is sufficient sediment for the waves to deposit it above sea level. On a global scale the estimate of sandy beach coast ranges from 11% (53 000 km; Inman and Nordstrom, 1971) to 34% (170 000 km; Hardisty, 1994). Beaches form essentially 100% of the coast of The Netherlands, 60% in Australia and 33% in the United States.

While beaches are totally dependent on waves and sediment, their presence is independent of most other surface processes, and consequently they occur in all latitudes, in all climates, in all tide ranges, and on all manner of coasts. They are, however, influenced and modified by other processes, particularly tide and wind, and parameters such as size and type of sediment, biota, air and water temperature, and water chemistry.

This book is about beaches, their nature and dynamics. Its aim is to provide an introduction to and understanding of beaches in most of the world's coastal environments. It is about the geography of beaches; about what controls their regional and global variation in time and space; about the dynamic processes that occur as waves and tides are transformed to both form and change beach systems; and about the different types of beaches that occur in response to changing environmental inputs and potential environmental change. It is also about the large scale longer term evolution of beach, dune and barrier systems; and about the application of this information to beach stratigraphy, beach ecology, beach stability and beach safety.

### 1.1.1 Beach systems

Beaches can be defined in a number of ways; however two ingredients are essential, namely waves and sediment. The simplest definition of a *beach* is a wave-deposited accumulation of sediment lying between modal wave base and the upper swash limit, where *wave base* is the maximum depth at which waves can transport beach material shoreward, and the *swash limit*, the landward limit of sub-aerial wave action and sediment transport. Between these two limits, waves and beaches reign supreme. While this definition includes all beaches, it does not take into account additional processes and parameters that can influence the character of a beach, nor does it adequately define the sub-systems that exist within beach systems. For the purpose of this book, the beach includes the nearshore zone of wave shoaling between modal wave base and the breaker zone, the surf zone between the breakpoint and the shoreline, and swash zone between the shoreline and limit of wave swash, together with all wave, tide and related processes acting thereon, and all surface and sub-surface morphology produced by the interaction of these processes with the sediment.

### 1.1.2 Beach dimensions – spatial

Beach dimensions can be defined in two-dimensional space using Figure 1.1, and in three dimensions with the inclusion of their longshore extent. The cross shore and longshore dimensions of beaches vary considerably. At the low energy end of the spectrum are narrow, shallow beaches (Figure 1.2a), with rip-dominated beach produced by moderate to high waves (Figure 1.2b) and wide multi-barred systems at the high energy end (Figure 1.2c). Furthermore, beaches range considerably in length. Along the south-east Australia coast the 1281

*Handbook of Beach and Shoreface Morphodynamics*. Edited by A. D. Short.
© 1999 John Wiley & Sons Ltd.

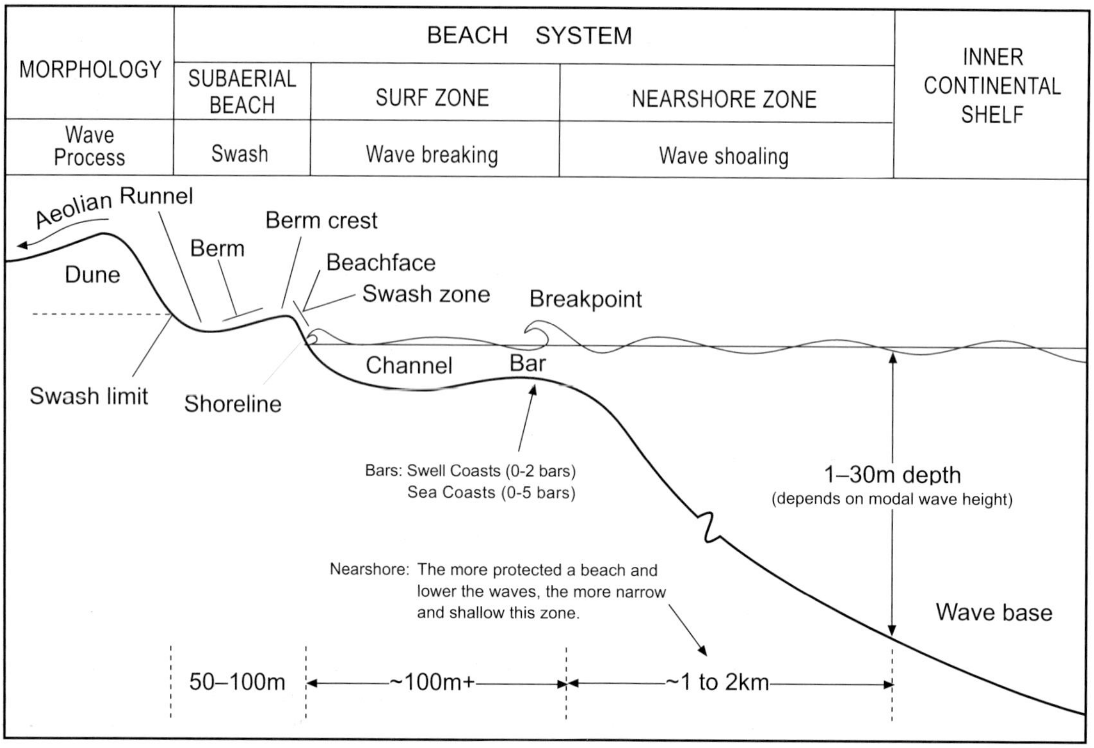


**Figure 1.1** Definition sketch of a high energy beach system including the zone of wave shoaling across the nearshore zone, wave breaking across the surf zone, and final wave dissipation in the swash zone. Low energy beaches are smaller in scale and have a small to non-existent surf zone.

beaches range in width from 0.1 to 3 km and in length from 20 m to 125 km with a mean length of 1.4 km ($s = 4.62$) (Short, 1993b, 1996).

### 1.1.3 Beach dimensions – temporal

In a temporal sense beaches range from the instantaneous to millennial. The instantaneous is represented by the beach boundary layer, the thin zone of wave–sediment interaction at and immediately above the seabed, that extends from wave base to the swash limit. The boundary layer reaches its greatest thickness in the breaker and surf zone, decreasing in thickness both shoreward to terminate in the thin swash zone, and seaward to wave base, beyond which waves do not normally interact with the seabed. The seaward extent and the thickness of the layer also vary temporally, increasing in thickness and extent with increasing wave height and period. At longer time scales, interactions within the boundary layer, coupled with sediment throughput, produce both beach change and ultimately the beach morphology and morphostratigraphy. This in turn may be subsequently eroded or preserved for millennia.

Beaches can therefore be examined at a range of time scales depending on the formative processes. These range from the instantaneous produced by prevailing processes; to tidal, diurnal, seasonal and increasingly longer periods produced by changing wave, tide, wind and climate regimes; to their long term erosion or preservation resulting from combinations of long term sediment budgets and processes including sea level, climate and tectonics. At the longer and larger scale beaches may include extensive prograded beach systems, kilometres in width and capped by beach ridges, foredunes and/or transgressive dunes. At this scale the beach becomes one part of a larger barrier system. This relationship is considered in Chapter 14.

At a time of ever-increasing specialisation and compartmentalisation of beach processes, driven by the desire to understand and predict the basic physical parameters that control boundary layer processes and sediment transport (the *instantaneous* scale) this book is primarily concerned with the *event*, *large* and even *geo-*

**Table 1.1** Characteristics of wave shoaling, surf and swash zones.

| Wave zone | Wave shoaling | Surf | Swash (wave breaking) |
|---|---|---|---|
| Beach subsystem | Nearshore zone | Surf zone | Sub-aerial beach (swash zone) |
| Outer limit | Modal wave base | Breaker zone | Wave/bore collapse |
| Inner limit | Breaker zone | Final wave collapse | Swash limit |
| Energy source | Shoaling progressive waves | Breaking waves | Broken waves/bores |
| Dominant process | • Wave orbital motions<br>• Velocity increases shoreward | Breaking waves, translatory waves, reformed waves, infragravity waves, surf zone currents | Laminar swash uprush<br>Uprush<br>Laminar to turbulent backwash |
| Surface morphology | Concave-upward slope $< 1^\circ$ | Varies: bar(s), troughs, channels | Seaward sloping 1–10° |
| Bedforms | Shoreward transition<br>• Symmetrical wave ripples<br>• Asymmetrical wave ripples<br>• Cross-ripples<br>• Lunate megaripples | Varies:<br>*Breaker zone*: lunate megaripples<br>*Bar*: plane bed<br>*Channel*: megaripples<br>*Troughs*: oscillatory and current ripples | Plane bed<br>Upper flow in lower swash zone |
| Vertical fluctuations | Decimetres to centimetres | Metres to decimetres | Metres |

*logical* scales. It is at this scale that our understanding of beach processes and ability to predict changes is perhaps best developed, and it is at these scales that the role of geographical variation in the dynamic processes that produce the broad spectrum of beaches and beach change can be most readily incorporated and comprehended.

## 1.2 TWO-DIMENSIONAL BEACHES

In two dimensions beaches consists of a number of sequential systems which can be grouped in three dynamic zones of wave shoaling, wave breaking and swash. Each zone has a characteristic set of wave processes and resulting morphological response and shape, as summarised in Table 1.1.

### 1.2.1 Wave shoaling zone

The *wave shoaling zone* lies between modal or average wave base and the breakpoint, and is also known as the nearshore zone. At its seaward limit, waves begin to interact with the seabed sufficiently to entrain sediment and move it shoreward. This limit is also called the *closure depth*, and was defined by Hallermeier (1981) as the depth below which seabed change is not detectable (Section 3.2.1). The depth of closure increases with increasing wave height and period, and on high wave energy beaches the depth may reach a maximum of about 30 m, and lie 2 to 3 km offshore (Figure 1.1). The same area is compressed in an estuary or laboratory wave tank to a depth of a few centimetres and width of a few metres. No matter what the dimensions the processes are essentially the same, namely, it is a zone over which wave orbital motions interact with the seabed to form a concave-upward slope between 0.1° and 0.8°. As waves move over this zone the decreasing depth results in increasing wave interaction with the seabed, leading to both higher velocity and longer orbital motions which transport more sediment, thereby building a steeper slope. This interaction also produces a predictable range of bedforms ranging from asymmetric wave ripple to ultimately a plane bed at the breaker zone. The morphodynamics of the shoreface, incorporating the nearshore zone, are presented in Chapter 3.

### 1.2.2 Surf zone

The *surf zone* extends from the breakpoint or area of wave breaking across the surf zone to the point where the wave or wave bore collapses on the lower beach face and becomes swash. Wave breaking usually commences when

$$EH_b = 1.5d \tag{1.1}$$

where $H_b$ is breaker height and $d$ is water depth. It can range in width from several hundreds of metres, with the outer breakpoint several metres deep, to a steep laboratory beach with the waves breaking and collapsing over a few centimetres.

a
b

**Figure 1.2** Refuge Cove (a) in southern Victoria is a low energy reflective beach a few tens of metres wide, while in western Victoria Johanna Beach (b) is a rip-dominated moderate to high energy beach, and Bridgewater Bay (c) a high energy dissipative beach, with a 300 m wide surf zone and more extensive nearshore zone (A D Short).

The surf zone is the more variable of the three zones and consists of a wide range of dynamic processes related to wave breaking and wave–current transformation. It is also strongly influenced by tide range and stage through tide control on water depth. The surf zone is the region where breaking gravity waves decrease in energy shoreward, while unbroken long (infragravity) waves increase in dominance. Both the breaking gravity waves and amplified infragravity waves contribute to the shape and character of the surf zone morphology, including the arrangement of bars, troughs and channels. The morphology must accommodate the shoreward-moving, though decaying gravity waves, the undulating infragravity waves, and the resulting currents that can move shoreward (wave bores), on–offshore (wave oscillations, infragravity currents), unidirectional longshore (rip feeder currents), bidirectional longshore (edge wave currents), and unidirectional offshore (rip currents). The dynamics of the surf zone are considered in Chapter 4.

### 1.2.3 Swash zone

The *swash zone* is the relatively narrow zone extending from the point of collapse of the wave or wave bore as it reaches the 'dry' beachface or shoreline, and the upper limit of swash action. As the wave collapses it immediately transforms from a thick wave form to a thin wave uprush, which runs up the beachface until loss of volume, by percolation into the beach surface, and gravity cause it to halt, then the remaining water runs back down the beach as backwash. This bidirectional flow is also influenced by a number of processes including wave type (wave or wave bore); wave height; tide range and stage; beach morphology, slope and width; sediment size, sorting and permeability; and water and air temperature. Swash processes are discussed in Chapter 5.

Beach systems therefore represent a dynamic interaction across a sloping seabed that causes waves to shoal, break and ultimately dissipate their energy. The shoaling, breaking and swash processes each represent a mechanism whereby wave energy is transformed from potential to kinetic energy, and in so doing does work

SCALE OF THREE DIMENSIONAL VARIATION IN BEACH MORPHODYNAMICS

| | SWASH ZONE | SHORELINE | SURF ZONE | SURF ZONE | NEARSHORE ZONE |
|---|---|---|---|---|---|
| **PROCESS** | Rhythmic swash | Swash (height & direction) | Cellular (Rip) Circulation | Breaker wave height | Wave refraction Wave attenuation |
| **MORPHOLOGICAL IMPACT** | Cusps | Rhythmic Shoreline (mega cusps) | Rhythmic Topography | Beach Type (Dissipative–Intermediate–Reflective) | Beach plan form |

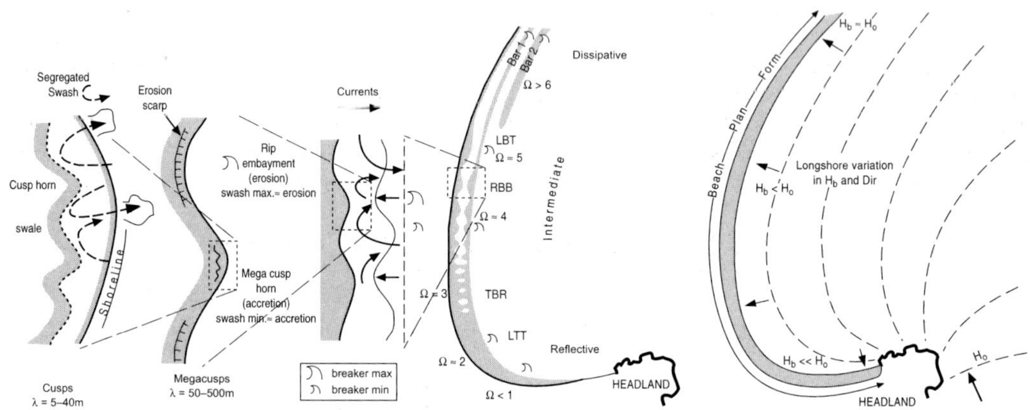


**Figure 1.3** Three-dimensional variation in beach topography begins with wave refraction and attenuation across the *nearshore zone* producing longshore variation in wave height and direction, leading to variation in beach plan form (shoreline) and beach type. Within the *surf zone*, the three-dimensional variation is related to both surf zone circulation and topography and includes cellular rip circulation and rhythmic bar and beach topography. Within the *swash zone* beach cusps are a manifestation of three-dimensional swash circulation. (Intermediate beach types are: LBT – longshore bar and trough; RBB – rhythmic bar and rip; TBR – transverse bar and rip; LTT – low tide terrace.)

on the seabed. The type of work is manifest in the boundary layer currents. In the wave shoaling zone the wave orbital motions produce a low, concave-upward gradient and lower flow regime bedforms. In the surf zone the breaking waves and surf zone currents produce a potentially undulating topography containing lower, transitional and occasionally upper flow regime bedforms; while wave uprush and backwash, an area of transitional to upper flow regimes, produce a planar through to low rippled, relatively steep surface (Table 1.1).

## 1.3 THREE-DIMENSIONAL BEACHES

The three-dimensional beach system refers to the cross- and longshore character of the beach. Cross-shore variation due to wave shoaling and breaking is discussed above. Longshore variations are driven by longshore changes in wave characteristics and related processes and can occur in the wave shoaling, surf and swash zones. During wave shoaling wave refraction and variable wave attenuation are the most common causes of longshore changes in wave direction and height. Their main impact is twofold. First, wave refraction causes the wave crest to bend which usually leads to a curvature in the adjacent shoreline (Figure 1.3). Second, the refraction and/or attenuation will produce a longshore variation in wave height, which will in turn influence the two-dimensional characteristics.

In the surf zone, cellular circulation induced by edge waves and associated undulating topography can induce further cross-shore variations in wave breaking and direction. At the shoreline any three-dimensional surf zone character will produce longshore variation in wave height and direction, which results in a reworking of the shore and the formation of rhythmic shoreline topography. Finally, even uniform waves, arriving at a straight shore, can produce three-dimensional swash circulation and the formation of high tide beach cusps.

The three-dimensional character of beaches, including the impact of bedrock topography and structures, is covered in Chapters 7, 8 and 9.

## 1.4 BEACH COMPONENTS AND BOUNDARY CONDITIONS

Beaches are dynamic sedimentary systems which require a number of boundary conditions and external inputs (Figure 1.4). While waves and sediment are essential for beach formation, a sub-surface boundary or base is required for the beach to rest on. The three-dimensional morphology of this surface or 'accommodation spacc' is critical to the formation and maintenance of the beach. During and subsequent to its formation, beaches also depend on the sediment budget, wave climate and tidal regime, together with other factors including wind regime, regional biota, water chemistry and temperature. Each of these is briefly discussed below (Table 1.2), and elaborated on in Chapter 2.

### 1.4.1 Geological boundary conditions

The geological substrate upon which a beach rests will influence the nature and size of a beach in three major ways. First, there is a very narrow range of *cross-shore gradients* upon which beaches can form. Roy *et al.* (1994) have developed a model that suggests beaches will only occur on gradients between 0.1° and 0.8°. Gentler gradients lead to excessive wave shoaling and formation of sub-aqueous sand ridges and flats, while steeper gradients induce offshore sediment transport and shoreline erosion usually leading to bedrock coasts (Figure 1.5).

Second, the overall three-dimensional morphology of the substrate determines the *accommodation space*, that is, the amount of sediment that can be accommodated

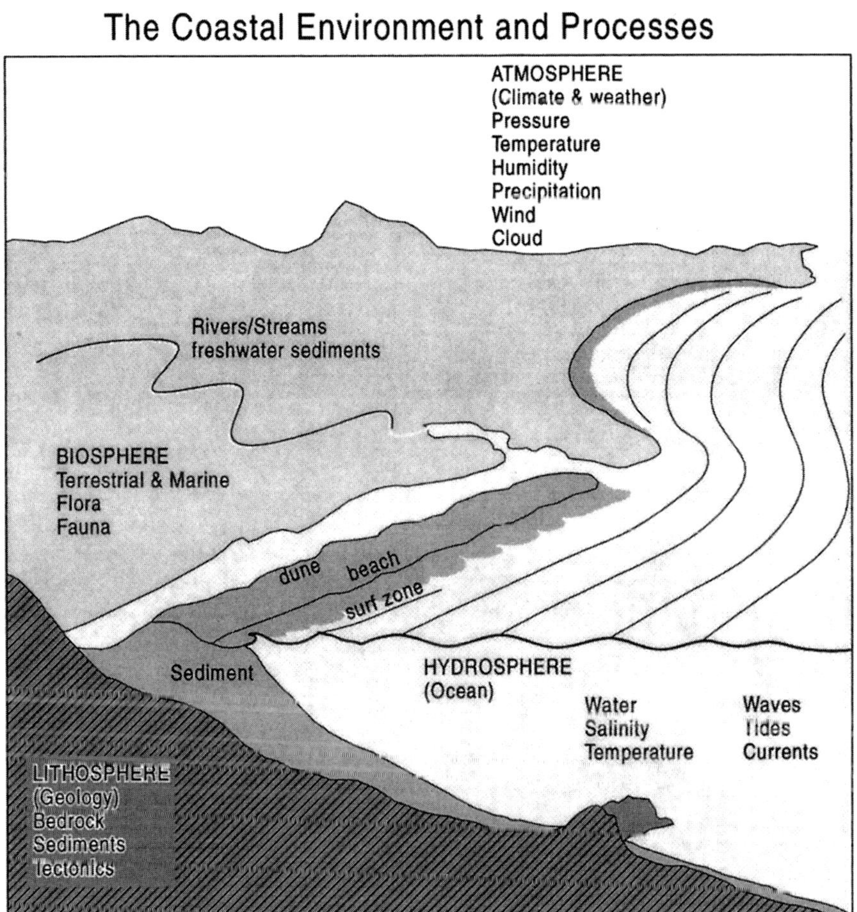


**Figure 1.4** The beach environment is the most dynamic part of the earth's surface. It contains elements of all four spheres that make up the earth, namely the atmosphere, the hydrosphere or ocean, the lithosphere or geology and the biosphere. The four spheres interact to produce a wide range of beach systems and types through the world (modified from Short, 1993b).

**Table 1.2** Beach components and impacts.

| Component | Presence | Impact |
|---|---|---|
| Geology | Cross-shore gradient | Beach gradient window |
| | Nearshore bedrock | Wave shoaling, refraction |
| | Longshore boundaries (Bedrock/headlands) | Beach dimensions, length |
| Sediment budget | Sand supply | Long term stability |
| Sediment type | Sediment size | Beach gradient/type |
| | Sediment mineralogy | Beach colour, chemistry |
| Wave type | Shoaling and breaking waves | Energy source |
| Tidal regime | Periodic tidal waves | Periodic shoreline oscillation |
| Wind regime | Wind waves | Local wave energy |
| | Aeolian processes | Beach sediment budget |
| Biota | Calcareous sediments | Increase sediment budget |
| | Calcareous structures | Modify wave shoaling, breaking |
| Water temperature | Sea ice | Retards processes |
| | Frozen beach | Retards swash percolation |

between the substrate and the beach surface for a given position of sea level and wave conditions. Together with sediment supply this will influence the long term and large scale evolution of the beach system.

Third, all beaches ultimately rest on and within a *pre-existing coastal topography* be it bedrock or older sedimentary deposits, commonly called the 'geological inheritance'. This topography will modify the nearshore zone as with reefs or hard surfaces, and thereby influence wave shoaling. This in turn will modify wave height and direction at the shore, and consequently beach shape and type. It will also influence the overall beach character through controls on beach length, orientation and curvature, while *headlands* or any shore-parallel structures, such as groynes and breakwaters, will modify the planar surf zone circulation and longshore sediment transport. These impacts are addressed in Chapter 9.

### 1.4.2 Sediment budget

A *supply of sediment* is an essential beach component. To form a beach there must be sufficient sediment to be moved by the waves over a suitable substrate to an accommodation area. The volume of sediment in and moving through a beach system can be defined by the sediment budget, which is equal to the sources of sediment, minus the losses, over a given period of time. To maintain a beach the sediment budget must be balanced or positive, as a negative budget will result in partial or complete beach erosion. Sediment budgets typically fluctuate at a range of time scales from storm–post-storm cut-and-fill, to seasonal and longer term. The spatial scale of sediment budgets can range from small pocket beaches with fixed, self-contained budgets, to massive barrier islands and spits supplied by adjacent river mouths. Therefore, while sediment budgets can be readily qualified, the quantification of the volume and rate of supply at a range of time scales are both highly variable and extremely difficult, if not impossible, to calculate.

### 1.4.3 Sediment type

Sediment type refers to sediment size, sorting and mineralogy. While beaches are commonly composed of fine to medium sand in the low to mid-latitudes, they are also commonly composed of gravel (cobble, shingle) in higher latitudes, and locally of boulders. Sediment size is a major contributor to beach slope, overall morphology and beach type (Chapters 7 and 8), as well as sediment mobility.

The mineralogy or composition of beach sands is dominantly quartz grains. However, regional and local variation in sediment type is produced by climate, source rocks, substrate and biota. Warm, humid climates produce well weathered quartz in the mid- to lower latitudes, as well as favouring the regional production of calcareous sands. In contrast, cold climates contribute coarse, poorly weathered rock fragments. Plate setting and regional geology also influence regional mineralogy, with passive margins dominated by quartz and active margins by more unstable minerals. Regional biota, together with local factors such as substrate, will determine the potential contribution of carbonate detritus to the beach sands, which can range as high as 100%.

### 1.4.4 Wave type

Wave type, together with sediment and the substrate, is the other essential requirement for beach formation. Most beaches are built by gravity waves, either sea or swell. However, beaches of a type can also be formed or modified by less periodic waves such as tsunamis, by

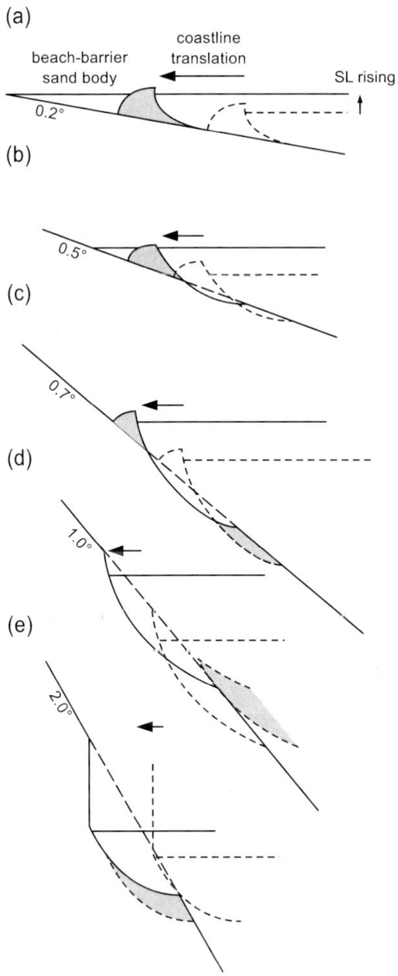


| Substrate Slope (degrees) | Barrier volume ($m^3/m$) | Sediment transfer ($m^3/m$) (+onshore) | Sediment transfer ($m^3/m$) (-offshore) | Coastline translation rate (m) |
|---|---|---|---|---|
| 0.1 | 23275 | +12120(180) | | 588(8.8) |
| 0.2 | 19788 | +5339(80) | | 258(4.2) |
| 0.4 | 12935 | +1966(29) | | 143(2.1) |
| 0.6 | 6053 | +823(12) | | 95(1.4) |
| 0.8 | 915 | +334(5) | -79(1) | 71(1.0) |
| 1.0 | 0 | | -253(4) | 57(0.85) |
| 1.5 | 0 | | -597(9) | 38(0.57) |
| 2.0 | 0 | | -804(12) | 29(0.43) |
| 4.0 | 0 | | -1144(17) | 14(0.21) |
| 6.0 | 0 | | -1277(19) | 9(0.13) |
| 10.0 | 0 | | -1372(20) | 4(0.06) |

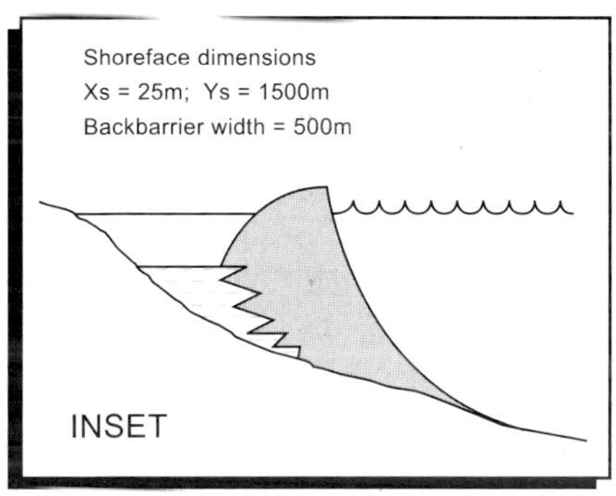


**Figure 1.5** Role of substrate slope on coastal deposition. Note how beaches and barriers will only form on slopes between 0.1 and 0.8° (source: Roy *et al.*, figure 4.6 in Carter, R W G and Woodroffe, C D (eds), *Coastal Evolution: Later Quaternary Shoreline Morphodynamics*, 1994, by permission of Cambridge University Press).

waves produced by collapsing glaciers, and in the laboratory by waves produced by mechanical means. Table 1.3 lists the major waves types and their morphological impact.

The modal (gravity) waves will determine the overall equilibrium beach gradient. Wave height influences the depth at which waves shoal and break, thereby the seaward extent of shoaling and breaker zones, and the beach type. Increasing wave period also increases the depth of wave shoaling and influences beach type. Wave energy, a function of wave period and the square of the wave height, determines the amount of potential energy available to do work across the beach and thereby the rate of beach change. Infragravity waves, which are a product of the wave climate and beach gradient, will, together with beach gradient, determine the two and three-dimensional dynamics and morphology of the swash and surf zones.

### 1.4.5 Tidal regime

Tides are not required for beach formation; however increasing tide range will, in combination with wave conditions, contribute substantially to beach morphol-

**Table 1.3** Wave influences on beaches.

| Wave type | Two-dimensional impact | Three-dimensional impact |
|---|---|---|
| Modal (gravity) waves | Equilibrium gradient | Plan shape of shoreline |
| Wave height | Beach size | Beach type |
| Wave period | Beach size | Beach type |
| Wave energy | Beach size | Beach type |
| Infragravity waves | Surf zone morphology | Beach/surf zone morphology |

ogy. As most beaches occur on open and sea coasts with a tide range, tides are a non-essential though ubiquitous component of most beaches.

Tides have three main impacts. Tides continually *shift the shoreline*, both horizontally raising and lowering it between each tidal wave crest and trough (high and low tide), and vertically, shifting the shoreline a distance depending on tide range and intertidal beach gradient. In areas of high tide range, the tidal variation in nearshore water depth can also modify breaker wave height by increasing wave shoaling at low tide. The shoreline mobility also *shifts the swash, surf and wave shoaling zones*. The impact of these shifts is to smear the potential impact of each. Masselink (1993) has shown that as tide range increases, beaches become wider, lower gradient and more featureless. Tides also contribute energy through *tidal currents* which regionally and locally can transport sediment and influence beach and barrier morphology (Davis and Hayes, 1984). The impact of tides on beach morphodynamics is discussed in Chapter 8.

### 1.4.6 Wind regime

Local winds have a number of potential impacts on beaches. In seas, lakes and enclosed water bodies the local wind is essential for *wave processes*. Wind *direction*, particularly longshore winds, will also contribute to nearshore currents and thereby sediment transport. Wind *velocity* will intensify all its impacts, and even strong onshore and offshore winds can contribute to sub-aqueous sediment transport through downwelling and upwelling currents. *Onshore winds* also contribute to aeolian sediment transport, which combined with other factors, such as sand size and coastal vegetation, can lead to loss of beach sand to coastal dunes. The role of winds in the beach backshore is presented in Chapter 6.

### 1.4.7 Biota

*Sub-aqueous biota* will contribute to the production of calcareous material (shells, algae, skeletons) and flora (seagrasses, seaweed, kelp) that can in turn be incorporated in the beach sediment as carbonate and organic detritus. The amount of skeletal carbonate material in beaches, while globally low, can over large regions reach a majority of the sediments, particularly in some tropical coral reef systems, and on temperate arid coasts rich in shelf carbonate production, such as southern Australia (Boreen *et al.*, 1993). Likewise the occurrence of seagrasses and kelps can increase dramatically in suitable environments. On sheltered open coast sections of the temperate southern Australian coast extensive seagrass beds supply dead seagrass to the sub-aerial beach. In places massive amounts of the grass can completely bury the beach to a depth of 1 to 2 m and for lengths of several kilometres. Likewise along the California coast the kelp beds increase bed roughness and substantially lower wave height in their lee, thereby affecting beach type.

In tropical waters coral and algal *reefs* can form solid shallow structures which will induce greater wave modification (shoaling, breaking, refraction) and thereby cause a decrease in breaker wave energy and lower energy beaches. The 2000 km long Great Barrier Reef off the north-east Australian coast effectively blocks all deep water waves, leading to a mainland shoreline dominated by low energy, fetch-limited seas generated in lee of the reef.

*Sub-aerial biota* will contribute to the beach indirectly through the impact of coastal dune vegetation on the presence, morphology and stability of dunes. Unstable dunes will encourage sand loss from the beach, while stable dunes will encourage beach stability or progradation, where the sediment budget is positive.

### 1.4.8 Water temperature

Seawater temperature has major impacts on beaches. In the high latitudes when seawater falls *below* $-4°C$ it freezes leading to the formation of sea slush, pancake ice and ultimately sea ice, which can thicken to a depth of 1 to 2 m. Against the shore, *shorefast ice* can freeze to the shoreline to a depth of 2 m. The presence of the ice impedes and finally ceases all wave and swash activity, as the beach and surf zone are literally frozen. Subsequent summer thawing will also lead to a predictable sequence of breakup and thaw beach features (Short, 1976).

When *air temperature falls below* $0°C$ the sub-aerial

**Table 1.4** Qualitative beach classifications.

| Parameter | Beach type | References |
|---|---|---|
| Waves height/energy | High energy | Davis and Hayes (1984) |
| | Low energy | Otvos (1982a,b) |
| Annual: wave climate | Storm wave | Davies (1964) |
| | West coast swell | |
| | East coast swell | |
| | Protected | |
| Seasonal: wave climate | Winter/storm/cut | Inman and Filloux (1960) |
| | Summer/swell/fill | Bascom (1964) |
| | | Davis and Schwartz (1982) |
| Wave type | Ocean/swell | Davies (1964) |
| | Sea | Nordstrom (1992) |
| | Estuarine | |
| Tide | Macro | Davies (1964) |
| | Meso | Hayes (1979) |
| | Micro | Davis and Hayes (1984) |
| Sediment size | Boulder | Davies (1980) |
| | Cobble/shingle | Nordstrom (1992) |
| | Sand | |
| Sediment composition | Quartz | Bird (1982) |
| | Carbonate | |
| Climate/latitude | Polar/high | Davies (1982a,b,c) |
| | Temperate/mid | |
| | Tropical/low | |

surface of the beach will freeze, impeding swash percolation and encouraging the accumulation of frozen swash on the beach. Over time, successive frozen swashes and tidal inundation build an *ice foot* on top of the intertidal beach.

At the other extreme, high seawater temperatures above about 26°C, encourage the take-up of calcium carbonate. In the intertidal zone of tropical beaches, particularly those rich in carbonate, this can lead to the precipitation of calcium carbonate in the interstitial pore spaces, leading to the cementation of the sand grains and the formation of beach calcarenite, commonly called *beachrock*. Beachrock in turn reduces swash percolation and encourages beach erosion.

The impact of the above factors on beaches is discussed more fully in Chapter 10.

## 1.5 BEACH CLASSIFICATIONS

Beaches are a product of waves and sediment, and are influenced by a range of additional parameters, some of which have been used to classify beach systems. The major qualitative beach classification systems are listed in Table 1.4.

Traditionally beaches have been classified by the major sources of energy, namely wave height, wave energy or the wave climate, the tide range, the sediment size, and impacts due to the atmospheric climate. All these classifications imply that the parameter used exercises such an overriding influence on the beaches that they can be distinguished by that parameter alone.

While such classifications are useful in explaining the geographical distribution and variation in beach types, they cannot be used to actually predict the beach type or behaviour. Such classifications are discussed in Chapter 2.

## 1.6 BEACH QUANTIFICATION

In their simplest form beaches are a function of waves and sediment, and measurable characteristics of these have been used by numerous authors to quantify beach type and behaviour. These include the dimensionless fall velocity (Gourlay, 1968); surf similarity (Battjes, 1974); surf scaling (Guza and Inman, 1975); (sand) transport direction (Sunamura, 1984a); breaker equivalent (Iribarren number); and bar number (Short and Aagaard, 1993). The various parameters used, their quantification and application to beach morphology and dynamics are discussed in Chapter 7.

## 1.7 BEACH MORPHODYNAMICS

The term morphodynamics was introduced into the coastal literature by Wright and Thom (1977). They defined it as the 'mutual adjustment of topography and fluid dynamics involving sediment transport'. On beaches this implies that the surface topography of the beach (swash, surf and shoaling wave zones) will adjust to accommodate the fluid motions produced by waves, tides and other currents, which in turn will influence the wave and tide processes. As Cowell and Thom (1994) state 'the essential properties of coastal morphodynamic processes are ... the feedback loops between topography and fluid dynamics that drive sediment transport producing morphological change'.

*Beach morphodynamics* therefore involves the mutual interaction of waves (tides, currents) with the beach topography, such that the wave processes modify the topography, which in turn will modify the waves and so on. The processes involve boundary layer interactions between wave orbital and translatory currents and the beach sediment, that produce sediment transport in the form of bed and suspended load. This in turn produces

**Table 1.5** Paradigms of beach studies.

| Paradigm | Scale (Figure 1.3) | Features | Major texts | Major international conferences (date of first conference) |
|---|---|---|---|---|
| Geological | Geological | Beach–barrier evolution | McCubbin (1982) | *International Sedimentology Congress* (1950s) |
| Geomorphological | Events<br>Geological | Beach change<br>Beach evolution | Guilcher (1958)<br>King (1972)<br>Zenkovitch (1967)<br>Komar (1976) | *Sediment Dynamics* (1977)<br>*Coastal Dynamics* (1991) |
| Engineering | Instantaneous<br>Events | Hard beach models<br>Numerical models<br>Field studies | Ippen (1966)<br>Wiegel (1964)<br>CERC (1973)<br>Silvester (1974)<br>Horikawa (1988) | *International Coastal Engineering Conference* (1951) |
| Mathematical | Instantaneous<br>Events | Numerical models<br>Field calibration | Massel (1989)<br>Hardisty (1990) | Above three |
| Large scale | Years–decades | Shoreline stability | List and Terwindt (1995) | *Large Scale Coastal Behaviour* (1993) |
| Management | Years–decades | Coastal zone management | Clark (1996) | *Coastal Zone* (1979) |

bedforms, and over time meso-scale features such as bars, troughs and beachfaces, and ultimately the equilibrium beach gradient from wave base to the swash limit. Beach morphodynamics can therefore be integrated from the instantaneous boundary layer dynamics through to Holocene barrier evolution.

The study of beaches at each of these scales is equally valid and important. However, as the aim of most studies is to understand or predict how beaches have or will change over time, there is some conflict as to the most appropriate scale at which to study beaches in order to answer this question, that is, to predict change.

Most studies of beach processes and change have been at the instantaneous and event scales. This has been for two reasons. First, the time period of most field and laboratory experiments prevents longer time scales being investigated; and secondly the mathematical parameterisation of boundary layer processes operates at instantaneous scales.

At the other end, Quaternary coastal studies include beach evolution at the geological scale. At this scale only those beaches that have been preserved can be studied; those eroded can at best be inferred if residual or remanent deposits remain, such as inner shelf sand bodies (Roy *et al.*, 1994) and clifftop dunes (Short, 1988b). These studies are necessarily selective and can at best provide a general time scale of beach evolution at the century to millennium level.

More recently the concept of 'large scale' coastal behaviour has attempted to understand beaches at perhaps the most important scale, that is the planning or engineering scale of years to decades. This scale has been bereft of past studies owing to the paucity of data and the acknowledged difficulty in predicting likely outcomes at this scale. Such prediction would need to 'involve composite evolution over many fluctuations in boundary conditions, each of which entails many cycles in the fundamental processes responsible for sediment transport' (Cowell and Thom, 1994). The status of large scale coastal beach behaviour is reviewed by List and Terwindt (1995).

## 1.8 THE STUDY OF BEACHES

### 1.8.1 Beach paradigms

The study of beaches involves both a range of possible time and length scales as discussed above, together with accompanying paradigms. These can be grouped into four schools, the geological, geomorphological, engineering and mathematical. The role of each in beach studies is illustrated in Table 1.5.

The *geological* approach involves the study of preserved beach and associated barrier deposits, which may range in age from decades to millions of years, incorporating both soft and hard rock (or lithified) beach systems. The prime concern here is to understand the nature of the deposit, and perhaps attempt to infer formative environments and processes.

The *geomorphological* approach is concerned with both more recent beach deposits and particularly contemporary beach change, at scales from events to decades. This approach involves monitoring both beach behaviour and the formative processes in order to help explain the morphological changes.

The *engineering* approach is concerned with predicting beach change at instantaneous to event scales, including the impact of structural modifications on beach behaviour. Much early work was conducted in wave tanks and scaled up to natural beaches. More recently engineers have recognised the need for *larger scale* predictive capacity, which is incorporating both field experiments as well as more consideration to the role of geomorphology/geology in beach behaviour.

The *mathematical* approach is concerned with explaining beach behaviour based on physical principles, involving rigorous mathematical parameterisation of morphodynamic interactions. Traditionally the interactions have been at the instantaneous scale, with attempts to scale up to longer term event and larger scales.

### 1.8.2 Major beach texts

The scientific study of beaches dates back to Cornaglia (1889) and is reviewed in Fisher and Dolan (1977). Cornaglia was concerned with the development of the beach and nearshore zone, and developed the concepts of wave base (the neutral line), equilibrium gradients balanced by wave and sediment parameters, graded shoreface sediments, as well as noting the impact of coastal structures on beach alignment.

The first major text was Johnson's (1919) *Shore Processes and Shoreline Development*, which includes chapters on wave and current processes, shoreline classification and shoreline development at geomorphological to geological scales. This was followed by Johnson's (1925) classic description of the New England–Arcadian coast. In England Ward's (1922) *English Coastal Evolution* provided the first regional description of the coast, including chapters on coastal deposition and beach systems. This was followed by Steers' (1964) *The Coastline of England and Wales*, and the *Coastline of Scotland* in 1973.

Following World War 2 Shepard's (1948) *Submarine Geology*, and particularly the second edition in 1963, was the first to present the new advances in the study of shallow water waves, surf zones and beaches. In France Guilcher's (1958) *Coastal and Submarine Morphology* devoted a chapter to beaches and coastal dunes including detailed beach terminology, while Williams' (1960) *Coastal Change* likewise devotes a chapter to wave transport and deposition. King's (1972) *Beaches and Coasts* was the first book devoted primarily to beaches, and covers processes, beach and coastal development at the geomorphological to geological scale, and reviews several attempts at coastal classification.

Bascom (1964) provided the first popular account of *Waves and Beaches*. In the same year the first *Shore Protection Manual* was published by the Coastal Engineering Research Center (1973). It contains a wealth of practical information on beach processes and related coastal engineering. Likewise Wiegel's (1964) *Oceanographical Engineering* is an excellent coastal engineering text. Russell (1967b) in *River Plains and Sea Coasts* presented a more personal view on many Quaternary coastal features including beach systems in a range of geographical locations. In Australia, Bird's (1968) *Coasts*, has a section on beaches, spits and barriers.

Perhaps the major publication of the 1960s was Zenkovich's (1967) *Processes of Coastal Development*, a translation of the Russian edition first published in 1962. This text provides a detailed overview of coastal processes, sediments and coastal evolution at the geomorphological to geological scale. He was particularly concerned with beach sediment transport and resulting accumulation landforms.

Since the 1970s there has been a series of books that have dealt entirely or partly with beaches. Meyer's (1972) *Waves on Beaches* is an edited book which presents a mathematical approach to wave processes and resulting sediment transport. Davies' (1972) *Geographical Variation in Coastal Development* was the first to record the geographical range in beach systems due to wave and climate impacts. Komar's (1976) *Beach Processes and Sedimentation* was the first text devoted exclusively to contemporary beach processes and geomorphology. The first encyclopedia covering *Beaches and Coastal Environments* was compiled by Schwartz (1982) and remains an excellent reference. In Japan Horikawa's (1988) *Nearshore Dynamics and Coastal Processes* compiled the impressive results of the Nearshore Environment Research Center into this book subtitled

*Theory, measurement, and predictive models* of beach and nearshore behaviour.

In 1979 Kaufman and Pilkey's *The Beaches are Moving* was the first book to deal primarily with coastal hazards, engineering and management, associated with eroding beaches. Since then Pilkey and Neal have edited the *Living with the Shore* series of similar books dealing with beaches and shoreline management in several American states, beginning with Neal *et al.* (1984) *Living with the South Carolina Shore*, and culminating with *The Corps and the Shore* (Pilkey and Dixon, 1996).

In the United Kingdom a series of coastal books dealing with beach geomorphology began with Pethick's (1984) *An Introduction to Coastal Geomorphology*, followed by the expansive *Coastal Environments* (Carter, 1988), subtitled *An introduction to the physical, ecological and cultural systems of coastlines*, and Hardisty's (1990) *Beach, Form and Process*, which deals with 'numerical experiments with monochromatic waves on the orthogonal profile', and is the first book devoted to numerical modelling of beach systems. More recently Viles and Spencer's (1995) *Coastal Problems*, addresses 'geomorphology, ecology and society at the coast'.

The increased specialisation of coastal books is highlighted in a number of regional and specialist studies, including McCann (1980) *The Coastline of Canada*; Thom (1984a) *Coastal Geomorphology in Australia*; Komar (1983) *Coastal Processes and Erosion*; McLachlan and Erasmus (1983) *Sandy Beaches as Ecosystems*; Nordstorm *et al.* (1990) *Coastal Dunes*; Brown and McLachlan's (1990) *Ecology of Sandy Shores*; Nordstrom's (1992) *Estuarine Beaches*, the first devoted to 'an introduction to the physical and human factors affecting use and management of beaches in estuaries, lagoons, bays and fjords'; Clark (1996) *Coastal Zone Management*; Bird (1996b) *Beach Management*; and Short's (1993b, 1996) *Beaches of the New South Wales* and *Victorian Coasts*. In the publications of Komar, Carter, Viles and Spencer, Nordstrom, Clark and Bird we see the increasing concern with coastal management as related to coastal and beach systems.

### 1.8.3 Beach scientific international journals

Traditionally papers dealing with beaches were located in the major journal of the discipline of the author(s). In geomorphology and geology the *Journal of Geology*, *Bulletin of the Geological Society of America*, *Journal of Sedimentary Petrology* and *Zeitschrift fur Geomorphologie* were the major international journal venues up to the 1970s. However the advent of more specialised journals, particularly *Marine Geology* in 1964, and the *Journal of Coastal Research* in 1985, have seen these become the major publishers of beach geoscience papers, with both journals publishing special issues devoted to beach topics including Greenwood and Davis (1984), Kraus and Pilkey (1988), Schwartz and Bird (1990) and Short (1993a). Papers dealing with beach geophysics have remained loyal to the *Journal of Geophysical Research*.

In the engineering field papers were traditionally published in the *Journal of the Hydraulics Division of the American Society of Civil Engineers*, and since 1977 in the journal *Coastal Engineering*. One of the earliest regional journals, *Coastal Engineering in Japan*, commenced publication in 1958.

At the same time the American Society of Civil Engineers has pioneered an increasing number of international conferences and proceedings which contain numerous beach papers. These began with the first *International Conference of Coastal Engineering* in 1951. The subsequent biannual conferences and proceedings cover a wide cross-section of engineering and geomorphological beach studies. Related specialised coastal conferences now include *Coastal Sediments* since 1977, *Coastal Zone* (Management) since 1979, and *Large Scale Coastal Behaviour* since 1993, and many regional and national coastal conferences such as the *Australasian Coastal and Ocean Engineering Conference*.

### 1.8.4 Beach studies 1940s to 1960s

The major impetus to the modern study of beaches commenced during World War 2, when beaches were increasingly used as landing zones. Because of problems encountered by landing craft and soldiers across the surf and swash zone, studies commenced in the United Kingdom (Williams, King) and the United States (Shepard, Inman) into the nature of beaches and bars.

Following the war, these studies were formalised in the USA within the Geography Program of the United States Office of Naval Research. This program between 1956 and 1975 funded beach studies in the United States, primarily at Scripps and the Louisiana-based Coastal Studies Institute (CSI), and also in the United Kingdom. Also the United States Army Corps of Engineers through its Beach Erosion Board funded coastal engineering studies, particularly in California. In the United Kingdom the Wallingford Hydraulics Research Laboratory was opened in 1951.

The beach studies undertaken in the late 1940s, 1950s and into the 1960s were directed by a range of individuals. It was a time when various techniques and ap-

**Table 1.6** Major pioneering beach studies 1940s to 1960s.

| Country | Institution/group | Major contributors | Type of beach research |
|---|---|---|---|
| USA, west | Scripps | Shepard, Bascom, Inman | Geomorphological, geophysical |
| | Beach Erosion Board | | Engineering |
| USA, east/Gulf | LSU-CSI | Russel, McIntire, Dolan | Geomorphological |
| USSR | Moscow | Zenkovitch | Geomorphological |
| UK | Various | Bagnold, King, Steers | Geophysical, geomorphological |
| | Wallingford | | Engineering |
| France | Sorbonne | Guilcher | Geomorphological |

proaches were used and tested, and when field equipment was often primitive and unreliable. Russell, commenting in 1968, stated that the field of coastal morphology was in its infancy. The major contributors to these studies are listed in Table 1.6.

### 1.8.5 Beach studies 1960s to 1990s

The turning point in the study of beaches commenced in the 1960s and took place in Japan and on both sides of the United States. In Japan, Homma and Sonu (1962) published the first study of rhythmic beach topography. In the USA, a CSI project to study beach changes was set up at Nags Head Pier, North Carolina (Figure 1.6a). Bob Dolan was the graduate student involved in taking the measurements. Sonu arrived at CSI in 1966 and was involved in the analysis of the Nags Head data, the results being published in a series of papers beginning with Sonu and Russell (1967).

Sonu's studies clearly highlighted a number of important features that had been ignored or left unrelated in previous studies. These were the Markovian behaviour of beaches (Sonu, 1969; Sonu and van Beek, 1971; Sonu and James, 1973), which has in turn led to the formalisation of beach morphodynamics; the three-dimensional nature of beaches (Sonu, 1973), now embodied in most beach studies; and integration of a range to new techniques based around field investigation of both beach processes and morphology. The last major field experiment conducted by Sonu was titled the Sea Air Land Interaction Study (SALIS, Figure 1.6b) expressing the fully integrated approach to the beach environment (Sonu *et al.*, 1973).

Sonu's work laid the foundation for a number of CSI and related field studies of beaches in Florida (Sonu *et al.*, 1973), Alaska (Short, 1975a; Figure 1.6c), Brazil (Suhayda *et al.*, 1977; Figure 1.6d), and spurred similar studies in Canada (Owens, 1977). It also highlighted the need for larger scale integrated studies. These were initiated in Japan (Figure 1.6e) with the establishment of the Nearshore Coastal Environment Centre (NERC) (Horikawa, 1988); in the USA with the Nearshore Sediment Transport Study (NSTS) (Seymour, 1989); in Canada with the Coastal Current Sediment Study ($C^2S^2$) (Willis, 1987); and finally the US Corps of Engineers' Coastal Engineering Research Center was suitably impressed by the original Nags Head studies to construct their own pier, which evolved into the Field Research Facility at Duck, North Carolina (Figure 1.6f), opened in 1982, the largest and most heavily utilised beach research station in the world (Birkemeier *et al.*, 1997), and just 25 km north of Nags Head where it all began. This book is also a derivative of Sonu's work. The editor first met Sonu and the CSI team at the 1971 Destin SALIS field experiments and went on to study under Sonu and his colleagues at CSI.

At the same time on the west coast the beachfront Shore Processes Laboratory at Scripps (Figure 1.8g), under the leadership of D L Inman, produced a continuous stream of geophysics-based wave and beach research and now well known graduates including P D Komar, A J Bowen, D L Aubrey and R T Guza. The graduates of Scripps, CSI and elsewhere have now developed an international network of beach research groups, some of which are listed in Table 1.7.

Following the lead from CSI the Australian group at the Coastal Studies Unit, University of Sydney, developed a highly mobile field capacity, choosing representative sites around the Australian (Figure 1.6h) and more recently New Zealand coast to undertake a wide range of beach experiments in low through high wave, and micro- through marco-tidal environments.

**Figure 1.6** Photos of some of the 1960s and 1970s beach field experiment sites: (a) Nags Head Pier, North Carolina; (b) SALIS experiment Destin, Florida; (c) Pingok Island, north Alaska (note the ice); (d) Aracaju, Brazil; (e) Ajigaura, Japan (T. Sasaki); (f) the old Scripps Pier, La Jolla, California; (g) Duck, North Carolina; (h) Goolwa, South Australia (all A. D. Short, except e).

**Table 1.7** Some major contemporary beach research groups.

| Country | Institution/group (year initiated) | Type of beach research | Major references | Major contributors |
|---|---|---|---|---|
| USA | Scripps (1940s) | Hydrodynamics | Shepard (1948) | Shepard, Inman and colleagues |
| USA | LSU-CSI (1950s) | Morphodynamics | Sonu *et al.* (1973) | Sonu and colleagues |
| USA | Oregon (1970s) | Hydrodynamics | Komar (1976) | Komar, Holman |
| USA | NSTS (1976) | Hydrodynamics | Seymour (1989) | Seymour |
| USA | CERC (Duck) (1982) | Hydrodynamics, engineering | Birkemeier *et al.* (1997) | Birkemeier and colleagues |
| Canada | Dalhousie<br>Toronto<br>$C^2S^2$ (1986) | Hydrodynamics | Owens (1977)<br><br>Willis (1987) | Bowen, Greenwood |
| UK | Various<br>Wallingford | Hydrodynamics<br>Morphodynamics | Carter (1988)<br>Hardisty (1994) | Hardisty, Huntley<br>Carter, Orford |
| Japan | Univ. Tokyo and others, NERC (1976) | Hydrodynamics, engineering, modelling | Horikawa (1988) | Horikawa, Sasaki, Sunamura |
| Australia | Univ. WA (1960s)<br>Univ. WA (1990s) | Engineering<br>Geomorphology | Silvester (1974) | Silvester<br>Eliot, Pattiaratchi |
| Australia | Sydney-CSU (1976) | Morphodynamics<br>Geomorphological | This volume | Wright, Short, Hesp, Thom, Nielsen, Masselink, Turner |
| The Netherlands | Utrecht/Public Works (1989) | Geomorphological<br>Hydrodynamics, engineering | | Terwindt<br>Kroon |
| New Zealand | Waikato (1995)<br>Auckland (1996) | Geomorphological<br>Morphodynamics | | Black, Osborne |

## 1.9 RATIONALE FOR BEACH STUDIES IN THE 21ST CENTURY

Russell (1967b) wrote that

> 'beaches . . . are natural objects of study to one who is interested in the morphology of sea coasts. Beach investigations rank foremost among the current research projects of the Coastal Studies Institute of Louisiana State University. We are interested in their origins and processes involved in their morphology, rather than attempts to categorise them by types. Taxonomic considerations are tempting because there is such a wide spectrum of appearances between the coarse shingle beaches of England and the silt beaches of Louisiana.'

At this time he was expressing both the lack of understanding of the origin of beaches and ensuing barriers, namely Quaternary coastal geomorphology; and the lack of knowledge of beach processes and their relation to beach morphology, namely beach morphodynamics. He did oppose the traditional taxonomic approach to coasts and beaches, at the same time alluding to the wide range of beach types. However, at the time the CSI was about to embark on coastal classification based on morphodynamic controls, which in turn led to a similar approach with beaches. It has been the morphodynamics that has helped to link the processes and morphology, and in doing so, to develop a far more useful approach to beach and coastal classification, as well as a greater understanding of the processes that drive these systems.

The scientific study of beaches is now just over a century old, with most major studies being undertaken is the past 30 years. Our knowledge of beaches has improved tremendously in the past few decades owing to:

- the increasing number of researchers;
- the increasing interdisciplinary nature of their studies;
- the increasing number of field sites in a wide range of coastal environments;
- increasingly sophisticated field equipment and ancillary hardware and software for data storage and processing;
- the improved mathematical formulation of beach processes and theoretical basis for an increasing number of beach processes;
- the increasing number of conferences and publications for cross-fertilisation of ideas.

The justification for all this effort is to improve our understanding of beach systems and in particular our ability to predict beach behaviour. This information is required for both basic and applied beach research. In the following chapters, the results of basic research are addressed in Chapters 3 to 6, which cover the shoreface, surf zone, beachface and backshore. Chapters 7 to 9 examine the impact of waves, tides and topography on beaches as well as presenting a morphodynamic classification of all beach types. The results of both basic and applied beach research are presented in Chapters 10 to 13 which address beach modification, beach ecology, beach stratification and beach safety. Finally, the longer term, large scale perspective of beaches and their associated barrier systems is addressed in Chapter 14.

While we have come a long way since Cornaglia's (1889) seminal work, and while tremendous progress has been made over the past three decades, there is still much we do not know or understand about beach systems. There is still no universal agreement on the existence and role of edge waves, on the breaking wave hypothesis, on the formation of beach cusps and three-dimensional beach topography, on the outer limit of beach processes, on the concept of equilibrium beach profiles, and the presence and role of phenomena such as shear waves. Our lack of understanding of these basic components of the surf zone make it exceedingly difficult, and essentially impossible at this stage, to scale up to larger scale, longer term predictions of beach behaviour, at a scale essential for beach and coastal management into the 21st century. We need to continue working at all four scales defined in Figure 1.3, so that ultimately we can scale up or down to make more robust predictions at all scales, and thereby really understand what is happening when waves interact with sand across the shoreface.

# 2

# Global Variation in Beach Systems

*Andrew D. Short*

## 2.1 GLOBAL VARIATION IN BEACH SYSTEMS

Beaches occur in all latitudes, in all tide ranges, in all climates, and may be composed of all grain sizes from sand through boulders. They occur wherever waves are able to move sediment to the shoreline. However, as discussed in Chapter 1, beaches, while a product of waves and sediment, are influenced by a range of other parameters. This chapter will examine the global and regional variation in beach sediments, wave climate and tide regimes and how in turn variation in these parameters is imprinted on the global and regional variation in beach morphodynamics.

## 2.2 COASTAL BOUNDARIES

The coastal boundary or the land–ocean interface refers to the overall structure, topography, geology, stability and orientation of a section of coast. These parameters will influence beaches through their impact on cross-shore gradients and topography, sediment nature and supply, embayment shape and length, orientation to marine and atmospheric processes, particularly waves and wind, and indirectly tide regimes.

At a global scale the nature of a coast can be defined relative to its position on a plate boundary. Inman and Nordstrom (1971) classified the world's coasts according to the type of plate boundary, with subsequent modifications by Davies (1980). Inherent on a plate boundary are a characteristic topography, structure, geology, stability, gradients and sediments (Table 2.1). In this section the two major types of plate boundaries, their location and characteristics are described with particular reference to their impact on beach systems.

*Convergent* coasts are predominantly mountainous (96%), and are the most dominant coastal type, occupying 40% of the world's coast. They usually have high relief mountainous shorelines, with beaches restricted to embayed areas and separated by headlands and rocky coastal sections. Beaches are characterised by relatively dark, coarse, unstable sediments in low quantity (Figure 2.1a). On high energy coasts with a range of sediment the beaches may have high energy, sandy, dissipative surf zones, backed by steep high tide cobble beaches. Because of the tectonic instability, uplifted beaches, barriers, reefs and marine terraces are commonly present.

*Passive* coasts, also called 'trailing edge' and in some areas 'marginal sea' coasts occur on the lee side of a convergent plate, usually well removed from the convergent boundary (Table 2.1 and Figure 2.1b). They consist of three types, Amero and Afro based on their location, and Neo for new passive margins. *Amero* coasts occur along the east coast of the Americas. They are well developed mature passive margins, and are characterised by wide, low coastal plains, deltas and wide continental shelves, and coral reefs in lower latitudes. They occupy 35% of the world's coast and are dominated by long sandy beach systems. *Afro* coasts are moderately well developed with moderate sediment yields, and tend to have coastal plains with narrow shelves, some hilly sections, some deltas and some reef systems. They occupy 7% of the world's coast and have alternating beaches and headlands. *Neo* coasts occur in new areas of plate spreading such as the Red Sea and tend to have hilly coasts with narrow shelves, few deltas and beaches, and occupy only 4% of the world's coast.

Beaches along low to mid-latitude passive margins are characterised by the long sandy beaches composed of fine to medium quartz sand, such as along the south-east and Gulf coasts of USA and much of Brazil. The high latitude north Alaskan passive margin has long beaches and several deltas; however beach sediment is predominantly coarse, owing to the impact of the polar climate. On Afro and Neo coasts, hills and embayments

*Handbook of Beach and Shoreface Morphodynamics*. Edited by A. D. Short.
© 1999 John Wiley & Sons Ltd

**Table 2.1** Coastal characteristics of convergent and passive margin coasts.

| | Convergent margin | Passive margin |
|---|---|---|
| Age | Young (1 to 10s millions years) | Old (100s million years) |
| Relief | Steep, mountainous | Low gradient plains |
| Landforms | High mountains and volcanoes | Coastal aggradation plains |
| | Narrow continental shelf | Wide, low continental shelf |
| | Deep sea trough | Continental slope |
| Tectonics | Active, earthquakes | Quiescent, stable |
| Weathering | Physical, mass movement | Chemical, fluvial, rivers |
| Drainage | Short steep streams | Long, meandering rivers |
| Sediments | | |
| Quantity | Low | High |
| Size | Fine to coarse | Fine |
| Sorting | Poor | Well |
| Colour | Dark | Light |
| Composition | Unstable minerals | Stable minerals |
| Coastal landforms | Rocky, few beaches | Extensive barriers and deltas |
| Wave attenuation | Low | Moderate to high |
| Tide range | Minimal amplification | Enhanced |
| Examples | West coast Americas | East coast Americas |
| | New Zealand | Southern Africa |
| | Iceland | Southern Australia |
| | Japan | North Alaska |
| | | India |

become more prevalent, interrupting the sandy beaches. All passive coasts are tectonically stable, which in areas of coastal progradation has led to the long term preservation of Quaternary shorelines, at or near present sea level (e.g. east Americas, southern Australia).

## 2.3 GLOBAL CLIMATES

Solar energy is the ultimate source of energy for many coastal processes. Ocean winds, currents and waves, as well as biological and chemical action, are all dependent on solar energy, while indirectly regional air masses through rainfall and temperature affect the denudation of the coastal hinterland and delivery of sediments to the coast. On a global scale the impact of climate is most evident in the latitudinal variation in temperature and its impact on a range of primary and secondary processes, as indicated in Table 2.2. For simplicity these impacts can be grouped by latitude into high, mid- and low latitude coasts and their corresponding polar, temperate and tropical climates.

Based on climate and location Davies (1980) classified the world's coasts into high latitude (polar), mid-latitude (temperate) and low latitude (tropical). Figure 2.2 illustrates the location of the world's major coastal types, while Table 2.2 defines the latitudinal boundaries of each and their general coastal processes and beach systems.

### 2.3.1 High latitude beaches

High latitude polar beaches are impacted by the dominance of physical weathering, the prevalence of sea ice, the resulting brief periods of summer open water, and the strong Coriolis effect. Because of the dominance of physical over chemical weathering, high latitude sediments and beaches are characterised by coarse, dark sediments composed of unstable minerals. Sediments are in relatively low abundance; however on passive margin trailing edge coasts, extensive barriers can form, as along the north Alaskan and Yukon coasts (Short, 1975a).

The most striking feature of these beaches however is the impact of sub-zero winter temperatures which freeze the beach and adjacent ocean surface, and cease all wave activity. The sea ice does however move with the tides and can build a shore linear *ice foot* against the shorefast ice. In the nearshore sea ice can ground on offshore bars and shoals building ice pressure ridges several metres high.

The overall impact of the frozen winter months is to retard coastal evolution, as normal coastal processes operate for a lower percentage of time, in some locations only a few weeks each year.

### 2.3.2 Mid-latitude beaches

Mid-latitude beaches have been the subject of most

**Figure 2.1** (a) A typical high relief convergent coast at Heceta Head, Oregon. The relief is steep, mass movement common, and the small beach composed of coarse, dark unstable sediments. (b) A low relief passive margin barrier at Little Talbot Island on the east Florida coast consisting of a series of low foredune ridges composed of fine, white stable quartz sand.

**Table 2.2** Impact of latitude/climate on high, mid- and low latitude coastal processes and beach systems.

| Latitude | High | Mid | Low |
|---|---|---|---|
| Climate type | *Polar* | *Temperate* | *Tropical* |
| Latitudinal range | 50/60–90° | 30/40–50/60° | 0–30/40° |
| Temperature characteristic | Surface < 0 °C<br>Seawater < −4 °C | Seawater > 4 °C and < 26 °C | Seawater > 26 °C |
| Climate | Cold winters, mild summers | Warm summers, cool winters | Hot summers warm winters |
| Winds | Light polar easterlies, some strong westerlies | Strong westerlies | Moderate trades, calm doldrums |
| *Waves sources* and size | *Polar easterlies*: low, short | *Westerlies*: storm waves, west coast swell, east coast swell – all moderate to high | *Doldrums*: low, *trade winds* moderate |
| Coriolis *strength* and impact | *High to maximum*<br>1–3 m set-up and set-down of sea level | *Moderate*<br>Moderate to minor set-up and set-down, upwelling and downwelling | *Low to zero*<br>Minor to negligible restricts tropical cyclones to > 10° latitude |
| Sediments | Coarse (cobbles, shingle), dark, unstable minerals | Quartz sand, temperate shelf carbonates | Fine quartz sand, coralline and algal reefs |
| Beach type | Reflective beachface, possible multi-bar | Reflective to dissipative | Reflective |
| Other climate impacts | Beach and sea/ocean frozen for part of year, retards coastal evolution | Water always open | Prolific growth of coral and algae may form reefs |
| | Permafrost in barriers | Dune calcarenite in arid regions lithifies beaches and dunes | Beachrock lithifies intertidal beach |
| Dunes | Low and poorly developed owing to: coarse sand, light winds, short season, frozen winter surface, poor vegetation cover, prone to overwash | Largest on west facing coasts in 40–50° latitude. Well developed: full range of dune types (foredunes through massive transgressive dunes) | Low to minimal owing to low waves and sediment supply and low winds. Predominantly foredunes, some dune transgression in trades |
| Dune vegetation | Stunted grasses | Grasses through temperate forests | Grasses through tropical rain forests |

beach research and are well known in the literature. They are typically composed of fine to medium quartz sand, exposed to low through high waves, moderate to strong winds, and have the full range of beach and barrier types. Some arid temperate coasts have extensive shelf carbonate sands, such as southern Africa and southern Australia, which impact long term beach evolution, and are discussed in Section 10.4.2.

### 2.3.3 Low latitude beaches

Low latitude tropical beaches tend to receive lower waves owing to the low regional winds, often protective coral reef systems and/or orientation away from mid-latitude wave sources. On exposed poleward-facing beaches, higher waves may arrive in winter from the mid-latitudes, as in the popular surfing destinations of the Hawaiian islands and southern Indonesia. The beaches are composed of well weathered quartz sands, which are diluted regionally and locally by carbonate sands. The beaches are usually lower energy and reflective, except where exposed to higher ocean waves, as in north-east Brazil (Suhayda *et al.*, 1975). Because of the higher ocean temperatures beachrock can form in a few decades, cementing the intertidal beach sands. This in turn has major secondary impacts, including accelerating beach erosion. When beaches are eroded, the beachrock remains as natural breakwaters to modify wave energy and thereby the retreating shoreline.

The major impact of latitude and climate on beaches is firstly, in the size and nature of the sediments; secondly, the level of zonal wave energy; and thirdly, the specialised physical impacts of sub-zero temperatures in the high latitudes, and the biological and chemical impacts of warm waters in the low latitudes.

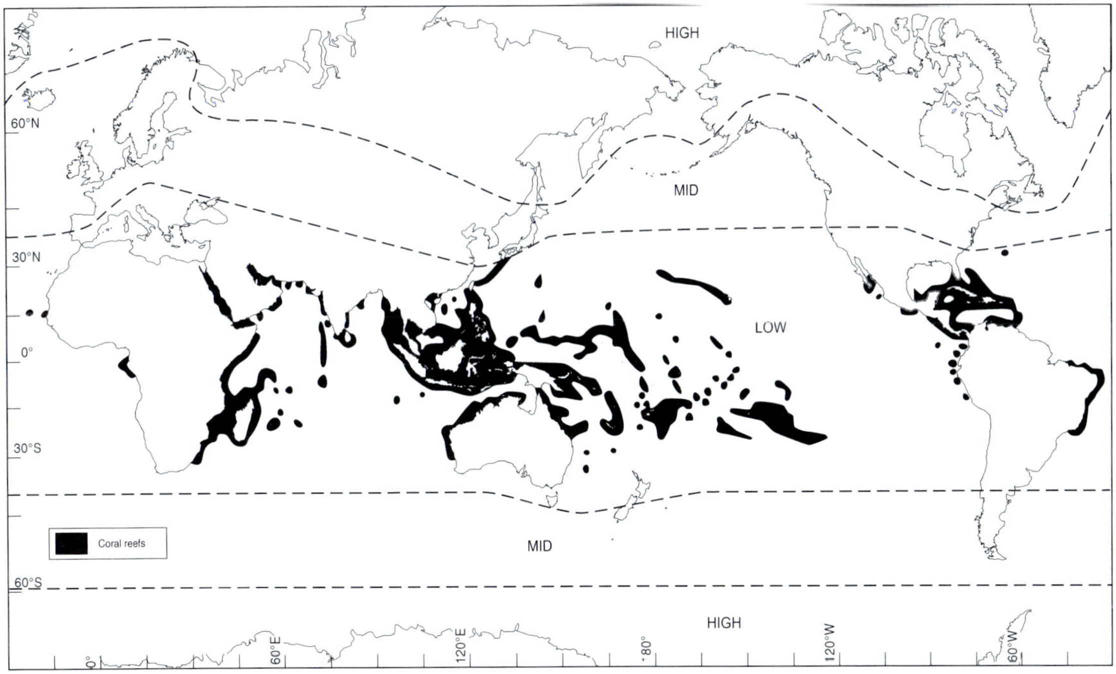


**Figure 2.2** Location of high, mid- and low latitude coasts and the extent of coastal reef systems (based on Davies, 1980).

## 2.4 GLOBAL COASTAL SEDIMENTS

The supply of terrigenous sediment to the coast is a function of the climate of the hinterland, particularly precipitation and temperature which control the rate of denudation; the geology which will determine the nature of sediments; the structure and topography which will influence the drainage pattern and location of coastal sediment sources including streams, rivers and glaciers; and the nature and extent of rocky coast exposed to wave erosion. The impact of the geology, topography and structure has been partly covered in Section 2.2 on plate boundaries. Most of the world's major rivers with associated high discharges and sediment yields are located in the humid mid- to low latitudes, while rivers in the higher mid- to high latitudes have both low discharges and sediment supply (Davies, 1980).

Based on the above, the following generalised comments can be made on the latitudinal variation in coastal sediments and their impact on beaches. In the *lower latitudes* sediments are abundant, though usually in association with major deltas. Low latitude deltaic bedload consists of well weathered fine quartz, while abundant suspended sediments lead to extensive muddy coasts, even on the open coast, as occurs for 2000 km downdrift of the Amazon between Brazil and Venezuela (Allison *et al*., 1995). There is a regionally significant input of carbonate material primarily derived from coral reefs. Tropical shores are therefore typified by extensive sedimentary deposits, exposed shores having quartz beaches, with carbonate sands adjacent to coral reefs, while more protected and low gradient deltaic shores are usually dominated by muddy tidal flats.

The *mid-latitudes* have a more regional variation in terrigenous sediment supply, with greater abundance on humid coasts, and little on arid coasts. At the coast weathered quartz sand dominates, with mud decreasing owing to fewer rivers and generally higher energy shorelines. Rock fragments increase locally adjacent to exposed eroding bedrock. On temperate arid coasts shelf carbonates tend to dominate in the absence of land supply, such as along the southern and western Australian and north-west and southern African coasts. The result is a range of shores alternating from rocky to sandy, and beach sediment systems, ranging from quartz to regional shelf carbonate and locally rock fragments (cobbles, shingles).

The *high latitudes* are dominated by physical weathering and pronounced seasonal (spring) river ac-

**Table 2.3** Sediment size and settling rates.

| Size | Grain diameter | Time to settle 1 m | Distance travelled per hour in 1 m/s current |
|---|---|---|---|
| Clay | 0.001–0.008 mm | Hours to days | 3.6 km |
| Silt | 0.008–0.063 mm | 5 min to 2 h | 3.6 km |
| Sand | 0.063–2.000 mm | 5 s to 5 min | 10s m |
| Cobble | 2 mm–6.4 cm | 1 to 5 s | $\ll$ 1 m |
| Boulder | > 6.4 cm | < 1 s | zero |

tivity. Consequently, sediment supply is low, coarse and composed of poorly weathered rock fragments. Shores are dominated by rocky coasts, with the coarse sediment (shingle, cobbles and boulders) favouring the development of steep reflective beaches.

The impact of global sediments on beaches can be summarised as follows. Coastal sediment *supply* is low in the higher latitudes favouring fewer beaches, and increases through the mid- into the lower latitudes, favouring a greater proportion of beaches. Sediment *size* decreases from cobbles and boulders in the high latitudes to fine quartz sands in the mid- to low latitudes, favouring reflective beaches in the high latitudes, and more intermediate to dissipative beaches on exposed mid- to low latitude beaches. Sediment *composition* is dominated by unweathered rock fragments in the high and glaciated coasts producing coarse, dark beaches, while the well weathered quartz and carbonate sands of the mid- to low latitudes produce lighter coloured, white and yellow beach sand.

## 2.5 BEACH SEDIMENTS

The size of beach sediment determines its contribution to beach dynamics. Unlike the case in air where all objects fall at the same speed, in water, sediment falls at a speed which is proportional to its size. Very fine sediment such as clays will simply not sink but stay in suspension for days or weeks causing turbid, muddy water. Silts, sediments coarser than clays but finer than sands, take up to 2 h to settle in a laboratory cylinder (Table 2.3). As a result these sediments will not settle in energetic wave-dominated environments. Sand, which composes most beaches, takes between 5 min for very fine sand to a few seconds for coarse sand to settle through 1 m of water.

As beaches may consist of sand to cobble and boulders, the size of the sediment is an indicator of the potential source or origin of the sediments. Mud and silts are usually precluded from beach environments owing to their low fall velocities which keep them in suspension through and beyond the energetic surf zone. Sand size sediments supplied to beaches are usually derived from river systems or reworked from the continental shelf during sea level transgression, and over time can travel considerable distances, but still have fast enough fall velocities to remain in the beach environment. Cobble beaches usually have a local source, either an adjacent bluff or a shallow nearshore source, and in some locations are adjacent to a coarse river supply. Boulder beaches only exist where there is an immediate local source, usually an eroding bluff, headland or glacier.

*Sediment size* will, in conjunction with waves, control the beach shape and dynamics. Fine sand produces a lower gradient (1 to 3°) swash zone, wider surf zone and potentially highly mobile sand. Medium to coarse sand beaches have a steeper gradient (5 to 10°), a narrower surf zone and less mobile sand. Cobble and boulder beaches are not only very steep, but have no surf zone, and have very low mobility. Therefore identical waves arriving at adjacent fine, medium and coarse sand beaches will interact to produce three distinctly different beaches.

*Sediment mineralogy* will on the other hand be a function of latitude and source. Quartz sand or silica dominates mid- to low latitude beaches, with rock fragments increasing into higher latitudes. Carbonate sediments increase into the low latitudes with regionally significant contributions on some arid temperate coasts. Heavy minerals on the other hand are purely a function of a suitable source rock and transport mechanism.

## 2.6 GLOBAL WAVE CLIMATES

Ocean waves are the major source of energy for coasts, with Inman and Brush (1973) estimating that half the energy budget of the world's coast is derived from waves. The wave characteristics of a particular location will, with sediment, determine the nature of the beach system. As waves are produced by wind blowing over a water surface, the nature of the world's waves is intimately linked to the size and shape of the ocean basins and seas, and the zonal and regional winds that blow over them. This section will briefly review wave generation, followed by an overview of the world's major wave regimes.

### 2.6.1 Wave generation

There are many forms of waves in the ocean ranging from small ripples, to wind waves, swell, infragravity waves, shelf waves, tidal waves and tsunamis. In this book the term 'waves' refers to the wind waves and swell, while other forms of waves are referred to by their full name.

#### *2.6.1.1 Wind waves or sea*

Wind waves are generated by wind blowing over the ocean. They are the waves that occur in what is called the area of wave generation, and as such they are called *sea*. Five factors determine the size of wind waves.

- *Wind velocity* – wave height will increase exponentially as velocity increases.
- *Wind duration* – the longer the wind blows with a constant velocity and direction the larger the waves will become until a fully arisen sea is reached, that is the maximum size sea for a given velocity and duration.
- *Wind direction* will determine, together with the Coriolis effect, the direction of wave travel.
- *Fetch* – the sea or ocean surface is also important, and the longer the stretch of water the wind can blow over, called the fetch, the larger the sea.
- *Water depth* is important as shallow seas will cause wave friction and possibly breaking; however this is not a problem in the deep oceans which average 4.2 km in depth.

Wave height and period are directly related to wind velocity and duration and fetch such that the *biggest seas* occur in those parts of the world where strong winds of a constant direction and long duration blow over a long stretch of ocean. The part of the globe where these factors occur most frequently is in the oceans between 40° and 60° latitude where the Roaring Forties and their westerly gales prevail. Satellite sensing of all the world's oceans has confirmed that the world's biggest waves occur in these latitudes in both hemispheres, with a winter maximum in the northern hemisphere and a year-round persistence in the southern hemisphere (Figure 2.3; Young and Holland, 1996).

#### *2.6.1.2 Swell*

Wind waves become swell when they leave or cease to be in the area of wave generation, either by travelling out of the area when the wind is blowing or when the wind stops blowing. Wind waves and swell are also called free waves and progressive waves. This means that once formed they are free of their generating mechanism, the wind, and can travel considerable distances without it. They are also progressive, as once formed they can move or progress unaided.

Once swell leaves the area of wave generation, it undergoes a transformation that permits it to travel great distances with minimum loss of energy. Whereas sea waves are highly variable in height and length, swell emanating from a sea decreases in height, increases in length, and becomes more uniform (Davies, 1980). As the speed of waves is proportional to their length, they also increase in speed. Figure 2.4 schematically illustrates the transformation of sea to swell to breaking waves.

A quick and simple way to accurately calculate the speed of waves in deep water is to measure their period, that is the time between two successive wave crests. The speed is equal to the wave period multiplied by 1.56 m. Therefore a 10 s wave travels at $10 \times 1.56$ m/s, which equals 15.6 m/s or 56 km/h. In contrast an 8 s wave travels at 36 km/h and a 12 s wave at 67 km/h.

Swell not only travels faster than seas, but also in what surfers call 'sets' or more correctly *wave groups*, that is, groups of higher and lower waves. Wave groups are also a source of long, low waves that are the length of the wave group. Wave group velocity is less than wave velocity in deep water, but equal to it in shallow water. As wave groups enter shallow water and the surf zone they become very important in contributing infragravity energy to surf zone processes.

### 2.6.2 Zonal wave climates

Wave climate refers to the average monthly and annual wave conditions at a location including height, period and direction, and the sources of those waves. The first global classification of waves was provided by Davies (1964) who used ship observations to derive four major wave climates with two sources of secondary waves. Davies' observations have been confirmed by more recent satellite imagery of ocean waves (Figure 2.3). Based on these data the following general comments can be made about the world's major deep water wave environments and their potential impact on beach systems.

The zonation of wave climates is based on their major generating sources, that is, the zonal wind systems. On a global scale these consist of the three major wind regimes: the trade winds of the lower latitudes; the wester-

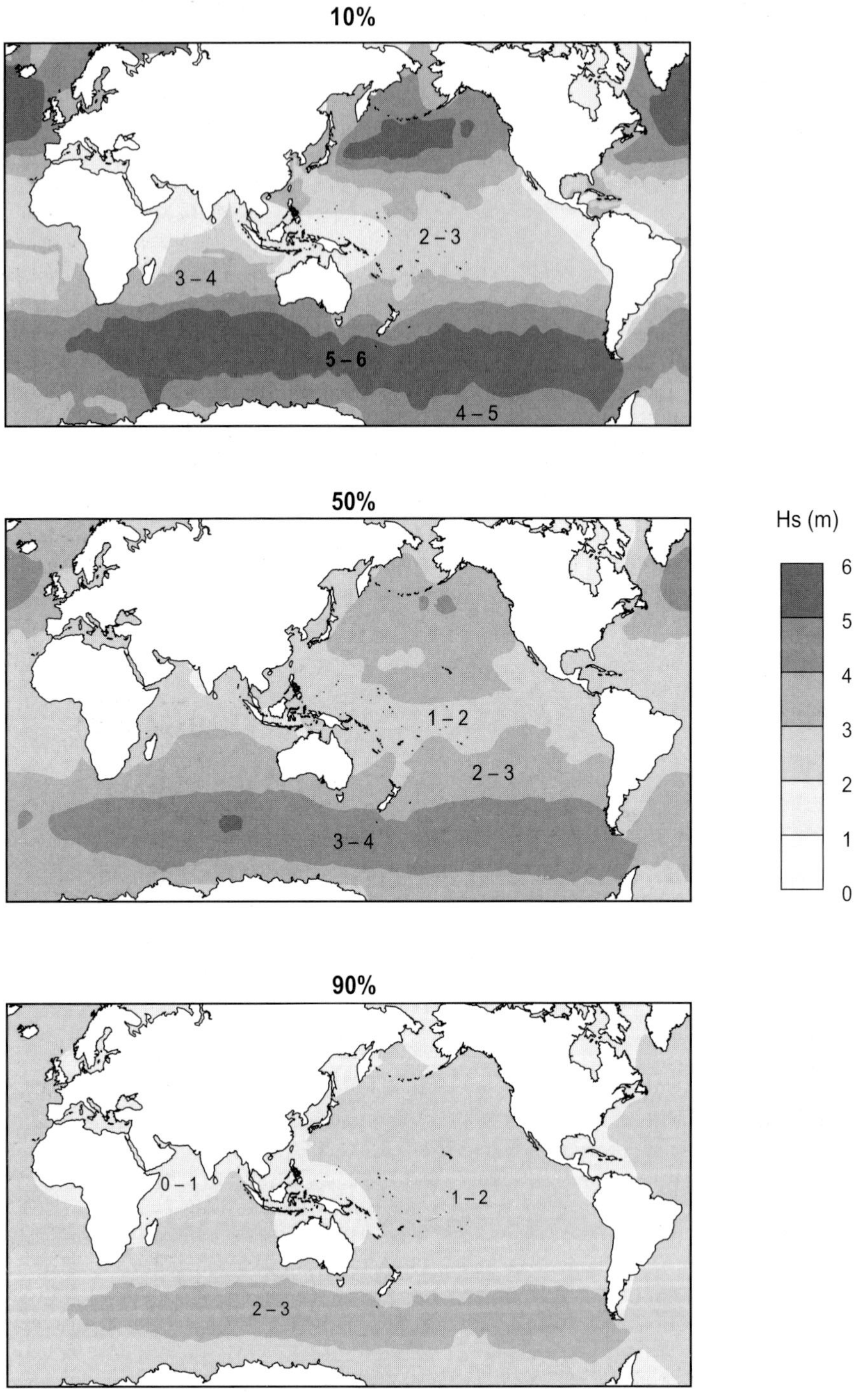


**Figure 2.3** Global values for significant wave height which will be exceeded 10% (top), 50% (middle) and 90% (lower) of the time. Based on Geosat radar altimetry (modified from Young and Holland, 1996).

**Ocean Wave Generation, Transformation and Breaking**

| Wave type | Breaking | Shoaling | Swell | Sea |
|---|---|---|---|---|
| Environment | Shallow water - surf zone | Inner continental shelf | Deep water >> 100m | Deep water >>100m<br>Long fetch = sea/ocean surface<br>Wind velocity waves<br>Wind duration waves<br>Wind direction = wave direction |
| Distance travelled | 100 m | 1 km | 100's to 1000's km | 100's to 1000's km |
| Time required | seconds | minutes | hours to days | hours to days |
| Wave profile | 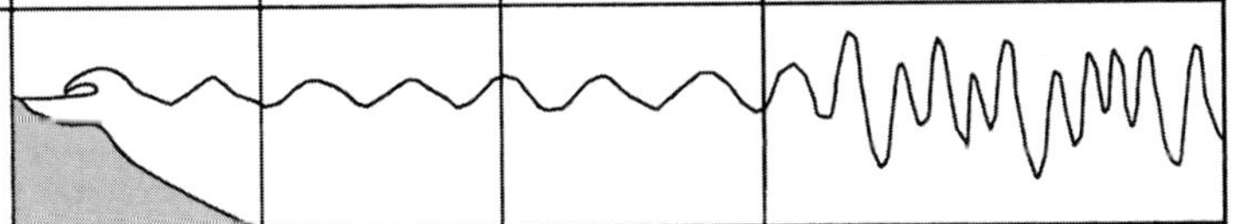 | | | |
| Water depth | 1.5 x water depth | < 100m | >> 100m | >> 100m |
| Wave character | wave breaks<br><br>wave bore<br><br>swash | higher<br>shorter<br>steeper<br>same<br>speed | regular<br>lower<br>longer<br>flatter<br>faster | variable height<br>high<br>short<br>steep<br>slow |
| Example;<br>height (m)<br>period (sec)<br>length (m)<br>speed (km/hr)<br>distance travelled (km/day) | <br>2.5 to 3m<br>12 sec<br>50 to 0m<br>15 to 0 | <br>2 to 2.5m<br>12 sec<br>220-50 m<br>66 to 15 | <br>2 m<br>12 sec<br>220 m<br>66 km/hr<br>1600 km/day | <br>3 to 5 m<br>6 to 8 sec<br>50 to 100 m<br>33 to 45 km/hr<br>800 to 1100 km/day |

**Figure 2.4** Ocean waves begin life as 'sea waves' produced by strong winds blowing over the surface of the deep ocean. When they leave the area of wave generation they transform into lower, longer, faster and more regular 'swell' which can travel for hundreds to thousands of kilometres. As waves reach shallow water they undergo a process called 'wave shoaling' which causes them to slow, shorten, steepen and finally break. This figure provides information on the characteristics of each type of wave (modified from Short, 1993).

lies of the mid-latitudes; and the polar easterlies in the high latitudes. The wave systems generated by these three major sources are presented first, followed by important regional and seasonal sources, with their characteristics listed in Table 2.4.

*2.6.2.1 Storm wave*

The world's largest waves are generated by the east to west tracking mid-latitude cyclones in the 40° to 60° N and S latitudes, the Roaring Forties. These are gener-

**Table 2.4** Characteristics of major world deep water wave environments.

| | Latitude/ location | Source | Seasonality | Deepwater height | Period | Direction |
|---|---|---|---|---|---|---|
| 1. Storm wave | 40–60° N and S/S and W facing coasts | Mid-latitude cyclones – sea (Roaring Forties) | N hem. – winter. S hem. – year round | High (2–5 m) | Long (10 to 14 s) | Westerly to south-westerly |
| 2. West coast swell | 40–0° N and S/west facing coasts | Mid-latitude cyclones – swell | N hem. – winter. S hem. – year round | Mod. – high (1.5–3 m) | Long (12–14 s) | N hem. – NW. S hem. – SW |
| 3. East coast swell | 40–0° N and S/east facing coasts | Mid-latitude cyclones – swell | N hem. – winter. S hem. – year round | Mod. (1–2 m) | Mod.–long (8–12 s) | N hem. – NE. S hem. – SE |
| 31. Trade winds + mons-oons + tropical cyclones (TC) | 25–0° N and S/exposed coasts | Trades Monsoons Tropical cyclones | Trades in winter Monsoons and tropical cyclones in summer | Moderate trades (0.5–1.5 m) Low monsoons (0.5–1 m) High TC | Trades – 6–9 s Monsoons – 4–6 s Short TC | Parallel winds NE and SE trades NW and SW monsoons Variable TC |
| 4. Protected coasts | 70–90° N and S Pole-facing coasts | Polar easterlies ( + sea ice) | Summer open water season | Low (0.5 m max) | Short (3–5 s) | NE |
| | Tropics 10–0° N and S | Doldrums | Year round | Low (< 0.5 m) | Short ( < 5 s) | Variable |

ated year round in the southern hemisphere, with a winter maximum, and during the northern winter in the northern hemisphere. Davies (1980) termed these 'storm wave' environments (Table 2.4). Geosat altimetry data show that in the storm centres, waves exceed 6 m 10% of the time, while they generally exceed 2 m 90% of the time (Figure 2.3). The waves track westerly, under the dominant westerly wind. In the southern hemisphere the cyclones have a frequency of one every 2 to 3 days, leading to a persistent series of high storm waves battering west-facing coasts. However, because of the vast southern oceans and little land, only the west coast of the South Island of New Zealand and southern Chile are exposed to this wave climate. In the northern hemisphere these waves arrive in winter battering the west-coast of North America and north-west Europe.

### 2.6.2.2 *West coast swell*

The high storm waves generated in the storm wave environment become swell on leaving the generation area. In both hemispheres, the Coriolis force deflects the waves towards the equator causing them to travel as south-west (southern hemisphere) and north-west (northern hemisphere) swell. West-facing coasts, from the mid-latitudes all the way to the equator and beyond, receive these waves as west coast swell environments. The height and period of the swell slowly decrease away from their generating area. Along the southern Australian coast the south-west swell takes up to 1 day to arrive and averages 1.5 to 2 m in height, with a slight winter maximum. Periods average 12 s year round. The same swell travelling from 40° latitude to the equator will take up to 4 to 5 days and will arrive as very regular, long crested, lower, longer swell. West coast swell arrives year round in the southern hemisphere, with a slight winter maximum, while in the northern hemisphere it is a distinctive winter wave, and usually non-existent in summer.

### 2.6.2.3 *East coast swell*

East coast swell environments face east and receive the same swell, originating from the storm wave environment, as the west coast swell. However, the swell is lower and less frequent for two reasons. Greater refraction is required to reach the east coasts, and the swell they receive originates from the lower energy, more equatorward sector of the cyclones. East coast swell environments, such as eastern USA, eastern South America and eastern Australia, tend to receive south-east (southern hemisphere) and north-east (northern hemisphere) low to moderate swell, year round in the southern hemisphere, and in winter in the northern hemisphere. Along the south-east Australian coast the swell averages 1.6 m with a period of 10 to 12 s and arrives 200 days a year (Short and Trenaman, 1992).

#### 2.6.2.4 *Trade Winds*

All east coast swell environments grade into trade wind environments below about 25° latitude. The south-east and north-east trades blow year round in their respective hemispheres, with a winter maximum in duration and velocity. The trades are the second great wave-generating area. They blow across huge expanses of oceans; however their low to moderate velocity (5 to 10 m/s) produces waves that average 1 to 1.5 m, rarely exceeding 2.5 m. Also, because the winds blow all the way to the coast, the waves remain in their generating area as sea, resulting in shorter periods, usually between 6 and 9 s.

#### 2.6.2.5 *Polar easterlies*

The polar easterlies blow all year around both poles. In Antarctica they tend to blow offshore generating little or low waves at the coast, and in the Arctic Sea they blow onshore; however in both locations they can only generate waves during the brief open water period, as for most of the year the coastal ocean is covered by sea ice. In the Arctic, the easterlies are persistent in direction and low in velocity, much like a lower velocity trade wind. These characteristics, combined with the short open water period (weeks to months), and the presence of sea ice over much of the sea year round, provide relatively small fetches between the ice and the Arctic coasts. Consequently the waves are low, usually less than 0.5 m, and short, 2 to 4 s, and only arrive during the late summer–autumn open water conditions (see e.g. Wiseman *et al.*, 1973).

#### 2.6.2.6 *Protected coasts*

A number of coasts are protected from ocean swell and rely on local winds for their wave energy. By definition these include all inland seas and lakes. On the oceanic coasts the protection can take a number of forms.

In *high latitudes* ice protects the Arctic and Antarctic coasts for much of the year (Section 10.3). In addition, the Coriolis-induced deflection of most of the mid-latitude waves away from the poles further protects the high latitude coasts, particularly Antarctica.

In the *low latitudes* the presence of coral reefs protects extensive stretches of coast, including much of the Queensland coast in lee of the Great Barrier Reef. Other coasts are protected by calcarenite reef systems, as along the tropical north-east beachrock 'reef' coast of Brazil, and the temperate calcarenite reefs of south-western Australia.

*Island arcs* and archipelagos protect many smaller seas, such as the Bering, Okhotsk, Japan, South China and Caribbean seas; while more *land-locked seas* such as the Gulf of Mexico and Mediterranean, North and Baltic seas, and inland Black, Caspian and Aral seas are all removed from ocean swell. As a consequence of the protection from swell, all protected coasts rely on local winds blowing over the available fetch to produce usually short, steep seas of short duration. Each will have a wave climate that will be a function of the size (fetch) and depth of the sea, combined with the prevailing winds. In the tropics, tropical cyclones can produce extreme wave events on usually protected coasts, such as in the Gulf of Mexico, south-east Asia and northern Australia.

### 2.6.3 Other wave sources

In addition to the zonal winds there are four other sources of waves, each with distinctive seasonal, latitudinal and/or locational characteristics.

#### 2.6.3.1 *Tropical cyclones, hurricanes, typhoons*

Tropical cyclones, called hurricanes in the Americas and typhoons in east Asia, originate around 10° N and S latitude, particularly in the warmer western tropical oceans, occurring during summer, with a late summer peak in frequency. Once formed, they move towards the west and north in the northern hemisphere, and west and south in the southern hemisphere, generally deteriorating between 30° and 40° N and S latitude. Tropical cyclones generate intense winds and rapidly build high seas, reaching several metres. The seas can emanate as swell arriving from the south-east in the northern and north-east in the southern hemisphere.

Waves generated by tropical cyclones are seasonal, infrequent and limited in area of impact. It is therefore not possible to characterise their likely annual impact for any one area. Areas receiving the brunt of such a cyclone receive high, short seas and usually accompanying storm surge, while adjacent areas receive moderate to high swell.

#### 2.6.3.2 *East coast cyclones*

East coast cyclones originate on the east coast of continents in the lower mid-latitudes. They are not associated with tropical or mid-latitude cyclones. Rather, they originate around 30° to 35° latitude, over or just off the east coasts of Australia (Holland *et al.*, 1987), South Africa and the USA where they are referred to as north-east storms (Dolan and Davis, 1992). They rapidly intensify and can reach the strength of a tropical cyclone. They can occur in all months, with a greater frequency in later summer and winter (Short and Trenaman, 1992; Dolan and Ferm, 1968). They produce short, high seas, and emanating high swell. In south-east Australia they occur on average 10 times a year and are responsible for the biggest waves, averaging 2–3 m, and cause the most severe beach erosion (Short and Trenaman, 1992), while along the eastern seaboard of the USA their frequency, intensity, associated wave heights and coastal impacts have been classified by Dolan and Davis (1992).

#### 2.6.3.3 *Monsoonal winds*

The south-west (northern hemisphere) and north-west (southern hemisphere) monsoons are seasonal onshore winds, occurring in the lower latitudes in respective summers. They are of generally low velocity and erratic in flow, and are accompanied by thunderstorms. They can, however, reach considerable strength in the western Indian Ocean, but are typically weak across south-east Asia, northern Australia and in the Americas. They generate low, short waves that arrive from the south-west or north-west during the monsoonal period (Scheffer *et al.*, 1994).

#### 2.6.3.4 *Sea breezes*

Sea breeze winds can occur on all coasts, but are more prevalent in the low to mid-latitudes, and reach their greatest strength wherever summer land–sea temperature differentials are great, and where they reinforce the zonal winds. Typically sea breezes occur on warm to hot summer days, building in strength from the late morning into the afternoon. The onshore winds produce accompanying short, steep seas. Their relative impact will depend on the competing waves. In many areas they are a small sea riding on top of dominant swell, as across southern Australia. However, where swell is absent, and the sea breeze strong, they can become the dominant waves. Along the Perth coast of Western Australia Masselink (1996) found that the sea breeze commonly reaches velocities of 10 m/s and more, and quickly generates low short wave, with a 4 s period and up to 1 m in height. The direction of sea breeze is dependent on regional coastal orientation and Coriolis deflection.

### 2.6.4 Breaker wave climate

The major zonal and regional wave climates described above refer to waves in deep water, before they interact with the seabed and transform to shallow water waves. In transforming from deep to shallow water waves, they undergo predictable changes including decreasing in velocity and length (Figure 2.4). In addition they are usually modified in height and direction, resulting in a breaker wave climate ranging from identical or slightly higher, to substantially lower than the deep water wave climate (Wright, 1976). The breaker wave climate therefore refers to the seasonal variation in wave height, period and direction at the breakpoint, rather than the deep ocean.

As waves move into shallower water and begin to interact with the seabed or feel bottom, four processes take place, affecting the wave speed, length, height, energy and ultimately the type of wave breaking.

- *Wave velocity* is depth dependent in shallow water and decreases shoreward.
- Variable wave depth and variable speed produces *wave refraction* which bends the wave crests, as that part of the wave in deeper water overtakes that part in shallower water.
- At the same time that waves are refracting and slowing they are interacting with the seabed, a process called *wave attenuation*, which causes a transfer of some wave energy to do work on the sea floor, such as moving sand, as well as decreasing the overall energy and therefore height of the wave.
- Finally, as the water becomes increasingly shallow the waves shoal which causes them to slow further, decrease in length and increase in height. The speed at which this takes place determines the type of *wave breaking*.

Whereas deep water wave climates are uniform over vast areas of the deep ocean and can be mapped as such, breaker wave climate is dependent on a range of local factors, particularly the orientation and gradient of the inner shelf and shoreface, and consequently may vary considerably over short distances. Such variation is illustrated in Figure 1.3, which shows changing wave height and direction owing to wave refraction. Compared to the deep water wave climate, the breaker wave climate will usually have lower waves, with possible changes in direction. In addition the actual waves will

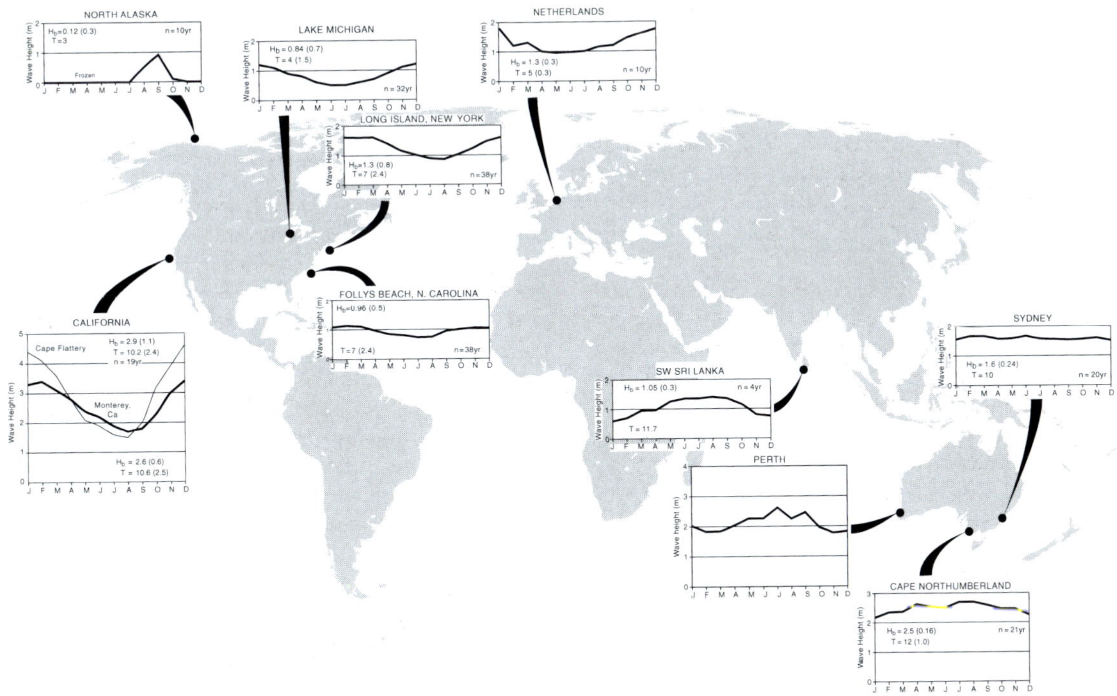


**Figure 2.5** Wave climates based on monthly mean wave height plots for 11 locations representative of west coast swell (California, Tuvalu, Sri Lanka, Perth and Cape Northumberland), east coast swell (Long Island, Follys Beach, Sydney), and three fetch limited 'protected' locations (North Alaska, Lake Michigan, The Netherlands) (sources: Wiseman *et al.*, 1973; Short, 1991b, Short and Trenaman, 1992; NOAA wave data).

be shorter, slower and steeper (Figure 2.4).

Four major factors influence the degree of deep water to breaker wave modification. Foremost are the inner shelf and nearshore gradients, which are a function of the bedrock boundary and sediment supply and size. The lower the gradient the greater the wave attenuation. Second is the orientation of the coast relative to the waves which will influence the amount of wave refraction. Third is the degree of embaymentisation, which will also affect the refraction and longshore extent of the headland influences. Finally, regional climate factors will influence the amount of biotic-induced bed roughness such as coral reefs and seagrass beds, and indirectly calcarenite reefs.

### 2.6.5 Regional wave climates

To illustrate the sources of waves and wave climates and their global and regional variation, the following sections present seven wave climates, each representative of its regional location.

#### *2.6.5.1 USA west coast (northern hemisphere west coast swell)*

The United States west coast receives north-west swell from mid-latitude cyclones located in the north Pacific. The swell arrives year round, but with a distinctive winter maximum, while in summer the coast receives lower swell or may receive no swell for extended periods. The two records (Figure 2.5) from Cape Flattery (48.6°N) and Monterey (35.8°N) both illustrate the clear winter maximum and the latitudinal decrease in wave height to the south, away from the swell source. This highly seasonal wave climate is more common in northern hemisphere locations and is the origin of the Californian seasonal wave and beach climate (e.g. Shepard, 1950).

#### *2.6.5.2 Southern hemisphere west coast swell*

In contrast to the northern hemisphere, the southern hemisphere experiences persistent mid-latitude cyclones year round, resulting in a year-round storm wave environment and more persistent year-round swell reaching west coast locations (Sri Lanka, Perth, Cape Northumberland). Note in Figure 2.5 how average height decreases away from the mid-latitude sources from 2.5 m at Cape Northumberland, to 2.1 m at Perth and 1.05 m at Sri Lanka.

#### *2.6.5.3 USA east coast (Long Island and Follys Beach)*

The United States east coast has a moderate wave climate with winter north-east swell originating from mid-latitude cyclones in the North Atlantic, and lower summer waves generated by onshore winds from high pressure cells in the mid-Atlantic. Overall wave height is moderate with mean deep water wave height ($H_0$) decreasing from 1.3 m at Long Island (41°N) to 0.96 m at Folly Beach (32.5°N) (Figure 2.5).

#### *2.6.5.4 NSW coast (east coast swell)*

The New South Wales coast in south-east Australia has one of the best documented wave climates in the world; the first Waverider buoy was established in 1971, and there is now one for every 200 km of coast. The coast receives waves from five sources which leads to a highly variable wave climate (Short and Trenaman, 1992), as illustrated by Sydney in Figure 2.5. Low to moderate south-east swell generated by mid-latitude cyclones deep in the Tasman Sea arrive year round, peaking in early winter; east coast cyclones averaging 10 a year produce the highest waves arriving as sea, then swell, from the east; tropical cyclones (three to four a year) produce moderate to high north-east swell in later summer; zonal light easterly winds produced by the dominant sub-tropical high pressure system produce low, short easterly waves; and summer north-east sea breeze winds add short, steep waves on 40% of summer days.

#### *2.6.5.5 North Sea (protected storm wave)*

The Netherlands coast of the North Sea is protected by the British Isles from most North Atlantic swell, but faces the mid-latitude cyclones that generate waves in the North Sea between Britain and the Dutch coast. Figure 2.5 illustrates the wave climate, which is typified by high seas during strong westerlies associated with the cyclones, and intervening calms. As the cyclones increase in frequency and intensity in the winter, so too does the wave height and the degree of variance. The wave climate is distinctly seasonal with a winter typified by high seas (3–5 m) every few days, with lower waves and calms in between, while in summer lower westerlies produce lower waves (1–2 m), also separated by calms. The mean wave height is 1.3 m, period 5 s and direct from the west (Short, 1991b).

#### *2.6.5.6 North Alaska (ice-protected – polar easterlies)*

The north coast of Alaska faces the Beaufort and Chukchi seas of the Arctic Sea. Figure 2.5 illustrates the wave climate on the north-facing Beaufort Sea coast, based on one year's wave data. For 8–10 months of the year the sea is frozen and there are no waves. Open water commences in June to July leading to a small fetch and low wave generated by the light polar easterlies. The peak in waves is due to strong south-west winds associated with the northerly penetration of a mid-latitude cyclone. Such events may only occur every few years. The average wave height for the open water season (3–4 months) was 0.3 m, with a period of 3 s, and predominantly north-east direction (Wiseman *et al.*, 1973).

#### *2.6.5.7 Great Lakes (protected – inland sea)*

The Great Lakes and all inland seas and lakes are dependent on local winds blowing over the fetch-limited lakes, with maximum wave height often also depth-limited. The example from the eastern shore of Lake Michigan (Figure 2.5) illustrates the overall low wave height ($H_0 = 0.84$ m) and the distinct seasonality, with higher waves during more severe winter storms associated with the passage of westerly cold fronts.

Other regional wave climates are provided for coasts of the United Kingdom (Draper, 1980), Brazil (Homsi, 1978), south-eastern Mediterranean (Goldsmith and Sofer, 1983), Nile delta (Naffa, 1995), Sri Lanka (Scheffer *et al.*, 1994) and Perth, Western Australia (Lemm, 1996).

## 2.7 GLOBAL AND COASTAL TIDE REGIMES

Tides are a non-essential beach component. However, as most beaches are affected by tides, it is important to understand the contribution of tides, particularly tide range, to beach morphodynamics. The contribution can range from nil in tideless seas and lakes, to a dominating factor in macro-tidal environments. While tidal range will always play a secondary role to waves in beach morphodynamics, increasing tide range will increasingly spread the impact of shoaling, breaking and swash wave activity. This section will briefly present the factors that control the absolute tide range and its global distribution, while Chapter 8 will examine the relative impact of wave and tide processes on beach morphodynamics. For a thorough description of tides and tidal processes see Pugh (1987).

*Tides* are the periodic rise and fall in the ocean surface due to the gravitational force of the moon and the sun, acting on a rotating earth. The amount of force is a function of the size of each and their distance from the earth. While the sun is much larger than the moon, the moon exerts 2.16 times the force of the sun because it is much closer. Therefore approximately two-thirds of our tidal force is due to the moon, and is called the lunar tides, and one-third is due to the sun and is called the solar tides.

Because the rotation and orbit of the earth and the orbit of the moon and sun are all rigidly fixed, the lunar tidal period, or time between successive high or low tides, is an exact 12.42 h, while the solar period is 24.07 h. Because these periods are out of phase it results in times when they are in phase, and their combined forces act together to produce higher than average tides, called *spring tides*, and 14 days later they are out of phase and counteract each other to produce lower than average tides, called *neap tides*. The whole cycle therefore takes 28 days and is called the lunar cycle over a lunar month.

The actual tide is in fact a wave, correctly called a *tidal wave*, and not to be confused with tsunamis. Tidal waves consist of a crest and trough, but are hundreds of kilometres in length. When the crest arrives it is called high tide, and the trough low tide. Ideally the tidal waves would like to travel round the globe. However the varying size, shape and depth of the oceans, plus islands, continents, continental shelves and small seas complicate matters. The result is that the tide breaks down into a series of smaller tidal waves that rotate around an area of zero tide called an amphidromic point.

*Tide range* in the deep ocean is zero at the amphidromic point and averages less than 20 cm over much of the ocean. However, tidal waves, because of their great length, always behave as shallow water waves, and as they move towards the coast and cross the relatively shallow continental shelves (< 150 m deep) they are amplified due to wave shoaling processes, and increase in height up to 1–3 m. In addition, a number of factors will further amplify the tides, including tidal resonance in certain large embayments, crossing of tidal waves, and coincidence of tidal waves. As a result, tides can reach heights of several metres, one to two orders of magnitude of their deep water height. Because of the shelf, and nearshore amplification of tides, they can increase dramatically in range and energy close to the shore and thereby become a dominant coastal process and a major contributor to beach morphodynamics.

The role of tides in coastal environments was first addressed by Davies (1964) who classified the world's tidal environments into micro (< 2 m), meso (2–4 m) and macro (> 4 m) tide range. In relation to beaches Davies observed that *micro-tidal* coasts behave like a tideless shore, while *macro-tidal* coasts are subject to strong tidal currents and a wide dispersal of wave attack.

## 2.8 SUMMARY

While beach systems can be generated in the laboratory devoid of natural input, natural beaches will to varying degrees be influenced by a range of parameters that will imprint a regional and/or latitudinal character. The major controls are exerted by regional tectonics and by latitude.

Tectonics, through plate boundaries, influences the size, mineralogy and quantity of sediment delivered to the coast, as well as overall coastal gradient and stability (Table 2.1). Latitude, through its influence on both land and ocean climate, influences the delivery of terrigenous sediment to the coast, its size, mineralogy and quantity, and the ocean wave climate (Table 2.2). As waves and sediments are the two key components of all beach systems, the imprint of tectonics and latitude is often dominating, and as such induces predictable regional variations in beach type and character.

# PART B

# BEACH MORPHODYNAMICS



In Part B the beach is segmented into its four dynamic sub-systems across which waves expend their energy in the process of shoaling and breaking, and in doing so transfer energy to the seabed to do work in shaping the beach system. This coupling of the dynamics with the beach morphology is at the core of the morphodynamic approach of the entire book. This section begins with the shoreface in Chapter 3 where wave shoaling dominates, then moves in Chapter 4 to the most energetic system, the surf zone, dominated by breaking waves and wave dissipation, followed by the swash zone in Chapter 5 where waves terminate on the beachface, and finally the episodically wave and wind impacted beach backshore in Chapter 6.

# 3

# The Shoreface

*Peter J. Cowell, David J. Hanslow and Justin F. Meleo*

## 3.1 INTRODUCTION: DEFINITION AND SIGNIFICANCE

The shoreface is a term coined by Barrell (1912) and used by Johnson (1919) to define the nearshore seabed extending from the low-tide shoreline offshore to a supposed break in slope, seaward of which the gradient was thought to be distinctly less steep (Figure 3.1). However, Johnson's definition is much narrower than that in use today. For example, Niedoroda *et al.* (1985a) and Niedoroda and Swift (1991) define the shoreface rather loosely as the transition zone between the surf zone and the inner continental shelf, whereas Wright (1995) regards the term as being synonymous with the inner continental shelf. Despite the semantic indifference, these works together provide a comprehensive overview on the shoreface. This chapter makes no attempt to duplicate the details contained in these previous reviews, especially in relation to hydrodynamic and sediment transport processes, which these reviews cover extensively. Rather, this chapter focuses on shoreface morphology and its behaviour on time scales ranging from days through to millennia. A systematic approach to this relationship is lacking in the literature, which has led to ambiguity in the science of shorefaces, especially in its application to coastal management.

Coastal management generally requires understanding and prediction of shoreface change over time scales of decades or longer, i.e. a class of change now formally termed *large-scale coastal behaviour* (LSCB) (Terwindt and Battjes, 1991; De Vriend, 1991, 1992a,b; Cowell and Thom, 1994). This time scale is much longer than those for which knowledge about the shoreface is most advanced, namely *instantaneous* and *event* scales (e.g. Cowell and Thom, 1994). These short scales apply from seconds to hours, and days to seasons respectively. Studies of shoreface processes on *instantaneous* and *event* scales predominantly comprise theory and measurement of the dynamics (i.e. hydrodynamics and sediment transport). Although these studies provide the most detailed insights into the shoreface (e.g. see Wright, 1995), knowledge here is still far from definitive. The uncertainty derives both stochastically and dynamically (Phillips, 1992) from the vast array of process combinations that are possible on the shoreface in nature. The stochastic uncertainty relates to the joint occurrence probabilities for each suite of flows with respect to their magnitudes, frequencies and directions. The dynamic uncertainty involves the consequences of non-linear interactions between the flows. Since combined wave–current boundary layers are ubiquitous on the shoreface, dynamic uncertainty compromises accuracy in sediment transport prediction (Nielsen, 1992). Stochastic and dynamic uncertainty underlie the magnitude–frequency problem, well recognised in geomorphology as the difficulty in determining the relative contribution toward overall geomorphic change attributable to each successive event (Schumm and Lichty, 1965; Wolman and Gerson, 1978). The magnitude–frequency problem is doubtless the main factor confounding attempts to scale up principles relevant to instantaneous processes for prediction and explanation of large-scale behaviour of the shoreface.

Until now, the most common approach used to model shoreface change at large scale (decades and beyond) involves concepts of shoreface translation in response to sea-level rise (Bruun 1962, 1988) and/or sediment supply or loss (e.g. Curray, 1964; Hands, 1983; Everts, 1987; Niedoroda *et al.*, 1985a; Swift *et al.*, 1991; Dubois, 1995, 1997). The main problems with this approach relate to assumptions concerning the shoreface geometry and definition of morphological scales associated with different classes of shoreface change. The approach usually assumes an invariant equilibrium

*Handbook of Beach and Shoreface Morphodynamics*. Edited by A. D. Short.
© 1999 John Wiley & Sons Ltd.

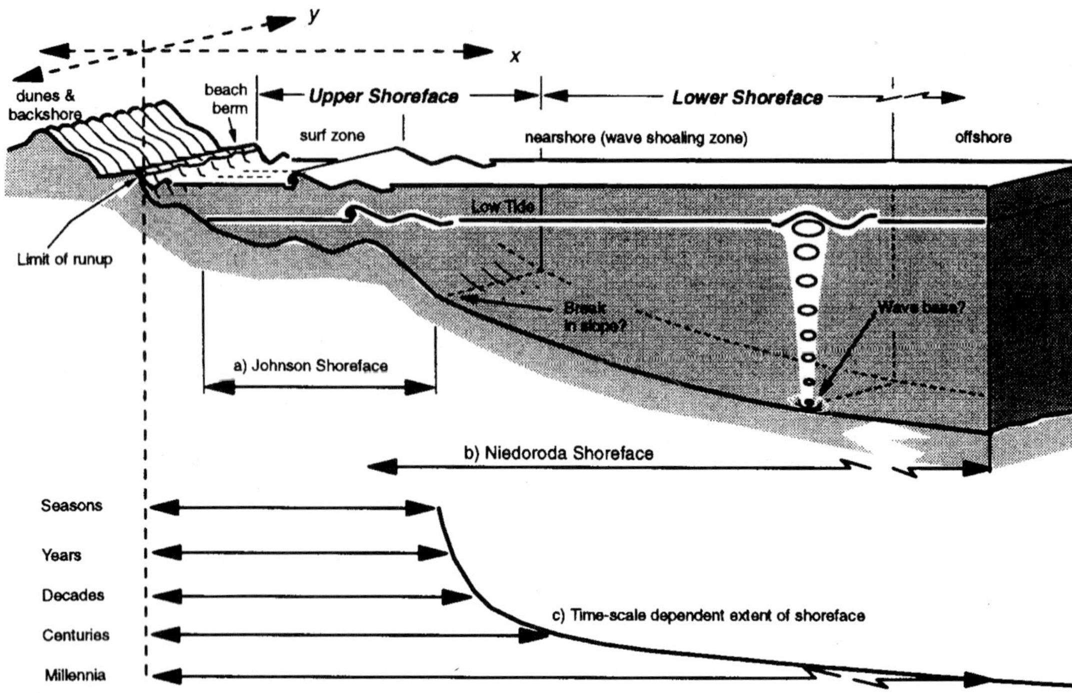


**Figure 3.1** The shoreface as defined (a) explicitly by Johnson (1919); (b) loosely by Niedoroda and Swift (1991); and (c) in this chapter by extending Johnson's definition in relation to the natural morphodynamic continuum. The curve at the bottom schematically depicts the relationship between time scale and scale of morphological change (? indicates dubious entity).

shoreface geometry, although this assumption can be relaxed somewhat (Swain, 1989; Cowell *et al.*, 1995). However, the issue of morphological scale remains unsolved (see curve in Figure 3.1). This situation is reflected in the widespread emphasis given in the literature and in coastal-management practice to the concept of annual *closure depth* in analyses of shoreface change (see Section 3.2.1). This emphasis is wholly inadequate with respect to LSCB and coastal management since depth of closure is now known to increase with time scale (Nicholls *et al.*, 1998). Annual *closure depths* are almost certainly less than those for profile *closure* over the long term (decades to millennia), although little is known quantitatively about shoreface change at this scale. In a general sense, the problem was expressed by Niedoroda *et al.* (1985a, p.534) as the need for 'matching dynamic length and time scales to the morphological scales' in LSCB.

The profile-translation approach to large-scale behaviour of the shoreface has evolved from concepts of shoreface equilibrium that go back a long way (Cornaglia, 1889; Fenneman, 1902; Bruun, 1962; Dietz, 1963, 1964; Moore and Curray, 1964). Equilibrium in its various forms (Schumm, 1991) is a fundamental product of morphodynamic adjustment (Wright and Thom, 1977; Cowell and Thom, 1994). The existence and character of any particular equilibrium form depends upon the time scale under consideration (Schumm and Lichty, 1965; Schumm, 1991). For example, shoreface equilibrium takes on a very different meaning over time scales greater than $10^4$ years since global sea level fluctuates through tens of metres during such intervals (i.e. associated with glacio-eustatic cycles). Under these conditions, only the uppermost part of the shoreface is likely to be in equilibrium (Roy and Thom, 1981; Roy *et al.*, 1994).

By comparison, under conditions of relatively stable or slowly varying sea level, typical of the last 6000 years, the shoreface may approach conditions of *steady-state equilibrium* (as defined by Schumm, 1991), although the lower shoreface appears to exhibit a significant *relaxation time* (i.e. the time required for morphodynamic adjustment toward the new equilibrium). Relaxation times for lower shorefaces on oceanic coasts appear to range from on the order of $10^2$ years (Stive and De Vriend, 1995) to $10^3$ years (Cowell and Thom, 1994). However, on *instantaneous* and *event* time scales, even the upper shoreface is subject to continual change with relaxation times on the order of days (Larson and Kraus, 1989, 1995). The upper shoreface therefore chases a *dynamic equilibrium* (Hack, 1960; Schumm, 1991) as boundary conditions vary stochastically, without ever quite making it (Inman *et al.*, 1993; Cowell and Thom, 1994).

The dependence of shoreface equilibrium on time scale and the implications of this dependence have been explored very little since an initial debate in the 1960s (Dietz, 1963, 1964; Moore and Curray, 1964). Yet this dependence would appear to provide an avenue for examining the nexus between various morphological scales and their associated time scales inherent in shoreface change (Figure 3.1). This chapter therefore exploits as its theme the link between these scales. The theme is advanced through a systematic identification of morphological components on the shoreface and their inherent scales, based on established conventions (i.e. in the literature) together with observed sedimentological and morphological patterns reported from contrasting shorefaces around the world (but especially from south-eastern Australia).

Such a theme naturally demands that we review mainstream concepts for shoreface equilibrium, since equilibrium provides theoretical reference conditions against which change and form can be compared. Equilibrium concepts enjoy far less than universal acceptance, especially with regard to practical application (Pilkey *et al.*, 1993) and the more traditional, simplistic theories may even be misleading if taken too literally. Nevertheless, the theoretical concepts do serve a valuable heuristic purpose in exploring shoreface morphodynamics, since equilibrium is a manifestation of morphodynamic feedback (Cowell and Thom, 1994). In reality, equilibrium processes and resulting morphologies probably depend very much on geography, as Pilkey *et al.* (1993), among others, strongly argue. An iconoclastic view might be that shoreface equilibrium does not exist at all in nature and that shorefaces constantly evolve in response to time-dependent dynamics. Support for this view exists in studies by the *International Geographical Union's Commission on the Coastal Environment* which have shown that at least 70% of the world's sandy coastline has retreated during the past few decades, with only 20% showing no net change (Bird, 1985). Nevertheless, the idea that shoreface form is in near equilibrium over time scales of years is evident in the manifest stability of what is, after all, merely a large pile of wet sand subject often to very violent ocean forces. At the very least therefore, we can use insights from equilibrium theory to develop a more general and unified understanding of shoreface variability between contrasting environmental settings that exist in nature.

The significance of the shoreface in coastal morphodynamics lies in the role it plays in the exchange of sand between the continental shelf and the backshore (beach and dunes). This process is of central concern in the management of coastal erosion due to storms, where sand is lost temporarily offshore from the sub-aerial beach, to be returned gradually during periods of beach recovery. The behaviour of the shoreface also determines whether the offshore zone is a net source or a sink for beach sands over time scales relevant to coastal management (e.g. Stive *et al.*, 1991). Over longer time scales (centuries to millennia) lateral translation of the shoreface governs the overall evolution of coasts, particularly (but not only) under conditions of rapidly changing sea levels related to the major cycles of glaciation (e.g. Roy *et al.*, 1994; Cowell *et al.*, 1995).

## 3.2 MORPHOLOGICAL SCALE

One problem with the original definition of the shoreface is that the offshore break in slope is not always present (as signified by the question mark in Figure 3.1). The seaward delineation of the shoreface is seldom clear-cut; it varies considerably depending upon the time scale of interest. An element common to all definitions is the dominant role played by gravity waves (wind waves and swell in the frequency band $0.5 < f < 0.033$ Hz). These waves are definitive agents in transporting sediments and shaping the shoreface into the distinctive morphology for which it is well recognised, i.e. a characteristically concave-up profile that is considered to comprise a response to the *regime* of shoaling waves (e.g. Fenneman, 1902; Johnson, 1919; Niedoroda and Swift, 1991; Thorne and Swift, 1991). On this basis, and in recognition of the natural continuum in wave-related sedimentation processes (Wright, 1995; Stive and De Vriend, 1995), the definition of the shoreface used in this book extends from the limit of wave runup on the beach face, seaward to the depth limit for effective influence by gravity waves in shaping seabed morphology (Figure 3.1). Hence, the innermost parts of the shoreface are dealt with in more detail in Chapters 4 and 5. In reality, the location of the landward and seaward limits of the shoreface (especially the latter) varies through time and with the time scale over which the shoreface is studied (see graph in Figure 3.1).

In adopting the shoreface definition in Figure 3.1, we continue the tradition of inconsistent terminology. For example, Price (1954), Reineck and Singh (1980), Niedoroda and Swift (1991) and Wright *et al.* (1991) all delineate the shoreface differently from each other and from Johnson. However, rather than distinguishing the shoreface as part of a simple classification scheme for coastal morphologies, our approach borrows the concepts of *hierarchy theory* from biology (e.g. O'Neill *et al.*, 1986). These concepts are now being adopted in earth science (e.g. Oost, 1995) to further the seminal ideas of Schumm and Lichty (1965) regarding scale. Accordingly, the shoreface is conceptualised as a large-scale geomorphic feature that has nested within it, and that is driven by the behaviour of, a set of smaller scale geomorphic features (i.e. the constituents identified in Figure 3.1). Accordingly, the surf zone and beach face are included as components of the shoreface not only because they form an important part of the wave-dominated inshore region. These components also are important morphodynamic elements involved in the annual changes to the cross-shore profile on the upper shoreface (e.g. Komar and McDougal, 1994), described below. At the next level up in the scale hierarchy, changes to bed elevations on the lower shoreface generally are detectable only over time spans of several years or more.

Drawing a distinction between the *upper* and *lower shoreface* means that, instead of representing the overall shoreface as a simple, concave-up curve in profile (e.g. as characterised by Niedoroda and Swift, 1991), the shoreface can be characterised more realistically as having a compound form. This compound form consists of two concave-up elements separated by a convex inflection, which occurs on the upper shoreface seaward of the surfzone (Figure 3.1). Until recently, the compound profile has been ignored in important equilibrium studies concerning the shoreface profile, partly perhaps because the seminal theories were tested against comparatively simple, low-gradient profiles from coasts of north-west Europe and the south-east USA (e.g. Bruun, 1954b; Dean, 1977, 1991). In 1993 however, Inman *et al.* published a model that accounts for these dual elements of the shoreface, which they termed the *shorerise* and the *bar-berm* (see Section 3.3.3). The bar-berm is analogous to the *shore terrace* referred to in the literature earlier this century (Fenneman, 1902) although the latter was commonly seen as a wedge of sediment that accumulated on the seaward side of a *submarine abrasion platform* that is eroded into bedrock. Dynamically however, the two features are probably equivalent (Dietz, 1963). For instance, Inman *et al.* (1993) present an illustration in which their bar-berm lies immediately seaward of an abrasion terrace associated with an eroded ('winter') profile on the southern Californian coast where rock headlands occur extensively and beach sand often forms only a thin veneer over bedrock. This setting is common on coasts fronted by steep, narrow continental shelves such as, for example, the southern part of the south-east Australian coastline (Roy and Thom, 1981).

### 3.2.1 The upper shoreface

The *upper shoreface* as defined here (Figure 3.1) is the region in which erosion and accretion result in significant (i.e. measurable) changes in bed elevation in any typical year (Figure 3.2). Because these changes are rapid compared to time scales associated with LSCB (decades or longer), Stive and De Vriend (1995) rather aptly term the upper shoreface the *active zone*. The changes in bed elevation are usually largest in the vicinity of the beach face and decrease progressively offshore forming an envelope of profile change that pinches out well beyond the outer bar (usually at several times the surf zone width). This definition of the upper shoreface therefore encompasses the surf zone. During episodes of beach erosion, sand is not only transferred from the beach face to the surf zone bars (see Chapters 4 and 5), but some distance beyond the surf zone as well. Thus the inner part of the upper shoreface is lowered and the offshore parts are raised. The process reverses during periods of beach recovery. The range of vertical movements in the bed decreases offshore for two reasons. First, the intensity of near-bottom water motions decreases away from the beach, especially seaward of the surf zone. Second, sand piled against the beach under conditions of full accretion is spread out over a much wider area offshore, and therefore more thinly, during periods of erosion. Eventually the envelope of profile change diminishes to a level that is less than the resolution of standard survey techniques, which Hallermeier (1981) defined for practical purposes as occurring where profile change is less the 0.3 m in the vertical during a typical year. Clearly therefore, two-way exchange of sand further offshore (i.e. with the lower shoreface) still occurs, despite the lack of a distinct morphological manifestation of these sand movements within a typical year.

Definition of the upper shoreface on this basis rests upon the work of Hallermeier (1981) who analysed the envelope of profile change, illustrated in Figure 3.2, and determined that the envelope converges at a *closure depth* of

$$h_c \cong 2.28H_{sx} - 68.5\left(\frac{H_{sx}^2}{gT_e^2}\right) \tag{3.1}$$

where $H_{sx}$ is the nearshore storm wave height that is exceeded only in 12 h each year, $T_e$ is the associated

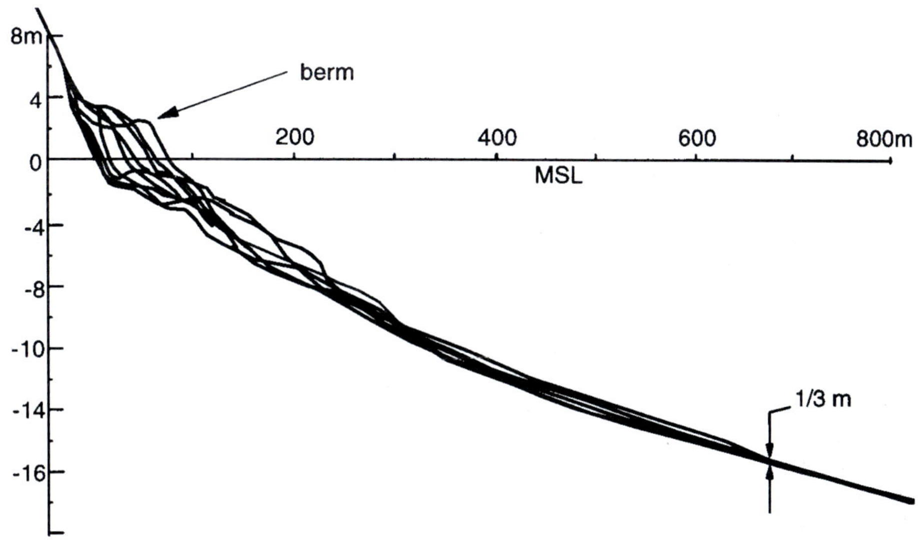


**Figure 3.2** Envelope of profile change on the upper shoreface during a 'typical year', following the conventions of Hallermeier (1981), with dimensions characteristic of the SE Australian coast.

wave period, and $g$ is acceleration due to gravity. The dependence of profile behaviour on wave climate is evident in this relationship. Statistical information on regional wave climates is not always available in terms of $H_{sx}$ and $T_e$. For practical use therefore, Hallermeier provided the simpler approximation

$$h_c \approx 2\bar{H}_{sig} + 11\sigma \quad (3.2)$$

which is expressed in the more conventional wave-climate statistics $\bar{H}_{sig}$, which is the mean annual significant wave height (defined as the annual mean of the highest one-third of waves measured each day), and $\sigma$, which is the standard deviation of $\bar{H}_{sig}$. Birkemeier (1985) re-evaluated Equation 3.1 using data from the CERC Research Facility, located at Duck on the Atlantic coast of the USA (North Carolina) and produced an alternative set of relationships. These alternatives also have been assessed for wave data from the coast of south-east Australia (Meleo, 1994) with the resulting estimates for $h_c$ from Equation 3.1 averaging 24% less than the approximation (Equation 3.2), whilst being on average 33% and 53% greater than Birkemeier's (1985) best-fit and simplified estimates respectively (with standard deviations of 6%, 1% and 8%). These discrepancies indicate that, even if detailed wave data are available, it is pointless to quibble over accuracy in prediction of $h_c$, and so simple approximations are deemed to suffice for practical purposes (e.g. Hanson, 1989).

Typical values of $h_c$ are listed in Table 3.1 for shorefaces on three coasts with moderately high-energy wave climates. These values for the seaward limit of the upper shoreface are in line with Johnson's (1919) concept of the shoreface proper, i.e. 'the narrow zone seaward from the low-tide shoreline permanently covered by water over which the beach sands and gravels actively oscillate with changing wave conditions' (Wiegel, 1953; Baker, *et al.*, 1966). However, it is important to remember that, in applying Hallermeier's closure depth convention (Equation 3.1) to definition of the upper shoreface, vertical movements of the seabed are negligible only during a 'typical' year. Over longer periods, or with the occurrence of severe storms (i.e. those with return intervals much longer than 1 year), vertical changes in bed elevation of more than 0.3 m are expected at a normal closure depth determined on the basis of Equation 3.1. In the case of gradual changes over periods much longer than a year, it is probably more useful to think of the profile changes in terms of lateral (shore-normal) migration of the 'theoretical' closure depth (Equation 3.1), along with the entire upper shoreface (Figure 3.3a). Under these circumstances, the upper shoreface has a time-averaged constant form that Inman and Dolan (1989) refer to as a *translational* profile.

Sudden changes in bed elevation that exceed 0.3 m at the normal closure depth (Equation 3.1), due to severe storms, are viewed as a temporary distortion of the upper shoreface (e.g. Inman *et al.*, 1993), entailing a horizontal stretching of this zone (Figure 3.3b) to produce what Inman and Dolan (1989) term a *transient* profile. The stretching involves both landward displace-

**Table 3.1** Comparison of the seaward extent of the upper shoreface ('closure depths', $h_c$) for the SE Australia, US Atlantic, US Pacific and Netherlands North Sea coasts (in metres). Wave data are from: Hallermeier (1981) and Birkemeier (1985) for the US; Kroon (1994) for the central Netherlands; and for south-east Australia wave data are based on means for seven beaches distributed fairly evenly between latitudes 30°S and 37°S (Meleo, 1994). The last column lists water depths, $h_t$, for the transition from nearshore sands to coarse sand based on data reported from the US Atlantic coast (North Carolina) by Swift *et al.* (1971), the US Pacific by Hunter *et al.* (1988), The Netherlands North Sea coast by van Straaten (1965), and in SE Australia by Hanslow (1989) (see text for explanation).

| | $H_{sig}$ | $\sigma$ | $h_c$ (Eqn 3.2) | $h_t$ |
|---|---|---|---|---|
| US Atlantic | | | | |
| Virginia beach | 0.7 | 0.4 | 6.2 | |
| Nags Head | 1.0 | 0.5 | 7.9 | |
| North Carolina | | | 7.0 | 10 |
| US Pacific | | | | |
| Huntington Beach | 0.9 | 0.3 | 5.3 | |
| La Jolla | 1.2 | 0.5 | 7.7 | |
| Point Mugu | 0.9 | 0.3 | 5.0 | |
| Monterey Bay | | | | 10–16 |
| SE Australia | 1.5 | 1.2 | 12.1 | 16–30 |
| Netherlands | 1.2 | 0.8 | 10.7 | 8–12.5 |

ment of the profile as well as a seaward shift in the point on the profile corresponding to the theoretical closure depth (Figure 3.3b). The most severe storms cause removal of the entire upper shoreface sand wedge (Niedoroda *et al.*, 1985a) which presumably is spread out over the lower shoreface. Inman *et al.* (1993) report stripping of the entire upper shoreface during a cluster of the most extreme storms in California this century, associated with the 1982/83 *El Niño* event. We have observed similar deepening of the entire upper shoreface in hydrographic surveys following a cluster of extreme storms in south-east Australia during 1974/75. The surveys were conducted as part of the work reported in Wright and Short (1983), and Thom and Hall (1991). In contrast to the western USA, the major erosion events in Australia (Foster *et al.*, 1975) are associated with the *La Niña* counter phase of the *El Niño Southern Oscillation* (ENSO) (Allan *et al.*, 1996). Following the 1974/75 storms beach recovery took at least a decade (Figure 3.4), although little is known about the details of the mechanisms and lower-profile changes involved.

The link between the behaviour of surf zone bars (described in Chapter 4), and overall changes in the upper shoreface profile at annual time scales, remains one of the many important but as yet unresolved questions in beach morphodynamics. This shortcoming is due to the scarcity of long-term profile data extending far enough offshore. Exceptions here are the data reported by Inman *et al.* (1993) from California and North Carolina (Duck) in the USA, and the Nile delta, Egypt, and by Lee *et al.* (1995), who also analysed data from Duck. Overall, these data demonstrate that during 'typical' (Hallermeier) years, the upper shoreface encompasses all the across-shore sediment exchanges that are large enough to produce measurable profile change. However, during periods subject to severe storms, especially when these occur in clusters, actual profile closure is not contained even within the data reported by Inman *et al.* (1993) and there may even be an increase in the total volume of the upper shoreface (Lee *et al.*, 1995). Then there is significant lateral shift in *closure depth* predicted on the basis of Equation 3.1. Paradoxically, the shift is landward (Figure 3.3c).

Thus, we adopt Equation 3.1 only as a reference convention in defining the upper shoreface. That is, the definition cannot be used to determine the seaward extent of either significant exchanges of sand volume across the shoreface or instability in seabed elevations in an absolute sense. Recent studies based on analysis of profile data spanning a decade at Duck on the east coast of the USA showed that closure depths increase with time scale (Nicholls *et al.*, 1998) and that clusters of storms within several weeks drive interannual fluctuations in profile morphology of the upper shoreface (Lee *et al.*, 1998). However, essentially nothing is known about fluctuations in profile closure depth over the longer term (decades or more). Similarly, little is known of how these fluctuations relate to the morphological effects of, and recovery from extreme storms, such as the 1991 'Halloween Storm' described by Wright *et al.* (1994a) on the US Atlantic coast, and those of 1974/75 and 1982/83 on the south-east Australian and Californian coasts, referred to above. The interdecadal profile behaviour evident in Figure 3.4 is consistent with observations of sub-decadal fluctuations in the upper shoreface data from Duck (Lee *et al.*, 1998). The Duck data indicate that gradual beach recovery occurs over a period of several years, between storm cluster events, during which time individual storms of moderate intensity had little effect on the recovery trend (Lee *et al.*, 1998). It seems likely that the fluctuations associated with extreme storms are attributable to variations in ENSO over decadal intervals, at least in the Pacific, but possibly beyond. For example, the Halloween Storm on the US Atlantic coast occurred during the 1991 *El Niño* event.

The offshore limit of the upper shoreface generally is

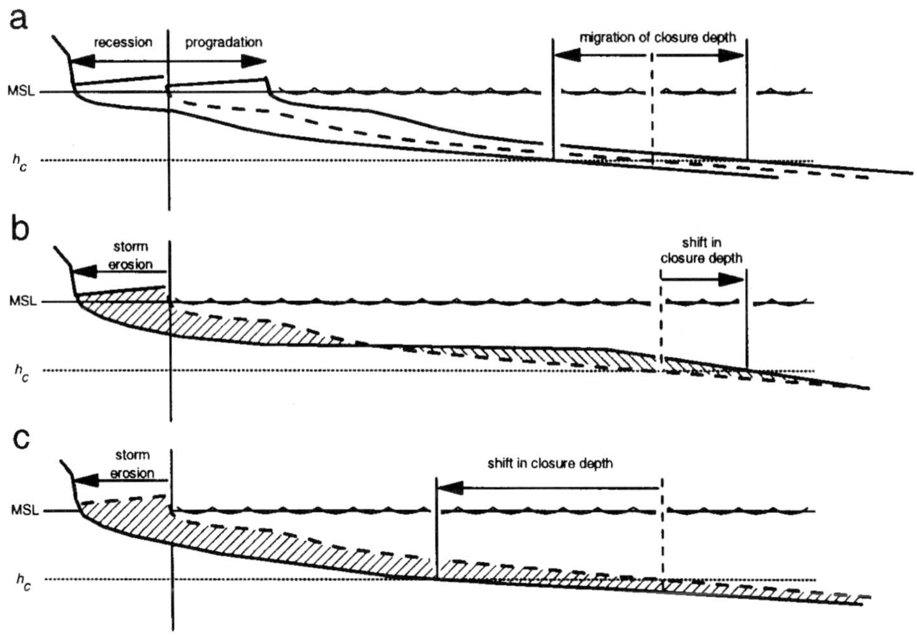


**Figure 3.3** Movements in the seaward boundary of the upper shoreface (i.e. the closure depth, $h_c$ estimated on the basis of Equation 3.1) at intervals significantly longer than the annual time scale, associated with: (a) a 'translational' profile; (b) a 'transient' profile due to a severe storm (return intervals greater than 1 year); and (c) a 'transient' profile due to removal of the entire upper shoreface sand prism, associated with normal wave conditions, during an extreme storm (return intervals on the order of decades). Only the region of shoreface erosion is shown in (c); the region of deposition is off the diagram to the right (i.e. on the lower shoreface).

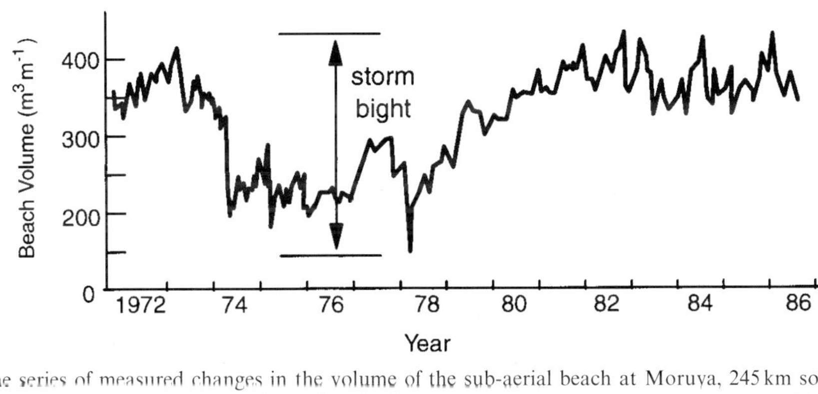


**Figure 3.4** Time series of measured changes in the volume of the sub-aerial beach at Moruya, 245 km south of Sydney, SE Australia, showing beach recovery over more than a decade after a series of very severe storms in 1974 (modified from Cowell and Thom, 1994; based on Thom and Hall, 1991). Much of the missing sand volume in the 'storm bite' was transferred to the lower shoreface.

not distinguished by any sharp break in the topography. However, the theoretical *closure depth* often (but not always) corresponds approximately to a distinctive break in sediment characteristics. *Upper shoreface sands* (Niedoroda *et al.*, 1985a) generally are well sorted and often (but not always) similar to the beach sediments, although there is usually (but not always) a distinct decrease in mean grain size in the seaward direction. In many places, seaward fining of sediments occurs down to an abrupt transition at depths, $h_t$, that are roughly comparable to $h_c$ (Table 3.1), as illustrated in studies on the US Atlantic coast (North Carolina) by Swift *et al.* (1971), The Netherlands North Sea coast by van Straaten (1965), and in south-east Australia by Hanslow

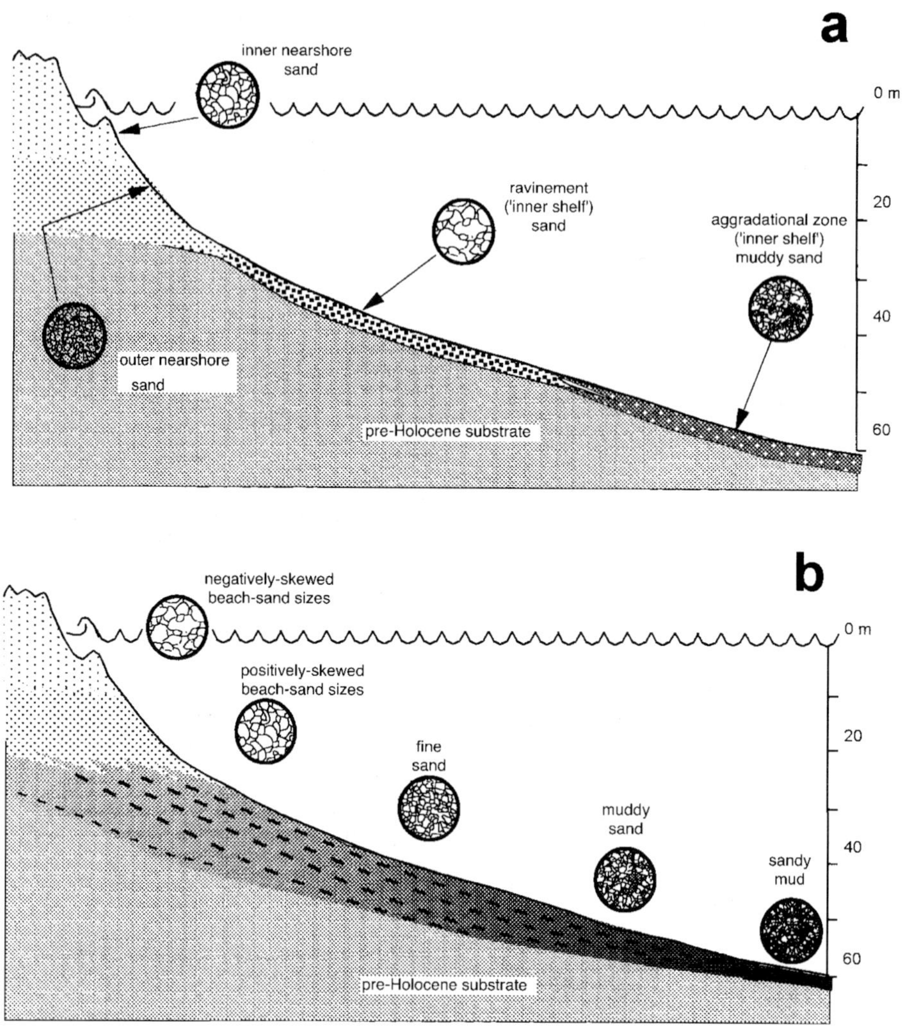


**Figure 3.5** Schematic illustration of the arrangement of sediment units on: (a) the shoreface in SE Australia, characteristic of a segmented shoreface (modified from Roy and Stephens, 1981); and (b) a graded shoreface in which sediments become progressively finer offshore until they pass into mud deposits.

(1989). Coarser, poorly sorted sands occur seaward of the transition. Niedoroda *et al.* (1985a) deem that this grain size discontinuity marks the offshore limit of the upper shoreface. However, a sedimentological definition of the seaward limit is not universally applicable since the coarse-grained sand zone does not occur on all coasts, as explained in the following.

On prograding, deltaic shorefaces, such as that of the low-gradient Mississippi shelf, sands fine progressively seawards from the upper onto the lower shoreface, through to fine silts and clays on the continental shelf (Niedoroda *et al.*, 1985a; Swift and Thorne, 1991; Swift *et al.*, 1991). In distinctly different environments, on parts of the south-east Australian coast where the inner shelf is steep ($> c.1^\circ$), very well sorted sand tends to maintain constant modal grain size from the shoreline out into water depths of 50 to 60 m where it mixes with muddy sediments of the mid-shelf (Field and Roy, 1984; Ferland, 1990; Roy *et al.*, 1994). The shoreface in these settings forms the surface of features that Roy *et al.* (1994) term *headland-attached sand bodies*. These deposits also occur seaward of beaches and their adjoining headlands along steep segments of the coast (Ferland, 1990). Steep, narrow shelves elsewhere also lack a coarse sand zone seaward of the upper shoreface. At La Jolla in southern California, the seaward-fining trend not only continues across the whole shoreface but even down the slope into the Scripps and La Jolla Canyons (Inman, 1953). This steep shoreface has a significant supply of sediments from onshore sources, a condition shared in common with the otherwise strikingly different shorefaces on low gradient deltaic coasts. In both cases, the supply of sediments from onshore precludes accumulation of coarser material lower on the shoreface.

Even on coasts with coarse-sand zones fronting the upper shoreface, abrupt grain-size transition occurs in water depths that vary considerably within and between coastal regions (see $h_t$ in Table 3.1). For example, while $h_t = 10$ m is typical of North Carolina (Table 3.1), coarse sand can occur in depths as shallow as 3 m on beaches impoverished of upper shoreface sediments (Thieler *et al.*, 1995). Furthermore, in extensive studies of shoreface sediments from south-east Australia, Roy and Stephens (1981) distinguish a zone of distinctly different sediments lying in intermediate water depths between the seaward-fining sequence and the coarse-sand zone (Figure 3.5a). Roy and Stephens refer to the sediments of this intermediate zone as the *outer nearshore sands* to distinguish them from the seaward-fining *inner nearshore sands*. The *outer nearshore sands* are finer than sediments further inshore, and tend to maintain constant mean grain size along the profile. The junction between the *inner* and *outer nearshore sands* occurs typically in water depths of 12 m (Roy and Stephens, 1981), which is roughly equivalent to the theoretical $h_c$ (Table 3.1). In contrast, the transition to coarse sand (where it exists) occurs at roughly 20 m. It would appear therefore, that this junction constitutes a sedimentological marker more tuned to hydrodynamic factors than the coarse sand discontinuity further offshore. The latter more likely depends upon the evolutionary history of the shoreface, including geological factors.

Overall therefore, sediment patterns seem not to provide consistent indicators of dynamic length scale on the upper shoreface. They do however provide evidence for interpretation of morphological scales inherent in upper shoreface behaviour. On the other hand, Equation 3.1 appears to offer useful generality in a crude definition of dynamic length scale on the upper shoreface, at least in terms of reference conditions related to the annual time frame. The definition depends only on wave climate, ignoring other dynamics that are undoubtedly important. Nevertheless, this emphasis highlights the importance of waves in shoreface morphodynamics.

A more general approach to time scale probably involves issues related to profile behaviour underlying the beach-volume changes depicted in Figure 3.4. Unfortunately, so far little is known about this aspect of the time problem. Thus, herein lies the most pressing need for future research into the upper shoreface. Without greater understanding of profile behaviour at time scales relevant to what Stive *et al.* (1991) term the *fluctuating component of LSCB*, little solid progress is possible in predicting the major beach erosion and recovery cycles evident in Figure 3.4.

### 3.2.2 The lower shoreface

Sediment-size distributions are more complicated on the *lower shoreface* than on the upper shoreface. Moreover, sediment size is inherent in the very definition of the lower shoreface (see Equation 3.3). The lower and upper shoreface are contiguous, so the inshore boundary of the former corresponds to the offshore boundary of the latter (i.e. the closure depth). However, the offshore limit to the lower shoreface is less distinctly defined and probably should be thought of as varying in relation to the time scale of interest (Stive and De Vriend, 1995), as indicated in Figure 3.1. However, in the absence of other clear-cut definitions, it is useful to turn to another of Hallermeier's (1981) criteria for a convention to delineate (at a first approximation) the seaward limit to the lower shoreface. Based on a synthesis of theoretical and field studies, Hallermeier estimated that the limiting depth for significant on–offshore transport of sand by waves throughout a typical year is

$$h_i = (\bar{H}_{sig} - 0.3\sigma)\bar{T}_{sig}(g/5000D)^{\frac{1}{2}} \tag{3.3}$$

where $\bar{T}_{sig}$ is the mean annual significant wave period and $D$ is the characteristic grain size for the shoreface. Hallermeier specified that the grain size used in Equation 3.3 should be set as the median diameter ($D_{50}$) determined from a sand sample taken in a water depth of $h \approx 1.5h_c$. Typical values of $h_i$ are listed in Table 3.2.

Equation 3.3 is a refined version of the old concept of *wave base* (Gulliver, 1899), which was defined as the water depth beyond which 'wave action ceases to stir the sediments' (Baker *et al.*, 1966). For the sake of simplicity, Davies (1980) equates wave base to the shoreward limit for so called *deep-water waves*. By convention, this limit is taken to be where the water depth is half the deep-water wave length (i.e. $h = L_o/2$). Strictly speaking, this convention relates more to where the bed starts to affect waves. However, Davies' somewhat arbitrary linking of wave base to the deep-water wave criterion may be of pragmatic value. The general expres-

**Table 3.2** Comparison of the seaward extent of the lower shoreface ('wave base') for the SE Australia, US Atlantic, US Pacific and Netherlands North Sea coasts (in metres). Data are from: Hallermeier (1981) and Birkemeier (1985) for the US; Kroon (1994) for the central Netherlands (Egmond) wave data and van de Meene (1994) for sediment data (Ijmuiden); and Meleo (1994) for SE Australia. The last two columns list water depths for 'wave base' alternatives (see text).

| | $H_{sig}$ | $\sigma$ | $T_s$ | $h_i$ (Eqn 3.3) | $h = L/2$ | $h = L/4$ |
|---|---|---|---|---|---|---|
| US Atlantic | | | | | | |
| Virginia beach | 0.7 | 0.4 | 8.3 | 21.1 | 53.7 | 26.9 |
| Nags Head | 1.0 | 0.5 | 9.8 | 31.1 | 74.9 | 37.4 |
| North Carolina | | | | | | |
| US Pacific | | | | | | |
| Huntington Beach | 0.9 | 0.3 | 13.2 | 43.9 | 135.9 | 67.9 |
| La Jolla | 1.2 | 0.5 | 12.0 | 52.6 | 112.3 | 56.1 |
| Point Mugu | 0.9 | 0.3 | 14.7 | 52.5 | 168.5 | 84.3 |
| SE Australia | | | | | | |
| Byron Bay | 1.6 | 1.2 | 9.6 | 46.2 | 71.9 | 35.9 |
| Sydney | 1.6 | 1.2 | 9.5 | 36.3 | 70.4 | 35.2 |
| Moruya | 1.5 | 1.2 | 9.5 | 36.0 | 70.4 | 35.2 |
| Netherlands | 1.2 | 0.8 | 5.0 | 13.4 | 19.5 | 9.7 |

sion for wave length according to *linear airy wave theory* (e.g. Komar, 1976) is

$$L = \frac{gT^2}{2\pi}\tanh\frac{2\pi}{L} \tag{3.4}$$

and for the special case of *deep-water waves*, $\tanh(2\pi/L) \rightarrow 1$, so that wave length simply becomes

$$L_0 = gT^2/2\pi \tag{3.5}$$

Equation 3.5 was used to compute estimates of wave base ($h = L_o/2$) and the results are compared to corresponding values of $h_i$ in Table 3.2. Clearly Davies' concept of wave base places the seaward limit in water depths that are too deep by a factor of 2 to 3. Komar (1976) argued that the $h \leq L_o/2$ convention for *deep-water waves* is, for most practical purposes, unnecessarily stringent in terms of the accuracy of the approximation provided by Equation 3.5. He observed that for $h = L_o/2$, the error in using Equation 3.5 is 0.37% whereas the error is still only 0.5% when $h = L_o/4$. Komar therefore redefined *deep water* as $h \leq L_o/4$ with respect to the limit of application for the deep-water approximation (Equation 3.5). Redefining wave base in line with Komar's reasoning gives water depths for the seaward limit of sediment disturbance by waves that are much closer to $h_i$ (Table 3.2). The agreement is reasonable, except for locations with extreme values of $T_s$ (Huntington Beach, Point Mugu, in the case of long-period wave climates, and The Netherlands in the case of short-period wave climates). The level of agreement indicates that it should be possible to derive an estimate for the seaward boundary to the lower shoreface that is independent of sediment size (unlike in Equation 3.3).

Estimation of wave base independently of sediment size is a desirable goal because of the geographic variability in the patterns of sediment texture that occur on the lower shoreface (alluded to in the previous section). Hallermeier's estimation of wave base ($h_i$) relies upon there being a single representative sediment size for the lower shoreface, which implicitly assumes that the size distribution across the lower shoreface forms a similar pattern everywhere. Unfortunately, this is by no means the case, as Hallermeier cautioned. The inconsistency is illustrated by the anomalous case of central Netherlands where $h_i$ (Table 3.2 ) is less than 3 m deeper than $h_c$ (Table 3.1).

The anomaly is explained by the discontinuity in the pattern of seaward-fining sand at about $h = 13$ m on the central Dutch shoreface, seaward of which $D_{50}$ increases from 0.19 mm to 0.25 mm at a water depth of $1.5h_c$ (van de Meene, 1994). Seaward of the discontinuity, sediments are medium sand, at least out to 10 km offshore, whereas fine sand occurs everywhere landward of the discontinuity (van de Meene, 1994). These coarser sediments lying offshore are referred to as the *shelf sand sheet* by Niedoroda *et al.* (1985a), or as *inner-shelf sands* (Figure 3.5a) by Roy and Stephens (1981), and form a veneer up to roughly 1.25 m thick over older sediments (Roy *et al.*, 1994). Typically these sediments are poorly sorted and dominated by the coarser fraction which may have modal values in the coarse sand range. Hallermeier (1981, p.265) refers to this type of anomaly in relation to wave and sediment data from the US New Jersey coast.

The estimates of $h_i$ for south-east Australia in Table

3.2 are based on grain sizes sampled from water depths typically of $1.5h_c \approx 18$ m, which generally lie landward of the sharp textural break of the *inner-shelf sands* (Roy *et al.*, 1994). The samples therefore are not always indicative of modal sediment sizes of the coarse-sand zone, within which the seaward limit of the lower shoreface occurs. Therefore, estimates for $h_i$ on the south-east Australian coast presumably are often too large in those areas featuring an *inner-shelf sand* zone. In reality, while inner-shelf sand of median size or larger may be immobile in water depths shallower than $h_i$, the sub-median portion of the grain-size population may well be mobile (Cowell, 1986a; Roy and Cowell, 1991). Moreover, the finer sediments may be more mobile in the presence of a dominant coarser sand fraction than for uniformly fine sands (Cowell, 1986a). Enhancement of fine-fraction mobility can be expected due to effects of the coarse sediment fraction in generating boundary-layer turbulence, especially where the geometry of wave-generated bed ripples (Nielsen, 1981) is governed by the dominant coarse fraction (Cowell, 1986a; Roy and Cowell, 1991). However, the degree to which this effect would tend to offset the over-estimation of $h_i$ based on Equation 3.3 is of course fortuitous.

Overall therefore, the existence and location of the grain size discontinuity introduces a degree of arbitrariness into estimating $h_i$ using Hallermeier's (1981) procedures. The cross-shore patterns of grain size in The Netherlands and south-east Australian examples are typical of a broad class of shorefaces thought to be associated with receding coasts (Swift *et al.*, 1971; Niedoroda *et al.*, 1985a). The coarser sediments of the *shelf sand sheet* comprise lag deposits of *ravinement surfaces* identified initially by Swift (1968), and considered to be the by-product of shoreface retreat due to sea-level rise (Swift *et al.*, 1991; Emery and Myers, 1996). This class of shoreface is found on coasts throughout the world, including south-east Australia (Figure 3.5a), north-west Europe (van Straaten, 1965; van de Meene, 1994), the US Atlantic (Swift *et al.*, 1971) and the US Pacific (Hunter *et al.*, 1988). They are referred to in this book as *segmented shorefaces* since they comprise a distinct sequence of sediment zones. The sequence in south-east Australia is described by Roy and Stephens (1981) as going successively from *inner* to *outer nearshore sands*, *inner shelf sands*, *mid-shelf muddy sands* and finally *muds* (Figure 3.5a).

The classical *graded shoreface* described by Johnson (1919) constitutes another class of sediment pattern. The graded shoreface entails fining of sand across the entire shoreface, eventually passing through to mud on the adjacent continental shelf (Figure 3.5b). This pattern has been associated with prograding coasts that have an abundant deltaic sediment supply (Swift, 1976), as in the archetypical case of the shoreface fronting Galverston Island in the Gulf of Mexico (Bernard *et al.*, 1962). Swift (1976) refers to such settings as *allochthonous* environments, in contrast to coasts where waves rework sediments faster than they are supplied, which Swift terms *autochthonous* coasts. *Autochthonous shorefaces* are often (but not always) receding (Roy *et al.*, 1994). As with *segmented shorefaces*, using a single grain size to estimate $h_i$ on a *graded shoreface* also introduces unwelcome arbitrariness since the progressive fining of sediments offshore means that, in reality, a significant level of sediment agitation by waves occurs until sediments grade through to muds on the mid-shelf.

The distinctly different character of sediment textures in these contrasting shoreface environments means that Equation 3.3 cannot be applied prudently without consideration of the textural patterns. Consideration of morphological scale on the lower shoreface also requires interpretation of the textural patterns. The way forward in this regard is not obvious, especially on graded shorefaces with their smooth continuum in sediment size from sands through to silts and muds on the mid-shelf. Equation 3.3 however provides a valuable reference condition, even if its absolute physical significance is limited under many conditions. That is, even if sub-sand size sediments are active further offshore than $h_i$ on any given shoreface, Equation 3.3 provides a standard against which the particular sediment patterns can be compared. This comparison may provide a basis for interpretation of dynamic length scale in relation to morphological scale and how this appears to manifest itself in the sediment pattern. Any direct relationship in principle should pertain to the annual time scale with respect to sediment agitation, because of the background to Equation 3.3. However, in terms of profile change, the relevant time scale would be considerably longer. Currently, these considerations are purely speculative. Nonetheless, they serve to highlight the uncertainties attaching to application of $h_i$ estimates in coastal management.

Hallermeier (1981) reasons that $h_i$ defines an appropriate morphological scale for prediction of shoreface translation in response to sea-level rise over the long term. While this may be reasonable in determining the seaward limit to sediment budgets involved in such responses (e.g. Bruun, 1962; Beets *et al.*, 1992), it is unlikely that the assumption of an invariant equilibrium profile, of the type associated with the Bruun Rule (Bruun, 1954b, 1988; Dean, 1997), can be applied to depths of $h_i$. For example, data on lower-shoreface profile change from the central Netherlands coast show a progressive deepening and seaward rotation of the shoreface over the past 100 years in association with a 0.15 m sea-level rise and a positive littoral-sediment supply (Stive *et al.*, 1991). Stive and De Vriend (1995) modelled this behaviour by applying an invariant profile to the upper shoreface, with the lower shoreface subject to dynamic change governed by wave-driven sand transport.

However, the overall problem of matching dynamic length and time scales for the lower shoreface is far from resolved. The problem falls into two parts involving evaluation of (i) the depth limit for application of the invariant profile assumption and (ii) seaward extent of significant (detectable) profile change over the long term. Both aspects of the problem undoubtedly depend on time scale, involving a seaward shift in both as the time scale of interest increases. Over time spans of decades, Stive and De Vriend's (1995) application of the invariant profile assumption to the upper shoreface seems reasonable. Over time spans of centuries to millennia, the assumption can probably be applied further seaward, but how far is unknown.

The seaward extent of significant profile change, and its dependence on time scale (Figure 3.1) so far remains essentially unknown. The question for future research here is how this dependence relates to $h_i$ and observed variations in sediment texture on the lower shoreface. One way to pursue this question is through an attempt to scale up the results from direct investigation of sediment dynamics. Another approach is to obtain estimates for rates of sediment accumulation on aggrading (allochthonous) shorefaces, based on shallow-stratigraphic surveys. This approach requires radiometric dating of sediment accumulations to provide gross quantification of profile change (e.g. Field and Roy, 1984; Cowell, 1986a). Furthermore, the results could be used to work backwards to calculate net across-shore transport rates following the method described by Larson and Kraus (1995). The method infers net transport from estimates of the sediment discharge required to translate successive profiles through the measured surface displacement.

## 3.3 MORPHODYNAMICS AND SHOREFACE EQUILIBRIUM

Knowledge of shoreface morphodynamics is far from definitive because the flow field is not well enough understood (Nielsen, 1992) and the combined effects of *deterministic* and *stochastic complexity* (Phillips, 1992; Cowell and Thom, 1994) govern sediment transport on the shoreface. The combined result is that shoreface sediment dynamics in nature is subject to a high degree of uncertainty (Huntley and Bowen, 1990; Kraus and Horikawa, 1990). For this reason, the development of shoreface theory seems to have progressed from one bad idea to another if judged on the basis of predictive accuracy. However, taken collectively, the body of theory on shoreface equilibrium that now exists does afford generalised insight into shoreface morphodynamics.

The uncertainty in shoreface sediment dynamics is reflected in the message to emerge from the previous section, i.e. that no clear-cut seaward limit is evident for long term shoreface change. However, the estimated depth limit for significant sediment transport at the annual time scale at least provides a reference condition for such change, and therefore a starting point for further work on this aspect of LSCB. On a similar basis, theoretical conditions for equilibrium provide a useful reference state for comparison of shoreface morphologies observed in nature. This is not to say that we expect observed morphology to mirror a theoretical equilibrium. It is likely that attainment of equilibrium in nature only occurs rarely, due to constant changes in governing factors on a variety of time scales (e.g. sea-level change, imbalance in littoral-sediment budgets, and stochastic variation in hydrodynamic forcings). The effects of geological constraints imposed by the substrate further complicate equilibrium forms (Pilkey *et al.*, 1993; Roy *et al.*, 1994; Riggs *et al.*, 1995; Thieler *et al.*, 1995). Moreover, the theoretical models for equilibrium are highly simplified as a matter of necessity and therefore cannot be expected to represent real-world conditions particularly well. However, these models have been a valuable aid in understanding shoreface processes, even if they are of limited practical value.

Therefore, comparison of observed shorefaces against a theoretical equilibrium may provide a means of evaluating geological effects and time-dependent shoreface changes. Long-term change in particular can be examined instructively in the context of disequilibrium conditions, e.g. with regard to the long (millennial) relaxation times evident for the lower shoreface referred to in the introduction to this chapter. Equilibrium concepts also provide insight into the morphodynamic coupling between processes responsible for shoreface form and change. The variation in this coupling with water depth is particularly relevant to the issue of scale. The most significant pattern here is covariation of shoreface slope and near-bed wave orbital velocities. Both decrease progressively offshore, with the reduction in near-bed wave action being attributable to the attenuating effect of increasing water depths. Since the waves are assumed to be responsible for the onshore transport of sediments (as explained later), one class of equilibrium theory (the *null-point* approach) explicitly includes the effects of shoreface slope as the counter-balancing agent responsible for offshore transport. Slope often is included in terms of the direct downhill effects of gravity in moving sediments. However, for reasons discussed later, slope may be considered more sensibly as a surrogate parameter for a host of processes that depend upon variation in water depth on the shoreface. Such processes include wave-induced sediment mobility and related concentration gradient-diffusion effects, and depth-dependent hydrodynamics. Moreover, a second class of equilibrium theory (the *entropy-maximising* approach) does not include shoreface slope at all. This approach involves only the effects of waves as they are transformed while crossing the shoreface.

The inclusion of wave effects in both classes of equilibrium theory is to be expected given that shorefaces are by definition a morphological response to waves. Consequently, shallow-water wave transformation is fundamental to shoreface morphodynamics. The general behaviour of wave length as water depth changes (Equation 3.4) is perhaps the most important aspect of this transformation since these two variables ($h$ and $L$) control the intensity of the interaction between waves and the seabed across the shoreface for any given input wave height. The importance of this relationship stems from the dependence on $L$ inherent in other wave parameters on the shoreface. Furthermore, at a practical level, these wave parameters (surface elevation, phase speed, group velocity, shoaling coefficient, velocity potential, orbital diameters and water velocities) can all be determined readily from Equation 3.4. Although *linear wave theory* is the simplest theory available, Equation 3.4 is clearly in an awkward implicit form. For practical use Equation 3.4 is more conveniently approximated (Nielsen, 1982) in terms of the deep water *wave number* ($k_o = 2\pi/L_o$), by

$$kh = \begin{cases} k_o h(1 + 2e^{-2k_o h}) & k_o h > 1.26 \\ \left[1 + \frac{1}{6}k_o h + \frac{11}{360}(k_o h)^2\right] & k_o h \leq 1.26 \end{cases} \quad (3.6)$$

The tendencies for shorefaces to approach an equilibrium form varies geographically as a function of coastal ocean climatology and geological setting. In principle at least, geological factors are included to a limited extent in all models in terms of the size (and sometimes the density) of sediments. However, other geological factors such as the bedrock substrate configuration and inherited patterns of grain-size distribution on the shoreface (attributable to former fluvial and tidal delta activity) introduce qualities that are not effaced by processes of sediment redistribution, even in high-energy environments (Roy and Stephens, 1981; Pilkey *et al.*, 1993). To a limited extent, the effects of ocean climatology are built into equilibrium models based on the *null-point* concept through inclusion of wave parameters, but are left out the main *entropy-maximising* models. The coastal ocean climatology can be expected to determine where the time-averaged balance lies between the type and magnitude of the various flow vectors contributing to onshore and offshore sediment transport at any given locality. Note here that since equilibrium must be viewed in terms of time-averaged profile characteristics in the real world, which are governed by stochastic flow forcing, the question of time scale becomes intrinsically important. Thus, Wright *et al.* (1991) observe that whether or not the shoreface is considered to be in equilibrium depends upon the time scale over which it is observed: by no means a new concept in geomorphology (Schumm and Lichty, 1965).

In evaluating concepts of profile equilibrium through field observations, grain size patterns provide a more sensitive indicator than profile shape for the interpretation of shoreface regime. This is illustrated by the distinction made between sediment units in the preceding discussion on the lower shoreface. For example, these textural differences were used there to interpret dynamic relationships, such as the manifestation of predicted closure depths on shorefaces in nature. Profile morphologies generally lack systematic irregularities that might otherwise provide the basis for such interpretations. Accordingly, this section not only reviews ideas on equilibrium profile morphology but it also emphasises equilibrium concepts concerning the distribution of sediment sizes across the shoreface.

### 3.3.1 Morphodynamic representation of the shoreface

Shoreface morphodynamics are usually represented for convenience in terms of across-shore and alongshore components ($x$ and $y$ respectively, Figure 3.1) of flows ($u$,$v$) and sediment fluxes ($q_x$, $q_y$) (Niedoroda and Swift, 1991). These quantities are related exactly to the rate of change in bed height, expressed in terms of water depth ($h = -z$) by the sediment continuity equation

$$\frac{\partial h}{\partial t} = \varepsilon\left(\frac{\partial C}{\partial t} + \frac{\partial q_x}{\partial x} + \frac{\partial q_y}{\partial y}\right) - \frac{\partial V}{\partial t} \quad (3.7)$$

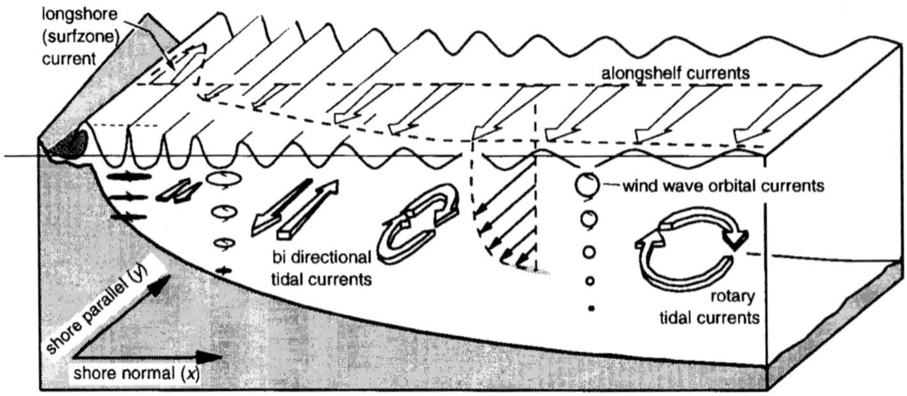


**Figure 3.6** Shallow water transformations due to the presence of the coastal bottom boundary, for flows across a wide range of length scales (adapted from Wright, 1995).

where $\varepsilon$ accounts for sediment density and porosity, $C$ is the total concentration of sediment in motion at any instant (and therefore not contributing to bed elevation), and $V$ (per unit surface area) provides for: (i) sediment gains and losses through autogenic and anthropogenic processes, and (ii) the effects of subsidence or uplift (Cowell and Thom, 1994). The sediment transport terms (containing $q$) mediate the mutual adjustment of flows and morphology (Wright and Thom, 1977; Cowell and Thom, 1994). The flow patterns determine sediment mobilisation and flux gradients in Equation 3.7 that in turn are the dominant cause of morphological change on the shoreface. Conversely, the morphology exerts control over the flows through morphodynamic feedback mechanisms involving: (i) the fluid continuity principle by which the shape of the shoreface at any time constrains the overall pattern of flow accelerations; (ii) the bottom (or solid) boundary conditions, according to which the normal and shear components of flow both cease at the bed; and (iii) the flow-attenuating effects of friction attributable to the granular and topographic roughness of the shoreface surface (Cowell and Thom, 1994).

In the context of the shoreface, these feedback mechanisms manifest themselves strongly in the shallow-water transformations that coastal flows undergo as water depths decrease toward the shore (Figure 3.6). These transformations include the flattening out of orbital motions within the immediate proximity of the coastal boundary, where 'proximity' is relative to the length scales of the flow (Figure 3.6), the spatial compression of *energy* that causes *wave shoaling*, and the growth of non-linearity in flow dynamics which produces, among other things, increasing distortions in wave motions (e.g. asymmetry). Similarly, wave *shoaling* causes shoreward increase in bed activity, both in terms of topographic change and sediment mobility (as evident in previous sections). Since these transformations can be regulated by topographic change, a mutually dependent coupling exists between changes in flows and morphology on the shoreface. This coupling is mediated by sediment transport (Wright and Thom, 1977; Cowell and Thom, 1994).

The instantaneous sediment fluxes themselves depend on the depth-integrated product of sediment concentration ($C$) and the flow-velocity components ($u$,$v$)

$$q_x = \int_{z=-h}^{0} (C_z u_z)\mathrm{d}z \quad \text{and} \quad q_y = \int_{z=-h}^{0} (C_z v_z)\mathrm{d}z \tag{3.8}$$

at any instant and any point on the shoreface. For time-averaged flows, the alongshore component of flow is commonly an order of magnitude larger than the corresponding across-shore flows (Wright, 1995). The spatial gradients in the alongshore components are usually much smaller than those for the across-shore flows, which allows $\partial q_x/\partial x$ to dominate in Equation 3.7 (Wright, 1995, p.18). The alongshore sediment flux gradients are considered to be weaker because the shoreface dimensions are much larger parallel to the shoreline, especially on well studied locations such as the north-west European and US Atlantic coasts where bedrock headlands are mainly absent. Even where prominent headlands interrupt the shoreface, such as in south-east Australia, headland bypassing by shoreface sands often allows effective continuity of the shoreface

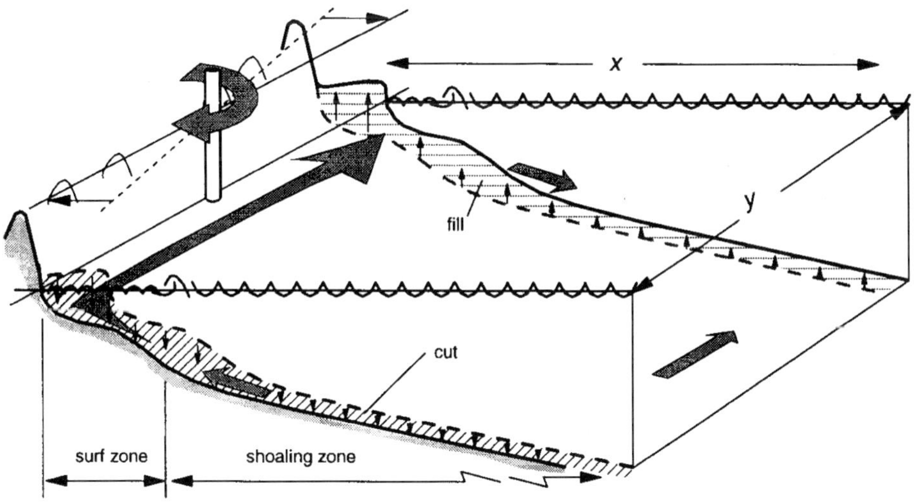


**Figure 3.7** Regional sediment dispersal system: rapid alongshore transport in the surf zone coupled with cross-shore redistribution of sediments through profile adjustment; slow alongshore fluxes on the lower shoreface. Differential translation of shoreface profiles along the coast causes rotation and realignment of the shoreline.

along many kilometres of bedrock-embayed coastline (Cowell, 1986a). Furthermore, cross-shore variation in water depth occurs systematically on all coasts. The offshore variation in water depth enhances the potential for strong variation in sediment mobility and transport gradients perpendicular to the coast, especially in relation to wave-driven sediment transport.

Alongshelf sediment fluxes therefore are more important to the regional characteristics of coastal evolution (shoreface translation), especially under conditions of slowly varying or stable sea level (Roy *et al.*, 1994), whereas local morphodynamics associated with changes in profile shape are attributable to across-shore transport (e.g. Stive and De Vriend, 1995). The alongshore dimension is then considered simply in terms of its role in coupling adjacent coastal cells through net sediment exchanges via the littoral sediment budget (Komar, 1973a; Hanson, 1989). The surf zone generally provides the main artery for these littoral sediment exchanges (Figure 3.7) since rates of alongshore transport are typically two to three orders of magnitude higher inside than outside the surf zone (Wright *et al.*, 1991). Littoral sediment feed occurring in the surf zone is compensated locally (i.e. at any point along the coast) by translation of the shoreface profile. Translation is landward where the littoral feed is negative and seaward where the feed is positive (Figure 3.7)

Profile adjustment involves redistribution of sediments through across-shore transport. Although across-shore transport occurs at rates that are much less than those typical for alongshore transport, across-shore transport generally entails much shorter length scales (i.e. $x \ll y$ in Figure 3.7). Generally therefore, shoreface change overall in $x$ and $y$ is dominated by rapid alongshore transport in the surf zone coupled with across-shore redistribution of sediments through profile adjustment (Figure 3.7). The process is evident in alongshore variations and changes in profile morphology. Consequently, most studies into shoreface dynamics have focused on across-shore sediment transport and profile development (Carter, 1988, pp.199–213).

### 3.3.2 Null-point equilibrium

Cornaglia (1889) was the first to suggest that a balance might be achieved between waves trying to move sediment onshore and gravity trying to pull the sediment downslope. Cornaglia recognised that, as the water depth decreases, the asymmetry between the onshore- and the offshore-directed flows at the bottom could result in a net onshore fluid force on sediments (Figure 3.8). This force, Cornaglia hypothesised, is opposed by the downslope component of gravity. At some point the forces would balance and the sand grain should simply oscillate back and forth. However, since wave asymmetry increases as the waves shoal, a stable equilibrium could be achieved only on a concave-up slope which confers a counter-balancing shoreward increase in the downslope gravity component. The location of the *null point* would vary depending on the size of the particle

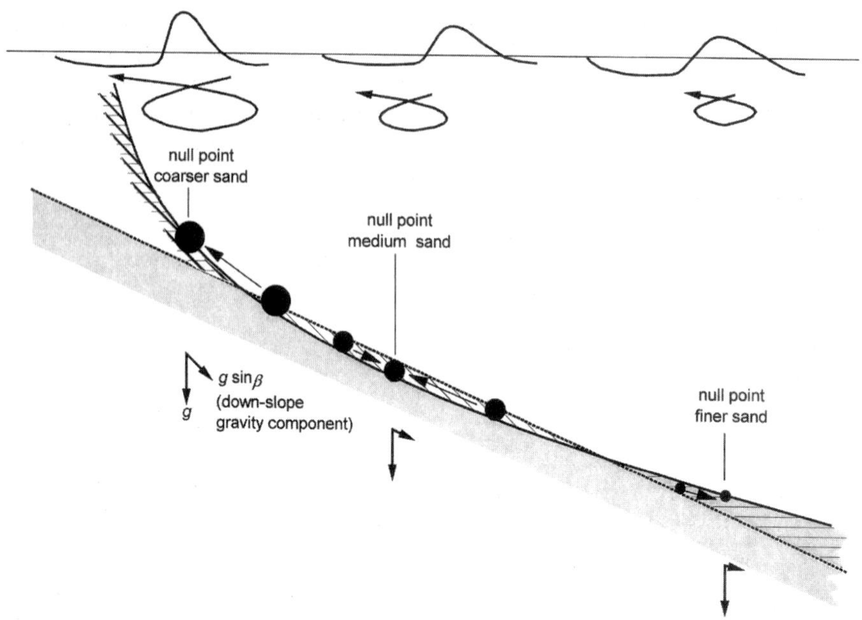


**Figure 3.8** Profile adjustment due to null-point equilibrium in which profile concavity develops in response to shoreward increase in wave asymmetry, with coarser sand grains moving into equilibrium on steeper slopes.

and the wave conditions; the coarser the grain size the shallower the *null point*, resulting in a seaward-fining profile if a mixture of sediment sizes were present. The *null-point* concepts as developed by Cornaglia (1889) were explored further by Ippen and Eagleson (1955), Eagleson *et al.* (1958, 1963) and Eagleson and Dean (1961).

Expressing Cornaglia's hypothesis in terms of the forces operating over one wave cycle, Ippen and Eagleson (1955) obtained the following expression for the net sediment velocity

$$(C_d + C_r)\rho\frac{\pi D^2}{8}(u_m - u_s)^2 = \frac{\pi}{6}(\rho_s - \rho)gD^3\sin\beta\cos\theta \tag{3.9}$$

where $C_d$ is the drag coefficient, $C_r$ is the coefficient of rolling friction, $\rho$ is the water density, $\rho_s$ is the sediment density, $u_s$ is the average velocity of the grain, $\beta$ is the slope of the bed, $\theta$ is the angle between the wave crests and the bottom contours, and $u_m$ is the onshore, near-bottom mass transport velocity, given by Longuet-Higgins (1953) as

$$u_m = \frac{5}{4}\left(\frac{\pi H}{L}\right)^2 C\frac{1}{\sinh^2 kh} \tag{3.10}$$

where $C$ is the wave phase speed. From this expression, Eagleson *et al.* (1963) obtained the following formula for the grain size, $D_e$, in oscillating equilibrium (units in metres)

$$D_e = \left[40.3\frac{H_o^2 v}{gTL_o}\frac{\rho}{\rho_s - \rho}\frac{F_e(h/L_o)}{\sin\beta}\right]^{\frac{7}{6}}\left(\frac{\pi}{vT}\right)^{\frac{2}{3}} \tag{3.11}$$

where $H_o$ is the deep-water wave height, $v$ is the kinematic viscosity of water, $L_o$ is the deep-water wave length, and

$$F_e(h/L_o) = \frac{\coth^2 kh}{\sinh^2 kh + k_o h} \tag{3.12}$$

This expression relates the grain size in oscillating equilibrium to the deep-water wave conditions. It applies to non-cohesive sediment in a laminar boundary layer (Jago and Barusseau, 1981), although rough-turbulent boundary layers usually prevail on the shoreface (Nielsen, 1992).

Field studies from south-east Australia (Hanslow, 1989), the Gulfe de Lion in France (Jago and Barusseau, 1981), and south-west Lake Michigan, USA (Graf, 1976), all found that grain-size variation with water depth, predicted on the basis of Equation 3.11, is vastly different to that of observed grain sizes, particularly in shallow water where grain size is over-predicted. The formulations of Eagleson and Dean, while qualitatively predicting a mutual adjustment between a concave-up profile and seaward fining in grain size, quantitatively do not predict the observed patterns of grain size distribution.

Bowen (1980) noted that the *null-point* hypothesis, as developed by Eagleson and Dean (1959), is seriously at odds with observations of grain motion. The hypothesis predicts that, if a range of grains of different sizes is placed on the seabed, then only one size can remain at the *null point* water depth, with coarser grains tending to move offshore and finer grains onshore, contrary to observation. Field tracer studies by Zenkovich (1946) and Murray (1967) generally show that coarser particles tend to move onshore. The experiments of Zenkovich (1946) suggest that a *null point* exists in nature, but the theoretical treatment of the motion of the grains is incorrect.

Bowen (1980) therefore adopted a different rationale for the *null-point* concept. Following an approach developed for the intertidal beach by Inman and Bagnold (1963), Bowen used Bagnold's (1963, 1966) so-called *energetics* approach for suspended and bedload transport to provide an expression for the local equilibrium slope defined by conditions under which net transport is zero. Bagnold's formulations essentially assume that the waves agitate the sediments and any excess shear stress causes the transport (Bailard and Inman, 1981). Bowen's approach further assumes, as did the original *null-point* hypothesis, that sediment transport averaged over several wave periods should be zero everywhere. Equilibrium results when the onshore movement of both the bedload and the suspended load fractions are balanced by the seaward transport of primarily the suspended fraction under the influence of gravity. Coarser grains move onshore while finer grains move offshore as suspended material, consistent with observations (Zenkovich, 1946; Murray, 1967). The overall result is the concave-up profile and progressive decrease in grain size with water depth, characteristic of Johnson's (1919) *graded* shoreface. As with the Eagleson and Dean (1959) model, the actual form of the profile in principle depends on the wave climate and the available range of sediment sizes.

The sediment transport rates given by Bagnold's *energetics* approach (Bailard and Inman, 1981) are, for suspended load transport

$$q_s = \frac{\varepsilon_s C_d \rho u^3 |u|}{w_s - u\beta} \tag{3.13}$$

and for bedload transport

$$q_b = \frac{\varepsilon_b C_d \rho u^3}{\tan\theta - u\beta/|u|} \tag{3.14}$$

where $\varepsilon_s$ and $\varepsilon_b$ are transport efficiency factors, $w_s$ is the sediment settling velocity (e.g. Raudkivi, 1976), $u$ is the horizontal water velocity near the bed (positive seawards), and $\tan\phi$ is the friction angle of the sediment, with the other variables as defined previously. The total transport is given by the sum of $q_s$ and $q_b$ integrated over one full wave period, with equilibrium manifesting as $q_s + q_b \to 0$. This occurs when the combination of grain size and slope give equal total-load transport in both the onshore and offshore directions over one wave period.

The onshore–offshore asymmetry in near-bed velocities, $u$, required in Equations 3.13 and 3.14 (Figure 3.9a) are commonly estimated from second-order *Stokes wave theory* (e.g. Martinez, 1987), according to which

$$u = \frac{\pi H}{T}\frac{\cos(kx - \omega t)}{\sinh(kh)} + \frac{3}{4}\left(\frac{\pi H}{L}\right)^2 C\frac{\cos[2(kx - \omega t)]}{\sinh^4(kh)}, z = -h \tag{3.15}$$

A similar approach to that underlying Equation 3.15 was developed by Bailard and Inman (1981) in deriving a simpler expression based on the second and third order velocity moments over a wave cycle, such that

$$\tan\beta = \tan\phi_s \langle u^3(t)\rangle / \langle u^2(t)\rangle^{3/2} \tag{3.16}$$

where the angular brackets denote time-average quantities.

Field studies indicate that the equilibrium condition $q_s + q_b = 0$ does not occur on the shoreface in relation to the observed grain sizes at different water depths, but that there is some correlation between grain sizes measured in the field and the point in reversal of total transport based on Equations 3.13 and 3.14 (Richmond and Sallenger, 1984; Hanslow, 1989). This transport reversal occurs when

$$w_s = u\beta \tag{3.17}$$

For grains smaller than this criterion (i.e. with lower settling velocities) the transport is onshore, while for larger grain sizes the transport is offshore. Therefore, while the grain sizes given by Equation 3.17 may agree with observations from nature better than the proper theoretical condition $q_s + q_b = 0$, it suffers from precisely the same shortcoming as the Eagleson and Dean (1957) *null-point* model.

Figure 3.9a illustrates the difference in the effect of wave orbital asymmetry on different grain sizes. The higher thresholds for movement of coarser sand fractions increase the dominance of onshore transport tendencies for these sediments compared to finer fractions. Heuristically, this appears to explain the observed pat-

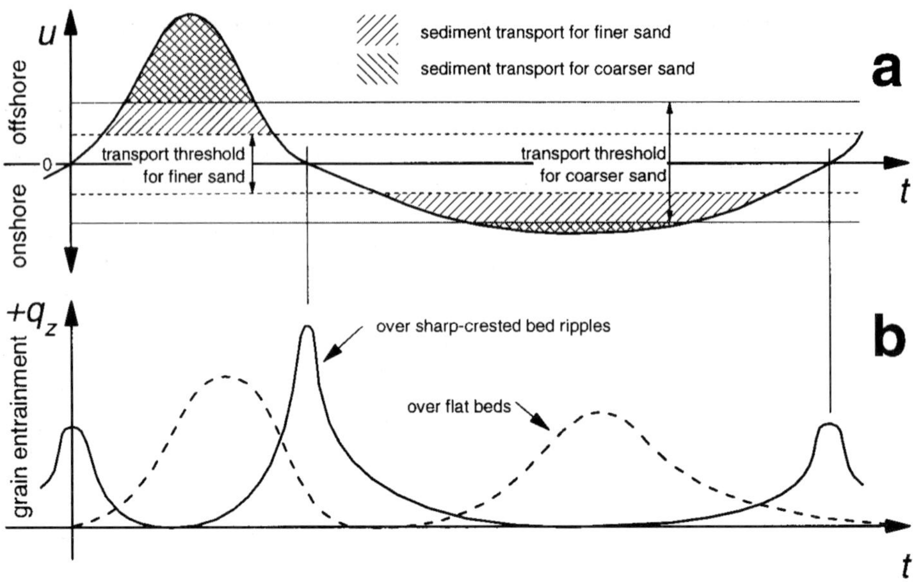


**Figure 3.9** Asymmetry in (a) near-bed water velocities under waves causing potential for onshore transport of sand (adapted from Komar, 1976); and (b) complicating effects of phase shifts between orbital velocity and sediment entrainment (based on Nielsen, 1992). The wave asymmetry increases in shallow water and has components of: (i) velocity magnitude asymmetry, and (ii) time asymmetry for flow in each direction. The potential effects for net sediment transport are amplified because transport depends on the velocity to a power of at least 3 and possibly 6 depending on the transport formula used (Wells, 1967; Bailard and Inman, 1981; Nielsen, 1988a). Note however that in reality the actual direction of net transport is known to be indeterminate due to the nature of mass transport in turbulent wave boundary layers (Trowbridge and Madsen, 1984) and the phase difference between instantaneous water velocities and suspended sediment concentrations over sharp-crested ripples (Inman and Bowen, 1963; Nielsen, 1992).

tern of coarser sand in shallowest depths, with progressive fining in the seaward direction, as observed in nature. However, Figure 3.9a portrays an over-simplification if consideration is given to the time-dependent nature of sediment entrainment under waves on the shoreface. Nielsen (1992) shows that, even over flat beds which predominate toward the surf zone on the upper shoreface (Figure 3.10), sediment entrainment volumes (dashed curve in Figure 3.9b) lag behind the oscillatory water velocities (which follow the curve in Figure 3.9a). Beds with sharp-crested vortex ripples experience an entrainment phase lag (solid curve in Figure 3.9b) that nears 90° with a significantly enhanced asymmetry between onshore and offshore entrainment volumes (Nielsen, 1992).

The result of these phase shifts in entrainment is that the direction, not just the magnitude, of net sand transport becomes uncertain under conditions of orbital asymmetry attributable to the effects of wave shoaling or a current superimposed on the wave field (Inman and Bowen, 1963; Wright *et al.*, 1991; Nielsen, 1992). This undoubtedly is one of the main reasons that sediment-transport prediction on the shoreface remains unreliable. The problem certainly violates the assumptions underlying the *energetics* model for sediment transport, which may go a long way to explaining the discrepancies between observed grain size patterns and those predicted from Bowen's (1980) *null-point* model.

The important role of bedforms in modulating sediment entrainment varies across the shoreface (Figure 3.10). In deeper water, where wave bed stresses are weaker, sharp-crested ripples prevail. Here, the effects of the lee vortices caused by the ripple crests produce the largest phase shifts in sediment entrainment (Figure 3.9b). Further inshore, the problem is reduced because ripple crests are rounded due to the effects of the higher wave-induced shear stresses (Nielsen, 1992). These stresses are indicated in Figure 3.10 in terms of the *grain roughness Shields parameter* (Madsen and Grant, 1976), which is expressed in terms of the median sand size, $D_{50}$, by Nielsen (1992) as

$$\theta = \frac{1}{2} f' \psi \tag{3.18}$$

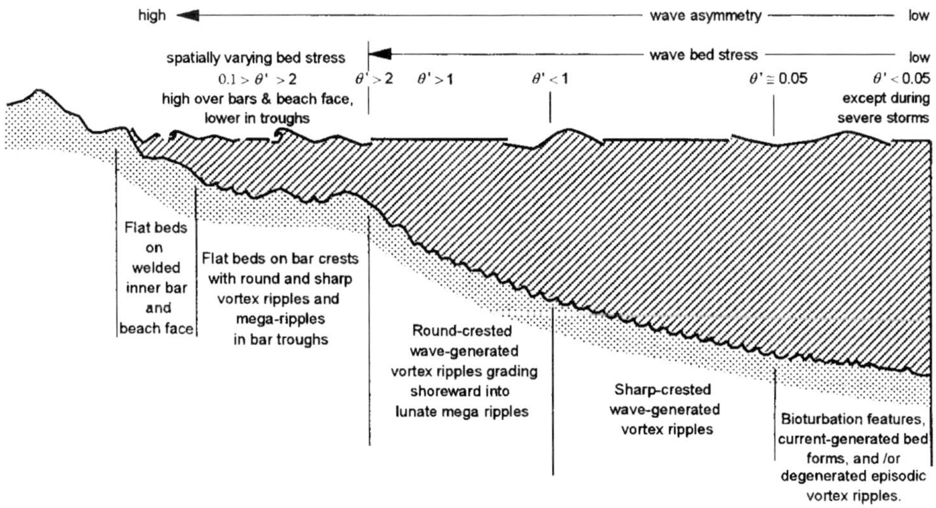


**Figure 3.10** Schematic distribution of bedform characteristics across the shoreface in relation to wave-related shear stress $\theta'$, or grain 'skin friction' (after Clifton *et al.*, 1971; and Nielsen, 1992).

where $f'$ is the grain-roughness friction factor

$$f' = \exp\left[5.213\left(\frac{2.5D_{50}}{a}\right)^{0.194} - 5.977\right] \tag{3.19}$$

and $\psi$ is the *sediment mobility parameter*

$$\psi = \frac{(a\omega)^2}{(\rho_s/\rho - 1)gD_{50}} \tag{3.20}$$

for waves with radial frequency $\omega = 2\pi/T$ and orbital semi-excursions $a$, immediately outside the boundary layer.

The zone dominated by vortex ripples occurs in shallower water while fair weather prevails. Estimation of across-shore sediment fluxes is most problematical under these conditions, as indicated in the field measurements of Wright *et al.* (1991). These measurements, from the lower shoreface on the US Atlantic coast (between 36° and 37° N), show that cross-correlations between suspended sediment concentrations and the instantaneous near-bed current speed were weak and out of phase during fair weather. During storms the measurements show that cross-correlations were high and in phase, with mean currents unambiguously suggesting offshore transport due to downwelling. Under these conditions, assumptions underlying the *energetics* transport model are more likely to be valid. On shorefaces subject to a *storm-wave climate* (Davies, 1980), it may be possible to assume that sediment transport is negligible under fairweather conditions. Then, the vortex ripple effects are perhaps less likely to confound attempts to predict the synoptic transport regime, such as that estimated for the central Netherlands coast by Roelvink and Stive (1991). Because the Netherlands shoreface is so shallow (see Figure 3.14b in Section 3.3.4), suppression of vortex ripples during storms may occur across the entire shoreface. On steeper shorefaces (e.g. south-east Australia and perhaps the US Atlantic), the zone of active vortex ripples probably shifts further offshore during storms (depending on storm intensity) into the area normally dominated by bioturbation features and current-generated bedforms (Figure 3.10).

Bagnold (1963) suggests that the condition in Equation 3.17 is that of *autosuspension* for bed sediments. Under these conditions, transport in the form of a *turbidity current* might be initiated. The importance of density-driven *turbidity currents* for shoreface sedimentation has been the subject of some debate. *Turbidity currents* are widely recognised as being important for the creation of graded beds in deeper water. Hayes (1967a) suggested that *turbidity currents* may be important for sediment transport in the shoreface environment during the ebb phase of *storm surges*, although it is now known that offshore transport can occur due to *downwelling* under geostrophic *storm-surge* conditions (Morton, 1981; Swift *et al.*, 1983; Niedoroda and Swift, 1991; Wright *et al.*, 1991, 1994a). Hayes argued that sediment-laden water generates density undercurrents that sweep down the shoreface forming graded beds. Morton (1981) cast some doubt on Hayes' (1967a) hypothesis, although

Walker (1979) had suggested that *turbidity currents* may also evolve from the *downwelling* mechanism advocated later by Morton (1981).

Swift (1984) suggested, on the basis of work by Cacchione and Drake (1982) and Clarke *et al.* (1982), that the threshold concentration required for autosuspension is several orders of magnitude higher than those commonly reported in storm-generated boundary layers. The term *autosuspension* here refers to the feedback nature of the transport mechanism. The sediment is suspended through fluid turbulence which itself is generated by the motion of the current. However, Wright *et al.* (1986a, 1991, 1994a) measured, during storms on the shoreface, boundary-layer concentrations of several grams per litre. These concentrations well exceed the *autosuspension* threshold of $C \approx 1$ g/l (Middleton 1966; Lowe, 1976, 1982) and are highly indicative of concentrations required in analytical models for *turbidity currents* applicable to the shoreface during storms (Seymour, 1986). Wright (1987) suggested that, where the shoreface is composed of fine-grained sediment, seaward transport may take place as gravity-induced movement of *hyperpycnal plumes* or underflows. He considered that these underflows need not maintain *autosuspension* in order to move sediment downslope, although this condition would permit the sediment movement to extend further seaward.

Overall, the existence of a *null point* is indicated empirically by the existence of *graded* shorefaces, and from the field tracer studies of Zenkovich (1946) and Murray (1967). It is also indicated in laboratory data of Schepers (1978) and Tilmans (1979) who measured sediment flux at various points down the length of a wave tank under conditions of varying waves, mean currents and for different grain sizes. These data (see also Nielsen, 1988a) suggest that for different combinations of the wave phase, and first and second harmonics of the velocity component, a *null point* is feasible. However, under conditions of continuously varying waves in nature (i.e. *non-stationarity*), the *null point* would be unstable (Sleath, 1984) and therefore must involve a *dynamic equilibrium* of the type described by Hack (1960) and Schumm (1991). Grains would be constantly dislodged from their old equilibrium positions as average wave conditions changed, and would migrate toward a new equilibrium site. Because of this, grain displacement necessarily must be toward equilibrium, rather than away from it as dictated by the Eagleson and Dean (1959) model of the process.

Furthermore, the reliance on the downslope gravity component (Figure 3.8) is another over-simplification in the standard *null-point* approach. The downslope gravity component provides the seaward-directed force required to balance against the onshore transport driven by near-bed orbital asymmetry induced through the shoaling of wind waves and swell. However, it is absurd to place too much physical meaning in the concept of sand being driven downslope by gravity on the shoreface (except in the case of *turbidity currents* described above), given that: (i) most shorefaces are of very low gradient, especially on the lower shoreface, such that shoreface slope $\beta \ll \phi_s$, the angle of repose for sand; (ii) the fluid drag in water overwhelms the small horizontal component of gravity where vortices characterise the boundary layer due the presence of bed ripples (Nielsen, 1992, p.162); and (iii) sharp-crested bed ripples (*vortex ripples*), that are ubiquitous outside the surf zone for much of the time (Figure 3.10), have average slope angles of 0.64 tan $\phi_s$ (Nielsen, 1992, p.128) which, again, is much greater than the mean local slope of the shoreface. This last effect of ripple steepness relates to the frequent periods of time when flow is not too vigorous in a particular range of water depths (Nielsen, 1992, p.128). At such times, but for the presence of ripples, an offshore-transport bias seems plausible due to enhanced downslope rolling of sand grains moving in traction mode.

During high-energy transport events on the other hand, when a considerable proportion of sediment movement involves suspension of sand within the wave boundary layer (Wright *et al.*, 1991, 1994a,b; Nielsen, 1992), the intensity of fluid drag and advection makes any tiny downslope gravity force seem irrelevant. If the downslope component of gravity does play a role in facilitating offshore transport, the effect is more likely to be indirect. For example, mild seaward slopes may exert an influence through their effect on concentration gradient diffusion (Inman, 1953; Niedoroda *et al.*, 1995). This process acts in the offshore direction due to the seaward decrease in time-averaged concentrations of mobile sediments ($C$ in Equations 3.7 and 3.8). The gradient, $\partial C/\partial x$ is due in turn to the reduction in near-bed wave activity as water depth increases since, $C = f(a\omega)$ in a general sense (Nielsen, 1992, p.217). It is the near-bed orbital motion that varies with water depth, as illustrated by *linear wave theory* in which

$$a = H/\sinh(kh) \tag{3.21}$$

Thus, the inclusion of slope in representation of equilibrium mechanisms may be thought of as a surrogate for other depth-dependent processes responsible for offshore transport of sediments. Regardless of how slope derives its physical significance, its inclusion in these *null-point* models is fundamental to the morphodynamic feedback responsible for the characteristic form of the shoreface. Furthermore, without this feedback, there would be no equilibrium (Cowell and Thom, 1994).

### 3.3.3 Entropy-maximising equilibrium

An alternative approach to profile equilibrium is based on the assumption that the co-adjustment of flows and morphology occurs in such a way as to produce uniform dissipation of energy across the shoreface. This assumption has precedents in broader concepts involving the analogy of thermodynamics in geomorphology (Strahler, 1952; Leopold and Langbein, 1962). The analogy was first drawn by Gilbert (1877) who introduced the concept of the 'graded' river in relation to channel equilibrium, a concept that Johnson (1919) borrowed for the shoreface. The thermodynamics analogy represents a macroscopic approach that sidesteps the need to scale up the sediment dynamics (Thorne and Swift, 1991) and thereby avoids the uncertainties inherent in sediment transport (as outlined above). The development of equilibrium then becomes an *entropy-maximising* process in which sediments diffuse to eliminate work gradients (Wright, 1987, 1995, p.18). In practical terms, this approach simply involves a representation of the system through equations of state that are solved for a given set of boundary conditions to yield the equilibrium form (Thorne and Swift, 1991). *Entropy-maximising* concepts have also been applied to the geomorphology of alluvial channel *thalwegs* (Leopold and Langbein, 1962) and shallow estuaries (Langbein, 1963; Wright *et al.*, 1973).

Proceeding from the assumption of uniform dissipation, and considering only wave energy, Bruun (1954b) and later Dean (1977, 1991, 1997) derived analytical solutions for the average cross-shore profile of the form

$$h = Ax^m \tag{3.22}$$

Inman (1953) suggested that bed transport seaward of the surf zone may take place in the absence of a net current if a gradient were to develop in the concentration of suspended sediment ($C$ in Equation 3.8). Transport under such conditions would occur as a result of a gradient-diffusion type process where grain movement is the result of essentially random motions (Crank, 1957).

Both Bruun (1954b) and Dean (1977) undertook their derivations for the surf zone and also compared their analytical solutions to a large number of upper shoreface profiles measured in the field within and just beyond the surf zone. Bruun used profiles from the coasts of California and the North Sea fronting Denmark while Dean analysed over 500 profiles from the Atlantic and Gulf of Mexico coasts of the USA. Nevertheless, Equation 3.22 is applied widely to the shoreface in general (i.e. well beyond the surf zone), probably as a result of currency gained through its applications in the prediction of coastal response to sea-level rise, i.e. the *Bruun Rule* (Bruun, 1962, 1988; Schwartz, 1967; Dean and Maurmeyer, 1983; Everts, 1987; SCOR, 1991; Cowell and Roy, 1988, 1992, 1995; Dubois, 1995, 1997; Dean, 1997). Both Bruun and Dean arrived at a value of $m = 2/3$ from analysis of the field data and their analytical derivations, despite use of different underlying assumptions; Bruun assumed uniform rate of energy dissipation per unit area of seabed (uniform *bed stress*), while Dean assumed uniform dissipation of energy per unit volume.

Dean's assumption makes physical sense inside the surf zone where energy dissipated by wave breaking is controlled, at first approximation, by variation in water depth (i.e. $H/h \approx \gamma$, a constant inside the surf zone). As Wright (1995, p.21) observes, outside the surf zone where wave energy dissipation is restricted to the effects of bottom friction, an assumption of constant dissipation rate per unit area of bed has greater validity. Dean's (1977) solution based on this assumption yields $m = 2/5$ which is consistent with best-fit values for field data from the lower shoreface analysed by Inman *et al.* (1993). However, upper shoreface profiles from the same study also indicate a best fit of $m = 2/5$, as do several other data sets from the surf zone cited by Wright (1995), despite the expectation that the assumption of uniform volume dissipation might be more appropriate because of wave breaking in the surf zone. Kotvojs and Cowell (1991) analysed upper-shoreface profiles along 1000 km of coast in south-east Australia and found a weak correlation between $m$ and the beach type of Wright and Short (1984), as Boon and Green (1988) had envisaged (Figure 3.11). *Reflective beach types* (see Chapter 7) gave the lowest values of $m \approx 2/5$, whereas profiles from more dissipative surf zones tended toward high values ($m \approx 4/5$), which is consistent with expectations based upon the two different assumptions concerning dissipation rates.

Dean (1991, 1997) plots $A$ graphically against grain size. Dean (1997) provides the empirical relationship

$$A = 0.067 w_s^{0.44} \tag{3.23}$$

in terms of sediment settling velocity, $w_s$, together with a more refined tabulation of recommended $A$ values for grain sizes in the range $0.10 \leq D \leq 1.09$ mm. However, field data from both surf zones and the upper and lower

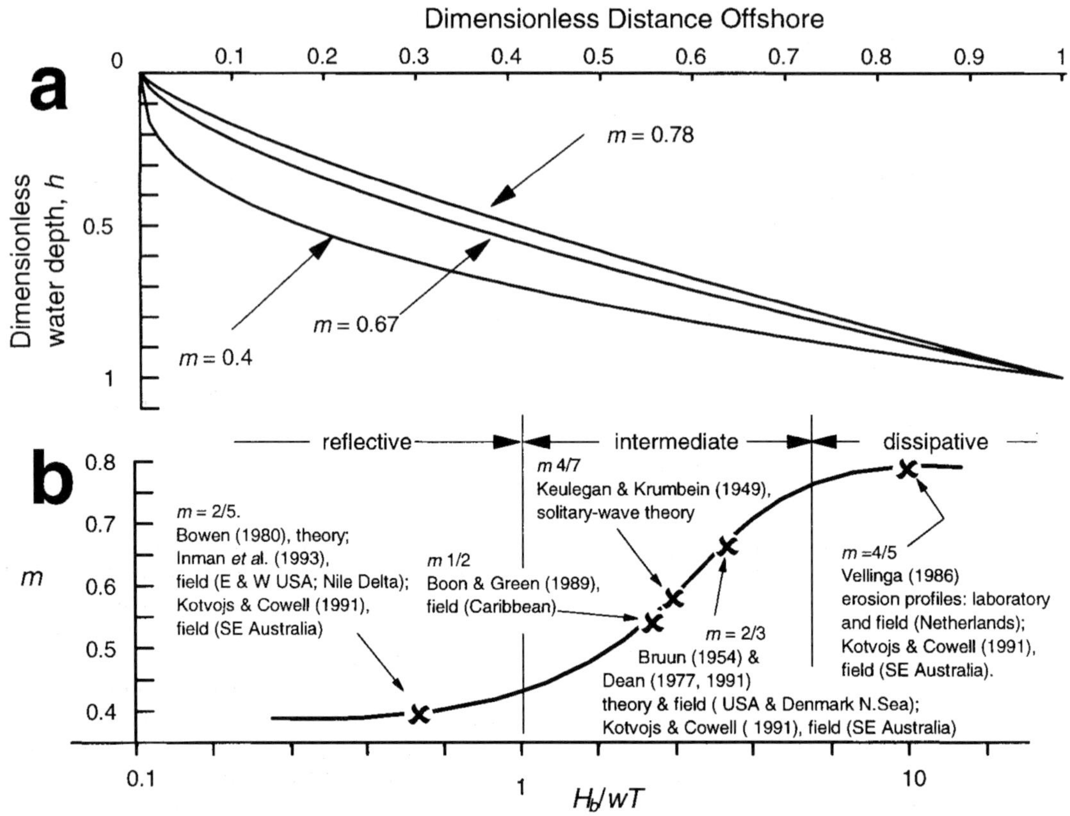


**Figure 3.11** Illustration for the application of Equation 3.22 using: (a) theoretical values of $m = 2/7$ and $m = 2/4$, corresponding to uniform dissipation rates with respect to unit water volume and bed area respectively; and (b) empirically tuned variation of $m$ in relation to beach type.

shoreface in south-east Australia indicate that Equation 3.23 is not particularly useful, and that $A$ is more reliably obtained from fitting Equation 3.22 to the measured profile (Cowell and Kotvojs, 1987; Kotvojs and Cowell, 1991; Meleo, 1994). While this does not help much in a predictive sense (e.g. in beach nourishment design using sand grades different from the original beach), the procedure has proved useful for shoreface modelling in general (Cowell *et al.*, 1995).

The main attraction of Equation 3.22 is its simplicity, so there have been several empirical attempts to improve on its representativeness, without adding too much complication (Bodge, 1992; Inman *et al.*, 1993; Komar and McDougal, 1994). Bodge reanalysed the profile data used by Dean (1991) and found that the exponential function

$$h = B(1 - e^{-kx}) \tag{3.24}$$

provided a consistently better fit than Equation 3.22, even when $A$ and $m$ were allowed to vary freely. The coefficients ($B$ and $k$) in Equation 3.24 are purely empirical. Komar and McDougal (1994) modified Equation 3.24 so that it could be defined in terms of conditions at its extremities (i.e. the slope of the beach face, $S_o$ and the closure depth, $h_c$ in Equation 3.1), whereby

$$h = \frac{S_o}{k}(1 - e^{-kx}) \tag{3.25}$$

in which

$$k = \frac{1}{x_c}\left[\frac{3}{2} - \left(\frac{6h_c}{S_o x_c} - \frac{15}{4}\right)^{1/2}\right] \tag{3.26}$$

with $x_c$ being the distance of $h_c$ from the shoreline. The explicit introduction of $h_c$ into this model means that it applies to the upper, but not the lower shoreface. In principle however, it is possible to include the lower shoreface either by: (i) extrapolating Equation 3.25 out

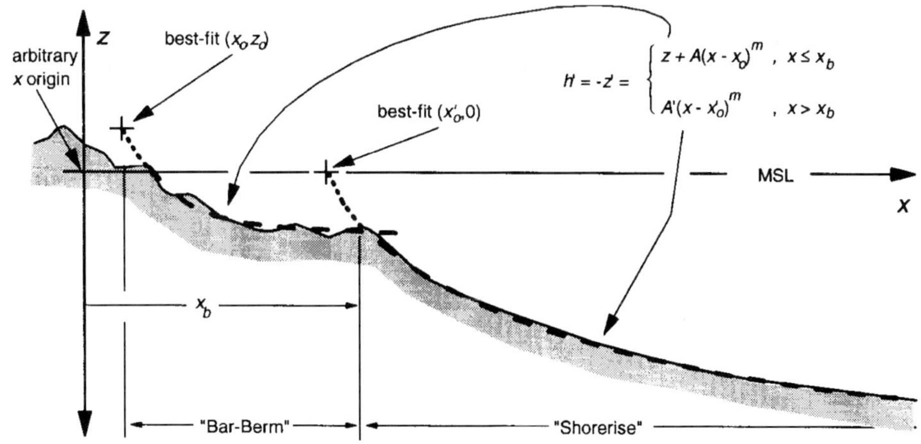


**Figure 3.12** Compound shoreface proposed by Inman *et al.* (1993).

to $h_i$, or (ii) substituting $x_i$,$h_i$ into Equation 3.26 in place of the closure depth coordinates. Neither possibility has been explored, although by increasing the scale out to $x_i$,$h_i$, the significance of the beach face $S_o$ boundary condition would be diminished considerably. Regardless, Equation 3.25 provides added justification for redefining the shoreface to include the surf zone, as suggested in Figure 3.1.

As it stands, Equation 3.25 corresponds to the *bar-berm* section of the compound profile (Figure 3.12), defined by Inman *et al.* (1993). Inman *et al.* apply Equation 3.22 to both the outer and inner profile (*shore rise* and *bar-berm*). They fitted their compound profile to data from a range of environments with markedly different wave climates and continental shelf settings, including California (where they principally used eight transects measured over 40 years), North Carolina (from the CERC research facility at Duck), and the Nile delta coast. They found that $m \approx 2/5$ applied nearly as well to both parts of the profile. Given the environmental variability inherent in the data sets, this study provides the most compelling evidence that the assumption of *uniform bed stress* is widely applicable, although $A$ (in Equation 3.22) varied widely within and between the two elements of the profile. This is consistent with the results obtained by Kotvojs and Cowell (1991) and Meleo (1994) and further indicates that site-specific values of $A$ obtained empirically are more reliable than that specified by Equation 3.23, even if such values have less physical meaning. It is worth noting that Dean (1977) originally referred to $A$ as the scale factor (and $m$ as the shape factor), which remains the only generally applicable interpretation of $A$. As such, $A$ probably incorporates not only the effects of regional variations in grain size, expressed in Equation 3.23, but also other geological controls such as the general gradient of the continental shelf, which is a function of the tectonic setting and depositional history on time scales of the Cenozoic (Inman and Nordstrom, 1971).

Having defined the equilibrium condition based on this general entropy-maximising approach, it is then possible to consider profile change in terms of a *disequilibrium stress* based on the deviation of conditions from either uniform bed stress or dissipation rate per unit volume, depending upon which assumption is applied (Kriebel and Dean, 1985; Larson and Kraus, 1989). Cowell and Thom (1994) demonstrate that adjustment of the shoreface on coastal barriers, following the onset of a sea-level stillstand, involves relaxation times on the order of $10^3$ years. In addition, most shorefaces can be expected to be undergoing adjustments to seasonal and other secular changes in wave climates. Dean (1997) claims that profile equilibrium takes from 1 to 5 years to re-establish following nourishment of the upper shoreface. This being the case, it is likely that many shorefaces will not have a form described by one of the above functions (Equations 3.22, 3.24, 3.26). However, no studies have been undertaken so far that explore long term evolution of the shoreface due to this type of *disequilibrium stress*, although the effects of storms on surf zones have been investigated using this approach (Kriebel and Dean, 1985; Larson and Kraus, 1989).

One element missing from initial versions of the Bruun–Dean model, but included in the *null-point* concept, is the offshore fining in grain size. However, Cowell and Kotvojs (1987), Dean (1991) and Kotvojs and Cowell (1991) showed that the effects of grain size variation with water depth could be included by variation in

$A(x)$. Jago and Barusseau (1981) have shown that, for a particular transect, the seaward-attenuating annual wave-power profile and the seaward trend in textural fining show very similar gradients. While the similarity of these two profiles need not indicate a mechanistic relationship between them, it does support the view that the seaward-fining grain size distribution may be subject to gradient diffusion.

This possibility was investigated by Bruun (1986) in relation to the entropy-maximising, constant bed-stress assumption for conditions outside the breaker zone. Bruun suggested that constant bed stress is most likely to hold true out to a depth exceeding $3.5H_{bs}$, where $H_{bs}$ is the significant wave height at breaking for severe storm conditions (i.e. return interval of decades). For example, the wave climate on the south-east Australian coast experiences severe storms with $H_{bs} \approx 10$ m. Here therefore, $3.5H_{bs} \approx 35$ m, which is comparable to regional values of Hallermeier's $h_i$ (Table 3.2). Bruun bases the constant bed-stress assumption on the reasoning that the decrease in wave height across the shoaling zone will be such that the near-bed velocity, and hence the bed shear stress, remains almost constant: 'The shoaling effect decreases the wave height, but the decrease in wave height in first as well as second order equations

$$u_{\max} = \pi H/T \cosh(2\pi h/L) + \ldots \tag{3.27}$$

is balanced by a decrease in $\cosh(2\pi h/L)$ so that $u_{max}$ stays almost constant' (Bruun, 1986, p.205). The decrease in wave height is presumably related to friction, not shoaling, but Bruun does not elaborate. The shear stress, which is given by

$$\tau_{\max} = \frac{1}{2} f_e u_{\max}^2 \tag{3.28}$$

will be constant if $(D/a)^{3/4}$ is constant, assuming: (i) the energy dissipation factor is

$$f_e \approx (\kappa/a)^{3/4} \tag{3.29}$$

and (ii) $\kappa \approx D$ for a near-bed, orbital semi-excursion of $a$ (Bruun, 1986). From this, Bruun derives the relationship

$$K = h^{5/12} D^{3/4} \tag{3.30}$$

for the variation of grain size with depth on the upper shoreface, which simplifies to the power function

$$D = K_1 h^n \tag{3.31}$$

where $K$ and $K_1$ are constants and $n = -5/9$. Sato and Kishi (1954), cited in Sleath (1984), also derived a similar relationship to Equation 3.31, but based on different physical reasoning. They assumed that cross-shore distribution of grain sizes is related to variation in the threshold of sediment motion with water depth for different grain sizes. Sato and Kishi arrived at a different value for the exponent in Equation 3.31 accordingly, i.e. $n = -3/2$. Sleath (1984) judged that Kishi and Sato's value of $n$ gives a faster decrease in grain size with water depth than supported by experimental data, a conclusion that is supported by analysis of upper shoreface data undertaken by Hanslow (1989), who obtained a modal value of $n = -0.1$. The analysis included grain-size data from the coast of south-east Australia, as well as data from North Carolina (reported by Swift *et al.*, 1971), and California (reported by Inman, 1953; Trask, 1955). Using Bruun's (1986) value of $n = -5/9$, Hanslow obtained a modal value of $K_1 = 0.2$, varying through a range between 0.15 and 0.45.

Bruun (1986) points out that the irregularity of waves on natural beaches, together with variations in intensity of different storm events, would mean that the theoretical distribution of grain sizes may never be produced. Smaller storms could carry grains to points of temporary deposition which would lead to a somewhat mixed distribution. This would simply cause a *disequilibrium stress* conceptually similar to that proposed by Kriebel and Dean (1985) in relation to profile morphology. The sorting of sediments across the shoreface may therefore be a response to departure of the grain-size pattern from that predicted by Equation 3.31. The validity of the uniform *bed-stress* assumption must become increasingly questionable as the water depth increases, especially beyond the domain specified by Bruun (1986), i.e. $h > 3.5H_{bs}$. However, this limiting depth in south-east Australia for example, is approximately 35 m, which is comparable to $h_i$ in Table 3.2.

### 3.3.4 Dynamic adjustment

In reality, shoreface morphodynamics are much more complicated than apparent from the perspective of *null-point* or *entropy- maximising* concepts outlined in the last two sections. Viewed in relation to instantaneous deterministic processes, sediment transport is the result of time-dependent advection of sediments by flows. Since many of the onshore–offshore flows undergo reversal on the shoreface over a wide range of time scales, the net transport is generally a small residual of the total transport. Then the time-average transport takes the form of a convective flux. For the most part, sediment advection by flows other than waves on the shoreface still involves agitation of sediments by waves, with the

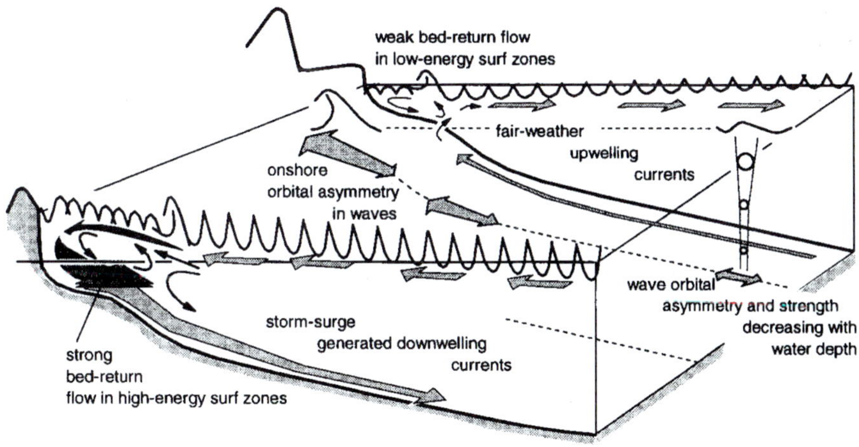


**Figure 3.13** Flows driving the primary components of cross-shore sediment transport on the shoreface (in the absence of inlets and estuaries where cross-shore tidal components also become important).

non-wave flows responsible for net drift of sediments, i.e. Bagnold's (1963) assumption, albeit subject to the uncertainties raised in the context of Figures 3.9 and 3.10. Equilibrium of the shoreface profile occurs only when all the counteracting mechanisms responsible for transport residuals reach a balance and the residuals cancel exactly. Then, profile equilibrium can be expressed in the context of Equation 3.7 and, considering across-shore sediment fluxes only, as

$$\frac{\partial h}{\partial t} = \varepsilon\left(\frac{\partial C}{\partial t} + \frac{\partial q_x}{\partial x}\right) - \frac{\partial V}{\partial t} = 0 \qquad (3.32)$$

Over time scales of seasons or years, generally $\partial h/\partial t \neq 0$ (Wright, 1995), and perhaps not even over millennia (Cowell and Thom, 1994). Over periods of days to seasons, the profile changes are by definition observable down to water depths of $h_c$, and in principle occur at a sub-detection level down to $h_i$ (as explained in Section 3.2). These changes relate to *dynamic equilibrium*, where morphodynamic adjustment occurs in response to 'small' variations in the forcing (waves and currents).

Over the previous two decades, significant advances have been made in knowledge of the important components of flow (Figure 3.13) that constitute the shallow-ocean climatology relevant to shoreface morphodynamics (Niedoroda *et al.*, 1985a; Niedoroda and Swift, 1991; Stive *et al.*, 1991; Wright, 1995). *Stochastic* and *deterministic complexity* of the flows and associated morphodynamic processes still defy a completely process-based approach to predicting shoreface change (Cowell and Thom, 1994; Stive and De Vriend, 1995; Niedoroda *et al.*, 1995). However, an understanding of these flows remains necessary for evaluation of more generalised models as well as the highly parameterised, multi-process models that are emerging (e.g. Roelvink and Stive, 1991; Roelvink and Brøker; 1993; De Vriend *et al.*, 1993). For instance, it is clear that the scope for geographic variability in shoreface characteristics and behaviours is expanded considerably if sediment transport is driven by a coastal ocean climatology that includes more than just wind waves and swell. The balance between counteracting sediment advection processes can be expected to depend on the relative contributions of each type of flow in Figure 3.13 (and possibly others) to the flow field. When this potential for variability is included with the other factors, such as the geological setting of the continental margin, then shorefaces can be highly idiosyncratic, as illustrated in Figure 3.14.

Onshore transport of sand remains dominated by the effects of wave asymmetry, as outlined in relation to the approximations contained in Equations 3.13 to 3.16. Other secondary but possibly significant flows include inshore *upwelling Ekman currents* (Figure 3.15a) associated with wind-driven geostrophic currents (Niedoroda *et al.*, 1985a) and possibly a small drift component due to tides (Roelvink and Stive, 1991). However, offshore sediment transport is now thought to be dominated by *downwelling Ekman currents* across the lower shoreface (Figure 3.15b), and by so-called *undertow* (the *bed-return flow* described in Chapter 4) driven by wave breaking on the upper shoreface (Niedoroda *et al.*, 1985a; Stive *et al.*, 1991). Swift *et al.* (1983) review work in which *downwelling* velocities of 60 cm/s were measured; velocities were highest on the upper shore-

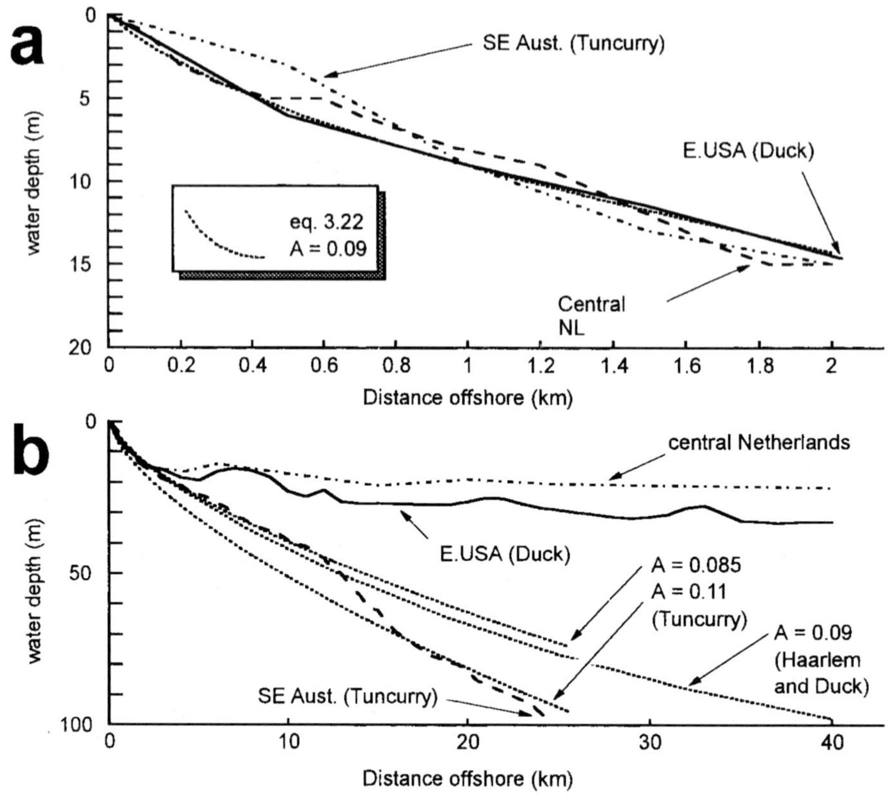


**Figure 3.14** Contrasting shorefaces from SE Australia (Tuncurry), The Netherlands (Haarlem) and the Atlantic Coast of the USA (Duck, North Carolina), together with theoretical profiles based on Equation 3.28 for (a) the upper shoreface, and (b) the lower shoreface. The environmental differences (coastal-ocean climatology and geological setting) clearly must be incorporated in A.

face where water surface slopes are steepest. Wright *et al.* (1991, 1994a) reports near-bed, *downwelling* velocities of up to 15 cm/s in water depths of 8 to 13 m during a storm on the US Atlantic coast. Presumably therefore, *bed-return flow* and *downwelling* are superimposed on the upper shoreface during storms, including inside the surf zone.

Rip currents contribute to offshore sand transport during storms (Cook, 1970; Cook and Gorsline, 1972; Reimnitz *et al.*, 1976), although their effects are probably limited to exchanges between the surf zone and upper shoreface (Cowell, 1986b; Wright *et al*, 1991). Moreover, *bed-return flow* is probably of more ubiquitous importance than rip currents. The alongshore location of the latter is often topographically fixed, especially on coasts with prominent bedrock headlands, but their position tends to be arrested even by surf zone morphology (Short, 1985). Johnson and Eagleson (1966) dismiss rip currents as being of major importance because of the shore-parallel nature of seaward-fining sediment distributions observed on the upper shoreface. Offshore sediment advection during storms appears likely to be assisted by the formation of turbidity currents initiated by the effects of strong wave agitation of the bed (Seymour, 1986; Wright *et al.*, 1986a; Wright, 1987), as outlined in Section 3.3.2.

Both *bed-return flow* and *rip currents* occur only when wave breaking is active and the strength of these flows increases with height of waves at breaking. Therefore, like *downwelling*, both are most active during storms. Wind waves and swell can occur more continuously, especially on *swell-dominated coasts* (Davies, 1980). Wright (1995) reports current- and sediment-mobility measurements that show that the dominance of offshore over onshore sediment transport is actually determined by the climatology of individual storm events, and may vary even during the rise and decay of a specific storm. Consequently, the morphodynamic behaviour of the

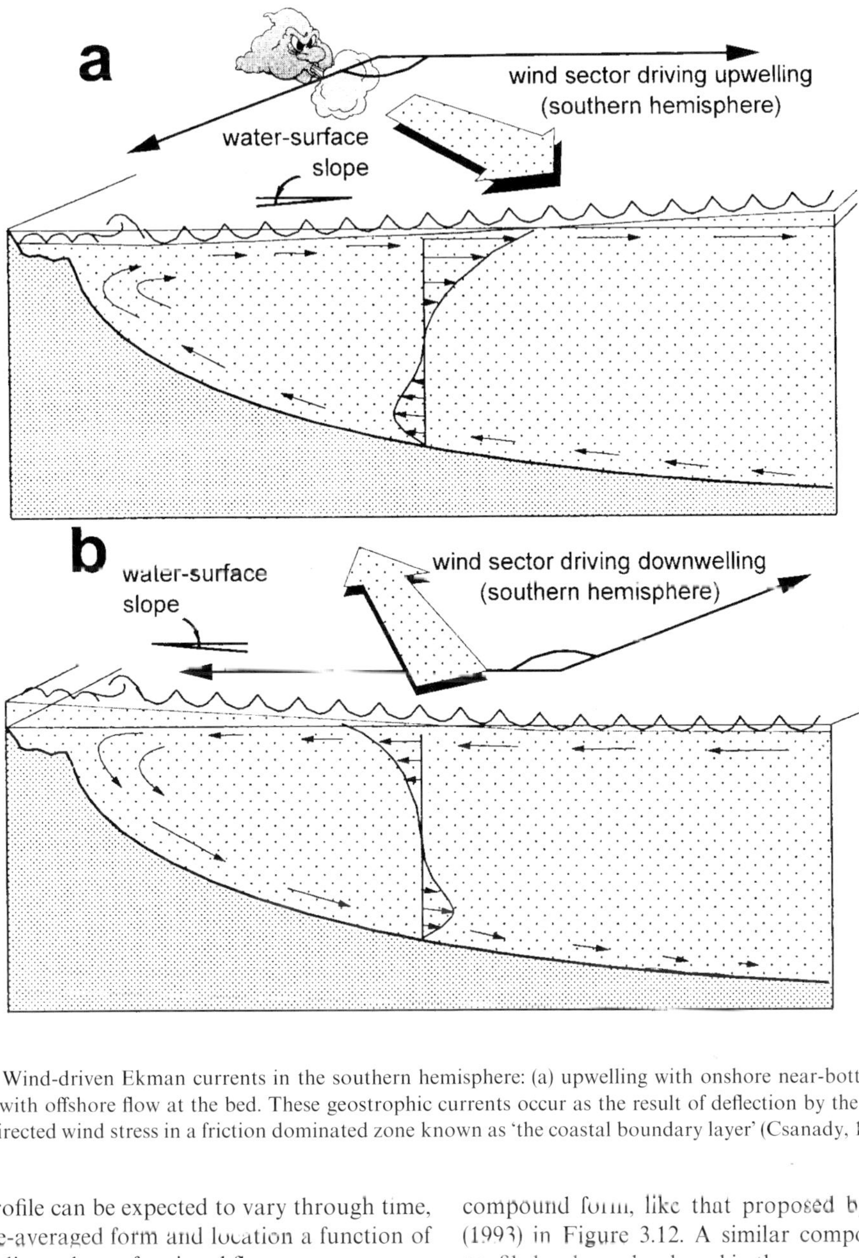


**Figure 3.15** Wind-driven Ekman currents in the southern hemisphere: (a) upwelling with onshore near-bottom flow; and (b) downwelling with offshore flow at the bed. These geostrophic currents occur as the result of deflection by the Coriolis force of alongshore-directed wind stress in a friction dominated zone known as 'the coastal boundary layer' (Csanady, 1977, 1982).

shoreface profile can be expected to vary through time, with its time-averaged form and location a function of the general climatology of regional flows.

The same principle should apply across the shoreface, as well as through time, since the suite of primary flows dominating the upper and lower shoreface are distinctly different (Inman *et al.*, 1993). Therefore, the equilibrium response (Equation 3.32) in terms of the net across-shore transport driven by these flows should also be distinctly different for the upper and lower shoreface. On this basis, it is reasonable to expect a profile with compound form, like that proposed by Inman *et al.* (1993) in Figure 3.12. A similar compound shoreface profile has been developed in the process of redesigning a model for simulating shoreface translation (Cowell *et al.*, 1992) which parametrically incorporates morphological features expected on the basis of the advective flow field described above. The profile, which takes the basic form of Equation 3.22, is defined by

$$h = WA_c x^{m_c} + (1 - W)A_i x^{m_i} \tag{3.33}$$

where the subscripts $c$ and $i$ denote the parameters

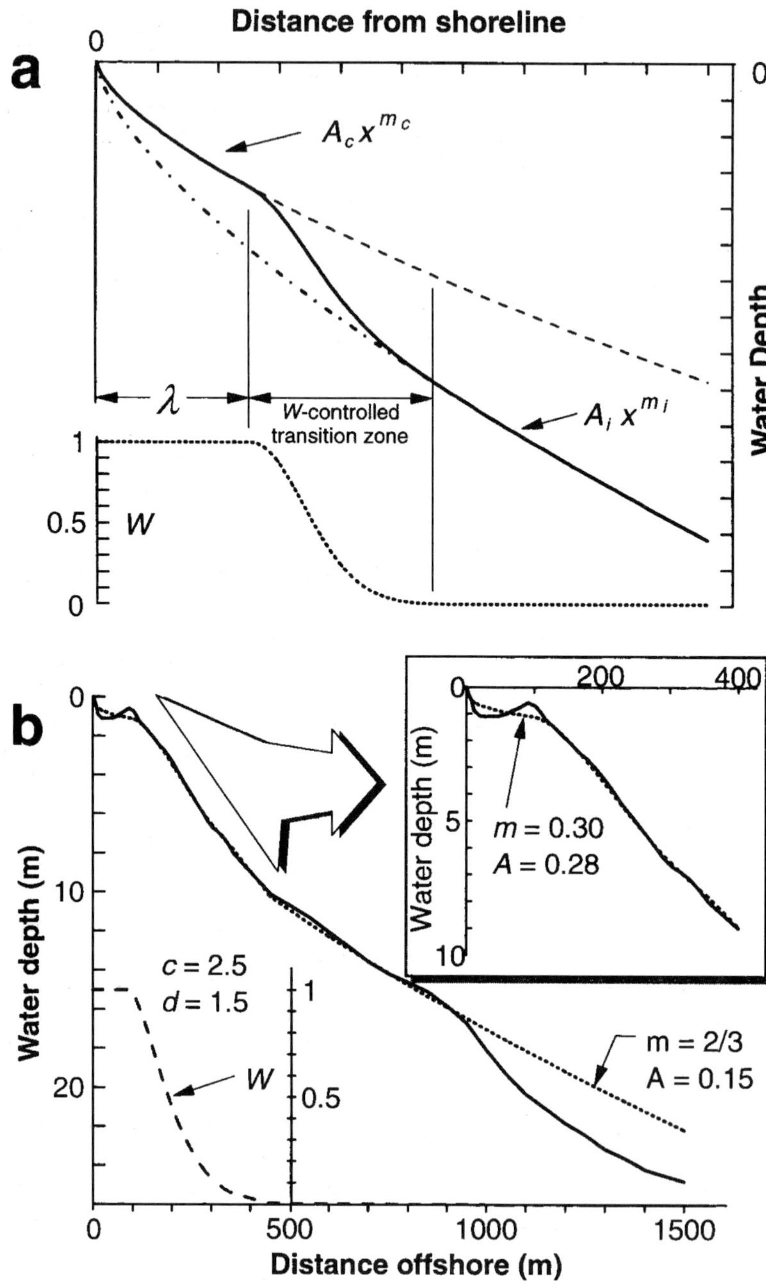


**Figure 3.16** Compound profile with transition zone that simulates the relative effects of upper and lower shoreface processes: (a) definitions; and (b) illustration.

relevant to the upper and lower shoreface respectively, and $W$ is a weighted transition function given by

$$W = \exp\left\{-\left[\frac{|\lambda - x| - (\lambda - x)}{\lambda c}\right]^d\right\} \tag{3.34}$$

in which $c$ and $d$ control the rate of roll-off from the upper to lower shoreface, and $\lambda$ is defined in Figure 3.16. The rationale behind Equation 3.33 is that the upper and lower shoreface respond to different processes all of which operate in combination with the shoreline as their landward boundary. This assumes that wave asymmetry and Ekman currents can contribute to sediment advection not only on the lower but also the upper shoreface, including in the surf zone. Thus the origin of the lower shoreface profile is the shoreline, which is the primary difference from the Inman *et al.* version. The extent and geometry of the transition zone are designed to reflect the degree to which *bed-return flow* dominates over *downwelling* activity on the upper shoreface. Increasing $c$ flattens the roll-off and extends the transition zone seaward, while $d$ affects the shape of the roll-off, such that increasing $d$ causes a seaward bulge in the transition profile (Figure 3.16).

Equation 3.33 attempts heuristically to capture a morphological response to the general flow field described above, but based on a parameter-fitting approach. The possibility of predicting the profile based on the wave-current flows themselves still seems a long way off. The stumbling block mainly relates to prediction of across-shore sediment transport over an important range of conditions. Quantitative estimation of onshore transport during fairweather conditions, when beach recovery occurs, is highly uncertain (Wright *et al.*, 1991), and it may be that lesser storm events are crucial in transporting sand from the lower shoreface toward the surf zone (Stive *et al.*, 1991).

## 3.4 MORPHODYNAMIC TIME SCALE AND EXTERNAL CONTROLS

Considered over intermediate time scales (years), shorefaces generally appear to be quite stable, except where there is significant gain or loss of sediments from alongshore. Even these changes might be seen as affecting only profile translation so that the shoreface may still appear to have a constant form. However, on longer time scales (centuries to millennia), there is increasing evidence that many shorefaces are continually but slowly evolving, even under conditions of prolonged sea level stillstand.

Morphodynamic time scales relate to a time lag inherent in the morphodynamic adjustment of the shoreface whenever forcing conditions change. In terms of adjustment toward a new equilibrium, this lagged response can be characterised by the *geomorphic relaxation time*

$$T_R = \frac{\Delta h}{\partial h / \partial t} \tag{3.35}$$

(another borrowing from thermodynamics) in which the numerator contains the total change in depth required to establish the new equilibrium at any point on the shoreface, and the denominator is given by Equation 3.7 (Cowell and Thom, 1994). *Geomorphic relaxation time* therefore depends on the sediment volume that must be displaced to achieve equilibrium. This volume increases with the horizontal dimensions of the bed area that is aggrading or degrading, and with $\Delta h$ averaged across that area. Because of this, *relaxation time* tends to increase with morphological length scale, given that sediment flux rates and gradients have upper limits (Cowell and Thom, 1994).

For a given change in bed height, such as Hallermeier's (1981) practical criterion of $\Delta h = 1/3$ m for profile closure, the denominator in Equation 3.35 decreases progressively with water depth down the shoreface. The decrease is expected due to the attenuation of wave motions in deeper water (Equation 3.21) and the consequent reduction in agitation of sediments by waves ($C$ in Equation 3.8) which limits the magnitude of sediment flux terms in Equation 3.7. The effect on the diminishing denominator in Equation 3.35 is to increase the time required for morphologically significant sediment transfers, as suggested in the plot at the bottom of Figure 3.1. The plot schematically depicts increasing morphological scale of the shoreface as the time scale increases. The longest of these time scales entail *morphogenetic* shoreface change, i.e. change involving the evolutionary history and hence the very formation of the shoreface. Only by considering shoreface behaviour at *morphogenetic time scales* (millennia) does it become possible to understand shoreface morphology and change in relation to the sediment patterns described in previous sections. This is the case especially for the lower shoreface since detectable changes here occur so slowly that direct observations are few. So far, these changes have been inferred mostly from sedimentological and stratigraphic investigations (e.g. Roy *et al.*, 1994). The large-scale behaviour of the lower shoreface also has significance for the upper shoreface because the two elements are kinematically coupled. This is an important consideration for coastal management in relation to issues such as offshore sand mining and predicting impacts of sea-level rise.

Relaxation times and shoreface behaviour in general also depend upon qualities inherited from the geological setting (Roy and Stephens, 1981; Pilkey *et al.*, 1993; Riggs *et al.*, 1995; Thieler *et al.*, 1995). Inheritance factors may affect *morphogenetic time scale* through (a) initial conditions set by the transgressive morphologies formed in response to the rapid rise in global sea levels that occurred at rates on the order of 0.01 m/a during the *post-glacial marine transgression* (e.g. Roy and

Thom, 1981; Roy *et al.*, 1994), or (b) the different response rates imposed by the resistance to movement for various materials in the substrate, which may range from rock, through sands in a cohesive sediment matrix, to cohesionless sediments whose mobility depends on grain size (Carter, 1988).

### 3.4.1 Kinematic coupling between the upper and lower shoreface

The separation of the shoreface into upper and lower units reflects changes in both the magnitude as well as the frequency of sediment-transport events (Wright, 1987). The boundary between the two units is to some extent arbitrary in that sediment transport does not suddenly stop at this point. There is however a difference in the suite of flows driving sediment transport processes within the two regions (Inman *et al.*, 1993), the upper shoreface being dominated by the highly active surf zone, and the lower shoreface dominated by shoaling-zone processes and wind-driven currents. In reality the boundary between these two hydrodynamic regimes varies through time, so that the distinction between the upper and lower shoreface is not absolute as sometimes assumed in coastal-management studies. Rather the definition is one of conceptual convenience, being made on the basis of the estimated maximum depth for nearshore erosion during annually recurring extreme-wave conditions (12 hours per year), following Hallermeier's (1981) approach. The upper shoreface so defined can be characterised as having *relaxation times* on the order of days in response to continual changes in wave conditions (Larson and Kraus, 1989, 1995), although responses to major disturbances are thought to be on the order of several years from experience gained through beach nourishment projects (Dean, 1997). On the lower shoreface, readily detectable movements in bed elevations (i.e. $\Delta h \geq 1/3$ m following Hallermeier's criterion) generally occur only over time scales of decades to millennia. As explained in preceding sections, these characteristic morphological time scales derive from the limiting effect that water depth has on rates of across-shore sediment transport and the larger length scales associated with the transport processes on the lower shoreface (Cowell and Thom, 1994).

Because of these limits to the rate of change, the lower shoreface exerts control over horizontal migration of the upper shoreface (Stive and De Vriend, 1995) and therefore governs macro-scale problems that underlie important issues in coastal management, such as (a) trends in shoreline advance and retreat due to changes in sea level over time scales of centuries to millennia (Cowell *et al.*, 1995; Stive and De Vriend, 1995), and (b) the rate of beach recovery, over time scales of years to decades, after beach erosion caused by the most severe storms (Thom and Hall, 1991; Lee *et al.*, 1998). Therefore, quantification of lower-shoreface change is an inescapable requirement for predicting large-scale behaviour of the upper shoreface (and therefore the all-important coastal-management question of shoreline position).

### 3.4.2 Inherited controls in shoreface response

During the rapidly rising sea levels of the *post-glacial marine transgression* (PGMT), it seems likely that the upper shoreface was capable of remaining in equilibrium while some of the sediments on the lower shoreface would have been left behind in the *shoreface ramp* (Stive and De Vriend, 1995). Therefore, the lower shoreface was probably well out of equilibrium at the end of the PGMT (Roy and Thom, 1981; Roy and Stephens, 1981). The site-specific morphological nature of this disequilibrium would have depended upon the morphology of transgressive deposits during the final stages of the PGMT. Computer simulation of shoreface evolution relevant to this period demonstrates the over-riding importance of regional slope in the coastal substrate under conditions of relative sea-level rise (Cowell and Roy, 1988; Cowell *et al.*, 1991, 1992, 1995; Roy *et al.*, 1994).

Regional slope is therefore a geologically inherited factor fundamental to the setting of initial conditions for ensuing *morphogenetic evolution* during the late Holocene. Simulation results relevant to the PGMT (Figure 3.17) show that on low-gradient substrates, *barrier rollover* (Swift *et al.*, 1991) dominates with a consequent net displacement of shelf sands landward. As a result, transgressive deposits are thickest in the landward direction, while a thinning lens of barrier sand extends seaward (Figure 3.17a). This lens overlies backbarrier sediments, but only if a net sediment volume is added to the transgressive deposits during the *barrier rollover* processes (Roy *et al.*, 1994). The old land surface forms the substrate to all these sediments. The fundamental behaviour over these low substrate slopes (i.e. ignoring added sediments) is in accordance with the *generalised Bruun Rule* (Dean and Maurmeyer, 1983).

Results from substrates that are steeper than the minimum slope on the translating shoreface indicate an opposite response to *barrier rollover* (Figure 3.17b). That is, the net displacement of sediments is seaward and occurs through excavation of the old land surface by the receding shoreface in accordance with the *stan-*

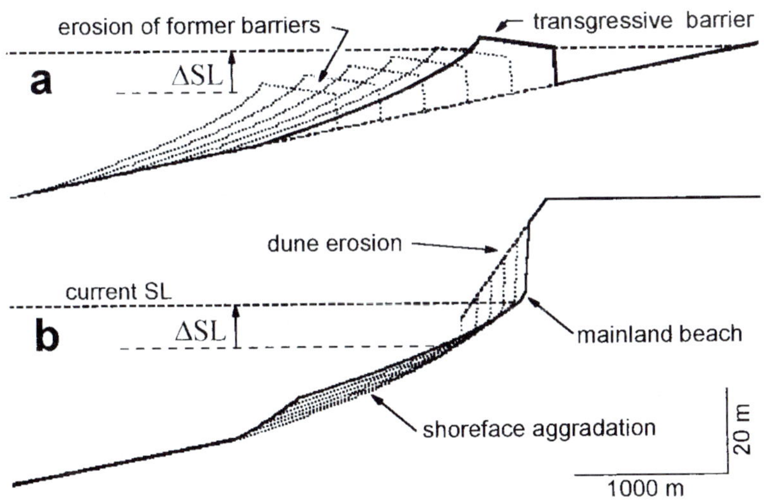


**Figure 3.17** Output plots from the Shoreface Translation Model (Roy *et al.*, 1994; Cowell *et al.*, 1995) characterising shoreface evolution under rising sea level over (a) a low-gradient regional substrate, and (b) a steep regional substrate. 'Substrate steepness' as used here is relative to the slope of the shoreface.

*dard Bruun Rule* (Bruun 1962, 1988; Dean and Maurmeyer, 1983). Consequently, initial conditions for ensuing *morphogenetic evolution* entail reworking by the shoreface of a sand body that is thicker in the seaward direction (Field and Roy, 1984; Roy *et al.*, 1994). Shelf substrate slope is an artefact of the regional geology and thereby an important source of geological inheritance contributing the initial morphological conditions that influence subsequent *morphogenetic evolution* of the shoreface. Other transgressive morphologies also affect initial conditions for shoreface evolution. For example, *shoal-retreat massifs*, produced in response to rising sea levels by the retreat of estuaries, river mouths and other convergences in the littoral transport system (Swift *et al.*, 1991; Wright, 1995), cause initially thicker deposits offshore at various places along most coasts. Initially, these areas of positive relief may be extensive in the vicinity of deltas such as that of the Rhine and Meuse rivers on the Netherlands coast where progradation occurred throughout the mid-Holocene as the *shoal-retreat massif* degraded by feeding sediments into the littoral transport system (Beets *et al.*, 1992).

Other sources of geological inheritance derive from the character of material that the shoreface must rework as it adjusts toward equilibrium. For example, although coarse, lag sands are generally associated with *ravinement surfaces*, the coarse zones in some other cases appear related to inheritance from former channel deposits (Cowell, 1986a; Riggs *et al.*, 1995; Thieler *et al.*, 1995).The effect of these deposits has been documented by studies from south-east Australia in Newcastle Bight and Shoalhaven Bight, about 120 km north and south of Sydney respectively (Roy and Crawford 1980; Roy and Stephens, 1981). Both embayments have long open beaches fully exposed to the dominant ocean swells and both show quite dramatic alongshore changes in sand size without corresponding alongshore variations in wave energy. These variations can be traced to former positions on the shelf of old river mouths (coarse sand) and aeolian dunes (fine sand) that presumably were reworked onshore as the present-day barriers formed (Figure 3.18). The presence of these grain-size trends in drill holes within the barriers is evidence for the inheritance of geological factors during the thousands of years of barrier growth (Roy and Crawford, 1980).

In morphodynamic response to these inherited sand populations, both embayments have developed a range of beach and shoreface morphologies: gentle offshore slopes with multiple bars in the case of fine sand areas, and steeper slopes with a single bar-trough system where the sediment is coarse grained. Presumably both morphologies are able to achieve dynamic equilibrium with a range of incident wave conditions without becoming effaced – something that process studies alone could not necessarily predict (Pilkey *et al.*, 1993; Roy *et al.*, 1994).

## 3.5 SUMMARY

The shoreface, as defined morphodynamically, extends from the limit of wave runup on the beach face, seaward to the limit of effective influence by gravity waves in

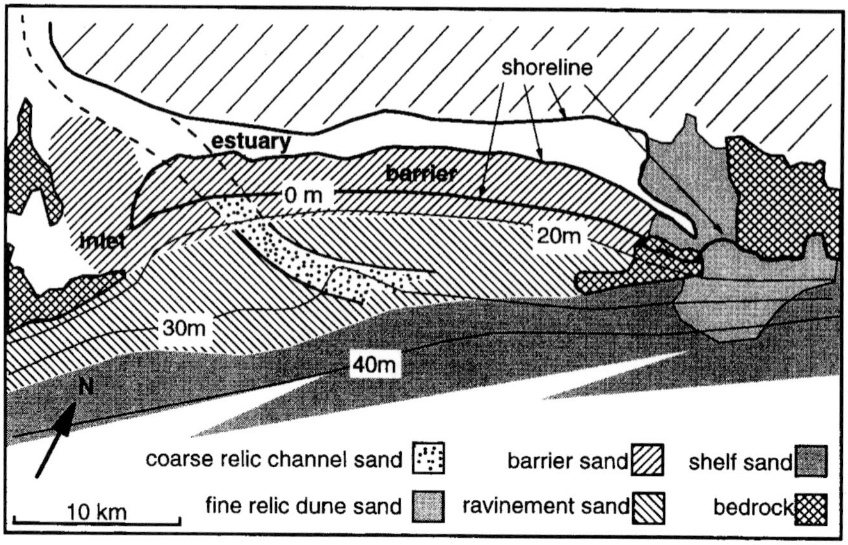


**Figure 3.18** Inherited sediment textures on the shoreface in Newcastle Bight, SE Australia (modified from Roy and Cowell, 1991).

shaping seabed morphology (Section 3.2). The offshore limit depends on time scale (Sections 3.2.2 and 3.4) but conventions concerning sediment disturbance during a typical year provide a reference point for this limit (Section 3.2.2). Surf-zone processes dominate the upper shoreface which extends out to the annual profile closure depth (Section 3.2.1). Profile change is greatest in amplitude and frequency on the upper shoreface. This means that the time-average profile form can be considered constant over decades to millennia. Surf-zone currents drive upper-shoreface morphodynamics, with wave shoaling and associated orbital asymmetry dominating onshore transport of sand. Breaking wave induced bed-return flows ('undertow') are primarily responsible for offshore sediment transport on the upper shoreface, although other lateral circulations involving rip currents also are important locally (Section 3.3.4).

The upper shoreface adjusts toward equilibrium in a matter of days in response to fluctuations in wave conditions. However, re-equilibration may take several years after a major disturbance such as beach nourishment (Section 3.3.1). Lateral and vertical translation dominate large-scale behaviour of the upper shoreface, depending on littoral sediment budgets, sea-level change and long-term changes on the lower shoreface. The morphological behaviour of the lower shoreface therefore controls the large-scale behaviour of the upper shoreface (Section 3.4.1). Other controls over upper shoreface behaviour (namely imbalance in littoral sediment budgets and sea-level change) may be more important under specific circumstances. However, the kinematic coupling between the upper and lower shoreface (Section 3.4.1) cannot be ignored in coastal management under any circumstances.

The general form of upper and lower shoreface profiles seem to be much the same (Section 3.3.3), but the morphological scales differ generally by a factor of 3 to 10, with the lower shoreface being the larger of the two (Section 3.2). This scale difference means that small profile changes on the lower shoreface can have disproportionate significance for the upper shoreface, given the constraints of mass continuity in sediment exchanges between the two zones. Unfortunately, little is known about the secular and fluctuating changes on the lower shoreface, so that equilibrium concepts provide the main point of reference (Section 3.3).

The lower shoreface requires millennia to evolve toward equilibrium, and probably many remain out of equilibrium today, even in high-energy environments such as south-east Australia, despite 6000 years of sea-level stillstand. Because of its slow rate of change, the lower shoreface tends to steer the upper shoreface in a landward or seaward direction, depending on whether evolution of the lower shoreface causes it to act as a net sink or source for sand-sized sediments (Section 3.3.1). Actual shoreline movements due to this coupling between the upper and lower shoreface also depends on other factors such as the littoral sediment transport budget. Similarly, inherited factors such as substrate slope and composition, as well as the style of antecedent coastal evolution, also exert control over behaviour and relaxation time of shorefaces in nature (Section 3.4.2). These inherited factors, together with the coupling between the lower and upper shoreface, remain poorly understood but cannot be ignored in long-term coastal management.

# 4

# The Surf Zone

*Troels Aagaard and Gerhard Masselink*

## 4.1 INTRODUCTION

The *surf zone* is that part of the nearshore where incident waves break and breaking-induced processes dominate the fluid motion, and hence sediment transport processes. This zone is located between the zone of wave shoaling and the swash zone.

The *seaward boundary* of the surf zone is at the onset of depth-limited breaking (thus disregarding offshore wind-driven whitecaps). On laboratory beaches with regular waves, the position of the *breakpoint* is constant since all waves have the same height and hence break at the same depth. On natural beaches, however, the wave field is irregular and the breakpoint moves onshore and offshore as a function of incident wave height. During a storm, the surf zone is significantly wider than under calm weather conditions. The *landward boundary* of the surf zone is the point closest to shore which is submerged during the entire wave cycle, i.e. the lower limit of the swash cycle. The location of this point is approximately at mean sea level, which in turn depends on tidal stage and wave/wind/storm set-up.

The surf zone is characterised by a complex mixture of wave and current motions operating over a range of frequencies. These motions are related to the breaking of incident waves and consist of high frequency turbulence, incident, infragravity and far-infragravity wave motion, and mean currents. Figure 4.1 represents a time series from a wave recorder which illustrates the irregularity of natural surf zone waves. A natural wave field possesses a range of wave heights and periods which can be described by histograms (Figure 4.1B, C) or more conveniently by a wave spectrum (Figure 4.1D). In this particular case, two predominant wave trains (swell and wind waves) were interacting to produce the complex time series in Figure 4.1A. A small amount of infragravity wave energy with periods longer than 15 seconds was also present.

This chapter discusses incident wave characteristics and transformations in the surf zone; various hydrodynamic phenomena generated by the wave breaking process (wave set-up, infragravity waves, surf zone currents); sediment dynamics and sediment transport; and finally the nearshore morphology generated by complex interactions between these phenomena, particularly the generation and migration of nearshore bars.

## 4.2 WAVE TRANSFORMATION AND WAVE SET-UP

### 4.2.1 Irregular wave field

Waves in natural surf zones are highly irregular (Figure 4.1A) and statistical techniques are required to properly describe the sea state in quantitative terms. Wave-by-wave analysis is one of the most common approaches to describe irregular waves. This technique consists of identifying the individual waves in the wave record using the zero-upcrossing or zero-downcrossing method (Figure 4.2), and applying statistical techniques to compute representative wave parameters. For surf zone waves the zero-downcrossing method is generally preferred (IAHR, 1989). A useful indicator of the *wave height* is the root mean square (r.m.s.) wave height

$$H_{rms} = \sqrt{\frac{\Sigma H_i^2}{N}} \tag{4.1}$$

where $H_i$ is the $i$th observation, $\Sigma$ indicates summation over $i$ from 1 to $N$ observations and $N$ is the number of waves in the data record. The r.m.s. wave height is an

*Handbook of Beach and Shoreface Morphodynamics*. Edited by A. D. Short.
© 1999 John Wiley & Sons Ltd.

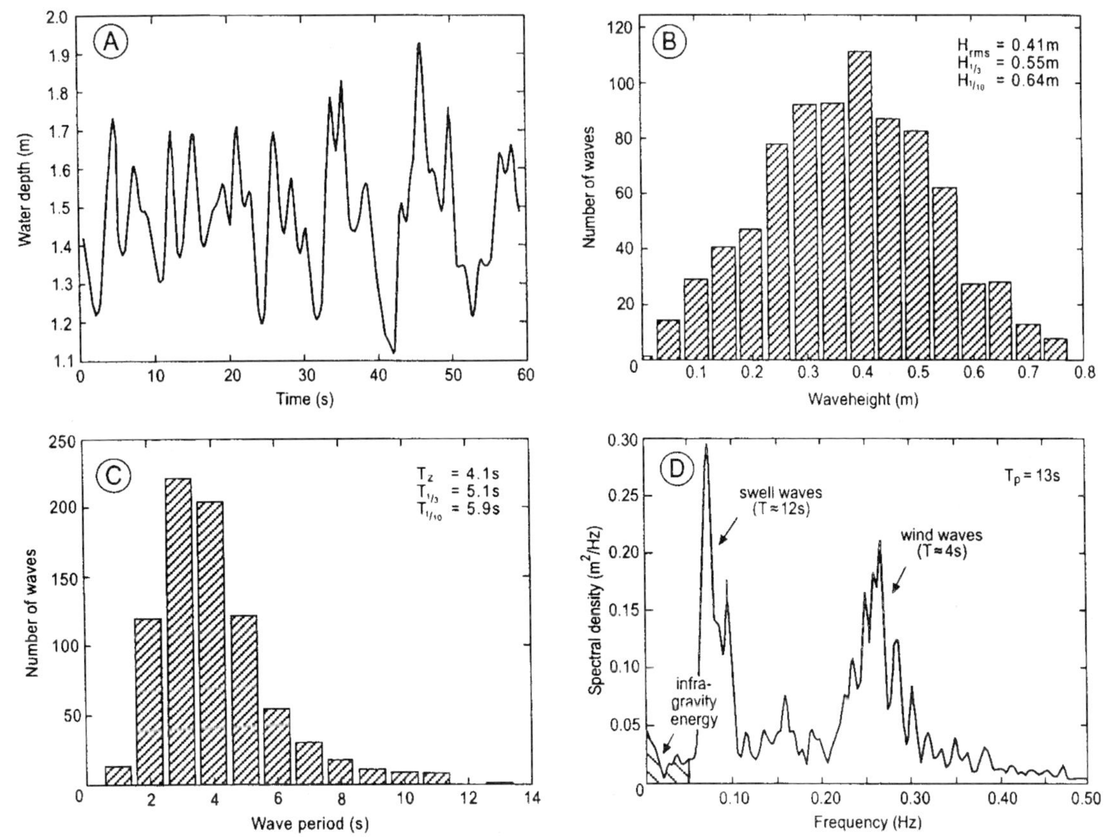


**Figure 4.1** Example of wave data collected in the surf zone: (A) time series of water surface elevation; (B) frequency distribution of wave heights; (C) frequency distribution of wave periods; and (D) wave spectrum.

appropriate measure of the wave energy level because the energy of a wave depends on the wave height squared. Other frequently used wave parameters to be derived using wave-by-wave analysis include $\bar{H}$ (mean wave height), $H_{1/3}$ (average wave height of one-third of the waves), $H_{1/10}$ (average wave height of one-tenth of the waves), $T_z$ (mean wave period), $T_{1/3}$ (mean wave period associated with $H_{1/3}$) and $T_{1/10}$ (mean wave period associated with $H_{1/10}$) (Figure 4.1B, C). $H_{1/3}$ is often referred to as the significant wave height ($H_s$) and approximately corresponds to visual estimates of wave heights. In the coastal engineering literature, $H_{1/3}$ or $H_s$ is the preferred wave height descriptor. Wave data collected from the surf zone often contain long-period fluctuations (Section 4.3). These motions need to be removed from the signal prior to wave-by-wave analysis using a high-pass filter.

Longuet-Higgins (1952) suggested that the probability distribution of wave heights in an irregular wave field is described by the *Rayleigh distribution*

$$P\left(\frac{H}{\bar{H}}\right) = \frac{\pi}{2}\frac{H}{\bar{H}}\exp\left[-\frac{\pi}{4}\left(\frac{H}{\bar{H}}\right)^2\right] \tag{4.2}$$

The Rayleigh distribution is characterised by an asymmetric shape (Figure 4.3) and can be used to compute the chance of a wave exceeding a certain height. For example, for a mean wave height of 1 m, the probability of a wave exceeding 3 m is 0.004 (0.4%).

Based on the Rayleigh distribution, various wave height measures can be correlated with the standard deviation of the water surface elevation record ($\sigma_\eta$)

$$\bar{H} = \sqrt{2\pi}\sigma_\eta \tag{4.3a}$$

$$H_{rms} = \sqrt{8}\sigma_\eta \tag{4.3b}$$

$$H_{1/3} = 4\sigma_\eta \tag{4.3c}$$

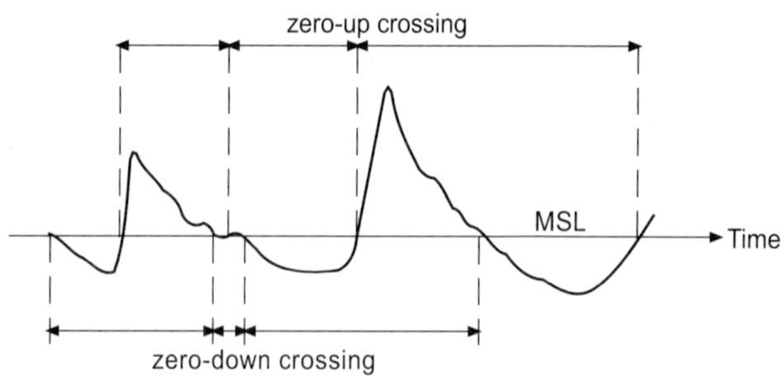


**Figure 4.2** Definition of individual waves by zero-upcrossing and zero-downcrossing method (after Horikawa, 1988).

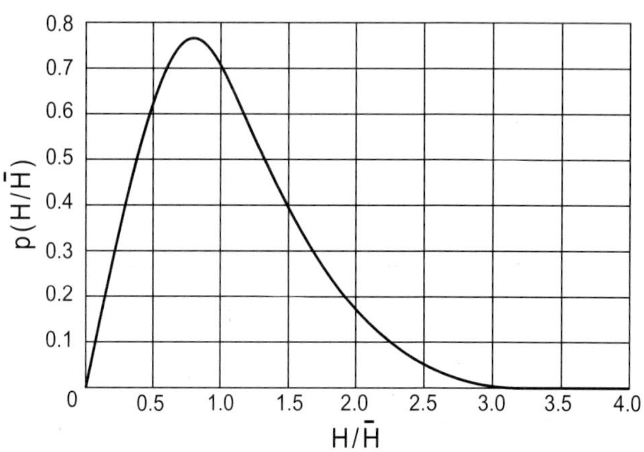


**Figure 4.3** Rayleigh distribution of wave heights normalised by the mean wave height ($H/\bar{H}$).

$$H_{1/10} = 5\sigma_\eta \tag{4.3d}$$

Equations 4.3a–d provide short-cuts to obtain measures of wave height and have been shown to hold reasonably well in the surf zone (Thornton and Guza, 1983).

An alternative method to describe the properties of irregular waves is spectral analysis (cf. Hegge and Masselink, 1996). The spectrum plots energy as a function of frequency and can be used, for example, to determine the contributions of wind waves and swell waves to the total amount of wave energy present (Figure 4.1.D). The peak period ($T_p$), defined as the wave period associated with the maximum wave energy, can be determined from the wave spectrum.

### 4.2.2 Wave transformation zones

A plane-sloping beach may be divided into four hydrodynamic regions on the basis of different stages of wave transformation (Figure 4.4): (1) the pre-breaking or shoaling region where waves are shoaling; (2) the outer surf zone region where the largest waves in the distribution break; (3) the inner surf zone region where the majority of the waves break or are broken; and (4) the run-up region or swash zone where the broken waves from the inner surf zone collapse on the beachface, resulting in swash motion (Svendsen *et al.*, 1978).

The different wave stages can be parameterised by the *Ursell number* (Ursell, 1953)

$$Ur = \frac{HL^2}{h^3} \tag{4.4}$$

where $H$ is the representative wave height, $L$ is the wave length and $h$ is the mean water depth. In deep water, waves are generally sinusoidal and $Ur \ll 1$. As the waves shoal, the Ursell number grows and the waves become increasingly deformed and non-linear until wave breaking occurs, followed by bore formation. Hence, the Ursell number provides a measure of wave non-linearity; the larger $Ur$, the more the wave shape deviates from a sinusoidal shape.

### 4.2.3 Wave breaking

In the surf zone, wave breaking occurs because the

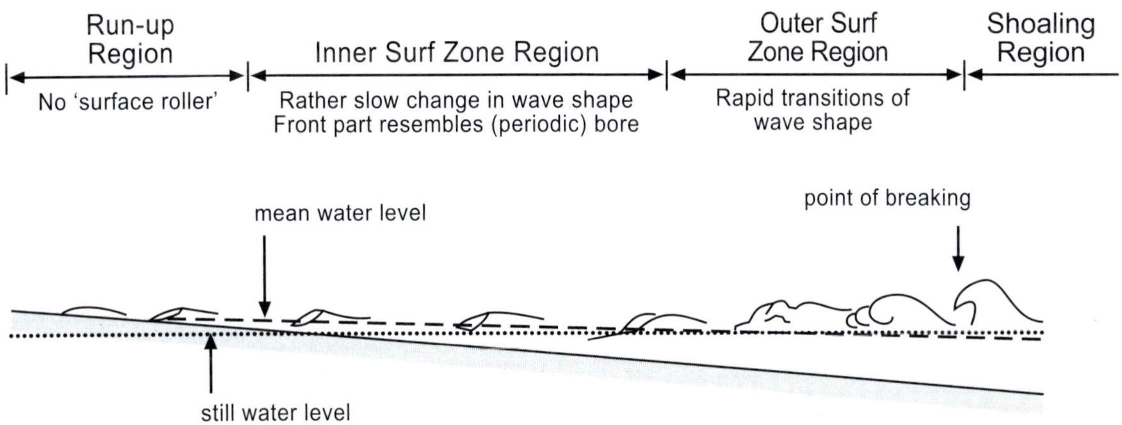


**Figure 4.4** Wave transformation zones on natural beaches (modified from: Svendsen *et al.*, 'Wave characteristics in the surf zone', *Proceedings of the 16th International Conference on Coastal Engineering*, 1978, by permission of the American Society of Civil Engineers).

water depth becomes too shallow for a stable wave form to exist and the wave height becomes depth-dependent. The mechanism of wave breaking is that the horizontal velocities of the water particles in the wave crest exceed the phase velocity of the wave. Consequently, the water particles leave the wave form, resulting in a disintegration of the wave into bubbles and foam. The breaking wave height is related to the water depth by the *breaker index* ($\gamma$) according to

$$H_b = \gamma h_b \tag{4.5}$$

where $H_b$ is the height of the breaking wave and $h_b$ is the mean water depth at the point of breaking. When the height-to-depth ratio of the wave exceeds the breaker index, breaking will occur. Solitary wave theory prescribes that $\gamma = 0.78$ (McGowan, 1894) and this provides a good first-order approximation.

Numerous laboratory studies have demonstrated that the breaker index increases with beach gradient and decreases with wave steepness (e.g. Weggel, 1972; Bowen *et al.*, 1968). It takes a finite amount of time for the waves to react to the decreasing water depth and on steep beaches this results in a delayed onset of breaking. Beach gradient and wave steepness can be combined in the *surf similarity parameter* (Battjes, 1974)

$$\xi_b = \frac{\tan\beta}{\sqrt{H_b/L_o}} \tag{4.6}$$

where $\tan\beta$ is the beach gradient, and the subscripts $b$ and $o$ indicate breaker and deep water conditions, respectively. The breaker index increases with the surf similarity parameter and according to Sunamura (1980)

$$\gamma = 1.1\xi_b^{1/6} \tag{4.7}$$

Over a range of $\xi_b$ values of 0.05–2, the breaker index calculated with Equation 4.7 varies from 0.67 to 1.23. It should be emphasised that Equation 4.7 is based on laboratory experiments and that a large amount of scatter was present in the data that were used to derive the formula.

Komar and Gaughan (1972) obtained an expression of the breaker height as a function of the deep water wave height and steepness by applying linear wave theory (and ignoring energy losses by bed friction) to evaluate the wave energy flux from deep water up to the breakpoint and using $\gamma = 0.78$

$$\frac{H_b}{H_o} = \frac{0.563}{(H_o/L_o)^{1/5}} \tag{4.8}$$

The value 0.563 was determined empirically from laboratory and field data. Due to wave shoaling, the height of the breakers always exceeds that of the deep water waves and the ratio $H_b/H_o$ increases with decreasing wave steepness. Hence, for a deep water wave steepness of around 0.01 which is typical of natural wave conditions and a deep water wave height of 1 m, the height of the breaking waves is predicted to be 1.41 m.

### 4.2.4 Breaker type

A continuum of breaker shapes occurs in nature. However, four types of breakers are commonly recognised:

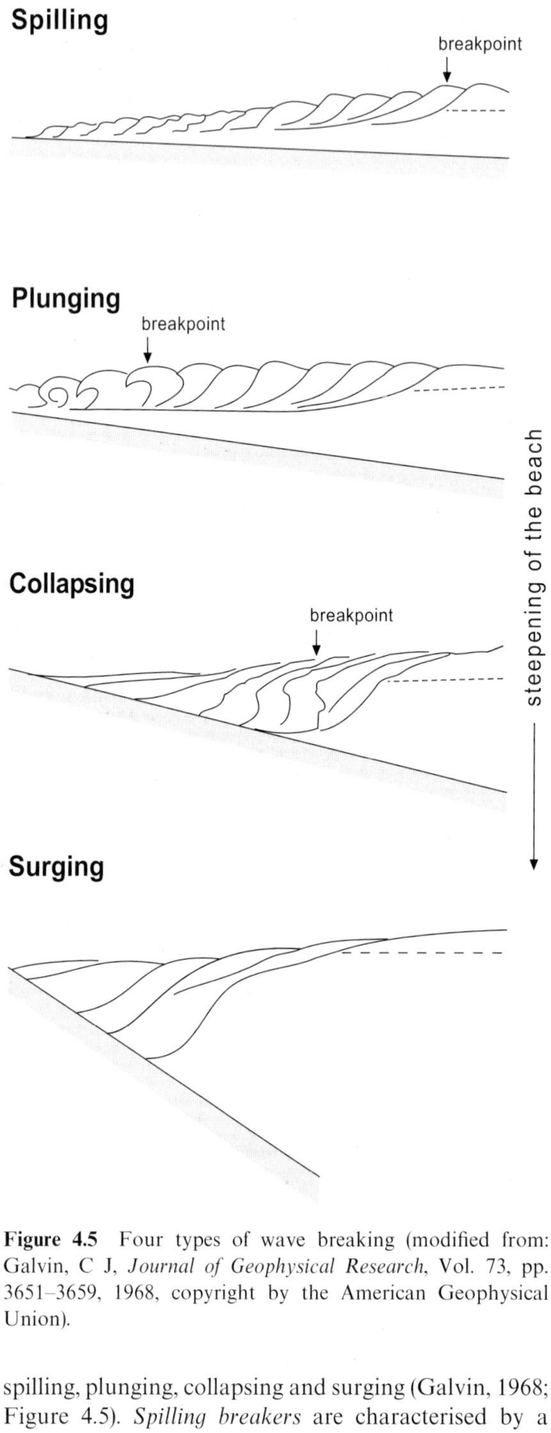


**Figure 4.5** Four types of wave breaking (modified from: Galvin, C J, *Journal of Geophysical Research*, Vol. 73, pp. 3651–3659, 1968, copyright by the American Geophysical Union).

spilling, plunging, collapsing and surging (Galvin, 1968; Figure 4.5). *Spilling breakers* are characterised by a gradual peaking of the wave until the crest becomes unstable, resulting in a gentle forward spilling of the crest. *Plunging breakers* are distinguished by the shoreward face of the wave becoming vertical, curling over, and plunging forward and downward as an intact mass of water. In *collapsing breakers* only the lower part of the front face of the wave becomes vertical and is thrust forward in an abbreviated plunging wave motion. In *surging breakers* the front face and crest of the wave remain relatively smooth and the wave slides directly up the beach without breaking. Spilling and plunging breakers occur in the surf zone, whereas collapsing and surging breakers are restricted to the swash zone. The transition from one breaker type to another is gradual and in natural surf zones a mixture of breaker types is often present.

The breaker type principally depends on the beach gradient and wave steepness which may be expressed by the surf similarity parameter defined for deep water conditions (Equation 4.6): spilling breakers occur for $\xi_o > 0.5$; plunging breakers occur when $\xi_o = 0.5$–$2.5$; collapsing breakers occur when $\xi_o = 2.5$–$3.7$; and surging breakers occur for $\xi_o > 3.7$ (Battjes, 1974; Okazaki and Sunamura, 1991). The continuum of breaker types from spilling to surging represents a decrease in the amount of wave energy dissipation and an increase in wave energy reflection. A slightly different but equivalent formula to describe breaker types in the surf zone is the surf scaling parameter (Carrier and Greenspan, 1958).

$$\varepsilon = \frac{a_b \omega^2}{g \tan^2 \beta} \tag{4.9}$$

where $a_b$ is the wave amplitude at the breakpoint and $\omega$ is the wave radian frequency ($2\pi/T$). Spilling breakers occur for $\varepsilon > 20$, collapsing and plunging when $\varepsilon = 2.5$–$20$, and surging for $\varepsilon < 2.5$ (Guza and Inman, 1975; Wright *et al.*, 1979). The local wind may also play an important role in determining the breaker type. Strong onshore winds favour spilling breakers, whereas strong offshore winds tend to be associated with plunging breakers (Galloway *et al.*, 1989; Douglas, 1990).

Depending on the mode of breaking, two types of *wave shapes* may be present in the inner surf zone (LeMehaute, 1962). Plunging breakers result in the development of saturated breakers (bores) where the rate of energy dissipation is at its maximum limit. *Saturated breakers* are characterised by a saw-tooth water surface profile with a steep, turbulent bore front and a gently sloping back. Spilling breakers initially result in the development of non-saturated breakers, which are symmetrical around the crest with breaking taking the form of a gentle spilling of water down the face of the wave.

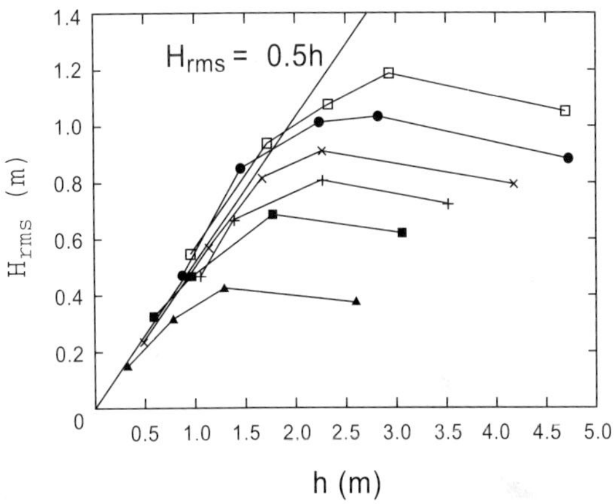


**Figure 4.6** Root mean square wave height versus mean water depth demonstrating that the wave height in the inner surf zone is controlled by the local water depth (modified from: *Marine Geology*, Vol. 129, Masselink, G and Hegge, B, 'Morphodynamics of meso- and macrotidal beaches: examples from central Queensland, Australia', pp. 1–23, 1995, with kind permission from Elsevier Science – NL, Sara Burgerhartstraat 25, 1055 KV Amsterdam, The Netherlands).

*Non-saturated breakers* develop into (saturated) bores when wave energy can no longer be dissipated by gentle spilling and bottom friction such as might occur when the bottom gradient increases and/or in very shallow water depths. A decrease in the bed gradient or an increase in the water depth (e.g. in the bar trough) may result in bores or non-saturated breakers losing their turbulent fronts and reforming into non-breaking waves.

### 4.2.5 Wave height in the surf zone

Where bore forms dominate, local water depth and beach slope determine the characteristics of flow and hence the energy dissipation, wave shape and wave decay. Numerous field studies (e.g. Thornton and Guza, 1982; Wright *et al.*, 1982a; Masselink and Hegge, 1995) have shown that in the surf zone, the r.m.s. *wave height* is a direct function of the local water depth (Figure 4.6), and may be given by

$$H_{rms} = \langle\gamma\rangle h \tag{4.10}$$

For natural beaches, $\langle\gamma\rangle$ varies between 0.3 and 0.6 and increases with beach slope (Sallenger and Holman, 1985)

$$\langle\gamma\rangle = 0.3 + 3.2 \tan\beta \tag{4.11}$$

For beach gradients approaching zero, Nelson (1994) found that $\langle\gamma\rangle$ approaches 0.39, slightly higher than predicted by Equation 4.11.

The parameter $\langle\gamma\rangle$ can be seen as the random wave equivalent of the breaker index and is considerably smaller than the latter for two reasons. First, the height-to-depth ratio of broken waves is smaller than for breaking waves and often decreases progressively in the onshore direction across the surf zone (Horikawa and Kuo, 1966; Dally *et al.*, 1985). Second, in natural surf zones a significant proportion of the waves have not yet broken and are still shoaling (e.g. Dally, 1990).

The existence of a direct relationship between wave height and water depth implies that at the shoreline the wave height is zero, suggesting that wave run-up and swash motion cannot exist. On natural beaches this is usually not the case, since there will always be some residual wave energy at the shoreline. The residual wave height is expected to increase with the beach gradient (Sunamura, 1984a).

### 4.2.6 Radiation stress and wave set-up

*Wave set-up* is a superelevation of the water level in the surf zone that enables waves to reach upper parts of the beach profile. During the 1938 hurricane which hit the US east coast, observations showed that the maximum mean shoreline elevation was about 1 m higher on exposed parts of the coast than in calmer waters (Guza and Thornton, 1981). Since this difference could not be explained by variations in storm surge height, it was suggested that the incident waves were responsible for the higher shoreline elevation on the exposed parts of the coast. Subsequent laboratory experiments (Savage, 1957; Saville, 1961) confirmed this suggestion and demonstrated that the superelevation of the mean water level near the shoreline constitutes a significant proportion of the incident wave height.

In popular terms, wave set-up is conceived as a piling up of water against the shoreline due to the waves. A theoretical explanation for wave set-up was provided by Bowen *et al.* (1968) using the radiation stress concept. Associated with wave propagation there is a transmission of momentum due to the pressure and velocity fluctuations under waves. This phenomenon was investigated by Longuet-Higgins and Stewart (1962, 1964) who defined *radiation stress* as the 'excess flow of momentum due to the presence of the waves' which is equal to the difference between the absolute pressure under the waves and the hydrostatic pressure

$$S = \int_{-h}^{\eta} (p + \rho u^2)\partial z - \int_{-h}^{0} p_0 \partial z \tag{4.12}$$

where $p$ is the total pressure, $\rho$ is the density of water and $p_0$ is the hydrostatic pressure integrated between the bottom and the mean water level ($\eta$ is the free water surface). Using the expressions for $u$ and $\eta$ from linear shallow water wave theory and integrating over a wave period, the cross-shore component of the radiation stress ($S_{xx}$) for waves approaching the beach with their crests parallel to the shoreline is given by (Longuet-Higgins and Stewart, 1964)

$$S_{xx} = \frac{3}{2}E = \frac{3}{16}\rho g H^2 \tag{4.13}$$

where $E$ is the wave energy density. Momentum balance prescribes that changes in the shoreward component of the radiation stress (due to wave breaking and shoaling) must be balanced by changes in mean water level and bed shear stresses (Svendsen, 1984a)

$$\frac{\partial S_{xx}}{\partial x} + \rho g h \frac{\partial \bar{\eta}}{\partial x} + \bar{\tau}_b = 0 \tag{4.14}$$

where $h$ is the mean water depth, $\bar{\eta}$ is the departure from still water level and $\bar{\tau}_b$ is the mean shear stress at the bottom due to currents. The bed shear stress term is usually neglected since even for a 1 m/s offshore-directed current, the reduction in wave set-up at the shoreline is at most 0.05 m (Nielsen, 1988b).

Ignoring the bed shear stress term in Equation 4.14, it can be seen how variations in the wave height and radiation stress affect the *water surface gradient* in the nearshore. Outside the surf zone, wave height and hence radiation stress increase due to shoaling. The resulting positive gradient in radiation stress must be balanced by a negative water surface gradient, resulting in a lowering of the water level, known as *wave set-down* (Figure 4.7). In the surf zone, wave breaking processes result in a decrease of the wave height, and hence a negative gradient in the radiation stress. This gradient is balanced by a positive water surface gradient, resulting in an elevated water level in the surf zone referred to as *wave set-up* (Figure 4.7). Wave set-down is always considerably smaller than wave set-up.

Assuming that the wave height in the surf zone is everywhere limited by the water depth (Equation 4.10), Bowen *et al.* (1968) obtained an explicit expression for the set-up gradient in the surf zone

$$\frac{\partial \bar{\eta}}{\partial x} = \frac{1}{1 + \left(\frac{8}{3\gamma^2}\right)} \tan\beta \tag{4.15}$$

Thus for a planar beach, the slope of the mean sea level is proportional to the beach gradient and is constant across the surf zone. However, in reality the set-up profile steepens considerably landward of the still water line (Figure 4.7), particularly if the mean water level in the swash zone is included in the computation of the wave set-up profile (Nielsen, 1988b; Gourlay, 1992). On beaches with irregular, barred morphology, the topographic control on wave breaking results in even more complex set-up profiles with a positive set-up gradient over the bar due to wave breaking and a negative set-up slope across the trough where broken waves reform (Greenwood and Osborne, 1990).

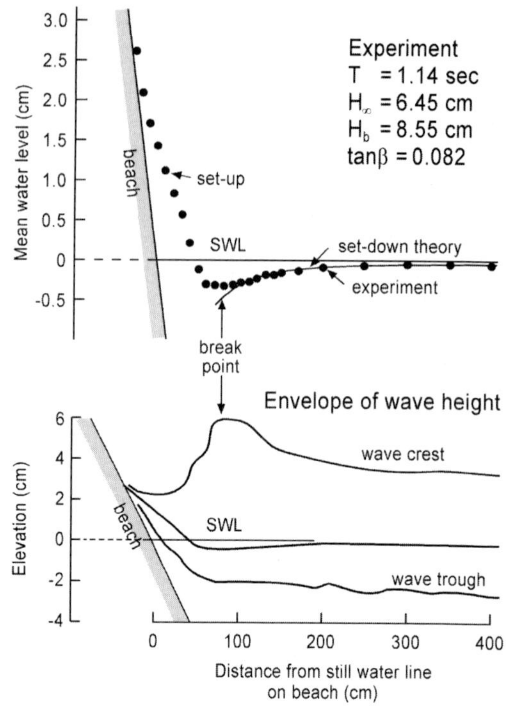


**Figure 4.7** Laboratory experiment showing the mean water level and the envelope of the wave height. The mean water level is depressed around the wave breakpoint (wave set-down) and elevated across the surf zone (wave set-up) (modified from: Bowen *et al.*, *Journal of Geophysical Research*, Vol. 73, pp. 2569–2577, 1968, copyright by the American Geophysical Union).

### 4.2.7 Wave set-up at the shoreline

The wave set-up at the shoreline is defined as the intersection of the set-up gradient with the beach profile. Ignoring wave set-down, the shoreline set-up ($\bar{\eta}_s$) can be approximated by (Nielsen, 1988b)

$$\bar{\eta}_s = \frac{3}{8}\gamma H_b \tag{4.16}$$

In a surf zone with regular waves ($\gamma \approx 0.8$), Equation 4.16 predicts a shoreline set-up of approximately $0.3H_b$ which is supported by laboratory measurements (Battjes, 1974). Under natural conditions with irregular waves only a proportion of the waves are breaking and contributing to the negative radiation stress gradient at any one position. Hence, wave set-up at the shoreline is expected to be less. In this case $\langle\gamma\rangle$ may be inserted for $\gamma$ (Guza and Thornton, 1981) and with $\langle\gamma\rangle = 0.4$ and $H_{b,rms}$ as the representative breaking wave height this yields

$$\bar{\eta}_s = 0.15H_{b,rms} \tag{4.17}$$

Equation 4.17 corresponds reasonably well with field measurements presented by Guza and Thornton (1981), Greenwood and Osborne (1990) and King *et al.* (1990).

Because the breaker index depends on the surf similarity parameter (Equation 4.7), primarily through the beach slope, the set-up at the shoreline is also a function of the surf similarity parameter. Using data from Yanagishima and Katoh (1990), Gourlay (1992) suggested that

$$\bar{\eta}_s = 0.35H_{o,rms}\xi_o^{0.4} \tag{4.18}$$

where both the r.m.s. wave height and the surf similarity parameter are determined for deep water conditions.

### 4.2.8 Wave transformation models

The transformation of incident waves across the nearshore zone is of fundamental importance in inducing radiation stress gradients, wave set-up, nearshore current circulation, sediment transport, and ultimately beach change. Therefore, considerable effort has been directed towards the development of models that predict the shoreward attenuation of the incident wave height. It is important to make a distinction between regular and irregular (random) wave transformation models. In *regular wave models* (e.g. Svendsen, 1984a; Dally *et al.*, 1985), the waves are of constant amplitude at each location and the waves in the surf zone are all broken. These models therefore predict the transformation of a single wave across the nearshore. In reality, the wave field is irregular and a distribution of wave heights exists (Section 4.2.1). The objective of *irregular wave models* is therefore to characterise the transformation of the wave height distribution (usually a Rayleigh distribution).

In the early irregular wave models, shoaling and breaking are only dependent on the local water depth (e.g. Collins, 1970; Goda, 1975). In these models, the deep water wave heights are first transformed into shallow water wave heights using linear wave theory. When the waves reach shallow water they begin to break, with the largest waves breaking farthest offshore. Wave breaking is simulated by truncating the tail of the Rayleigh distribution. The various methods that employ the above approach differ in their techniques to determine the wave height cut-off and the equations used to define the breaker index and the wave height after breaking. The major shortcoming of these models is that the height of the broken waves is only dependent on the local water depth. This causes problems on barred beaches where the predicted wave height will increase over the trough. However, this requires an increase in wave energy which is physically inappropriate.

More recently, wave transformation models have been developed which use a similar approach for the wave shoaling process, but the wave height decay within the surf zone is modelled differently (e.g. Battjes and Janssen, 1978; Thornton and Guza, 1983; Battjes and Stive, 1984; Dally, 1990). The height of the broken waves in the surf zone is calculated using the wave energy flux balance

$$\frac{\partial EC_g}{\partial x} + D_* = 0 \tag{4.19}$$

where $E$ is the wave energy density, $C_g$ is the wave group velocity, $x$ is the onshore coordinate and $D_*$ is the average energy dissipation. The various types of models differ in their definition of the energy dissipation. A simple approach is to model the dissipation rate in broken waves after that of a bore of corresponding height (LeMehaute, 1962). These recent models can properly predict wave transformation over barred beach profiles (Figure 4.8) such that wave energy lost by breaking over the bar is not regained when the wave is propagating across the trough.

## 4.3 INFRAGRAVITY WAVE MOTION IN THE SURF ZONE

A significant part of the wave energy in the nearshore zone is at frequencies that are lower than those of the incident swell and wind waves (Figures 4.1D and 4.9). This motion is referred to as *infragravity wave* motion and is generally characterised by periods ranging from 20 s (0.05 Hz) to several minutes. These limits are some-

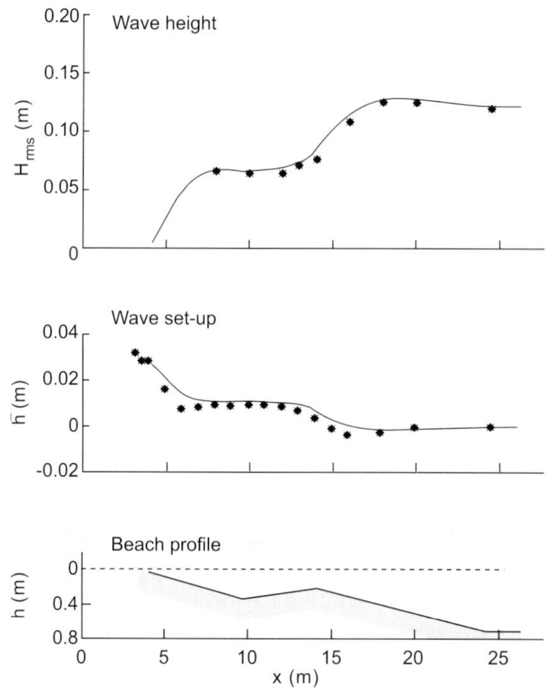


**Figure 4.8** Comparison of random wave transformation model (solid lines) with laboratory experiment (data points) using random waves. Regions of wave breaking and height attenuation, such as on the seaward-facing side of the bar and the beachface, are characterised by relatively steep wave set-up gradients. Waves reform and reshoal in the trough leading to a local decrease in set-up (modified from: Battjes, J A and Stive, M J F, 'Calibration and verification of a dissipation model from random breaking waves', *Proceedings of the 19th International Conference on Coastal Engineering*, 1984, by permission of the American Society of Civil Engineers).

what arbitrary. Particularly, the high frequency limit is dependent on incident wave frequencies and in short period, fetch-limited environments this limit may be more reasonably set at 10–15 s (0.1–0.067 Hz; Bauer, 1990). Infragravity wave motion is also known as surf beat, a term introduced by Munk (1949) who made the initial observations of low frequency motion in the nearshore region. The infragravity wave motion in the surf zone is generally standing in the cross-shore direction and the water surface elevation and the cross-shore current records show a 90° phase difference (Suhayda, 1974).

Standing infragravity waves provide a simple and quantitative theoretical explanation of many nearshore morphological features, including rip currents (Bowen and Inman, 1969), crescentic bars (Bowen and Inman, 1971), multiple longshore bars (Short, 1975b; Bowen, 1980) and welded sand bars and more complex morphologies (Holman and Bowen, 1982). In addition, infragravity waves may contribute to shoreline erosion during storms since a large part of the offshore suspended sediment transport can occur at infragravity frequencies (Beach and Sternberg, 1988; Section 4.5). Furthermore, infragravity waves periodically raise the mean water level at the beach, thus enabling the incident waves to reach the back beach and the toe of the dunes (Wright, 1980).

### 4.3.1 Infragravity energy level

Infragravity wave motion is forced directly or indirectly by the incident wave field. Consequently, the surf beat energy level in the surf zone is dependent on the incident wave energy level (Holman, 1981; Huntley *et al.*, 1993). Typically, the *infragravity wave height* in the surf zone is 20–60% of the offshore wave height (Goda, 1975; Guza and Thornton, 1985a; Howd *et al.*, 1991). Both the relative and absolute importance of infragravity wave motion over incident wave fluctuations increases in the onshore direction (Figure 4.9). Hence, wave run-up motion is often dominated by infragravity frequencies (Holman and Sallenger, 1985). The relative infragravity energy is related to the surf similarity parameter $\xi_o$ (Holman and Sallenger, 1985). On beaches with $\xi_o > 1.5$, swash excursions are dominated by the incident wave frequencies, whereas for smaller values of $\xi_o$, swash becomes increasingly dominated by infragravity frequencies.

Guza and Thornton (1982) conducted measurements of *swash oscillations* on the beachface and related the swash excursion to the offshore wave conditions. They divided the total swash signal into motions at incident wave frequencies and infragravity wave frequencies (separated at 0.05 Hz), and found that whereas incident swash energy was independent of the offshore wave conditions, infragravity swash energy was proportional to the offshore wave height (Figure 4.10).

The observations of Guza and Thornton (1982) have major ramifications for the behaviour of beaches during storm conditions (Figure 4.11). The incident wave energy level in the surf zone and the swash zone has an upper limit due to wave energy saturation (Section 4.2.5). During storm conditions, the surf zone will become wider, leading to greater wave breaking and similar incident wave levels in the inner surf zone. Infragravity energy, on the other hand, is not dissipated in the surf zone because the long wave length of the infragravity wave inhibits wave breaking. Because the infragravity energy level is proportional to the offshore wave height,

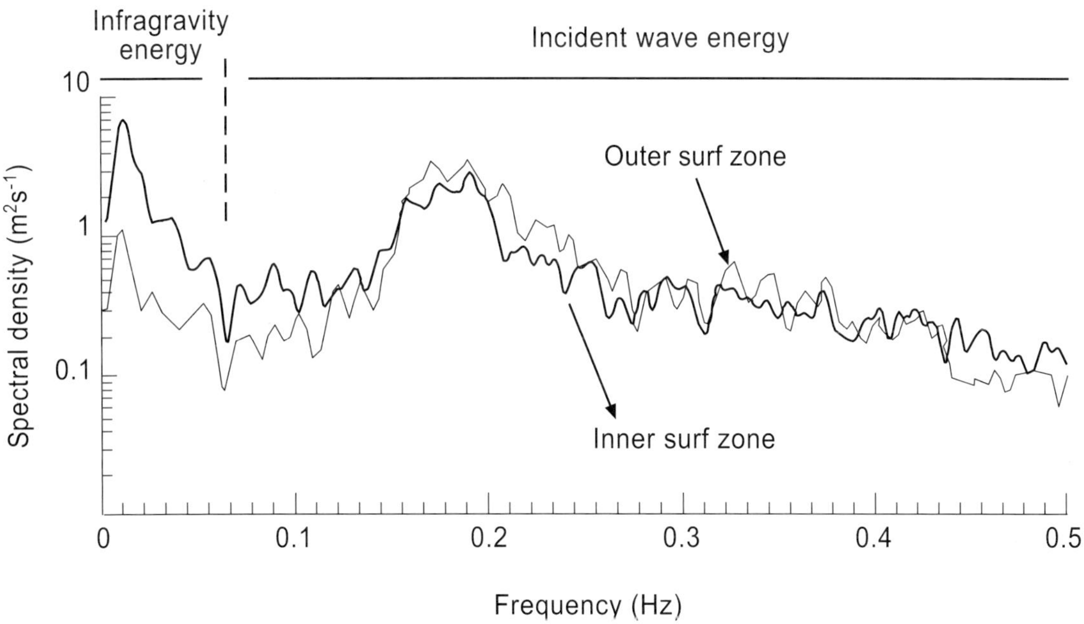


**Figure 4.9** Examples of cross-shore velocity spectra from the surf zone demonstrating the presence of large amounts of infragravity energy at frequencies below 0.05 Hz. The two spectra were recorded simultaneously in the outer and inner surf zone. Note the decrease in gravity energy and increase in infragravity energy between the outer and inner surf zone (after Aagaard and Greenwood, 1995a).

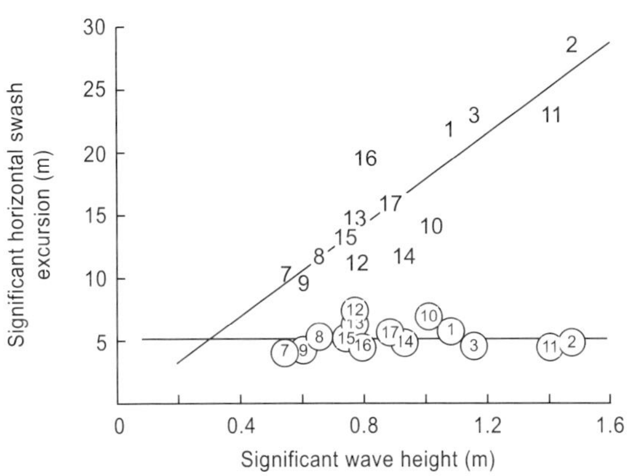


**Figure 4.10** Significant horizontal swash excursion at incident (circled) and infragravity frequencies (uncircled) as a function of offshore significant wave height. The division between incident and infragravity wave periods was taken at 20 s. The numbers refer to individual data runs (modified from: Guza, R T and Thornton, E B, *Journal of Geophysical Research*, Vol. 87, pp. 483–491, 1982, copyright by the American Geophysical Union).

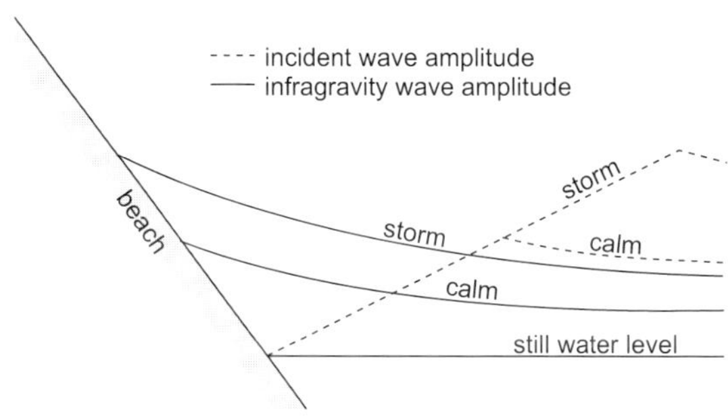


**Figure 4.11** Schematic representation of the variation in incident and infragravity wave energy across the nearshore during storm and calm weather conditions (reproduced with permission from: Holman, R A, in Komar, P D (ed.), *CRC Handbook of Coastal Processes and Erosion*, 1983, pp. 21–33. Copyright CRC Press, Boca Raton, Florida).

infragravity energy may dominate the water motion in the inner surf zone during storm conditions (Holman *et al*., 1978). This may result in cross-shore current velocities at infragravity frequencies that exceed 2 m/s (Beach and Sternberg, 1988).

### 4.3.2 Types of infragravity wave motion

Three types of infragravity waves can be identified (Figure 4.12):

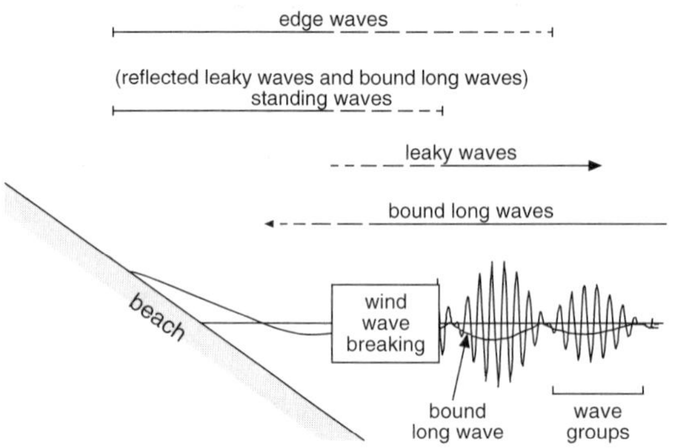


**Figure 4.12** The different types of infragravity wave motion identified in the nearshore zone (modified from: Huntley, D A and Kim, C S, 'Is surf beat forced or free?' *Proceedings of the 19th International Conference on Coastal Engineering*, 1984, by permission of the American Society of Civil Engineers).

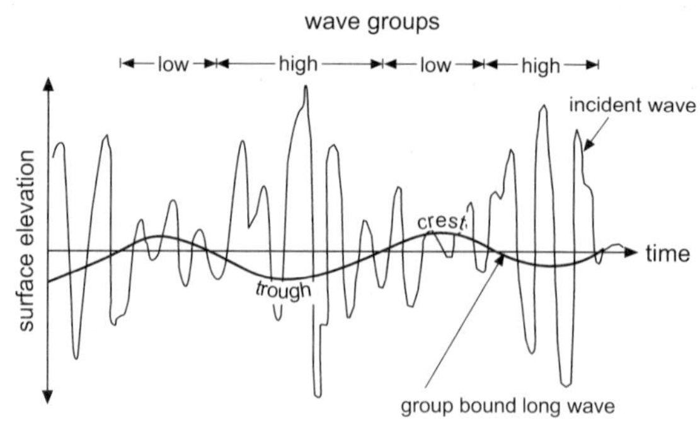


**Figure 4.13** Schematic representation of the bound long wave with the crest of the long wave associated with the low waves and the trough of the long wave with the high waves (after Barthel and Funke, 1984).

1. *Bound long waves* occur seaward of the breakpoint. They are forced locally by the wave groups and travel at the group velocity of the wind waves.
2. *Leaky waves* are forced non-locally and propagate and escape out to deep water upon reflection at the shoreline. Superposition of an incoming infragravity wave and the outgoing leaky wave may result in a surf zone standing wave.
3. *Edge waves* are free waves that travel along the beach and are trapped in the nearshore by reflection and refraction. They are standing in the cross-shore direction but may also be standing in the longshore direction.

Bound long waves must remain with the forcing (i.e. the wave groups), whereas leaky waves and edge waves are free waves and can propagate away from the generation source. Due to their long periods, infragravity waves will always be reflected from the shoreline, regardless of the beach gradient. Recently, fluctuations in the longshore current velocity have been discovered with periods beyond that of infragravity waves. These motions are referred to as *shear waves* and are manifestations of a shear instability in the longshore current. In addition to the organised forms of infragravity motion mentioned above, additional sources of infragravity energy are bore and swash interactions (e.g. Emery and Gale, 1951; Carlson, 1984; Mase, 1995).

### 4.3.3 Bound long waves

The *bound long wave* is generated by the incident waves and arises from the amplitude modulation of natural waves, with alternating groups of high waves and low waves. Longuet-Higgins and Stewart (1964) demonstrated theoretically that a small lowering of the mean water level accompanies a wave group. Groups of high waves are associated with larger radiation stresses than groups of low waves (Equation 4.13) and as a result radiation stress gradients cause fluid to be expelled from the high wave groups to the low wave groups (Equation 4.14). This leads to the development of a set-down wave which has the same wave length and period as the incident wave group, but is 180° out of phase with the group (Figure 4.13).

The amplitude of the bound long wave is positively related to the wave groupiness (Sand, 1982) and incident wave height (Mansard and Barthel, 1984). Under high wave energy conditions, the bound long wave may dominate the total infragravity energy outside the surf zone (Elgar *et al.*, 1992). The amplitude of the bound long wave increases significantly during shoaling (Mansard and Barthel, 1984), even after the incident waves have started breaking (List, 1992). However, during shoaling, the bound long wave progressively lags behind the wave groups and develops an asymmetrical form, such that a rise in the water level precedes the radiation stress maxima and a depression in water level follows the radiation stress maxima (Elgar and Guza, 1985; List, 1992; Masselink, 1995b).

### 4.3.4 Leaky waves

Strictly speaking, the term *leaky wave* should be reserved for a wave escaping to deep water upon reflection at the shoreline, thus 'leaking' energy back offshore. However, the term is typically used to refer to two dimensional standing wave motion in the surf zone

**Box 4.1** *Computation of the wave form of standing leaky waves.*

For a plane beach with uniform slope tan $\beta$, the offshore wave form of a standing leaky wave is given by the zero-order Bessel function (Suhayda, 1974)

$$\phi_x = J_0(4\chi)^{1/2}$$

where $\chi$ is the non-dimensional cross-shore distance

$$\chi = \frac{\omega^2 x}{g \tan \beta}$$

and $J_0$ is the first zero-order Bessel function

$$J_0(x) = 1 - \frac{(x/2)^2}{(1!)^2} + \frac{(x/2)^4}{(2!)^2} - \frac{(x/2)^6}{(3!)^2} + \frac{(x/2)^8}{(4!)^2} - \frac{(x/2)^{10}}{(5!)^2} + \frac{(x/2)^{12}}{(6!)^2} - \frac{(x/2)^{14}}{(7!)^2} \cdots$$

Inserting $(4\chi)^{1/2}$ for $x$ in the Bessel function $J_0$ will yield the standing leaky wave form. However, many more terms are required for the correct solution than shown in the above equation and the use of polynomial approximations is recommended (e.g. Press *et al.*, 1989).

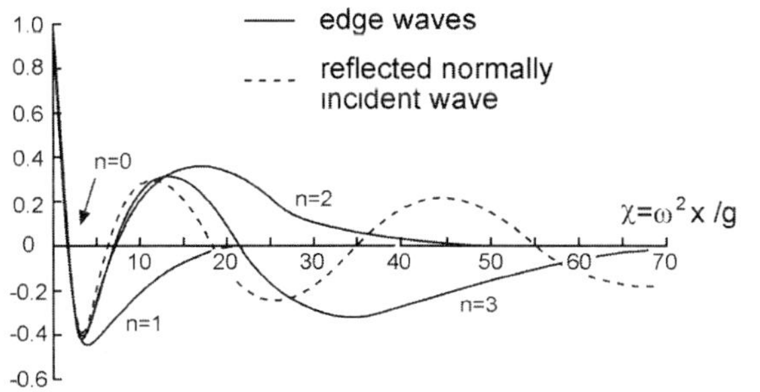


**Figure 4.14** The cross-shore surface elevation structure of a leaky mode standing wave and modes ($n$) 0–3 edge waves plotted in terms of non-dimensional offshore distance ($\chi$). The $y$-axis represents the amplitude of the wave form normalised by the shoreline amplitude.

caused by the superposition of an incoming and outgoing infragravity wave. In contrast to the three dimensional edge waves, the standing wave pattern associated with leaky waves does not show any variation in the longshore direction and they are therefore referred to as two dimensional infragravity waves.

Leaky waves occur when $\omega^2 > gk_y$, where $\omega$ is the wave radian frequency ($2\pi/T$) and $k_y$ is the longshore wave number ($2\pi/L$). If $\omega^2$ is smaller than $gk_y$, the reflected wave will be trapped by refraction and cannot escape to deep water. This results in edge waves (Section 4.3.5). If $\omega^2$ is larger than $gk_y$, the reflected wave will not be trapped and energy can leak back offshore in the form of leaky waves. Thus, for the long periods associated with infragravity waves, leaky waves require a near shore-normal incidence angle of the incoming wave ($L$ is very large and $k_y$ approaches zero).

The *formation of leaky waves* requires an infragravity wave entering the surf zone from offshore, either through the release of the group bound long wave at the breakpoint (Longuet-Higgins and Stewart, 1964) or the generation of a long wave at the breakpoint through the time-varying breakpoint mechanism proposed by Symonds *et al.* (1982). The *bound long wave* may be released when the short period, incident waves break and the wave groupiness is destroyed (Longuet-Higgins and Stewart, 1964). The released bound long wave then propagates shoreward as a free wave. At the shoreline this free wave reflects and sets up a standing wave pattern in the surf zone (Guza *et al.*, 1984; List, 1986; Masselink, 1995b). Outside the surf zone the reflected wave travels offshore as a progressive free leaky wave. The *time-varying breakpoint model* of Symonds *et al.* (1982) is based on the premise that alternating groups of high and low waves generate a time-varying wave set-up and hence infragravity waves. High waves break further offshore and generate a large set-up, and vice versa; the wave height modulation functions as a large wavemaker. This model predicts the generation of infragravity waves at the wave group frequency and higher harmonics in both seaward and shoreward directions. The shoreward propagating set-up wave reflects

**Table 4.1** Distance from the shoreline to surface elevation nodes and antinodes for edge waves and leaky waves. The distance is expressed in the non-dimensional distance ($\chi$; Equation 4.20).

| Wave type | Node 1 | Node 2 | Node 3 | Node 4 | Node 5 |
|---|---|---|---|---|---|
| Edge wave mode 1 | 1.5 | | | | |
| Edge wave mode 2 | 1.5 | 8.5 | | | |
| Edge wave mode 3 | 1.5 | 8.0 | 22.0 | | |
| Edge wave mode 4 | 1.5 | 7.9 | 20.4 | 42.3 | |
| Edge wave mode 5 | 1.4 | 7.8 | 19.8 | 39.0 | 69.5 |
| Leaky wave | 1.4 | 6.6 | 17.6 | 33.8 | 54.8 |
| | Antinode 1 | Antinode 2 | Antinode 3 | Antinode 4 | Antinode 5 |
| Edge wave mode 1 | 4.5 | | | | |
| Edge wave mode 2 | 3.9 | 16.1 | | | |
| Edge wave mode 3 | 3.8 | 13.6 | 35.1 | | |
| Edge wave mode 4 | 3.7 | 13.0 | 29.6 | 61.6 | |
| Edge wave mode 5 | 3.7 | 12.8 | 28.1 | 52.2 | 95.7 |
| Leaky wave | 3.5 | 12.3 | 25.8 | 44.5 | 67.8 |

at the shoreline resulting in a standing wave across the surf zone. Shoreward propagating long waves generated by this mechanism should show a positive correlation with the wave envelope function, in contrast to bound long waves that exhibit a negative correlation. List (1992) concluded, following comparison of field observations with model results, that long waves generated by the time-varying breakpoint are of secondary importance compared to those derived from the group bound long wave.

The offshore wave form of leaky waves is described by the zero-order Bessel function (Box 4.1) and is shown in Figure 4.14 as a function of the non-dimensional cross-shore distance $\chi$ (Holman, 1983)

$$\chi = \frac{\omega^2 x}{g \tan \beta} \tag{4.20}$$

where $\omega$ is the radian frequency, $x$ is the offshore distance and $\tan\beta$ is the beach gradient (note that some publications use the notation $\chi = 4\omega^2 x/g \tan \beta$). The amplitude of the leaky wave varies from a maximum at the shoreline, to zero at a succession of nodes which are separated by antinodes of progressively diminishing amplitude. As surface elevation and velocity are shifted in phase by 90° in standing waves, the maximum orbital velocities occur at surface elevation nodes. The distance from the shoreline to the first five antinodes and nodes is given in Table 4.1 (ignoring the first antinode occurring at the shoreline). For example, for a standing leaky mode wave with a period of 40 s on a beach with a gradient of 0.02, the first node and antinode are located at 11 m and 28 m from the shoreline, respectively. As will be seen in Section 4.3.5 the spacing of nodes and antinodes increases exponentially in the offshore direction and will also increase with wave period and beach slope.

### 4.3.5 Edge waves

When $\omega^2 < gk_y$, waves reflected from the shoreline will become trapped close to shore due to refraction and may result in the formation of edge waves. Under these conditions there is no offshore radiation of reflected energy and the reflected wave orthogonals bend back towards the shore where the wave is re-reflected. Thus, edge waves can be generated when the angle of wave incidence is different from zero (or if the shoreline is irregular), in contrast to leaky waves that require near-normal wave incidence. A special class of high frequency edge waves generated directly by the incident waves may occur when waves approach the coast with their crests parallel to the shoreline (harmonic and sub-harmonic edge waves). These edge waves will be separately discussed in Chapter 5.

By definition, *edge waves* are free waves that are standing in the cross-shore direction with their crests normal to the shoreline. They may travel along the beach as progressive edge waves or they may be standing in the longshore direction as standing edge waves. Standing edge waves are formed by the superposition of two progressive edge waves propagating in opposite directions. This may occur if, for example, a progressive edge wave is reflected from a longshore boundary (e.g. groyne, headland).

The *offshore structure* of edge waves is similar to that of leaky waves and is described by an antinode at the shoreline, a number of zero crossings equal to the mode number ($n$), and an exponential decay in the offshore

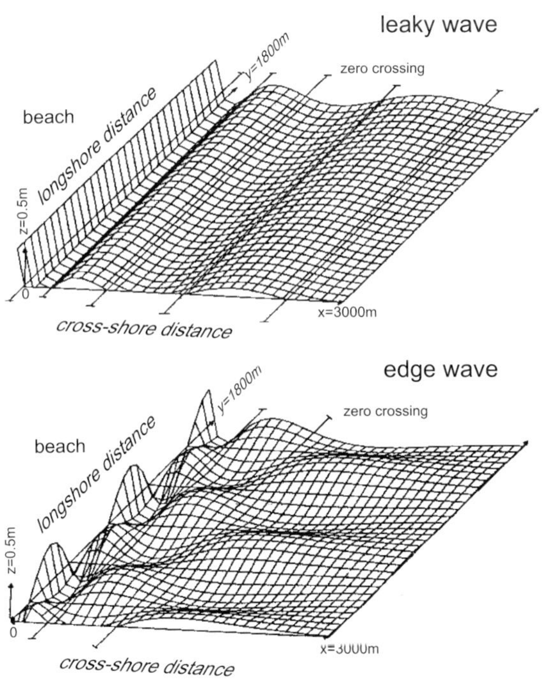


**Figure 4.15** Perspective view of the three-dimensional water surface of a leaky mode standing wave and a mode 3 edge wave computed for an infragravity wave period of 100 s on a slope of 0.011 (modified from: *Marine Geology*, Vol. 45, Wright *et al.*, 'Dynamics of a high-energy dissipative surf zone', pp. 41–62, 1982a, with kind permission from Elsevier Science – NL, Sara Burgerhartstraat 25, 1055 KV Amsterdam, The Netherlands).

direction (Figure 4.15). Thus, a mode 0 edge wave has only one antinode at the shoreline whereas a mode 3 edge wave has three additional antinodes. For progressive edge waves, the longshore sinusoidal wave form propagates along the beach, whereas for standing waves the wave form remains fixed at the same longshore location. The offshore form of edge waves and leaky waves with identical wave period is very similar close to the shoreline (Figures 4.14 and 4.15). This is also apparent from the locations of nodes and antinodes for the first five edge wave modes and leaky waves listed in Table 4.1. The resemblance of the offshore structure of edge waves and leaky waves has been one of the main difficulties in distinguishing between these two types of infragravity standing wave motion.

The edge wave period ($T_e$) and longshore wave length ($L_e$) are related by the edge wave dispersion relationship. For planar beaches with a uniform beach gradient of $\tan\beta$, the shallow water edge wave dispersion relationship is given by (Eckart, 1951)

$$\omega^2 = gk(2n + 1)\tan\beta \tag{4.21}$$

leading to the equation for the longshore wave length

$$L_e = \frac{g}{2\pi} T_e^2 (2n + 1)\tan\beta \tag{4.22}$$

The relation between the edge wave period and wave length is sensitive to changes in the beach morphology, and for non-planar beaches, the edge wave dispersion relation should be determined numerically (Holman and Bowen, 1979).

There is a limit to the number of edge wave modes that may exist on a beach with uniform slope and the *cut-off mode number* is given by (Ursell, 1952)

$$(2n + 1)\tan\beta \leq \pi/2 \tag{4.23}$$

The steeper the beach slope, the smaller the number of edge wave modes that can be trapped in the nearshore. Thus, whereas on a beach with $\tan\beta = 0.1$ only seven edge wave modes can remain trapped, for $\tan\beta = 0.01$ the number of theoretically possible edge wave modes is 78. It is clear that on natural beaches with typical gradients in the order of 0.025 a large number of edge wave modes can occur.

In addition to the edge wave cut-off mode, *cut-off periods* may be given for certain nearshore profiles. For beach profiles approximated by a linear slope $\tan\beta$ terminating at a constant depth at a distance of $x_s$ from the shoreline (Figure 4.16), a series of edge wave cut-off periods are described by Huntley (1976)

$$T_e = 2\pi\sqrt{\frac{x_s}{3.5n(n + 1)g\tan\beta}} \tag{4.24}$$

For example, on a beach with $\tan\beta = 0.025$ going out until a distance $x_s = 500$ m, a mode 3 edge wave has a cut-off period of 44 s. Three mode 3 edge wave profiles are shown in Figure 4.16, one with the cut-off period and the other with edge wave periods of 34 and 54 s. The edge wave at the cut-off appears to occupy the entire nearshore zone and may be preferentially excited because: (1) edge waves with longer periods extend beyond the nearshore slope, resulting in offshore leakage of edge wave energy due to inefficient refractive trapping (Ursell, 1952); and (2) edge waves with shorter periods have smaller offshore length scales and may be more dampened by surf zone processes (Bowen and Guza, 1978). Indications that edge waves on natural beaches are present at the cut-off frequency were first given by Huntley (1976) and later by Gerritsen and van Heteren (1984). It is these cut-off edge waves that are in some cases held responsible for the formation of nearshore

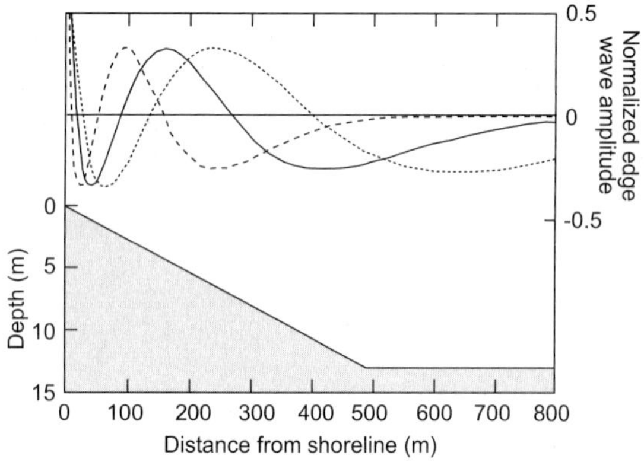


**Figure 4.16** Idealised beach with a linear slope of 0.025 terminating at a constant depth at a distance of 500 m from the shoreline, with the normalised wave amplitude of mode 3 edge waves with periods 34 s (dashed line), 44 s (solid line) and 54 s (dotted line). The cut-off period for this bathymetry is 44 s (Equation 4.24).

bar morphology (Aagaard, 1990). However, the cut-off mechanism does require a specific nearshore morphology.

The *generation and growth* in edge wave energy may occur as a result of non-linear wave interactions leading to resonant amplification of free modes through a triad mechanism (Gallagher, 1971; Bowen and Guza, 1978). A triad is composed of three waves, where the 'beating' of two waves results in the growth of the third. The wave radian frequencies and the longshore wave numbers of the waves constituting the triad must satisfy $\omega_e = \omega_1 - \omega_2$ and $k_e = k_1 - k_2$ (note that waves which approach the coast normal to the shoreline cannot contribute to a triad since the longshore wavelength is infinite and $k = 0$). An edge wave may be generated through a triad mechanism when $\omega_e$ and $k_e$ satisfy the edge wave dispersion relation. The edge wave frequency is then given by (Bowen and Guza, 1978)

$$\omega_e^2 = |\omega_1^2 \sin\alpha_1 - \omega_2^2 \sin\alpha_2| \sin(2n+1)\tan\beta \quad (4.25)$$

where $\alpha$ is the angle of wave incidence. The two waves resulting in the growth of the third (the edge wave) are usually considered to be obliquely incident waves propagating from different directions (Gallagher, 1971; Bowen and Guza, 1978).

Edge waves can also be generated in a manner similar to leaky waves, i.e. through the reflection of a bounded long wave or a time-varying breakpoint long wave for the case where the incident waves (and the wave groups) are non-normally incident and satisfy the edge wave dispersion relationship (Symonds *et al.*, 1982; Herbers *et al.*, 1995a).

### 4.3.6 Shear waves

Recently, a new class of progressive, wave-like, alongshore oscillations has been identified in the surf zone (Oltman-Shay *et al.*, 1989). These waves have relatively short alongshore wave lengths in the order of 100 m, but have periods ranging from 100 to 1000 s. The low frequency motion is apparent in the cross- and longshore current record, but generally not in the water surface elevation. These low frequency motions were initially termed far infragravity waves because of their extremely long periods. Bowen and Holman (1989) suggested that these oscillations are manifestations of a shear instability in the longshore current, the shear arising through horizontal velocity gradients of the mean current. This was subsequently confirmed by later studies (Dodd *et al.*, 1992; Putrevu and Svendsen, 1993), which also showed that bottom friction is important to the instability. Therefore, these motions are being referred to as shear waves. *Shear waves* exhibit an approximately linear relationship between the period of the motion and longshore wave length. Another distinguishing feature of the shear waves is that they are meandering alongshore and the relative strength of the cross-shore and longshore velocities varies as a function of the cross-shore location (Deigaard *et al.*, 1995).

### 4.3.7 Distinguishing between the different types of long wave motion

It is not possible to distinguish between the different types of infragravity motion using only a single sensor installed in the surf zone because the standing wave structure strongly depends on the distance from the shoreline. However, using data from cross-shore and/or longshore arrays of instruments with co-located water surface elevation sensors and current meters, separation of the different infragravity waves is feasible. *Shear waves* are relatively easy to identify because of their long periods and the near-absence of wave motion in the water surface elevation record.

Several techniques are available to differentiate between leaky waves and different edge wave modes using data collected using a cross-shore array of instruments.

1. *Cross-shore wave form.* Comparison of the measured cross-shore infragravity wave structure with the theoretically predicted wave form can indicate the type of wave motion. However, close to shore, the offshore

**Table 4.2** Phase relationships between water surface elevation and longshore and cross-shore current velocities for progressive waves, cross-shore standing waves, progressive edge waves and standing edge waves (after Huntley and Bowen, 1978).

| Wave type | Water surface versus cross-shore current | Water surface versus longshore current | Cross-shore current versus longshore current |
|---|---|---|---|
| Progressive waves | In phase | In phase | In phase |
| Cross-shore standing wave | Quadrature | In phase | Quadrature |
| Progressive edge wave | Quadrature | In phase | Quadrature |
| Standing edge wave | Quadrature | Quadrature | In phase |

forms are very similar for leaky waves and edge waves, whereas far from the shoreline the infragravity wave amplitude is rather small (Figure 4.14). In addition, the cross-shore structures of progressive and standing edge waves are identical. A further confounding factor associated with the identification of standing edge waves is the location of the instrument array in relation to the longshore edge wave nodes/antinodes. For example, if the instruments are closely located to the longshore edge wave node, the edge wave will not be represented in the water surface elevation record.

2. *Longshore current energy*. Another method to separate standing wave modes is by investigating the energy ratio of longshore current to cross-shore current velocity. Edge waves are three-dimensional, whereas leaky waves are two-dimensional. Therefore, the ratio of longshore to cross-shore current energy is larger for edge waves than for leaky waves.
3. *Phase relationships*. Comparison of the phase relationships between the water surface elevation, cross- and longshore current velocities is a relatively straightforward way to extract the type of long wave motion (Huntley and Bowen, 1978). Table 4.2 shows the various theoretical phase differences between the signals and indicates that it is possible to differentiate between standing edge waves and leaky waves/progressive edge waves. However, it is not possible to distinguish between progressive edge waves and leaky waves because they exhibit identical phase relationships.

The most successful way to identify confidently the type of long wave motion is through longshore arrays of instruments which enable the determination of the wave length of the infragravity wave. Holman and Bowen (1984) measured synchronous run-up records over a 7 km stretch of coastline and demonstrated the presence of a (longshore) standing edge wave with a wave length between 5 and 10 km. Katoh (1981) used aerial photographs of wave run-up and showed evidence of standing edge wave motion with antinodes located against a breakwater.

When longshore arrays of current meters are deployed, two-dimensional spectra can be computed (Huntley *et al.*, 1981). In these so-called *f-k* spectra (Figure 4.17) the amount of energy is plotted as a function of the wave frequency ($1/T$) and the reciprocal of the wave length ($1/L$). The amount of spectral energy is plotted in a normalised way, referred to as the enhancement factor, which is the ratio of the observed energy to the energy which would occur if the total energy at each frequency were evenly smeared over the wave length axis, usually up to limits of $0.025\,m^{-1}$ (40 m wave length). Reciprocal wave lengths can be positive or negative, depending on the direction of wave travel. The edge wave dispersion relationship dictates that the edge wave period is related to the edge wave length (Equations 4.21 and 4.22) and edge wave energy should therefore plot along the dispersion curves. Shear waves are characterised by the lowest frequencies and the shortest wave lengths and plot below the mode 0 edge wave dispersion curve. Shear waves are further characterised by a linear relationship between the wave length and frequency. For leaky waves, the longshore wave length is infinite and leaky wave energy should plot around the *f*-axis. High mode edge waves have a very large wave length, and their reciprocal wave length is close to zero. Therefore, even with *f-k* spectra it is still difficult to distinguish between high mode edge waves and leaky waves.

The two dimensional spectrum of the longshore current velocity (Figure 4.17A) indicates that a large part of the spectral energy is concentrated along the low mode (0–2) edge wave dispersion curves, demonstrating the presence of predominantly southward propagating low mode edge waves. The presence of southerly propagating shear waves (plotting under the mode 0 edge wave dispersion curve) is also clear. The *f-k* spectrum of the cross-shore current (Figure 4.17B), on the other hand, suggests a mixture of higher mode edge wave and possibly leaky wave energy. The reason for the seemingly

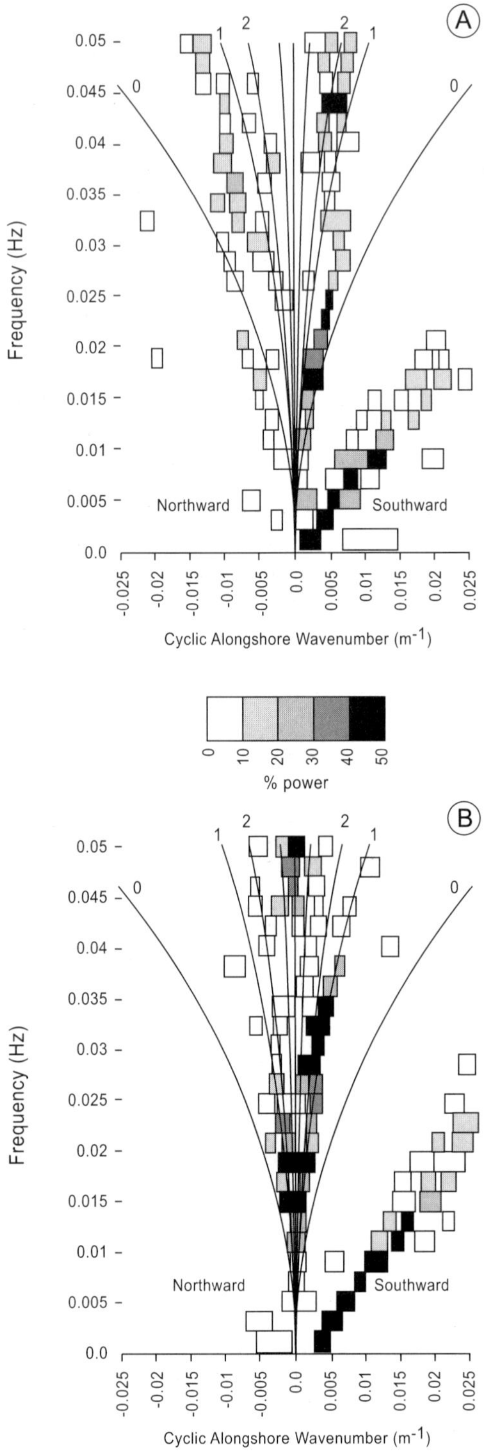


contrasting edge wave modes in Figure 4.17 is that the ratio of cross-shore to longshore velocity increases with edge wave mode number; this ratio scales as $(2n + 1)$ for progressive edge waves. The lower mode edge waves are therefore more likely to show up in the *f-k* spectra of the longshore current.

Due to the difficulties associated with the separation of the different types of standing wave motion, controversy still exists as to which infragravity type is more important. Numerical models generally assume leaky waves, partly because these two-dimensional waves are easier to model. However, the widespread occurrence of rhythmic topography suggests that (longshore) standing edge waves may be a frequent occurrence along barred shorelines. Unfortunately, the available data are not conclusive. Frequency–wave number spectra presented by Oltman-Shay and Guza (1987) and Huntley (1988) indicate that (progressive) edge waves may dominate the long wave spectrum. However, *f-k* spectra presented by Holman *et al.* (1990) suggest that leaky infragravity waves are significant contributors to infragravity energy in the surf zone. With a shore-normal wave approach, leaky modes and/or high mode edge waves may dominate the infragravity band (Howd *et al.*, 1990). Recent studies of Herbers *et al.* (1993, 1995a,b) on the shelves of the US Atlantic and Pacific coasts show that free infragravity energy generated by reflection of bound long waves was predominantly refractively trapped in the nearshore/shelf region as edge waves, except for very low energy conditions when leaky modes dominated. Additionally, Elgar *et al.* (1992) demonstrated that (forced) bound long wave energy may dominate the total infragravity energy in shallow water outside the surf zone under high wave energy conditions. This is consistent with a scenario where the majority of the incoming infragravity energy is trapped in the nearshore upon reflection, instead of re-radiating back offshore.

## 4.4 SURF ZONE CURRENTS

Surf zone currents are defined as currents generated by wave action in and near the breaker zone (Shepard and Inman, 1951) and are almost exclusively due to energy dissipation in breaking waves. While mass transports

**Figure 4.17** Example of *f-k* spectra of: (A) alongshore; and (B) cross-shore current velocity. The dispersion curves of the first four edge wave mode (modes 0, 1, 2 and 3) are drawn for comparison. Shear wave energy plots below the mode 0 edge wave dispersion curve (modified from: Oltman-Shay *et al.*, *Journal of Geophysical Research*, Vol. 94, pp. 18 031–18 042, 1989, copyright by the American Geophysical Union).

due to asymmetric incident waves and tidal flows still exist close to shore, such flows are generally of lesser importance in terms of velocities and sediment transport capacity in comparison to surf zone currents.

Three types of *surf zone currents* exist: (1) bed return flows (undertow); (2) rip currents (cell circulation); and (3) longshore currents. While three-dimensionality is an inherent characteristic of natural surf zones, rips and bed return currents are principally directed perpendicular to the beach, while longshore currents flow predominantly parallel to shore. In reality, cross- and longshore flows often coexist resulting in flows oblique to the shoreline, but for purposes of conceptual and modelling simplifications, a distinction is usually made between cross- and longshore components. All three surf zone current systems are due to cross- and/or longshore components of radiation stress (i.e. wave height) gradients associated with wave breaking. This also implies that when waves are not breaking, these currents will not be generated. However, flow inertia and lateral mixing allow surf zone currents to extend some distance beyond the surf zone.

### 4.4.1 Bed return currents

The *bed return current* is an offshore-directed mean flow near the bed and has been recognised for some time (Johnson, 1919; Evans, 1938; Bagnold, 1940). The current is part of a vertically segregated circulation of water which is directed onshore in the upper part of the water column and seaward near the bottom. Subsequent to early reports, bed return currents were more or less neglected and their existence and significance was challenged (e.g. Shepard *et al.*, 1941), probably because most field studies at that time were undertaken on beaches with horizontally segregated flows in the form of rip cell circulation systems. However, more recent quantitative investigations on dissipative beaches (Wright *et al.*, 1982a; Sallenger *et al.*, 1983) have reconfirmed the existence of bed return currents.

The bed return current is part of a simple stress balance (Svendsen, 1984b)

$$\frac{\partial S_{xx}}{\partial x} = -\rho g h \frac{\partial \bar{\eta}}{\partial x} - \bar{\tau}_b \qquad (4.26)$$

where $S_{xx}$ is the cross-shore component of onshore directed radiation stress, $\rho$ is the mass density of water, $h$ is the mean water depth, $\bar{\eta}$ is the wave-induced set-up and $\tau_b$ is the bed shear stress generated by the current (Section 4.2.6). Equation 4.26 is the depth- and time-averaged momentum balance and indicates that the depth-averaged momentum flux gradient ($\partial S_{xx}/\partial x$) is balanced by the set-up gradient ($\partial \bar{\eta}/\partial x$) and the bed shear stress induced by the mean current ($\bar{\tau}_b$). The cross-shore gradients in radiation stress and wave set-up gradients, however, are not equally distributed within the water column. The orbital velocity and total pressure are related to the elevation above the bed (Equation 4.12), and in surf zones by far the largest contribution to $S_{xx}$ comes from the region between the wave crest and the wave trough (Svendsen, 1984b). On the other hand, the pressure gradient from the set-up is uniformly distributed over depth. Consequently, there is a mismatch at each point in the vertical between the radiation stress contribution and the pressure gradient from the set-up (Figure 4.18). If a fluid element below trough level is considered, there will be (on average, over a wave period) a net seaward-directed horizontal force on that element corresponding to the local difference between the $S_{xx}$ term and the $\bar{\eta}$ term. A small fluid element above wave trough level, however, will experience a net landward force. These local imbalances in the stresses drive the vertically segregated mean flows with offshore-directed currents below, and onshore-directed flows above wave trough level (Figure 4.18).

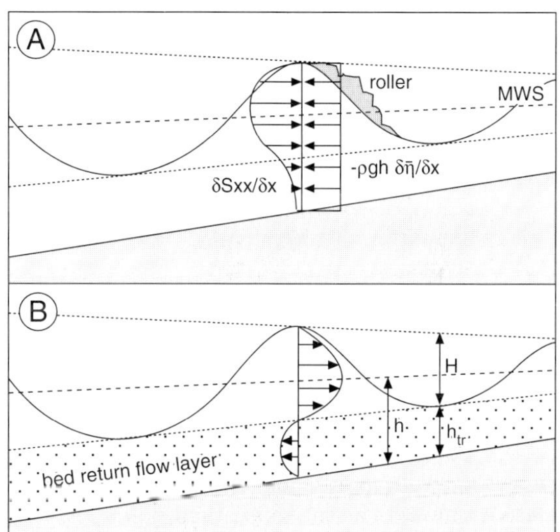


**Figure 4.18** (A) Schematic representation of the stress balance involved in the generation of bed return currents; (B) the resulting vertical velocity profile. MWS is the mean water surface. $h_{tr}$ is water depth beneath the wave trough (modified from: *Coastal Engineering*, Vol. 25, Masselink and Black, 'Magnitude and cross-shore distribution of bed return flow measured on natural beaches', pp. 165–190, 1995, with kind permission from Elsevier Science – NL, Sara Burgerhartstraat 25, 1055 KV Amsterdam, The Netherlands).

The bed return current is being fed by the onshore mass transport associated with the waves in the upper part of the water column. Svendsen (1984b) considered the onshore mass transport between the crest and the trough to consist of two contributions: (1) the effect of wave asymmetry (Stokes drift); and (2) the surface roller in breaking waves. This is in contrast to, for example, Dally and Dean (1986) who only considered the onshore discharge associated with Stokes drift. The contributions from Stokes drift and the surface roller can be written as (Fredsøe and Deigaard, 1992)

$$Q = Q_{drift} + Q_{roller} = \frac{\sqrt{gh}B_0H^2}{h} + \frac{A}{T} \tag{4.27}$$

where $Q$ is the net onshore discharge, $h$ is the water depth, $H$ is the wave height, $B_0$ is a shape factor which describes the peakedness of the waves, $A$ is the area of the roller in vertical projection (approximated by $A = 0.9H^2$; Svendsen, 1984b) and $T$ is the wave period. The value of $B_0$ can be expressed as a function of the Ursell number (Equation 4.4) using cnoidal wave theory and for surf zone waves ranges from $B_0 = 0.08$ for saw-tooth bores to 0.125 for sinusoidal waves (Fredsøe and Deigaard, 1992). The mean velocity of the bed return flow below trough level ($U$) is then found as

$$U = \frac{Q_{drift} + Q_{roller}}{h_{tr}} \tag{4.28}$$

where $h_{tr}$ is the water depth below wave trough level. This formulation has been found to compare reasonably well with measured current velocities on plane-sloping laboratory beaches with regular waves (Stive and Wind, 1986; Svendsen *et al.*, 1987), where maximum current velocities are found near the bed and are in the order of $(gh)^{1/2}$.

Field data on bed return currents are still scarce and efforts have not kept pace with modelling developments. Under natural conditions in the field, measured velocities are typically around 0.2–0.3 m/s (e.g. Wright *et al.*, 1982a; Greenwood and Osborne, 1990; Aagaard and Greenwood, 1994a; Haines and Sallenger, 1994; Masselink and Black, 1995). During an extreme situation with a maximum breaker height of 5 m, Sallenger *et al.* (1983) reported offshore-directed current velocities of 0.66 m/s near the bed. On *barred beaches*, maximum bed return current velocities have been recorded at or slightly landward of the bar crest (Greenwood and Osborne, 1990; Osborne and Greenwood, 1990; Smith *et al.*, 1992; Smith, 1994). On planar beaches, the cross-shore distribution of the bed return flow is characterised by a maximum in the mid-surf zone (Masselink and Black, 1995).

The bed return current is mainly dependent upon the local set-up gradient (e.g. Svendsen and Buhr-Hansen, 1988; Haines and Sallenger, 1994), which in turn is strongly related to the local wave height gradient, which again depends on the local beach slope ($H = \gamma h$; $\partial H/\partial x = \gamma \tan \beta$; Sections 4.2.5 and 4.2.6). Therefore, the bed return current velocity is expected to be larger on steeper beaches (Longuet-Higgins, 1983; Dibajnia *et al.*, 1994). This is supported by detailed field measurements of wave set-up and mean currents by Greenwood and Osborne (1990) that demonstrated a topographic control on set-up gradients, with the largest set-up gradients and bed return current velocities occurring over the steepest part of the nearshore profile.

Recently, more elaborate field-based numerical models on bed return currents have become available (e.g. Smith *et al.*, 1992; Haines and Sallenger, 1994). These models include the effects of wave incidence angle, boundary layer streaming (i.e. the residual circulation due to boundary layer expansion and contraction over a wave cycle) and wind stress, and also compute the vertical distribution of the bed return current. These models generally use the output of a random wave transformation model to compute radiation stress gradients and therefore allow for wave reformation to occur in the trough. As a result, computed bed return current velocities over the bar crest and seaward slope are larger than in the trough, in correspondence with field measurements.

### 4.4.2 Rip currents

Surf zone circulation is frequently dominated by horizontally segregated flows instead of the vertically segregated bed return current, in particular when longshore gradients in the wave set-up occur (Bowen, 1969a), for example on intermediate-state beaches. These flows consist of strong, narrow rip currents flowing seaward through topographic depressions in a bar and a weak net onshore flow due to mass transport in regions away from the rip. This cell circulation is completed by longshore flows increasing towards and converging at rips.

Rips are relatively easy to observe visually due to increased turbulence generated by wave–current interaction, gaps in the breaker line, streaks of darker water due to increased water depths and/or foam patches seaward of the breakers (Figure 4.19). The *rip current per se* consists of two converging feeder currents, the rip neck which occupies the rip channel across the bar, and

**Figure 4.19** (a) A low energy rip current (highlighted by dye) fed by two feeder channels and moving through the surf zone and seaward of the nearshore bar (Palm Beach, New South Wales, Australia; photo by R W Brander); and (b) a moderate energy rip head indicated by the fine calcareous suspended sediment (Fowlers Bay, South Australia, Australia; photo by A D Short).

the rip head. The largest velocities are encountered in the rip neck while flows expand and decelerate in the rip head. Because of their large flow speeds, rips may be important in the transport of sediment away from the beach and thus in beach erosion. Rips also pose a recreational hazard as swimmers may be caught in the strong offshore-directed currents and flushed out to sea (Shepard *et al.*, 1941; Short and Hogan, 1994).

Since the early scientific descriptions of rip currents (Shepard *et al.*, 1941), qualitative observations have been rather numerous while quantitative measurements are scarce. Mean rip velocities are significantly stronger than bed return flows and frequently exceed 1 m/s during high energy conditions (Short, 1985; Short and Hogan, 1994). During low/moderate energy conditions velocities of around 0.5 m/s are typical (e.g. Wright and Short, 1984; Bowman *et al.*, 1988). Figure 4.20 illustrates a typical example of rip current velocities with flow speed increasing from the shoals into the feeder current and peaking in the rip neck between bars. While bed return currents are maximum near the bed, rip velocities should be maximum close to the surface even though wave–current interactions may decrease flows at the top of the water column (Sonu, 1972). Velocities in the rip channel frequently pulse at low frequencies (30 seconds to several minutes; e.g. McKenzie, 1958; Chappell and Wright, 1978) as a response to infragravity waves and/or wave groups. Tidal fluctuations also have a marked influence on rips. At low tide, wave dissipation over bars and thus rip forcing increases and the flow is concentrated in troughs and rip channels resulting in increased flow velocities (McKenzie, 1958; Sonu, 1972; Short and Hogan, 1994; Masselink and Hegge, 1995; Aagaard *et al.*, 1997).

Rip currents are usually more or less evenly spaced along the shoreline with wave lengths ranging from approximately 50 m to 1000 m and more (Short, 1985). The ratio of wave length to surf zone width varies from 1.5 to 8 (Bowen and Inman, 1969; Sasaki, 1977; Short, 1985) and the spacing appears to increase with incident

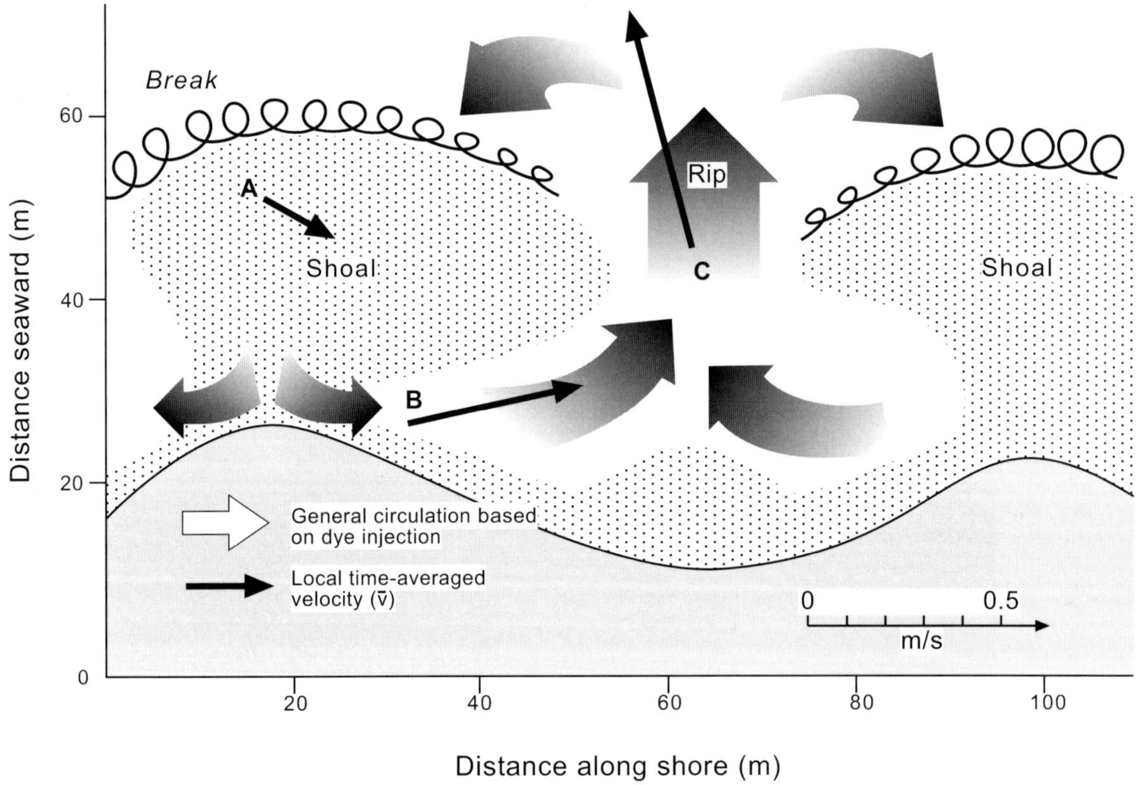


**Figure 4.20** Rip circulation observed and recorded in a transverse bar and rip system, Palm Beach, Australia (modified from: *Marine Geology*, Vol. 56, Wright L D and Short, A D, 'Morphodynamic variability of surf zones and beaches: a synthesis', pp. 93–118, 1984, with kind permission from Elsevier Science – NL, Sara Burgerhartstraat 25, 1055 KV Amsterdam, The Netherlands).

wave height (and thus surf zone width) and with decreasing beach gradient (Short, 1985; Huntley and Short, 1992). Hence, on a given beach, rips will attain the longest wave lengths under more dissipative conditions. Short (1985) identified three basic rip types: (1) erosion rips associated with increasing wave height and/or changing wave direction; (2) accretion rips associated with decreasing wave height; and (3) topographically controlled rips which are relatively permanent features located adjacent to headlands, reefs or coastal structures.

Cell circulation is driven by longshore gradients in set-up and/or radiation stress. On a uniformly sloping beach, gradients in set-up ($\partial\bar{\eta}/\partial y$) may occur because of *longshore variations in wave height* (Bowen, 1969a; Bowen and Inman, 1969). With longshore-varying wave heights, there will also be longshore gradients in radiation stress. Even though the orbital velocity is negligible in the direction parallel to the wave crests, a longshore flux of momentum will exist as the pressure departs from the hydrostatic (Equation 4.12). Longuet-Higgins and Stewart (1964) demonstrated that the longshore flux of momentum parallel to the wave crests is

$$S_{yy} = \frac{E}{2} = \frac{1}{16}\rho g H^2 \tag{4.29}$$

The forces arising from set-up and radiation stress gradients will combine to drive a current directed away from regions with large breakers. While these forces act in the same direction within the surf zone, it can be shown (Bowen, 1969a) that the forces act in opposition and exactly balance each other outside the surf zone where longshore gradients in mean water level (and thus pressure) are due to set-down under large waves (Section 4.2.6). Such a hydrodynamically controlled rip circulation can occur because of wave refraction patterns (Shepard *et al.*, 1941) or two intersecting wave trains (Dalrymple, 1975). A forcing exclusively due to set-up gradients can occur if a longshore uniform set-up is unstable to perturbations (Hino, 1974). In the analytical model derived by Hino (1974) such perturbations will result in rips with a wave length of four times the surf zone width.

However, rips usually exist under conditions with uniform longshore wave heights and an alternative mechanism to account for rip cell generation is *topographic control*. Sonu (1972) observed rip circulation occurring in association with a well developed rhythmic nearshore topography where waves were spilling across shoals and plunging in (pre-existing) rip channels. He explained the rip circulation as being driven by set-up gradients; over the shoals, a constant wave energy dissipation (and thus a large shoreline set-up) occurred, while set-up in the lee of the channels was less due to less intense wave breaking. Subsequently, Noda (1974) demonstrated analytically that rip circulation can occur under longshore uniform wave conditions with a variable bottom topography. However, a pre-existing topography cannot, in itself, explain a regular longshore rip spacing, or the initial construction of this topography.

A mechanism which can satisfactorily explain the common regularity in alongshore rip current/channel spacing is the presence of longshore *standing edge waves* (Bowen and Inman, 1969; Holman and Bowen, 1982). The drift velocity (mass transport) field in the bottom boundary layer under standing edge waves may rearrange the bed topography with shoals at longshore antinodes and depressions at nodes (Figure 4.21). To account for the longshore length scales involved in rip spacing, these edge waves would have to be of infragravity frequencies. A mode 1 edge wave with $T_e = 60$ s on a slope of 0.02 would generate a topographic wave length ($\lambda$) of 335 m, as $\lambda = L_e/2$ (Equation 4.22). Once these small bed perturbations are formed, the larger wave energy dissipation over shoals will initiate the cell circulation according to the models by Sonu (1972) and Noda (1974).

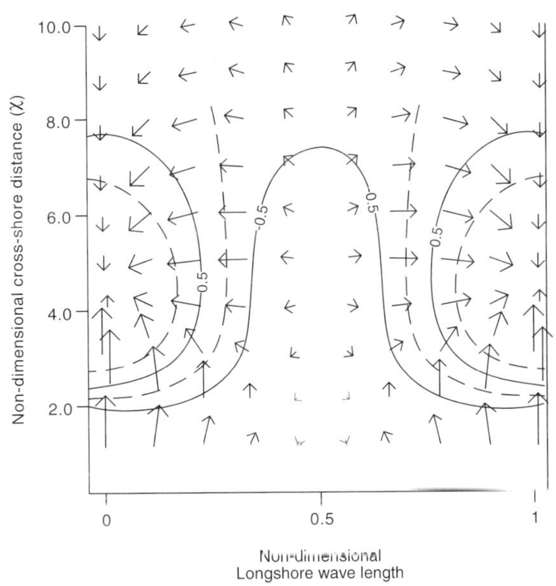


**Figure 4.21** Non-dimensional drift velocities in the bottom boundary layer under a standing mode 1 edge wave, as well as the resulting topographic perturbation. The horizontal axis is longshore rhythmic wave length (after Holman and Bowen, 1982).

A conceivable model for rip current generation involves standing edge waves, occurring under high energy conditions. The edge waves provide the initial bottom perturbation and the resulting spatially varying wave breaker patterns initiate a cell circulation leading to a positive topographic feedback (i.e. rip channels are being hollowed out, while shoals are accreting) resulting in a stable circulation pattern. The tendency for larger rip current spacings with larger incident wave heights (and decreasing nearshore gradients) has been hypothesised to be a result of edge wave dimensions in some way being constrained by surf zone width (Huntley and Short, 1992).

While numerical models (e.g. Bowen, 1969a; Noda, 1974; Sasaki, 1977) have been successful in reproducing the qualitative aspects of cell circulation, quantitative correpondence has been rather poor (Basco, 1983; Massel, 1989). Furthermore, quantitative field investigations on rips are few due to logistical problems associated with monitoring a rip current system. Nevertheless, mean rip velocity should depend on the amount of water carried onshore by the breaking waves, the distance between the rip channels and the channel cross-sectional area. Recently, Aagaard *et al.* (1997) tested the simple model

$$U_{rip} = \frac{Q\lambda_{rip}}{A_{rip}} \tag{4.30}$$

where $U_{rip}$ is mean rip current velocity, $A_{rip}$ is the cross-sectional area of the rip channel, $\lambda_{rip}$ is distance between rips and $Q$ was computed from Equation 4.27 with the water depth set equal to the depth over the bar crest. With bores propagating across a bar, Equation 4.30 predicts that rip velocity will increase with the square of the wave height and with the distance between rips. Increasing wave period and rip channel cross-section will decrease rip velocities.

### 4.4.3 Longshore currents

Longshore currents are continuous shore-parallel flows within the surf zone. These currents, which may reach velocities exceeding 1.5 m/s, are a powerful sediment transport agent and they are associated with problems concerning long-term beach erosion and harbour/inlet shoaling. A satisfactory theory capable of predicting the velocities of the longshore current and its horizontal velocity distribution did not emerge until Bowen (1969b) and Longuet-Higgins (1970a,b) investigated the role of the longshore component of radiation stress. When waves are obliquely incident on a beach, there will be a flux towards the shoreline of longshore-directed radiation stress. In shallow water, this component is given by (Bowen, 1969b)

$$S_{xy} = E \sin\alpha \cos\alpha = \frac{1}{8}\rho g H^2 \sin\alpha \cos\alpha \tag{4.31}$$

where $\alpha$ is the angle of wave incidence.

Longuet-Higgins (1970a,b) balanced the longshore gradient of this *radiation stress* against the bed shear stress generated by the longshore current (note that there are no pressure gradients associated with the longshore flow)

$$\frac{\partial S_{xy}}{\partial x} = -\bar{\tau}_b \tag{4.32}$$

Consequently, Longuet-Higgins (1970a,b) obtained a formula for the longshore current velocity assuming a plane slope and a saturated surf zone, and neglecting the $\cos\alpha$ term in Equation 4.31

$$V = \frac{5\pi}{8}\frac{\tan\beta}{c_f} u_m \sin\alpha_b \tag{4.33}$$

where $c_f$ is a friction coefficient in the order of 0.01, $u_m$ is the maximum orbital velocity and $\alpha_b$ is the breaker angle. The horizontal velocity profile predicted with Equation 4.33 is characterised by a linear increase from zero to its maximum value from the shoreline to the breakpoint, followed by a sharp discontinuity at the breakpoint because outside the surf zone there is no longshore current. To smooth this velocity profile and obtain a more satisfactory configuration, Longuet-Higgins (1970a,b) introduced a lateral mixing due to turbulent eddies.

The model developed by Longuet-Higgins (1970a,b) has since become the accepted basic theory. However, when applied to natural surf zones it suffers from (at least) two weaknesses: (1) the assumption of regular, monochromatic waves; and (2) the assumption of a plane-sloping beach. The former assumption necessitates the introduction of the lateral eddy viscosity (mixing) term. However, in natural surf zones waves are irregular and there is not a single breakpoint, but a wide breaker zone. As a result, the energy dissipation and the longshore current forcing will be more evenly distributed across the surf zone. Thornton and Guza (1986, 1989) eliminated the need for an eddy viscosity term by assuming a Rayleigh distribution of wave heights and obtained a formula for the longshore current velocity under *random waves* on a plane beach

$$V = \frac{3\pi}{8}\frac{B_0^3 f_p^{1/2}}{c_f\langle\gamma\rangle^4}\frac{\sin\alpha}{C}\frac{H_{rms}^6}{h^{9/2}} \tag{4.34}$$

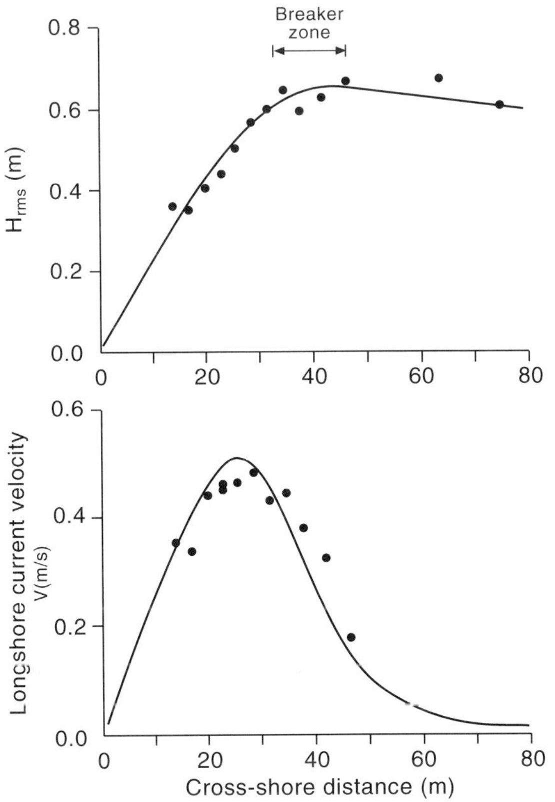


**Figure 4.22** Comparison of measured wave heights and longshore current velocities (circles), and predictions from an analytic random wave model solution for a planar beach (solid line) (modified from: Thornton and Guza, 1986, *Journal of Physical Oceanography*, by permission of the American Meteorological Society).

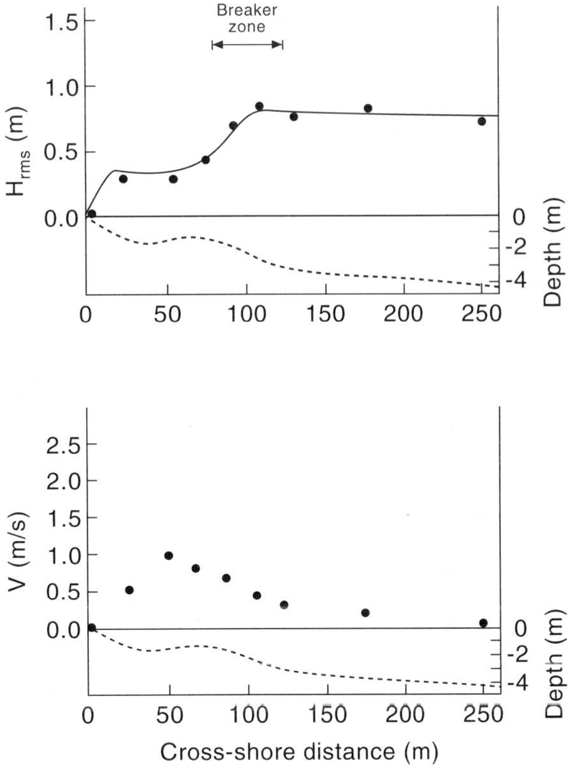


**Figure 4.23** Example of cross-shore distribution of wave heights and longshore current velocities on a barred beach (modified from: *Coastal Engineering*, Vol. 20, Church and Thornton, 'Effects of breaking wave induced turbulence within a longshore current model', pp. 1–28, 1993, with kind permission from Elsevier Science – NL, Sara Burgerhartstraat 25, 1055 KV, Amsterdam, The Netherlands).

where $f_p$ is peak wave frequency and $B_0$ is the wave shape factor, $H_{rms}$ and $h$ are evaluated locally, while $\alpha$ and $C$ are the wave incidence angle and wave celerity seaward of the surf zone. Figure 4.22 shows a comparison of model-predicted $H_{rms}$ and $V$, and measured values for an (approximately) plane beach. Good agreement was obtained with $\langle\gamma\rangle = 0.42$ and $c_f = 0.008$. Introducing an eddy viscosity term did not improve predictions, thus demonstrating that this term is redundant for random waves.

While the model of Thornton and Guza (1986, 1989) has proved successful on planar beaches it breaks down when applied to a barred beach. The reason for this is that wave height and water depth are evaluated locally. Predicted velocity maxima may be reasonable, but observations show that the velocity distribution of the current tends to be skewed landward and maximum velocities are often found in the trough between bars where radiation stress gradients are small and/or positive due to wave reforming (e.g. Greenwood and Sherman, 1986; Smith *et al.*, 1993) (Figure 4.23). Apparently, several mechanisms contribute to the lateral mixing and the landward displacement of the velocity maximum (Holman, 1995). Firstly, the onshore advection of turbulent kinetic energy associated with the wave bore may result in a spreading and delay in radiation stress decay (Smith *et al.*, 1993; Lippman *et al.*, 1995). This would displace the current maximum landwards as well as smoothing the longshore current profile. Secondly, shear waves which appear to be stronger on barred beaches than on planar beaches provide a mechanism for cross-shore momentum redistribution and can be an important source for lateral mixing (Church *et al.*, 1994). Thirdly, Putrevu and Svendsen (1993) suggested that a coexistence of cross-shore and longshore mean flows will produce a lateral shear and thus broaden the longshore current profile.

Longshore currents are not only dependent on incident waves but also on wind and tides. At high tide, longshore currents may increase due to increased local wave heights (Thornton and Kim, 1993). Furthermore, tidal currents can modulate the longshore current. Wind effects were disregarded for a long time, probably because the majority of field studies on longshore currents were conducted on swell-dominated (planar) ocean beaches. However, Nummedal and Finley (1978) and Whitford and Thornton (1993) demonstrated that under conditions with strong winds blowing alongshore, wind forcing of longshore currents may indeed be significant, particularly on gently sloping beaches.

## 4.5 SEDIMENT TRANSPORT

Sediment transport refers to the combined movement of sediment as bedload and in suspension. In the surf zone the velocity field is oscillatory, with constant changes of direction and velocity magnitude due to the superimposition of waves over a range of frequencies, and unidirectional, non-stationary currents. This results in consistent shifts between velocity accelerations and decelerations and the response of the sediment particles to these conditions is still poorly understood.

Figure 4.24 illustrates an example of cross-shore velocity ($u$) and sediment concentration ($c$) near the bed, under breaking waves in a moderate energy surf zone, as well as the resulting instantaneous sediment flux computed as the cross-product of velocity and concentration. The oscillatory nature of the velocity field is evident but the wave motion is superimposed on a mean offshore directed current of approximately 0.3 m/s. Initially, sediment concentrations (and the resulting sediment transport) are low, but at around $t = 3$–$4$ min large waves occur, coincident with a relatively strong offshore current pulse. Sediment concentrations and transport rates increase significantly and it is obvious that at this time some correlation exists between the maximum velocity over a wave cycle, sediment concentration and thus transport rate. However, these relationships are not straightforward and $u$ and $c$ are not in phase. Such characteristics have confounded the establishment of simple relationships between fluid velocity and sediment transport rate.

### 4.5.1 The bottom boundary layer and bed roughness

The *bottom boundary layer* is the zone in the immediate vicinity of the bed where the fluid motion is significantly influenced by the frictional resistance of the bottom. The velocity in the boundary layer grades from zero at the bed to the velocity of the free stream at some distance above the bed. The thickness of the wave boundary layer ($\delta$) depends on the wave period and is approximately (e.g. Grant and Madsen, 1986)

$$\delta \approx \frac{\kappa u_{*,w}}{\omega} \tag{4.35}$$

where $\kappa$ is von Karman's constant ($\approx 0.4$), $u_{*,w}$ is the wave-induced shear velocity and $\omega$ is the wave radian frequency ($2\pi/T$). The thickness of the wave boundary layer is typically in the order of a few centimetres. While an *oscillatory* boundary layer more or less breaks down and reforms twice every wave cycle, a *current* boundary layer is significantly thicker, as this layer has ample time to develop.

The most important physical aspect of the boundary layer is the shear stress ($\tau_b$), i.e. the force which the fluid motion exerts on the bed. The *shear stress* can be defined as $\tau_b = \rho u_*{}^2$ and determines the velocity gradient close to the bed and the mobilising forces applied to the sediment grains on the bottom. This quantity is of fundamental importance to sediment entrainment and transport.

Under unidirectional currents, the bed shear stress can be computed from the vertical velocity profile. In the lower 1–2 m of the water column this profile is generally well described by the von Karman-Prandtl equation (Wright, 1995)

$$u_z = \frac{u_*}{\kappa} \ln(z/z_0) \tag{4.36}$$

where $u_z$ is the (mean) velocity at elevation $z$ above the bed, $u_*$ is the shear velocity associated with the current and $z_0$ is the zero intercept of the velocity profile.

However, sediment entrainment and concentration in surf zones are mainly determined by the shear stress due to the waves. This is because the wave boundary layer is much thinner than the current boundary layer. Therefore, velocity gradients (the shear velocity) due to waves are significantly larger than for the mean current. An appropriate way of determining shear velocity and bed shear stress is through the use of the definition of *bed shear stress* under waves

$$\tau_{b,w} = \rho u_{*,w}^2 = \frac{1}{2} \rho f_w u^2 \tag{4.37}$$

where $\rho$ is the density of the water, $u$ is the free stream wave orbital velocity and $f_w$ is a wave friction factor, i.e. a coefficient of proportionality describing the relation

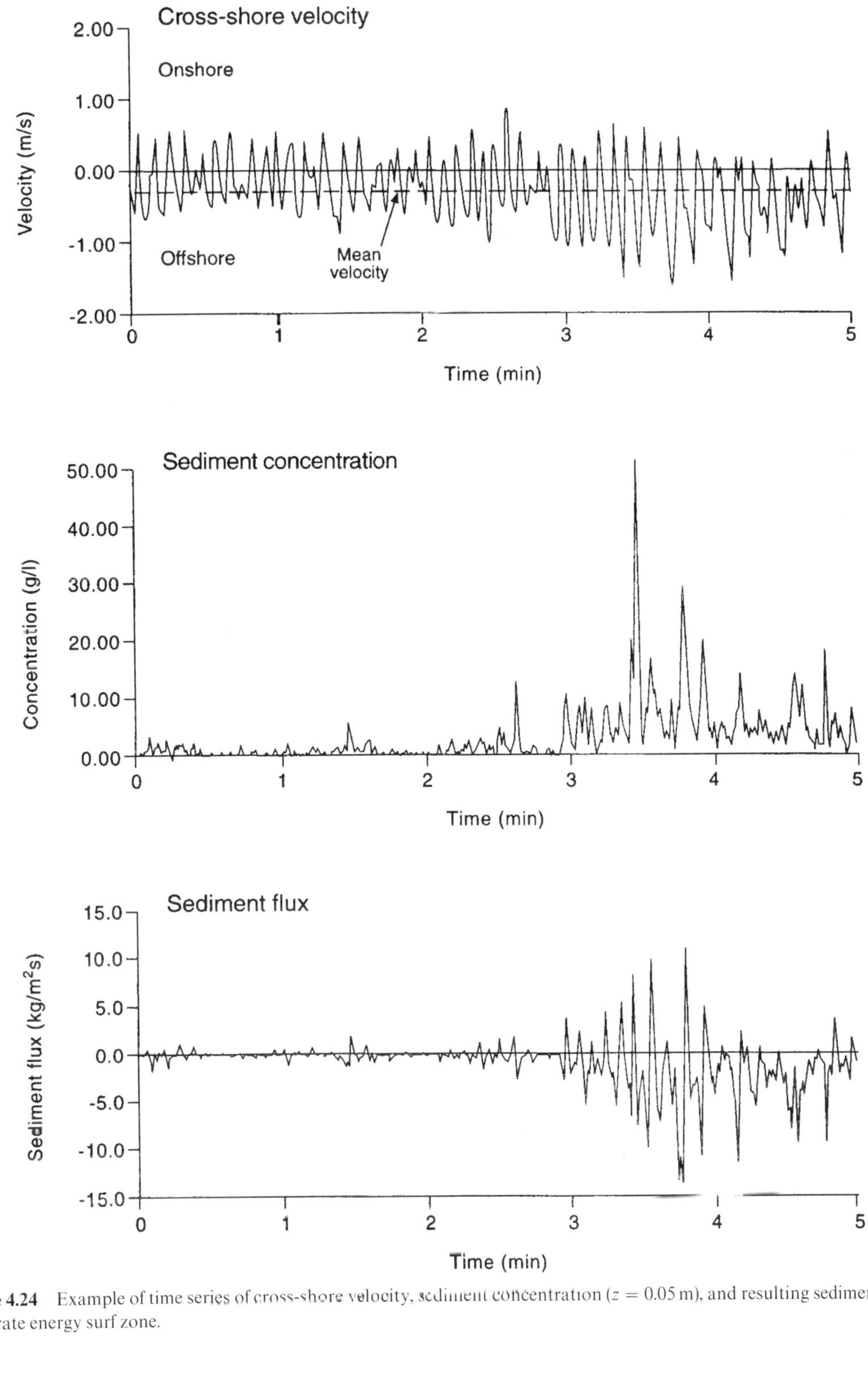


**Figure 4.24** Example of time series of cross-shore velocity, sediment concentration ($z$ = 0.05 m), and resulting sediment flux in a moderate energy surf zone.

between shear velocity and the free stream velocity. From Equation 4.37, $u_*/u = (f_w/2)^{1/2}$. A commonly accepted expression for the (non-dimensional) *wave friction factor* is due to Swart (1974)

$$f_w = \exp(5.213(k_s/A)^{0.194} - 5.977) \quad k_s/A < 0.63 \tag{4.38}$$

where $k_s$ is the bed roughness and $A$ is the water particle semi-excursion (i.e. orbital amplitude, $A = u_m T/2\pi$). For $k_s/A > 0.63, f_w = 0.30$.

The bed roughness (or hydraulic roughness) is usually defined as (e.g. Nielsen, 1992)

$$k_s = 30z_0 \tag{4.39}$$

Hence, the rougher the bed the steeper the velocity gradients and the more stress is exerted by the fluid motion. However, in sediment transport modelling, the specification of $k_s$ remains a problem. When waves and currents coexist, the bed roughness is mainly linked to the wave motion. *Total bed roughness* is composed of the *grain roughness* (roughness due to the individual particles on the bed, $k_d$), roughness exerted by *bedforms* (form drag roughness, $k_r$), and roughness due to sediment movement ($k_m$). The latter two terms are called *moveable bed* roughness (Wright, 1995). Thus

$$k_s = k_d + k_r + k_m \tag{4.40}$$

Nielsen (1983) suggested that for a flat fixed bed, $k_s = k_d = 2.5\ D_{50}$ (where $D_{50}$ is the mean grain size). However, in the surf zone the bed is rarely immobile and for a moveable bed under waves, Nielsen (1992) proposed

$$k_s = \frac{8\eta_r^2}{\lambda_r} + 170D_{50}(\theta' - 0.05)^{1/2} \tag{4.41}$$

where $\eta_r$ is the bedform height, $\lambda_r$ is the bedform wavelength and $\theta'$ is the skin friction Shields parameter

$$\theta' = \frac{\tau'}{\rho(s-1)gD_{50}} = \frac{0.5\rho f'_w u^2}{\rho(s-1)gD_{50}} = \frac{0.5 f'_w u^2}{(s-1)gD_{50}} \tag{4.42}$$

where $s$ is relative sediment density ( $\approx 2.65$ for quartz particles in water) and $f'_w$ is found from Equation 4.38 with $2.5D_{50}$ substituted for $k_s$. In Equation 4.41, the first term represents $k_r$, whereas the second term accounts for $k_d$ and $k_m$.

### 4.5.2 Thresholds of sediment motion

The definition of incipient grain motion involves some conceptual difficulties. How many grains in a population should be required to move by how much in order for this threshold to be exceeded? Nevertheless, such thresholds are reasonably well defined and they depend on the ratio between mobilising and stabilising forces on the grain. This ratio can be quantified by the *Shields parameter* which was initially based on a study of incipient motion in steady flows (Shields, 1936)

$$\theta = \frac{\tau}{\rho(s-1)gD_{50}} \tag{4.43}$$

In wave motion, this equation is replaced by Equation 4.42 with $\theta$ computed from the maximum bed shear stress over a wave cycle. Van Rijn (1993) compiled laboratory and field data on the initiation of motion under waves and concluded that the original Shields curve is also applicable in this case (Figure 4.25). According to Nielsen (1992), for sand-sized particles the critical Shields value can be given by $\theta_c \approx 0.05$. This is the threshold value appearing in the last term of Equation 4.41 for the total bed roughness.

A difficulty with this threshold is that it is based mainly on laboratory data and applies to regular waves and to a single grain size. In nature, a range of grain sizes occurs and the smaller grains in a population will be mobilised earlier and entrained longer. Also, these criteria are only valid for flat beds. If relict ripples are present, the thresholds are significantly lower due to flow concentration at ripple crests, and vortex shedding from the ripples. However, such problems are of minor importance in the surf zone where the threshold of motion is almost always exceeded.

### 4.5.3 Sediment transport modes

Sediment may be transported in the surf zone as either bedload or suspended load. *Bedload* is defined as sediment grains moving in more or less continuous contact with the bed, by rolling, sliding or jumping. A commonly accepted physically based definition of the distinction between bedload and suspended load is due to Bagnold (1956). He stated that grains contributing to the bedload are supported by intergranular forces (i.e. individual grains are in contact with the bed), while suspended load is supported by fluid drag, i.e. turbulence. While it may be difficult to distinguish visually between bedload and suspended load, particularly close to the bed where sediment concentrations are large, the transport mechanisms are different for these two modes. Bedload transport is determined by the effective bed shear stress on the grains and transport rates are often computed as a function of the skin friction Shields par-

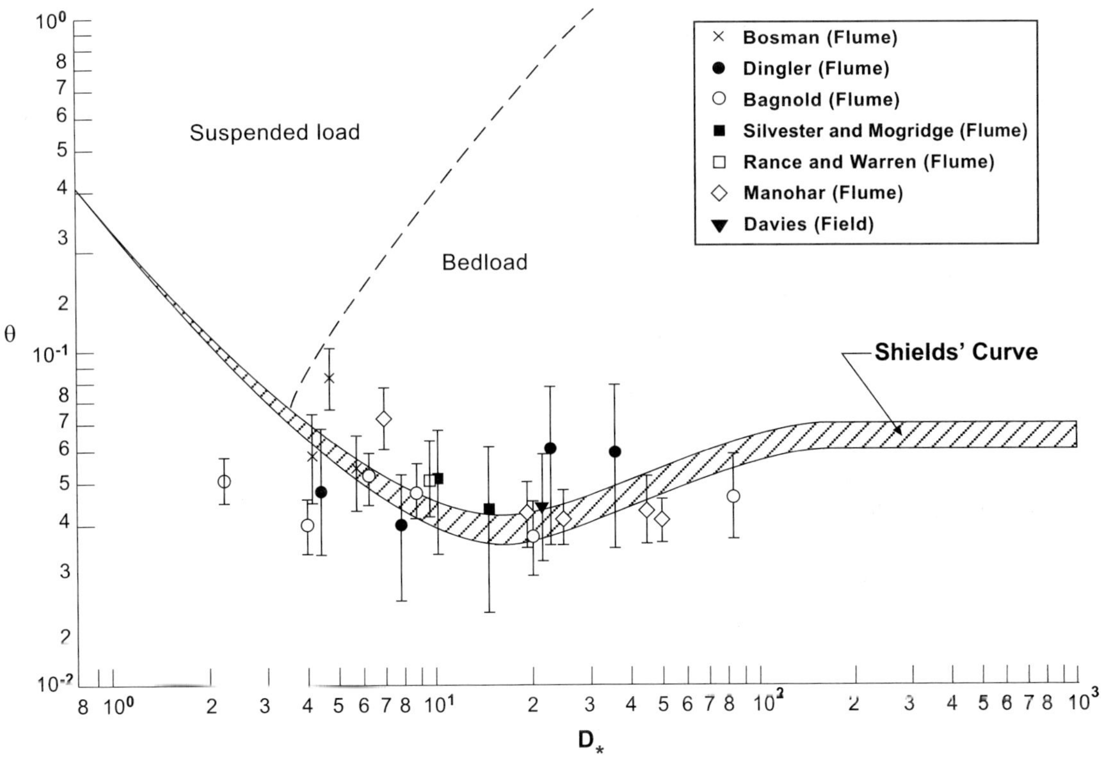


**Figure 4.25** Initiation of sediment movement on a plane bed under waves. $\theta$ is the Shields parameter and $D_*$ is the dimensionless particle size $((s-1)g/v^2)^{0.33}D_{50}$, where $v$ is the kinematic viscosity of the water. The cross-hatched area is the original Shields threshold curve for unidirectional flow. More recent data on the threshold under waves have been added (modified from: van Rijn, L C, *Principles of Sediment Transport in Rivers, Estuaries and Coastal Seas*, 1993, by permission of Aqua Publications).

ameter ($\theta'$; Equation 4.42). On the other hand, *suspended load* transport is generally computed as the cross-product of fluid velocity and sediment concentration.

It is therefore important to determine the relative magnitudes of bedload and suspended load. However, a state of confusion has reigned on this subject. First of all, it is difficult to define when a grain leaves the bed *per se*. This is further complicated by the fact that sediment grains in the surf zone are usually intermittently suspended, subsequently falling back to the bed (Figure 4.24). Empirical measurements in the laboratory have been hampered by scale problems and such measurements (as well as field measurements) have often been very pragmatic in their definitions. Instead of following the physically based definitions given above, bedload has often been considered as that part of the transport taking place below a certain level, or caught in traps. Finally, while reliable electronic recording devices now exist for suspended load, bedload is very difficult to measure in the surf zone.

Komar (1976, 1978) considered the relative proportions of bedload and suspended load by computing the theoretical longshore transport rate and comparing this rate with the cross-product of (computed) longshore current velocity and previously published values of mean sediment concentration in the surf zone, determined from pump samples. Komar (1976, 1978) further demonstrated that tracer dispersal was far slower than the advection speed of the suspended sediment, and concluded that ratios of suspended load to total load were in the order of 0.2–0.25. However, apart from the fact that theoretically predicted longshore transport rates may not be very reliable and pump sampling has its problems (Bosman *et al.*, 1987), the field data on sediment concentration were all collected well away from the bed. As suspended sediment concentrations are largest close to the bed, with rather steep concentration gradients, the computed values of suspended load were too low. Subsequently, Hanes (1988) reanalysed Komar's data and showed that the results were consistent with transport in intermittent suspension. Hanes (1988) suggested that the unsteady forcing due to the waves in the surf zone provides intervals during which suspended sediment grains are allowed to resettle at the bed and the average velocity of a grain in a tracer population that is partly suspended, partly moved as bedload and partly buried, will be far less than the longshore current velocity.

Bagnold (1956, 1966) suggested that sediment suspension becomes dominant when shear velocity approaches grain settling velocity ($u_*/w_s = 0.8$). Substituting this criterion into Equation 4.42 yields for quartz grains

$$\theta' = \frac{0.4 w_s^2}{g D_{50}} \qquad (4.44)$$

Most recent field studies (e.g. Hanes, 1988; Thornton and Abdelrahman, 1991; Beach and Sternberg, 1992) have argued that suspended sediment transport dominates in the entire surf/swash zone. On the other hand, bedload must be of a certain magnitude in the surf zone as bedload is probably a significant contributor to bedform migration.

### 4.5.4 Bedforms

When near-bed flow velocity exceeds the threshold, the bed is deformed and various *bedforms* appear. Initially, *wave ripples* having wave lengths ($\lambda_r$) in the order of $10^1$ cm and heights ($\eta_r$) in the order of a few centimetres are formed, later (with increasing flow velocity) giving way to *megaripples* with $\lambda_r \approx 10^2$ cm and $\eta_r \approx 10^1$ cm. Finally, when $\theta' > 1$, the bedforms are erased and the bed becomes flat (Nielsen, 1981; Hay and Bowen, 1993).

Studies on bedform sequences (e.g. Clifton *et al.*, 1971; Clifton and Dingler, 1984) have been of interest from a geological point of view as an aid in interpretations of facies relations in the geological record. From geomorphological/engineering perspectives, the bedforms are important to sediment transport as they: (1) increase local shear stress, resulting in increased suspended sediment concentration; and (2) increase bed roughness, thus influencing fluid velocity gradients and the vertical scales of the sediment concentration profile. Working on high energy planar beaches in Oregon with asymmetric incident waves, Clifton *et al.* (1971) identified a bedform sequence (Figure 4.26) progressing from asymmetric ripples over lunate megaripples to flat bed with increasing flow velocity (decreasing water depth). Ripples and megaripples were identified seaward of the breakpoint while *flat bed* was dominant in the surf zone. At the transition between the surf and swash zones, large scale bedforms reappeared. Davidson-Arnott and Greenwood (1974) confirmed this sequence for barred beaches with the qualification that sequences may repeat when waves break over a bar and then reform in the trough; ripples reappeared in troughs. Subsequently, later investigations (e.g. Osborne and Vincent, 1993) have identified more detailed and elaborate bedform types.

In wave-dominated environments, *ripple wave length* appears to be at least initially determined by the wave orbital diameter $d_0$ (Figure 4.27). Such ripples are termed *orbital ripples* (Clifton and Dingler, 1984) and are characterised by the formation of vortices in the lee of ripple crests. Based on laboratory flume data, Miller and Komar (1980) found that $\lambda_r = 0.65\ d_0$. However, when a certain threshold is reached ($d_0/D \approx 5000$; Clifton and Dingler, 1984), the intensified flow velocities erode sediment from the ripple crests. The ripple wave length becomes independent of $d_0$, with sediment grain size becoming the main determinating factor, and $\lambda_r \approx 400$–$600\ D$ (Clifton and Dingler, 1984; Wiberg and Harris, 1994). Clifton and Dingler termed these bedforms *anorbital ripples*. Ripples occurring when $1000 < d_0/D < 5000$ are considered *sub-orbital ripples* and for these ripples the wave length becomes less dependent on $d_0$ and increasingly dependent on $D$. These ripples are therefore forms transitional between orbital and anorbital ripples.

Where wave ripples are initially *two dimensional*, they may evolve into *three dimensional* forms with increasing orbital velocity, and further into megaripples (e.g. Osborne and Vincent, 1993; Hay and Wilson, 1994), or vice versa. Megaripples usually appear when waves are strongly asymmetric (Figure 4.26) or associated with significant currents, for example in rip channels. However, the thresholds for megaripple formation are as yet undetermined, partly because these large bedforms are difficult to generate in the laboratory. Furthermore, ripples and megaripples frequently coexist.

Measurements of *ripple migration* rates in the surf zone vary widely from 1 to 65 cm/h (Brander, 1991; Osborne and Vincent, 1993; Hay and Bowen, 1993) while migration rates of megaripples tend to exceed those of ripples under similar environmental conditions (Sherman *et al.*, 1993). So far, no predictive relationships have yet been proposed, but the rates of migration appear to depend on maximum orbital velocity, velocity asymmetry, orbital diameter and mean flow velocity. Until recently, bedform studies in the field largely relied on diver observations which restricted quantitative estimates to relatively calm weather conditions. However, with the appearance of acoustic measuring devices (e.g. Greenwood *et al.*, 1993; Hay and Bowen, 1993; Vincent and Osborne, 1993), it is to be expected that the data

| Offshore | Near shore | Surfzone | | Swash |
|---|---|---|---|---|
| | Outer rough | Outer planar | Inner rough | Inner planar |
| Wave shoaling | Wave built up | Wave breaking | Surfzone current | Uprush backwash |
| | | (bar) | (trough) | |

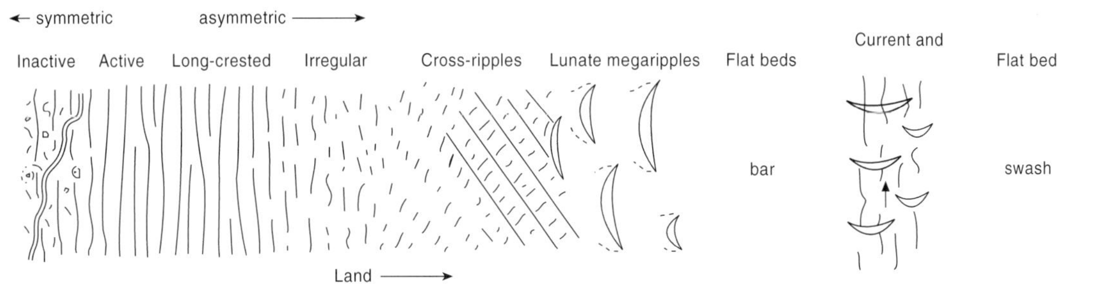


**Figure 4.26** Bedforms under waves. Seaward of the breakpoint, asymmetric wave ripples change into lunate megaripples with increasing wave asymmetry. Under bores inside the surf zone, flat bed dominates (modified from: Clifton *et al.*, 1971, *Journal of Sedimentary Petrology*, by permission of the Society for Sedimentary Geology).

base will significantly expand and more reliable estimates of ripple dimensions and migration rates will appear.

Ripple migration is at least partly a reflection of sediment transport as bedload. While transport associated with ripple migration may be small in comparison to total suspended sediment transport, there are indications that the migration of megaripples may involve a significant proportion of the net sediment flux (Dyer and Soulsby, 1988; Hay and Bowen, 1993). However, the latter authors acknowledged that it is difficult to separate the contributions from suspended load and bedload to bedform migration rate, as their observations showed intermittent suspension events over these bedforms. Therefore, the processes of bedform migration and suspended sediment transport cannot be separated at present.

### 4.5.5 Sediment resuspension and concentration profiles

*Sediment resuspension* is intimately linked to the configuration of the bed. Over a flat bed, concentrations close to the bed are in phase with the fluid motion (e.g. Ribberink and Al-Salem, 1994). As indicated in Figure 4.28, under asymmetric waves sediment concentrations remain low under the wave trough and concentration magnitudes display only one peak per wave cycle. Close to the bed, the result would be sediment transport in the direction of the waves. With increasing elevation above the bed, the phase lag between velocity and concentration increases and the direction of the sediment transport over a wave cycle may reverse. The increasing phase lag is due to the finite amount of time it takes for the sand to move upwards to higher elevations.

While the concentration magnitude over a flat bed is characterised by a single peak over a wave period, this may not be the case over a *rippled bed* where concentrations can display two major peaks (Figure 4.29; Tunstall and Inman, 1975; Nakato *et al.*, 1977). These major peaks generally occur close to fluid velocity reversals, i.e. velocity and concentration are phase shifted by approximately 90°. Sediment entrainment over rippled beds can be described as follows (Inman and Bowen, 1963; Figure 4.30). Under the wave crest (Figure 4.30A), a vortex is formed in the trough between ripple crests. Sediment from the trough becomes entrained in this vortex; the vortex is ejected when the velocity reverses (Figure 4.30B), and is transported backwards across the ripple crest under the wave trough while it expands and sediment settles to the bed. Simultaneously, a new vortex is formed under the wave trough (Figure 4.30C) and ejected during the next velocity reversal (Figure 4.30D). This entrainment/ejection mechanism is mainly a *convection* process, in contrast to the entrainment over a flat bed which approximates a *diffusion* process (Figure 4.31; Nielsen, 1992).

When fluid velocities are asymmetric (due to asymmetric waves, or a superimposed mean current), there will be a tendency towards a single major concentration peak with the vortex being ejected during the velocity reversal following the velocity maximum. Depending

BOURNEMOUTH UNIVERSITY
86

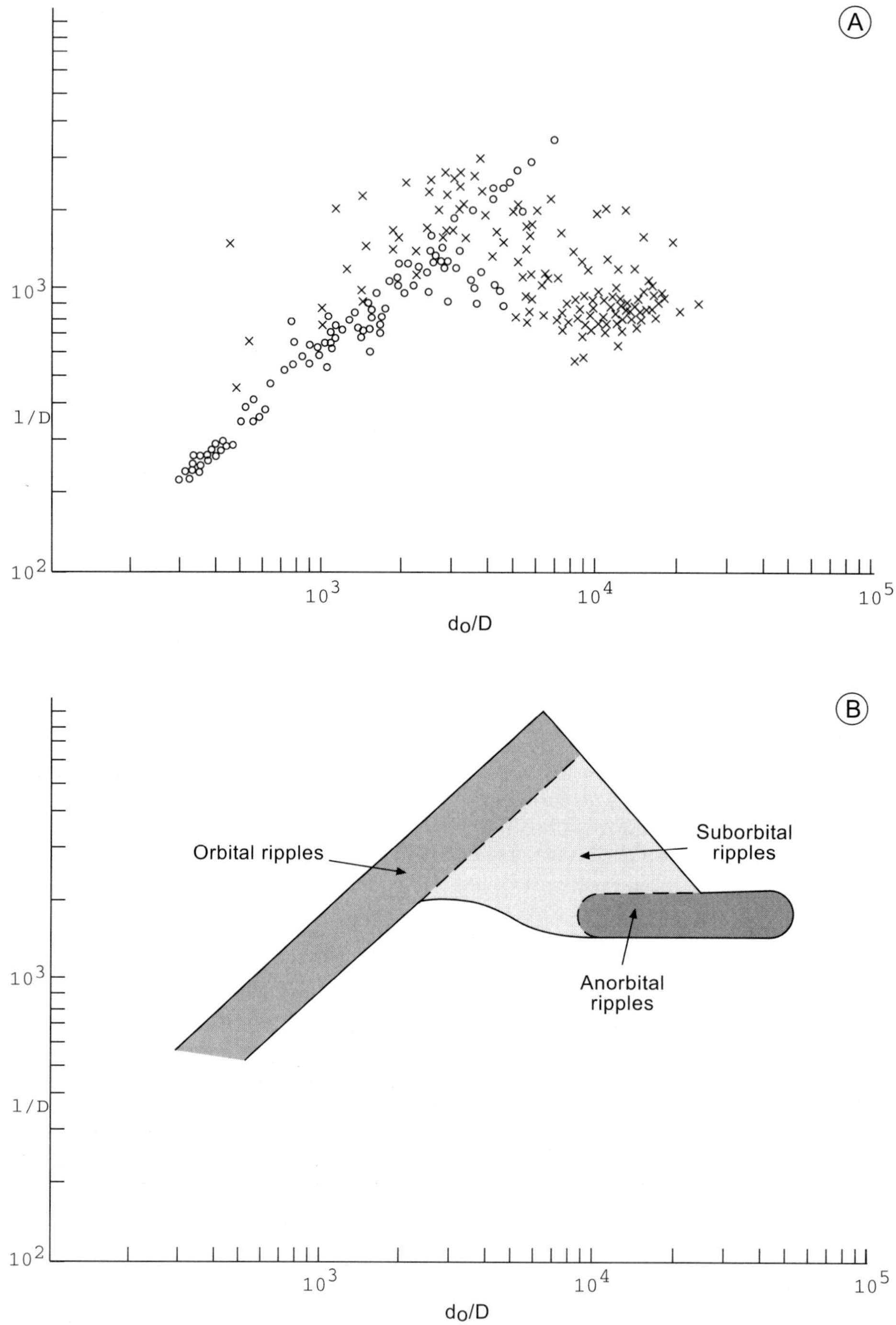


**Figure 4.27** (A) Field observations (crosses) and laboratory data (circles) indicating the relation between non-dimensional ripple spacing ($\lambda/D$) and non-dimensional wave orbital diameter ($d_0/D$); and (B) classification of ripple shapes based on the distribution shown in (A) (modified from: *Marine Geology*, Vol. 60, Clifton, H E and Dingler, J R, 'Wave-formed structures and palaeoenvironmental reconstruction', pp. 165–198, 1984, with kind permission from Elsevier Science – NL, Sara Burgerhartstraat 25, 1055 KV Amsterdam, The Netherlands).

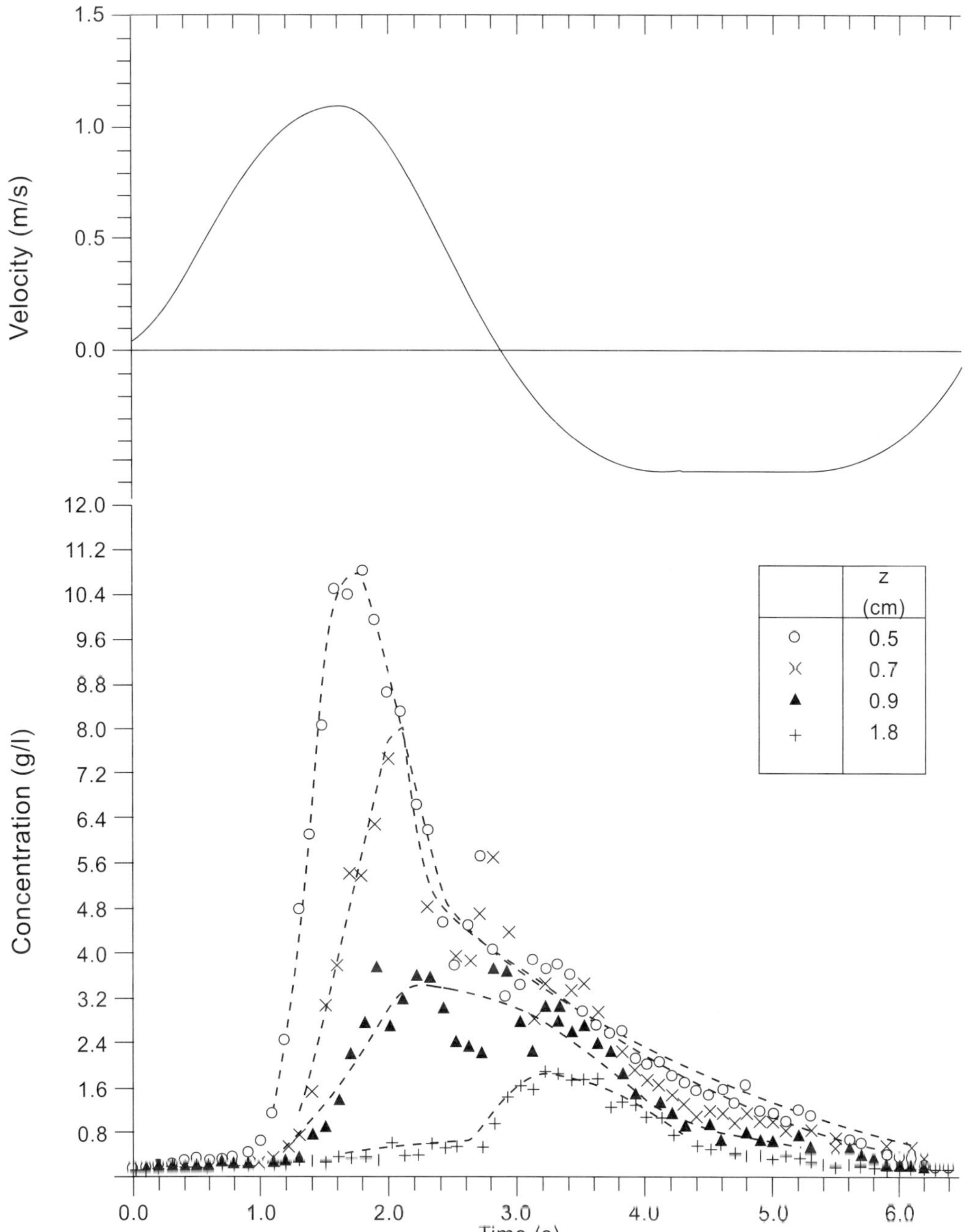


**Figure 4.28** Instantaneous sediment concentrations at different elevations under asymmetric waves on a flat laboratory bed (after Ribberink and Al-Salem, 1994).

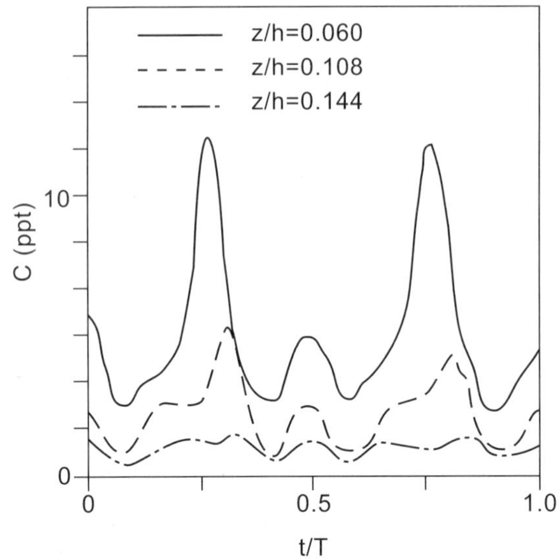


**Figure 4.29** Instantaneous sediment concentrations at different elevations above a ripple crest, as a function of non-dimensional wave phase (after Nakato *et al.*, 1977).

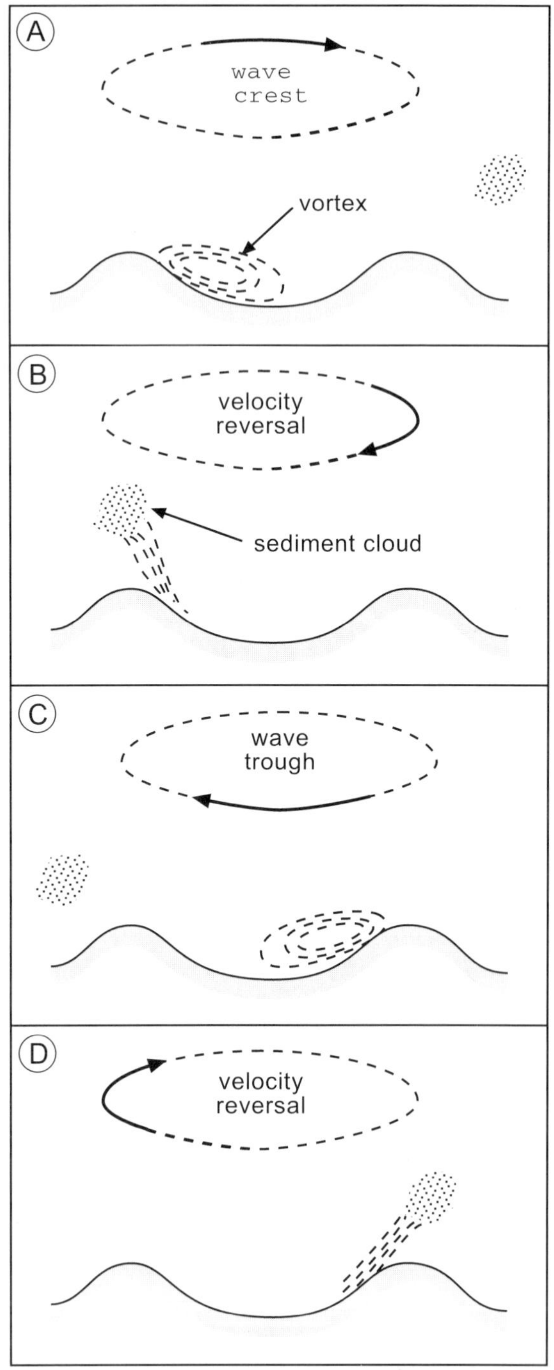


**Figure 4.30** Sediment resuspension under symmetric waves on a rippled bed. The arrow describes the wave orbital motion (see text for further explanation) (after Inman and Bowen, 1963).

upon the relationship between turbulence level at the bed and sediment fall velocity, which determines the amount of time the sediment remains in suspension, the net flux may be against the waves, or in the wave direction. Furthermore, it is obvious that the phase between velocity and concentration changes as a function of elevation (Vincent and Green, 1990; Osborne and Greenwood, 1993), similar to the flat bed case. While net sediment flux may be, for example, onshore at the bed, the transport direction will frequently reverse at higher elevations. Figure 4.32 is an example from Green and Vincent (1990) where suspended sediment transport contributions due to mean currents and oscillatory waves were separated. Transport due to the wave motion reversed a short distance above the bed, resulting in a directional change of the net transport.

### 4.5.6 Sediment concentration

*Concentration profiles* vary continuously over a wave cycle (e.g. Black and Rosenberg, 1991; Osborne and Greenwood, 1993); however, when averaged over a wave cycle the mean concentration profile usually approximates an exponential shape. Therefore, looking at sediment resuspension in a more quantitative manner, this can be modelled for a *flat bed* using the diffusion approach which states that in the case of a steady state situation, the downward flux of sediment ($w_s c$) must be balanced by the upward flux given by the product of the

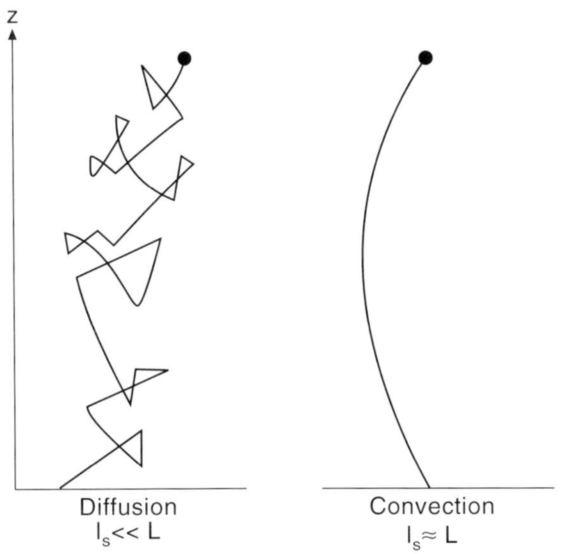


**Figure 4.31** Gradient diffusion and convective entrainment of sediment from the bed. $L$ is the vertical length scale of the concentration profile and $l_s$ is the mixing length (modified from: Nielsen, P, *Coastal Bottom Boundary Layers and Sediment Transport*, 1992, by permission of World Scientific Publishing Co. Pte Ltd).

concentration gradient ($\partial c/\partial z$) and an eddy diffusion coefficient ($\varepsilon_s$)

$$w_s c + \varepsilon_s \left(\frac{\partial c}{\partial z}\right) = 0 \tag{4.45}$$

Solving this equation, and assuming that $\varepsilon_s$ is constant through the entire water column yields

$$c_z = c_0 \exp\left(-\frac{w_s z}{\varepsilon_s}\right) \tag{4.46}$$

where $c_z$ is the concentration at elevation $z$, $c_0$ is a reference concentration at the bed relative to which $c_z$ is measured, and $(\varepsilon_s/w_s)$ is termed the mixing length ($l_s$) which can be determined as $l_s = \partial z/\partial \ln c$. The mixing length can be viewed as a typical distance (elevation) over which the sediment mixes with the surrounding fluid. This length scale can be measured empirically by using two (or more) concentration sensors deployed at different elevations and the shape of $c_z$ can then be approximated. However, in the absence of empirical measurements, $c_z$ must be predicted and one is faced with two problems: (1) defining a value for $c_0$; and (2) determining $l_s$. Experimental evidence (Nielsen, 1984; Black, 1994) indicates that

$$c_0 = 0.005\theta^3 \tag{4.47}$$

where $c_0$ is the non-dimensional sediment concentration. In the case of a flat bed, $\theta = \theta'$ (Equation 4.42 with $k_s = 2.5D_{50}$). The other problem is to determine the vertical length scale of the resuspension process which varies over a wave cycle and with elevation. However, Black and Rosenberg (1991) and Black (1994) obtained reasonable results for the concentration profile at peak flow velocity by assuming $\varepsilon_s$ (and hence $l_s$) is constant over the water column. For non-breaking waves on a flat laboratory bed, $l_s \approx 30D_{50}$ (Nielsen, 1986) whereas prototype scale laboratory data in the sheet flow regime, reported by Ribberink and Al-Salem (1994) indicate $l_s \approx 1.5$–$2$ cm.

Over *rippled beds* where convection is likely to be the dominant mechanism for sediment entrainment, Nielsen (1984) found that Equation 4.46 may still be applied with the reference concentration near the bed found from Equation 4.47 and $\theta = \theta'/(1 - \pi\eta_r/\lambda_r)^2$. The mixing length is approximately equal to the dimension of the vortex. Field measurements reported by Osborne and Greenwood (1993) indicated that $l_s = 3$–$14$ cm with the smaller values associated with two dimensional straight crested ripples and the larger values occurring under three dimensional irregular ripples. Van Rijn (1993) contains a useful summary of various models for the computation of $\varepsilon_s$ and $l_s$. It seems likely that this mixing length is critical in determining the transport direction associated with incident waves. Apart from the configuration of the bed, the mixing length will also depend on whether the waves are breaking or shoaling/reforming (Black and Rosenberg, 1991) and on the breaker type. Turbulence levels and vertical mixing are considerably larger for plunging breakers than for spilling breakers (Miller, 1976).

*Absolute concentration* magnitudes vary spatially across the surf zone. In the inner surf/swash zone, Beach and Sternberg (1991) measured concentrations exceeding 100 g/l (> 4% by volume) on time scales of several wave periods at $z = 2.8$ cm. Over barred topography, sediment concentrations may be an order of magnitude larger over bars where waves are breaking than in troughs with reforming waves. Another factor which is important is the breaker type with plunging breakers suspending more sediment than spilling breakers (Kana, 1978; Yu *et al.*, 1993) due to the larger levels of turbulence. Finally, a succession of large waves occurring in groups will cause increased mean concentrations as sediment is successively 'pumped up' into the water column (e.g. Hanes and Huntley, 1986; Osborne and Greenwood, 1993). This pumping effect due to wave

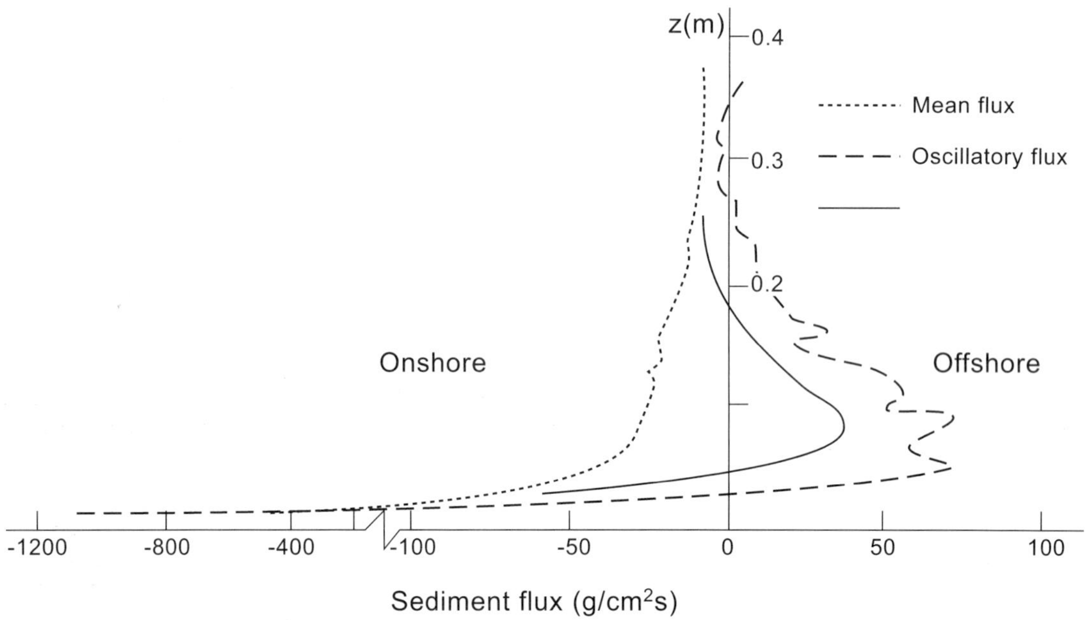


**Figure 4.32** Field measurements of sediment flux due to mean flows and oscillatory waves, and the resulting net flux, plotted as a function of elevation above the bed (modified from: Green, M O and Vincent, C E, 'Wave entrainment of sand from a rippled bed', *Proceedings of the 22nd International Conference on Coastal Engineering*, 1990, by permission of the American Society of Civil Engineers).

irregularity significantly hampers predictions of sediment concentration and sediment flux under natural conditions.

### 4.5.7 Sediment transport in the surf zone

Until recently, quantitative measurements of cross-shore sediment transport rates in the surf zone proved rather difficult as measurement techniques in the presence of significant oscillatory motions were inadequate. Tracer studies were conducted in order to attempt an explanation for morphological changes in the profile (e.g. Greenwood and Hale, 1982). However, in such studies sediment transport loads can only be viewed over individual (storm) events and interpretations of sediment transport patterns are problematic as the tracer will disperse seaward as well as landward. Measurements of longshore transport rates using tracers (Ingle, 1966; Komar and Inman, 1970), or pump samplers (Nielsen, 1984) and sediment traps (Kraus and Dean, 1987) have been considerably more successful due to the dominance of the (unidirectional) longshore current.

Fast-response sensors such as the Optical Backscatter Sensor (OBS; Downing *et al.*, 1981) are, in conjunction with electromagnetic current meters, the standard instruments for measuring sediment transport in the nearshore environment. Through recording simultaneous time series of velocity and concentration, the net transport rate can be determined by

$$\langle q_s \rangle_n = \frac{1}{N} \sum uc \tag{4.48}$$

where $\langle q_s \rangle_n$ is the net suspended sediment flux and $N$ is the number of data points in a time series. Moreover transport can be separated into components associated with mean flows and oscillatory currents (Jaffe *et al.*, 1984)

$$\langle q_s \rangle_n = \bar{u}\bar{c} + u'c' \tag{4.49}$$

where $\bar{u}$, $\bar{c}$ are mean velocity and mean concentration, and $u'$, $c'$ represent the oscillatory components of the velocity and concentration field, respectively. The oscillatory component of the sediment transport can be determined by subtracting the mean transport from the total, or alternatively, by integration of the cospectrum of velocity and concentration (Huntley and Hanes, 1987)

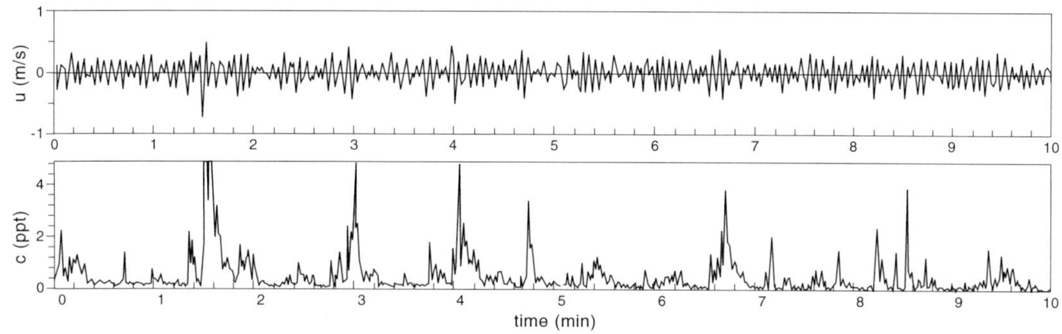


**Figure 4.33** Time series of cross-shore velocity and sediment concentration measured under shoaling waves on a non-barred beach (modified from: *Marine Geology*, Vol. 106, Osborne, P D and Greenwood, B, 'Frequency dependent cross-shore suspended sediment transport. 1. A non-barred shoreface', pp. 1–24, 1992, with kind permission from Elsevier Science – NL, Sara Burgerhartstraat 25, 1055 KV Amsterdam, The Netherlands).

$$u'c' = \frac{\Delta f}{F}\int_0^T C_{uc(f)}\partial t \qquad (4.50)$$

where $\Delta f$ is the spectral resolution, $F$ is the cospectral frequency range, $C_{uc(f)}$ is the cospectral density at a given frequency and $T$ is the length of the record. The cospectrum thus yields information on the transport rates and directions associated with different components of the velocity spectrum.

### *4.5.7.1 Cross-shore suspended sediment transport*

Cross-shore sediment transport on a non-barred beach was studied by Osborne and Greenwood (1992). Sediment resuspension and transport under shoaling waves was dominated by oscillatory currents at incident wave frequencies (Figure 4.33). The time series of velocity and concentration in Figure 4.33 illustrate that sediment resuspension occurs at time scales of the incident waves with large suspension events occurring in association with the wave groups where a sequence of large waves successively increase sediment concentration. Cospectra recorded under such conditions (Figure 4.34) show that oscillatory transport mainly occurs at incident wave frequencies with the transport rate rapidly decreasing with elevation. In most cases, cospectra recorded in the nearshore indicate a net onshore sediment flux. However, at times the net transport is directed offshore. The mixing length ($l_s$) is probably involved in determining the direction of the net flux, particularly where ripples are the dominant bedform in the nearshore.

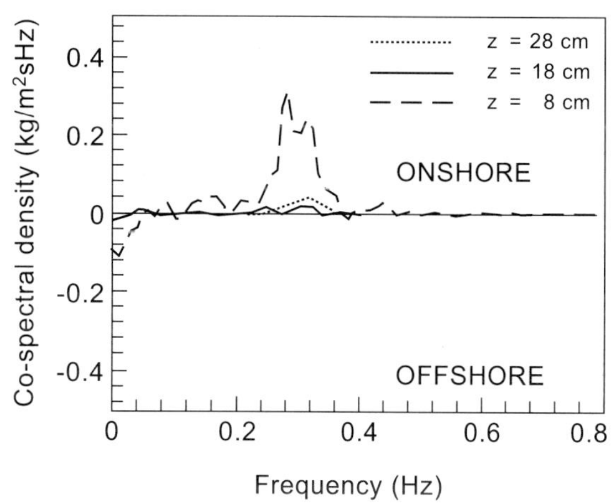


**Figure 4.34** Cospectra of cross-shore velocity and sediment concentration at different elevations recorded under shoaling waves on a non-barred beach (modified from: *Marine Geology*, Vol. 106, Osborne, P D and Greenwood, B, 'Frequency dependent cross-shore suspended sediment transport. 1. A non-barred shoreface', pp. 1–24, 1992, with kind permission from Elsevier Science – NL, Sara Burgerhartstraat 25, 1055 KV Amsterdam, The Netherlands).

Another common characteristic in cospectra recorded under shoaling wave conditions is a small offshore transport at low (infragravity) frequencies. This transport is due to group-bound long waves which have wave troughs (offshore velocity component) under the large incident waves (Shi and Larsen, 1984; Section 4.3.3). As mean flows are weak under non breaking waves the sediment transport is predominantly due to oscillatory flows. Similar characteristics were reported by Huntley and Hanes (1987) and Doering and Bowen (1988). On reflective beaches, waves break close to the beach face and in the surf zone a landward transport due to incident waves may be balanced by offshore transport due to bed return flows (Osborne and Greenwood, 1992).

On intermediate-state beaches with waves breaking on nearshore bars and reforming in troughs, sediment transport rates and magnitudes are considerably more complicated (Davidson *et al.*, 1993; Aagaard and Greenwood, 1994b). Sediment transport varies spatially and depends on the relative position within the surf zone. Figure 4.35 shows cross-shore suspended sediment transport rates associated with incident and (cross-shore standing) infragravity waves as well as mean currents throughout a moderate storm event on a coast with distinct and relatively steep nearshore bars. When integrated over time and space, transport rates due to wave motions and mean currents were approximately equal. In terms of the individual transport constituents, sediment transport due to incident waves was almost exclusively onshore at all instrument stations, but decreased as the shoreline was approached. However, spatial and temporal reversals of transport associated with infragravity waves were a dominant characteristic. Moreover, mean flux was directed offshore in the inner part of the surf zone ($x = 130, 210$ m) due to bed return flows, but onshore in the outer surf zone ($x = 220$, 245 m), probably due to bore-induced mass transport and/or infragravity wave drift velocities. The overall transport pattern appeared very complicated. In the troughs, sediment transport rates are rather low as reforming waves only suspend relatively small amounts of sediment (Yu *et al.*, 1993) and thus sediment exchange between the bar and the beach may be small until the bar welds to the shoreline. Intermediate-state beaches are also characterised by longshore process gradients, and net sediment transport rates and directions are expected to vary in the longshore direction as well. The three-dimensional nature of sediment transport on these beaches has therefore not yet been quantitatively investigated, primarily because vast amounts of instruments are required to monitor comprehensively the sediment transport process.

On dissipative beaches, suspended sediment transport becomes increasingly dominated by mean flows and (standing) infragravity waves, whereas sediment transport due to broken incident wave bores is small (Beach and Sternberg, 1988; Aagaard and Greenwood, 1995a). Offshore mean flows due to bed return currents cause an offshore-directed sediment transport which is superimposed by transports due to standing infragravity waves. In some cases, however, infragravity transports may exceed transport due to mean flows by a factor of two or more (Russell, 1993). In contrast to shoaling wave conditions on reflective beaches, sediment resuspension tends to occur at infragravity wave time scales. Figure 4.36 illustrates an example where suspension events are closely correlated with the infragravity wave velocity field. This low frequency modulation of the sediment concentrations is related to infragravity wave-induced modulations of incident wave heights, alternately increasing/decreasing the water depth at a given position, and/or sediment resuspension by infragravity wave velocities.

Whereas infragravity wave sediment transport close to the beach is generally directed offshore and may be the principal component in causing beach erosion (Russell, 1993), oscillatory (infragravity) transport reversals may occur within the surf zone (Figures 4.35 and 4.37), possibly related to the (anti)nodal structure of these standing waves. Aagaard and Greenwood (1995a) found that such reversals, which were superimposed on an offshore-directed mean transport due to bed return currents, resulted in a significant net transport divergence, causing rapid erosion and migration of a nearshore bar.

In *summary*, longshore uniform (reflective and dissipative) beaches tend to display relatively simple characteristics in terms of sediment transport direction; the former are usually dominated by onshore transport due to incident waves and the latter are generally associated with offshore transports due to mean flows, with standing infragravity waves a possible source of transport gradients and/or reversals within the surf zone. Intermediate-state beaches are significantly more complex with spatial net transport reversals, both in the longshore and cross-shore direction. The reversals depend on the relative magnitudes of the various transport constituents. These described patterns are of course extremely generalised. As discussed earlier, bedforms may impose significant constraints on the transport directions associated with oscillatory motions. Bedforms are very common, particularly on reflective and intermediate-state beaches, and their presence or absence may result in temporal changes of net transport rates and directions (Davidson *et al.*, 1993). Moreover, transport reversals can occur with elevation above the bed (Figure 4.32).

#### *4.5.7.2 Longshore sediment transport*

Longshore sediment transport is of great importance to problems associated with, for example, beach erosion, harbour shoaling and inlet stability, and quantifications of the transport rate have been pursued for a considerable length of time (e.g. Galvin, 1967). Transport can often be considered almost unidirectional due to (strong) longshore currents and the fact that oscillatory

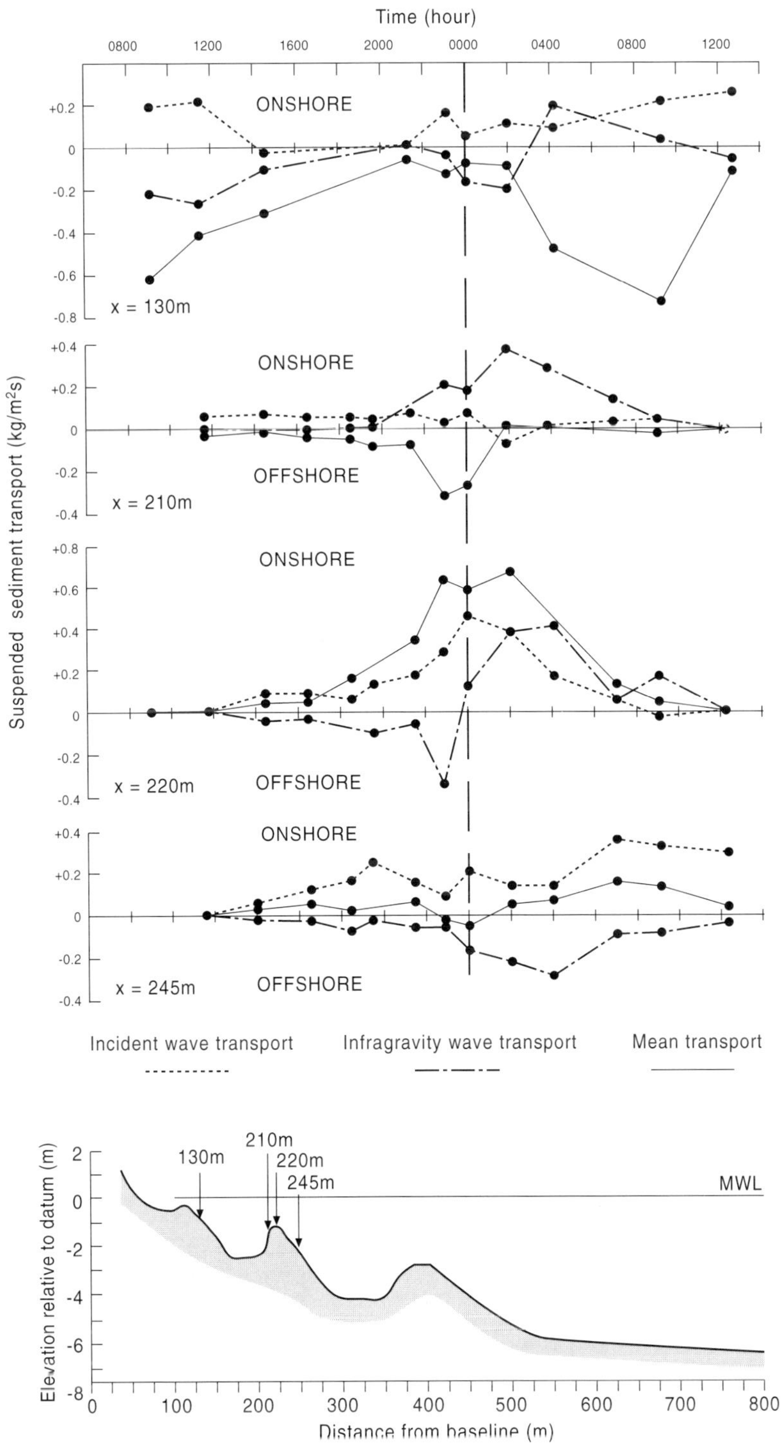


**Figure 4.35** Cross-shore sediment flux measured on an intermediate beach during a moderate storm event in a multiple barred system. The flux has been separated into the different transport constituents (mean currents, infragravity waves, incident waves). The positions of the measurements are indicated on the cross-shore profile (bottom panel) (modified from: *Marine Geology*, Vol. 118, Aagaard, T and Greenwood, B, 'Suspended sediment transport and the role of infragravity waves in a barried surf zone', pp. 23–48, 1994, with kind permission from Elsevier Science – NL, Sara Burgerhartstraat 25, 1055 KV Amsterdam, The Netherlands).

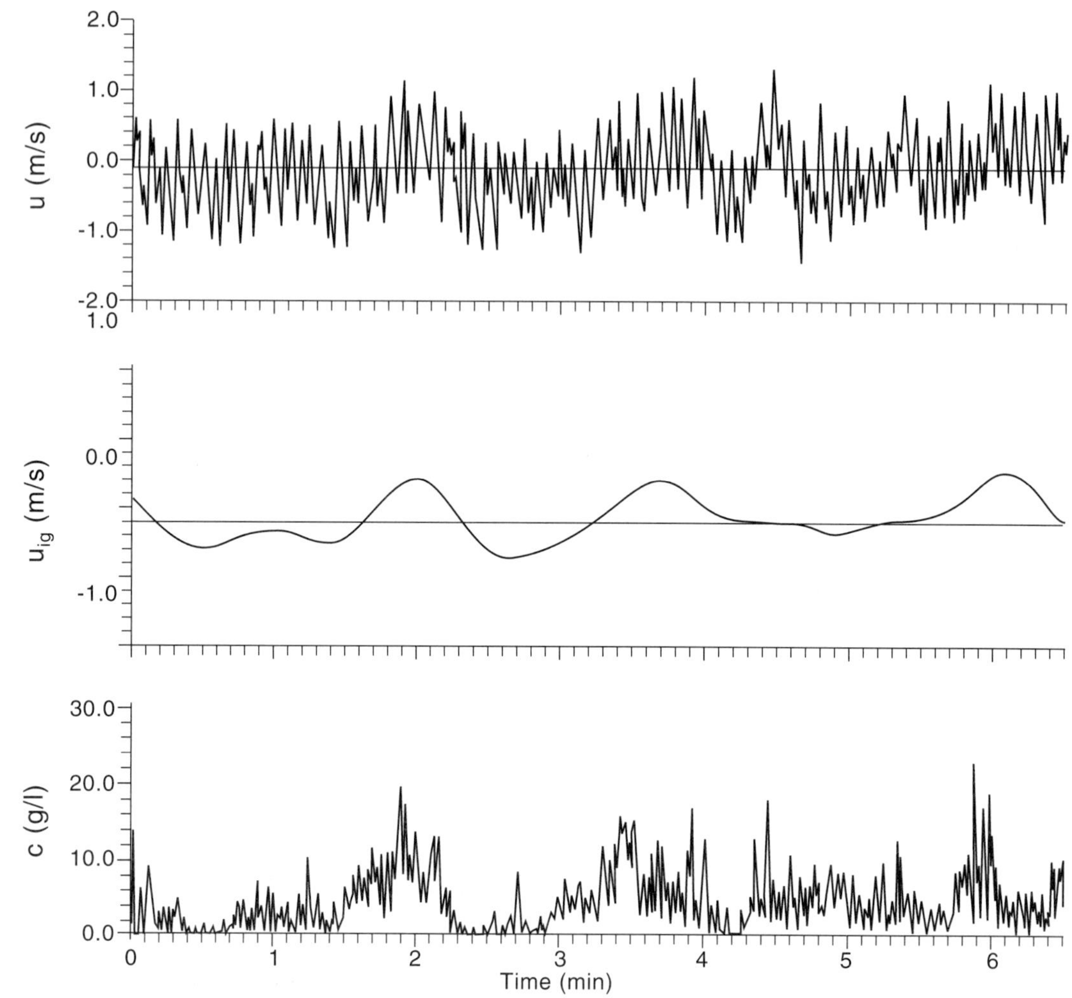


**Figure 4.36** Time series of cross-shore velocity ($u$), low-passed velocity ($u_{ig}$) and sediment concentration ($c$; at $z = 0.05$ m) on a dissipative beach. Concentration maxima are in phase with the onshore stroke of the infragravity waves (after Aagaard and Greenwood, 1995a).

flows in the longshore are generally relatively weak. Therefore $\langle q_s \rangle_n \approx \bar{v}\bar{c}$ (Sternberg *et al.*, 1984; Masselink and Pattiaratchi, 1998). Even though there are still significant problems with interpretations (Kraus, 1987), tracer studies have been rather more successful when applied to longshore transport than to cross-shore transport. Using this method, Komar and Inman (1970) derived the following simple empirical relation

$$I_l = K(ECn)_b \sin \alpha_b \cos \alpha_b \tag{4.51}$$

where $I_l$ is the immersed weight transport rate, $K$ is an empirical constant with a value of 0.77, $E$ is the wave energy, $C$ is the wave celerity, $n$ is the ratio of wave group velocity to wave celerity and $\alpha_b$ is the angle of wave incidence at the breakpoint. The immersed weight transport rate ($I_l$) is related to the volume transport rate ($Q_l$) by

$$I_l = \rho(s - 1)ga'Q_l \tag{4.52}$$

where $a'$ is a correction factor for pore space ($\approx 0.6$). Equation 4.51 can be restated as

$$I_l = K(CS_{xy})_b \tag{4.53}$$

where $S_{xy}$ is given by Equation 4.31. The general form of this empirical equation was subsequently confirmed by studies using tracers (Komar, 1983a,b) and streamer

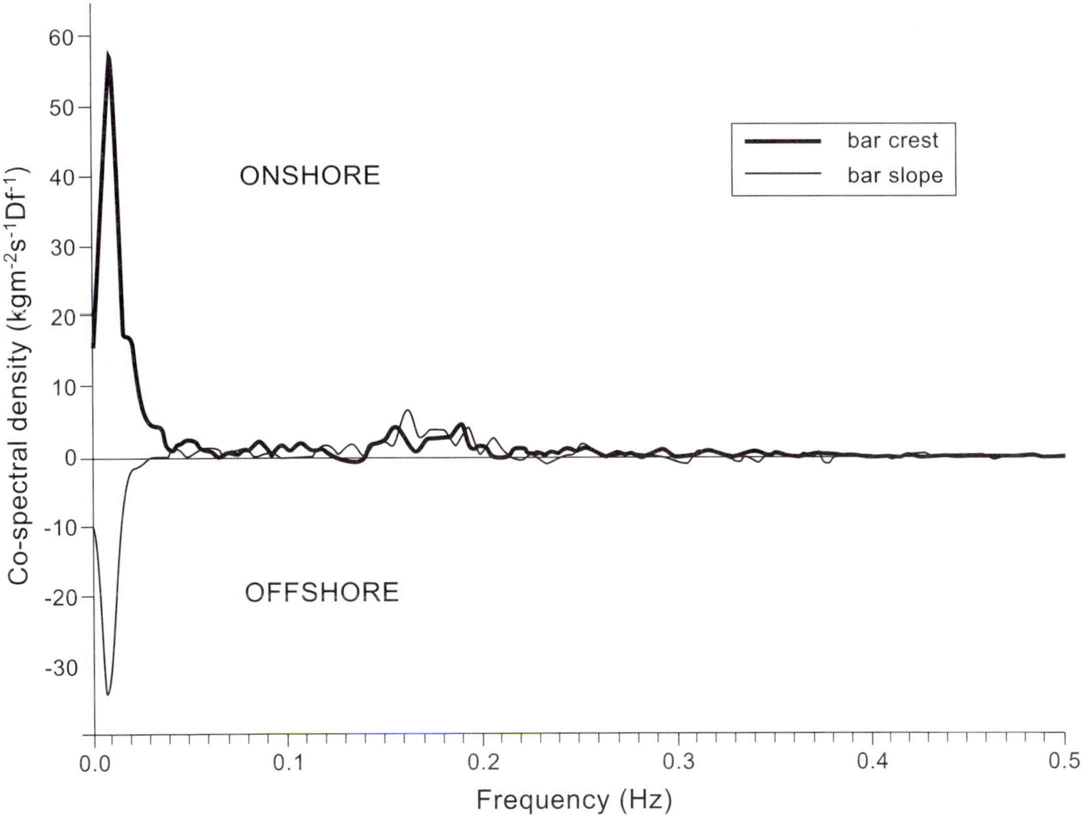


**Figure 4.37** Reversals of sediment flux associated with infragravity waves on a dissipative beach. Oscillatory flux is directed offshore over the lakeward bar slope, and onshore on the bar crest (after Aagaard and Greenwood, 1995a).

traps (Kraus and Dean, 1987; Kraus *et al.*, 1988). The model, or similar expressions, has been used extensively as it is easy to apply and provides transport estimates within a reasonable error range, assuming that sufficient amounts of sediment are available for transport. However, the cross-shore distribution of the longshore transport is characterised by a maximum transport rate in the outer surf zone, with no transport in the swash (Komar, 1971a). Furthermore, the model was derived for plane-sloping beaches, whereas on barred beaches transport rates are maximum over the bar and at the beachface. In other words, maximum transport rates are found where the waves break and sediment concentrations are large (Figure 4.38; Kraus *et al.*, 1981; Bodge, 1989). Equation 4.53 assumes that oscillatory motions are unimportant to the longshore sediment flux which is not always the case, as low-frequency oscillations of the longshore current may occur due to shear waves and these motions can constitute a significant part of the net longshore flux (Aagaard and Greenwood, 1995b).

There is evidence to suggest that the constant of proportionality in Equations 4.51 and 4.53 depends on environmental conditions. Bodge and Kraus (1991) suggested

$$K = 0.22 \ln \xi_b + 0.62 \qquad (4.54)$$

thus indicating smaller $K$ values on more dissipative beaches.

### 4.5.8 Sediment transport models

Due to the difference in the mechanics of sediment entrainment, sediment transport models generally distinguish between bedload and suspended load. Bedload is often modelled as a function of the (excess) shear stress acting on the sediment grains, while transport in suspension can be modelled through a '*uc* integral' approach.

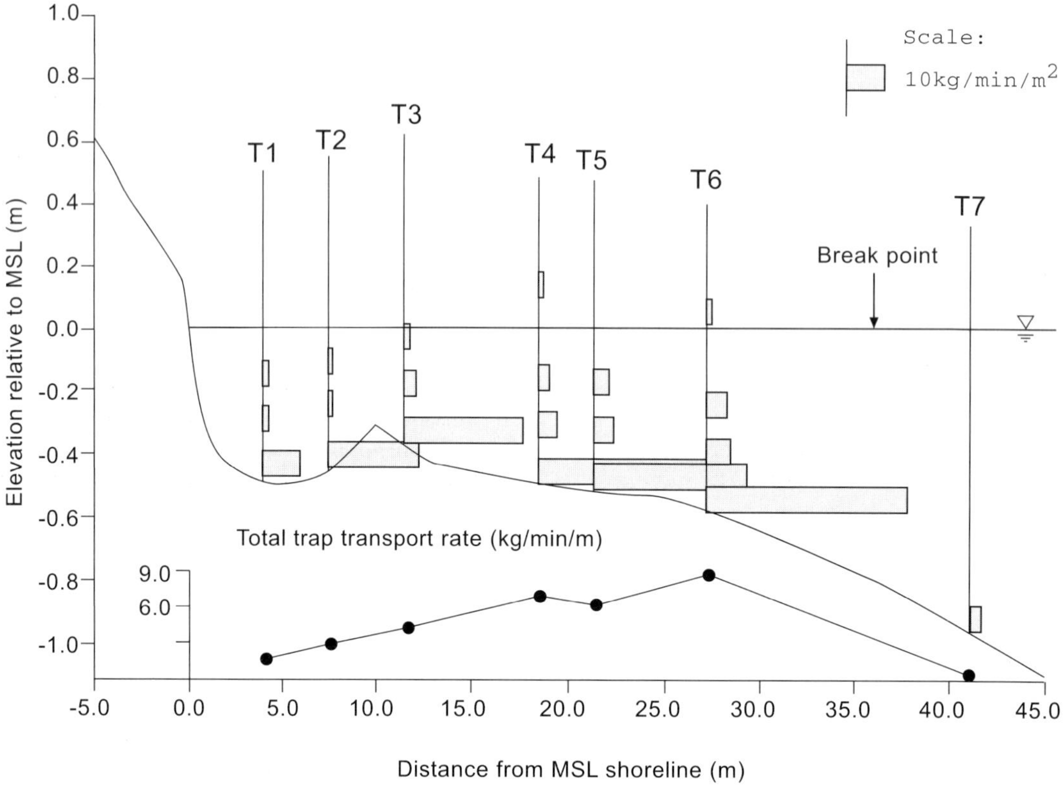


**Figure 4.38** Cross-shore and vertical distribution of longshore sediment transport rate determined from trap measurements (modified from: Kraus *et al.*, 'Toward an improved empirical formula for longshore sand transport', *Proceedings of the 21st International Conference on Coastal Engineering*, 1988, by permission of the American Society of Civil Engineers).

A large number of *bedload transport* models exist and they all rely to a significant extent on empiricism. As bedload transport is very difficult to measure under oscillatory wave motion, these models all use laboratory data on transport rates. One of the most widely cited models is due to Madsen and Grant (1976) who related the instantaneous bedload transport rate to the instantaneous skin friction Shields parameter (Equation 4.42)

$$q_b(t) = 40w_s D(\theta'(t))^3;\ \theta' > \theta_c \tag{4.55}$$

This model assumes a flat bed and the net transport rate is obtained by integrating $q_b(t)$ over a wave cycle (or longer). Other bedload models possess more or less the same general structure with transport rates related to various powers of velocity. Field tests of such bedload models have not been very successful (e.g. Seymour and King, 1982; Hay and Bowen, 1993).

Models for *suspended sediment* flux generally rely on the diffusion approach (Equations 4.45 and 4.46) to predict the sediment concentration profile. Sediment transport is then computed as the product of velocity and concentration, integrated over depth. The majority of these models have viewed the oscillatory motion as a stirring agent incapable of causing a net transport, but mobilising the sediment which is then available for transport by mean flows (e.g. Dally and Dean, 1984; Stive and Battjes, 1984). Quite apart from the difficulties involved in the specification of the appropriate reference concentration for the concentration profile, and a proper eddy diffusion coefficient (mixing length), these models compute $q_s = \bar{u}\bar{c}$ (or $\bar{v}\bar{c}$), and do not take into account the oscillatory sediment transport caused by asymmetric incident waves and/or infragravity waves. Further, they do not consider phase changes between velocity and concentration induced by bedforms. As field measurements have demonstrated, such models cannot generally be successful for cross-shore transport. However, for longshore transport the diffusion approach may be a more realistic model as oscillatory sediment flux is rather small under most conditions (Sternberg *et al*., 1984; van Rijn and Kroon, 1993).

More sophisticated intrawave models using the diffusion approach have recently been developed (e.g. Deigaard *et al*., 1986; Hedegaard *et al*., 1991). These models compute the transport due to all combinations of currents and waves but they require a detailed description of the hydrodynamics over a wave cycle.

A third type of model is the energetics approach, initially developed by Bagnold (1956, 1963, 1966). This approach, which considers bedload as well as suspended load, relates sediment transport to fluid power, or in other words, to the rate at which the fluid performs work at the bed. This rate is proportional to $\tau u$ (and thus $u^3$). The model does not explicitly consider suspended sediment concentration but uses the ratio between shear velocity and sediment fall velocity as a surrogate.

Bagnold's original model was extended by Bowen (1980) and Bailard and Inman (1981) to apply to combined wave and current flows in the nearshore environment. Bailard and Inman's formulation of the instantaneous immersed weight sediment transport rate at a point is

$$i(t) = \rho c_f \frac{e_b}{\tan\phi}\left[ |u(t)|^2 u(t) - \frac{\tan\beta}{\tan\phi}|u(t)|^3 \right] + \rho c_f \frac{e_s}{w_s}\left[ |u(t)|^3 u(t) - \frac{e_s \tan\beta}{w_s}|u(t)|^5 \right] \qquad (4.56)$$

where $\rho$ is the fluid density, $c_f$ is a drag coefficient ($= 1/2 f_w$; Ribberink and Al-Salem, 1994), $e_b$ and $e_s$ are bedload and suspended load efficiency factors (not all the energy dissipated at the bed is available for the transport of sediment), $\tan\beta$ is the bed gradient, $\phi$ is the internal angle of friction for the sediment ($\approx 33°$) and $w_s$ is the sediment fall velocity. The numerical values for $e_b$ and $e_s$ are given as 0.2 and 0.025, respectively, for the surf zone environment (Bailard and Inman 1981). The net transport is then computed by integrating over a wave cycle.

*Field tests* by Guza and Thornton (1985b) demonstrated that the most important terms in the equations involve the expressions $|u|^2 u$ and $|u|^3 u$. Foote and Huntley (1994) separated the latter into velocity components associated with incident waves, infragravity waves and mean flows, and the computed transport rates were correlated with measured cross-shore suspended sediment flux. Predictions involving infragravity waves and mean flows were modelled quite well. However, with regard to incident waves, predicted transport rates were largely correct in magnitude but in the opposite direction to the measured transport. This rather odd result was ascribed to the presence of bedforms, altering the phase relationship between velocity and concentration at the bed. On a flat bed in the laboratory, Ribberink and Al-Salem (1994) tested Equation 4.56 and found a correspondence within a factor of two between measured and computed total transport rates, with the suspended load term being dominant.

## 4.6 BAR MORPHOLOGY

Nearshore bars are ubiquitous morphological features along many of the world's beaches. They are the main expression of hydrodynamic and sediment transport gradients, with bars developing as a result of sediment convergence and troughs forming in areas of sediment divergence. Nearshore bars are the most dynamic morphological element of the surf zone and hence they are a major topic of coastal research.

### 4.6.1 Barred versus unbarred beaches

Long-term monitoring of beaches on the west coast of the United States resulted in the identification of 'winter' and 'summer' profiles (Shepard, 1950), and a seasonal cycle of beach morphology (Hayes and Boothroyd, 1969). Winter storms result in the removal of sand from the berm and the formation of a breakpoint bar, whereas calmer weather conditions in the summer induce landward migration of the bar, subsequent welding of the bar to the beachface and the generation of a non-barred cross-shore profile with a wide berm (Fox and Davis, 1973). Later work indicated that the terms 'winter' and 'summer' are somewhat misleading and demonstrated that cyclic beach response, as opposed to seasonal, may result in several cycles of barred/non-barred profiles within a year depending on the storminess of the wave climate (Short, 1979b; Nordstrom, 1980). Therefore, the terms 'barred' and 'non-barred' are more appropriate (Greenwood and Davidson-Arnott, 1979).

The separation between barred and non-barred profiles is generally related to the direction of cross-shore sediment transport. Offshore sediment transport results in erosion of the beach and the formation of barred profiles, whereas onshore sediment transport causes beach accretion and non-barred profiles (Figure 4.39). However, once a certain nearshore morphology has developed (barred or non-barred) and the prevailing hydrodynamic conditions remain relatively constant, the morphology can be maintained under conditions of no net sediment transport.

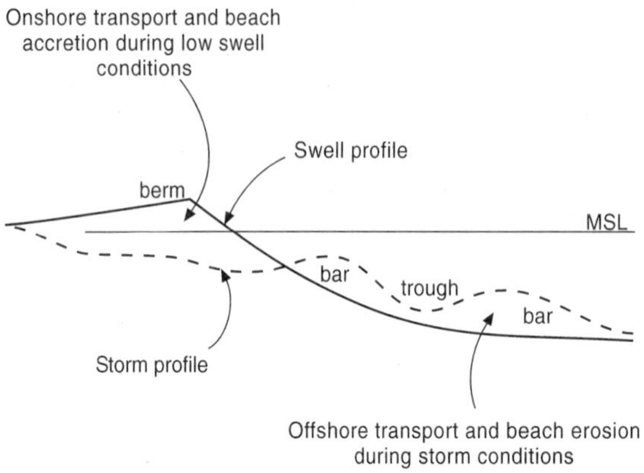


**Figure 4.39** Idealised barred and non-barred beach profiles. Calm weather conditions result in onshore sediment transport, beach accretion and the formation of a berm. Storm conditions induce offshore sediment transport, beach erosion and the formation of a bar.

When beach profile changes at one location are considered, the wave height appears to be relatively effective in distinguishing between barred and non-barred profiles. For the south-east coast of Australia, for example, Short (1979b) found that if the offshore significant wave height exceeds 1.2 m, barred profiles are formed, whereas offshore wave heights smaller than 1.2 m result in onshore sediment transport and eventually non-barred profiles. However, such relations are site-specific and factors such as wave period and sediment size must be included in a more universal model.

Early small-scale laboratory investigations (Johnson, 1949; King and Williams, 1949) correlated the cross-shore sediment transport direction, and hence the occurrence of bar morphology, with the deep water *wave steepness* ($H_o/L_o$). Steep waves (large wave height and/or short wave period; $H_o/L_o > 0.02$) were found to result in offshore transport and the formation of a barred profile. Waves with a low steepness (small wave height and/or long wave period; $H_o/L_o < 0.02$), on the other hand, promoted onshore transport and the formation of a non-barred profile. Field studies (King, 1972) and large-scale laboratory experiments (Saville, 1957) tended to find significantly lower threshold values and this discrepancy was primarily ascribed to the dependence of the cross-shore sediment transport direction on the beach sediment size (Rector, 1954; Iwagaki and Noda, 1962).

Based on laboratory experiments, Gourlay (1968) introduced the *dimensionless fall velocity*

$$\Omega = \frac{H_o}{w_s T} \tag{4.57}$$

to distinguish between barless beaches ($\Omega < 1$) and barred beaches ($\Omega > 1$). A similar parameter to the dimensionless fall velocity was derived by Sunamura (1984a)

$$K = \frac{H_b^2}{gT^2 D_{50}} \tag{4.58}$$

where $D_{50}$ is the median sediment grain size. A *K value* of 10 separates barred and non-barred beaches, with $K > 10$ resulting in bar formation (Sunamura, 1989). Using large-scale experiments, Kriebel *et al.* (1986) recommended an $\Omega$ value of 2–2.5 separating barred and non-barred beaches. Further analysis of large-scale laboratory data, and also field data (Kraus and Larson, 1988; Larson and Kraus, 1989), resulted in the formulation of a new *profile parameter P* (Dalrymple, 1992)

$$P = \frac{gH_o^2}{w_s^3 T} \tag{4.59}$$

For $P > 9000$ barred profiles develop, and non-barred profiles form when $P < 9000$. It should be noted that the offshore wave height is used to compute $P$. When the breaker height is used, the threshold value for $P$ is 22 200. The deep water parameters, however, appeared to be more successful in delineating the different beach profile types (Dalrymple, 1992).

One of the major shortcomings in applying the dimensionless parameters $\Omega$, $K$ and $P$ to field data is that their derivation is based on laboratory experiments with regular waves. Apart from the obvious problems associated with the scaling and the lack of three-dimensionality in a laboratory environment, it is also unclear which measure of the wave height and period should be selected when computing the parameters for random waves ($H_{rms}$ or $H_s$ and $T_z$ or $T_p$; Section 4.2.1). In addition, in laboratory experiments with constant wave conditions an equilibrium morphology is always reached, whereas this is rarely the case on natural beaches, necessitating the inclusion of some sort of lag period for field data (Allen, 1985). The application of these parameters to natural beaches and the problem of lag period are addressed in Chapter 7.

### 4.6.2 Bar morphologies

Nearshore bar morphology can assume a wide variety of configurations in which the number of bars present and their shapes may vary considerably. Multi-barred morphology (two to five bars) is primarily found in storm wave dominated sea and lacustrine environments where long periods of low wave energy conditions alternate with short intense storms (Evans, 1940; Goldsmith *et al.*, 1982a; Greenwood and Sherman, 1984; Aagaard, 1988; Short, 1992). The outer bar(s) are only active during the storms and can be considered relict features during periods of lower energy conditions. However, multiple bars (usually two bars) may also be found in environments that consistently experience high energy swell conditions (Wright *et al.*, 1982a).

Multi-barred nearshore morphology exhibits a tendency towards an increasing number of bars with decreasing beach slope (Dolan and Dean, 1985; Davidson-Arnott, 1988b). In addition, the bar spacing increases exponentially in the offshore direction. Short and Aagaard (1993) reviewed the literature on multi-barred beaches and introduced a *bar parameter* ($B_*$) to predict the number of bars on nearshore profiles that can be approximated by a linear slope $\tan\beta$, terminating at a constant depth at a distance $x_s$ from the shoreline

$$B_* = \frac{x_s}{gT^2 \tan\beta} \tag{4.60}$$

On the basis of field data, Short and Aagaard (1993) suggested that no bars occur when $B_* < 20$, one bar occurs for $B_* = 20$–$50$, two bars for $B_* = 50$–$100$, three bars for $B_* = 100$–$400$ and four bars for $B_* > 400$. The formulation of the bar parameter ($B_*$) is based on the premise that standing infragravity waves are responsible for bar formation.

A large number of different *bar types* occur in nature. The bar topography may consist of alternating transverse bars and rips, whereby the bars are attached to the shoreline and rip currents flow between the bars. Alternatively, the bar may have a crescentic or sinusoidal shape or it may be linear. The primary variable controlling the bar type appears to be the wave energy level. Transverse bar morphology is found at the low energy end of the spectrum, whereas linear bar morphology characterises high wave energy levels. Crescentic bars develop under intermediate wave energy conditions. The different types of bar morphology are discussed more extensively in Section 4.6.4 and in Chapter 7.

### 4.6.3 Bar formation

Early laboratory investigations conducted with regular waves attributed the formation of bars to the action of plunging breakers which scoured a depression in the beach sediment leaving a bar just offshore the breakpoint (Keulegan, 1948; Miller, 1976). On natural beaches with irregular waves the breakpoint mechanism of bar formation is not valid since waves do not break consistently at the same location. Bar formation and spacing on natural beaches is most likely the result of sediment transport patterns associated with: (1) infragravity standing waves; and/or (2) the interaction between steady offshore flows and onshore wave asymmetry.

#### *4.3.6.1 Bar formation by infragravity standing waves*

Laboratory experiments have demonstrated that linear bars can form as the result of reflection of the incident wave field at the shoreline (Carter *et al.*, 1973). Reflection of incident waves results in cross-shore standing waves which induce a horizontally segregated drift velocity field close to the bed, with the drift converging at nodes at the bottom of the boundary layer, while the convergence is towards the antinodes somewhat higher in this layer (Figure 4.40). Consequently, longshore bars may form at antinodes of the standing wave motion when suspended sediment transport prevails and at the nodes when bedload is the dominant mode of transport. Under high energy surf zone conditions (when bars are formed), suspended transport is the dominant mode and hence, theoretically, bars will form at the antinodal positions of standing waves (Bowen, 1980).

The distance ($x$) from the shoreline to consecutive nodes/antinodes of the standing wave motion is given by

$$x = \frac{g \tan\beta\chi}{\omega^2} = \frac{gT^2 \tan\beta\chi}{4\pi^2} \tag{4.61}$$

where $\omega$ and $T$ are the standing wave radian frequency and period, respectively, and $\chi$ is the value of the non-dimensional distance from the shoreline to a given node or antinode (Section 4.3.4, Equation 4.20, Table 4.1). The standing wave model of (multiple) bar formation predicts an exponential increase in bar spacing in the offshore direction (Figure 4.14) which corresponds to observations (Bowen, 1980).

The length scale of incident waves (tens of metres) does not correspond with that of nearshore bar morphology (hundreds of metres) and hence the predicted bar spacing using incident wave periods is significantly smaller than observed in nature. Larger, more realistic bar spacings are predicted when infragravity standing waves are considered. Short (1975b) found a good agreement between the observed locations of multiple longshore bars on the Alaskan coast and those pre-

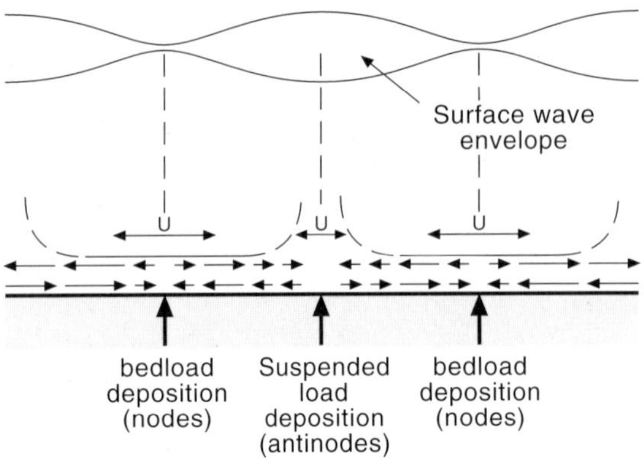


**Figure 4.40** The drift velocity field (single arrows) and the movement of sediment on a flat bed under the influence of a standing wave. Oscillatory velocities ($u$) are indicated by double arrows (after Carter *et al.*, 1973).

dicted using infragravity standing waves. Because of the similarity in offshore standing wave structure between leaky waves and edge waves (Figure 4.14), both may be responsible for multiple longshore bar formation (Bowen, 1980).

Standing infragravity waves do not only provide a mechanism for (multiple) longshore bar formation; standing edge waves may also be responsible for *rhythmic topography*. Bowen and Inman (1971) demonstrated that crescentic bars can form as a result of standing edge wave motion. In addition, Holman and Bowen (1982) showed how many complex rhythmic topographies can be quantitatively explained through edge wave interactions. The incident (breaking) waves stir the sediment which subsequently reorganises itself in rhythmic patterns associated with the net drift velocities associated with the standing edge wave field (Figure 4.21). Although the formation of rip currents is not explicitly included in the standing edge wave models, it can be inferred that once a certain rhythmic bar topography is initiated, the wave breaking patterns associated with the morphology may give rise to set-up gradients that drive rip current circulation (Section 4.4.2). The rhythmic wave length of the bar morphology ($\lambda$) is then given by half the edge wave length ($L_e$) which is related to the edge wave period ($T_e$) according to

$$\lambda = \frac{L_e}{2} = \frac{g}{4\pi} T_e^2 (2n + 1)\tan\beta \qquad (4.62)$$

where $n$ is the edge wave mode number.

It is apparent why so much interest has been directed towards edge waves, because they are one of the major explanations for the formation of nearshore morphology. Because in theory a wide range of edge wave modes and frequencies is possible, a variety of bar morphologies can be conceived. In order to produce a distinct bar morphology, a number of distinct edge waves are required. The edge wave cut-off mechanism discussed in Section 4.3.5 may be one way by which some edge waves may be preferentially excited. Consider a beach with a planar slope $\tan\beta = 0.015$ going out 1000 m before terminating at a horizontal bottom. From Equation 4.24 the edge wave cut-off periods can be calculated for different mode numbers (Table 4.3). Subsequently, the positions of the antinodes can be computed using the non-dimensional cross-shore distances listed in Table 4.1. Finally, using Equation 4.62, the wave length of the rhythmic morphology can be determined, assuming standing edge waves at the cut-off period are responsible for its formation. Aagaard (1990) used the above approach to demonstrate that along a particular beach, the longshore and cross-shore length scales of edge waves at the cut-off frequency corresponded to those of the nearshore bar morphology at that site, suggesting that standing edge waves were instrumental in the formation of the nearshore bar morphology.

It is apparent that even with the use of cut-off periods, a large number of potential positions exist where bars may form. In order for edge waves to form bars, ideally only one edge wave frequency/mode combination should be present, and an additional selection mechanism is required. One possibility is suggested by Aagaard (1990), who postulated that the selection of the edge wave mode may take place on the basis of the *wave group period*. The edge wave mode whose cut-off period is closest to the wave group period is then responsible for the bar morphology. An additional mechanism through which only a small number of edge wave modes and frequencies may be selected is through *topographic control*. It has been shown experimentally (Huntley, 1980) and numerically (Kirby *et al.*, 1981; Bryan and Bowen, 1996) that preferential excitation of certain edge waves may occur in relation with bar morphology. Bars located close to antinodal positions of normal standing waves tend to trap the antinodes of resonant edge wave modes, leading to maintenance or even continual growth of the bars even when the hydrodynamic conditions change. The presence of a *longshore current* may have a similar effect on the edge wave structure, by influencing the way in which a progressive edge wave, travelling either with or against the current, perceives the planar morphology (Howd *et al.*, 1992). The wave rays of an edge wave travelling with the current diverge as if travelling in the trough of the bar, whereas the wave

**Table 4.3** Edge wave cut-off periods, antinodal position and wave length of rhythmic morphology for the first five edge wave modes on a beach with a planar slope $\tan\beta = 0.015$ going out 1000 m before terminating at a horizontal bottom.

| Edge wave mode | $T_e$ (s) | Distance shoreline to antinodes (m) 1 | 2 | 3 | 4 | 5 | $\lambda$ (m) |
|---|---|---|---|---|---|---|---|
| $n = 1$ | 196 | 497 | | | | | 1349 |
| $n = 2$ | 113 | 165 | 548 | | | | 747 |
| $n = 3$ | 80 | 83 | 293 | 613 | | | 523 |
| $n = 4$ | 62 | 50 | 176 | 368 | 636 | | 404 |
| $n = 5$ | 51 | 34 | 119 | 249 | 431 | 656 | 333 |

rays of an edge wave travelling against the current converge as if it were a bar. Howd *et al.* (1992) further demonstrated that a positive feedback occurs between the effective topography (the topography experienced by the edge wave) and the true profile (initially planar). The cross-shore component of the edge wave drift field is modified in such a way that it promotes the evolution of the true beach profile into the effective profile, hence resulting in bar formation

Perhaps the most compelling evidence that infragravity waves affect the nearshore morphology was presented by Bauer and Greenwood (1990) They observed that a pre-storm linear bar–trough system migrated onshore during a storm and became rhythmic in response to the length scales and relatively fixed locations of nodal and antinodal lines of a mode 3 standing edge wave which dominated the infragravity spectrum. A well defined crescentic bar, as predicted by the edge wave drift velocity field, did not evolve because the hydrodynamic forcing ceased before an equilibrium condition could be attained. Nevertheless, the location of scour and accretion features demonstrated the tendency towards such a crescentic form. Similar results were obtained by Aagaard and Greenwood (1995a) who monitored the offshore migration of a nearshore bar in accordance with the perturbation introduced by infragravity standing waves on the cross-shore mean flows.

#### 4.6.3.2 *Bar formation by cross-shore flow mechanism*

The classic response of a planar beach to storm waves consists of erosion of the beachface and inner surf zone and deposition around the wave breakpoint resulting in the formation of a storm bar (Figure 4.39). The first explanation of bar formation was given as early as 1863 by Hagan (in Komar, 1976) who explained bar formation by 'a seaward going undertow meeting the shoreward moving waves'. This intuitive idea has now been formalised where the 'undertow' is represented by the bed return flow discussed in Section 4.4.1 and the intuitive expression 'shoreward moving waves' refers to the shoreward flow asymmetry associated with the highly non-linear shoaling waves. Thus, offshore of the breakpoint, the residual flow close to the bed is shoreward, whereas onshore of the breakpoint, the near-bed flow is seaward. This results in a region of *flow convergence* and hence sediment accumulation close to the breakpoint and the formation of a bar (Dyhr-Nielsen and Sorensen, 1970; Dally and Dean, 1984; Aagaard *et al.*, 1998).

For bars formed according to the cross-shore flow mechanism, the distance from the shoreline to the bar crest ($x_{bar}$) is described by (Holman and Sallenger, 1993)

$$x_{bar} = \frac{h_b}{\tan\beta} = \frac{\gamma H_b}{\tan\beta} \tag{4.63}$$

where $h_b$ is the depth at the break point, $\tan\beta$ is the nearshore gradient, $\gamma$ is the breaker criterion and $H_b$ is the breaker height. For natural beaches with random waves, the breaker criterion can be replaced by $\langle\gamma\rangle$ (Section 4.2.5).

The cross-shore flow mechanism of bar formation does not account for the formation of multiple bars because only one breakpoint or breaker zone will occur. However, several explanations can be given for the formation of multiple bars according to the cross-shore flow mechanism. The first involves the presence of multiple breakpoints as illustrated by Davidson-Arnott (1981). Once waves break on an outer bar of a gently sloping beach they may recover and reform as they travel across the deeper shoreward trough. These reformed waves may break for a second or perhaps a third time before they eventually reach the shoreline, resulting in multiple bar morphology. The analytical model of Dally (1980) accounts for repeated wave breaking and reformation, and consequently multiple bar formation. Alternatively, the presence of several distinct wave climates, each one responsible for a single bar, may result in multiple bar morphology (Evans, 1940). This is supported by findings of King and Williams (1949) who studied a double barred beach in North Africa and correlated the position of the outer bar with the breakpoint of storm waves, and the inner bar with wave activity during calm weather conditions. During calm wave conditions, the waves were found to pass the outer bar with little effect upon it. A third mechanism may involve varying water levels and breaker position due to tides (Komar, 1976). Tidal action can cause a bar formed at low tide to be at too great a depth to be affected by high tide surf zone waves, whereas a bar formed during high tide may be exposed during low tide.

#### *4.6.3.3 Standing waves versus cross-shore flow mechanism of bar formation*

Two mechanisms are proposed to explain the formation of bars on natural beaches. The major shortcoming of the standing infragravity wave hypothesis is the lack of a proper selection criterion for the edge wave mode (and frequency) responsible for the bars even though some sort of selection sometimes does exist. The major weakness of the cross-shore flow model is that it cannot account for rhythmic topography and requires some modifications to explain multiple bars.

Dally (1987) conducted laboratory experiments, specifically contrived to favour the standing wave mechanism of bar formation, but found evidence in support of the cross-shore flow mechanism. Sallenger *et al.* (1985) and Sallenger and Howd (1989), on the other hand, monitored the storm response of a nearshore bar system and concluded that the inner bar was in no way related to processes which would cause convergence of sediment in the breaker zone. The inner bar even became better developed and migrated in the offshore direction while being in the inner surf zone.

In the literature, bar formation by progressive edge waves and by leaky waves are often lumped together, and often the antinodal positions calculated for leaky waves are directly applied to those of edge waves (e.g. Aagaard, 1990). However, it is important to note that whereas an edge wave explanation requires a proper selection of the frequency (perhaps through a cut-off mechanism) and the edge wave mode (possibly through the wave group period or the antecedent topography), leaky waves only require the presence of a narrow banded frequency. Moreover, a recent leaky wave model proposed by O'Hare and Huntley (1994) predicts the formation of an inner surf zone bar whose location is relatively insensitive to the wave group period, incident wave height and the width of the infragravity spectrum.

Whatever mechanism initiates the bar(s), the morphology in some cases tends to reinforce itself, either through fixing the antinodal positions of the standing infragravity waves (Kirby *et al.*, 1981), or through inducing gradients in energy dissipation, mean flows and sediment transport (Sonu, 1972). It is perhaps appropriate to quote Holman and Sallenger (1993) who after an evaluation of the two mechanisms of sand bar generation concluded that bar formation is not 'simply a result of breakpoint generation nor of transport toward nodes or antinodes of a single standing infragravity waves, but is probably a hybrid'.

### 4.6.4 Bar dynamics

Nearshore bars respond to changing wave conditions and tend to move onshore under conditions of onshore sediment transport and migrate in the offshore direction when offshore sediment transport prevails. Due to the inherent lagged response of morphology to changing hydrodynamic conditions, nearshore bars often represent disequilibrium features that are continuously readjusting their configuration and location.

Short (1979b) presented daily observations of the nearshore morphology and hydrodynamic conditions in a single bar system on a modally high energy, swell-dominated beach in New South Wales (Australia) over a two year period and proposed an accretion and an erosion sequence. These sequences indicate that the bar morphology moves through a number of Markovian linked stages under the influence of falling and rising wave conditions. The bar sequences and their linkages are presented in Chapter 7.2.

# 5

# The Beachface

*Michael Hughes and Ian Turner*

## 5.1 INTRODUCTION

The *beachface* is located immediately landward of the surf zone, extending from the elevation of low tide to the upper limit of swash action. The beachface is the region where the shoreward propagation of ocean waves is finally halted either by dissipation or reflection, and despite it being probably the most familiar feature to many beach users, it remains one of the least well studied. The *swash zone*, where waves run up and back across the seaward-sloping beachface (Figure 5.1), is a notoriously difficult location to undertake morphodynamic research.

Processes operating in this relatively narrow region control the exchange of sediment between the sub-aerial beach and sub-aqueous surf zone during beach erosion, and the surf zone and beachface during beach accretion. There are three basic features of the swash zone that make it morphodynamically unique from the rest of the beach. The first is the land–water boundary, which shifts horizontally back and forth across the beachface over a broad spectrum of frequencies, ranging from the run-up of individual waves (2–15 s) to the semi-diurnal (twice daily), diurnal (daily) and lunar (fortnightly) tide cycles. This moving boundary condition makes modelling difficult. The second unique feature is that water depths in the swash zone are very small, particularly in the backwash (typically less than a few centimetres). As a result, sediment is often transported as complicated single-phase, granular-fluid flows, rather than as the better understood two-phase flows where the fluid and sediment in transport are readily distinguishable. The third unique feature is that at least some part of the beachface is usually unsaturated. This makes water infiltration a significant and influential factor in controlling the transport of sediment up- and downslope by swash flows.

The scope of this chapter is largely restricted to microtidal beaches, although many of the processes discussed in the following sections are generic. Limited reference is made to macrotidal beaches, but detailed discussion of the swash zone processes that are specific to such beaches is provided in Chapter 8. While the focus of this book is on the mutual interaction of fluid and sediment dynamics, and the resulting morphology (i.e. morphodynamics), to simplify their explanation swash kinematics and sediment transport are discussed first. The truly morphodynamic nature of the beachface is then highlighted in later sections where morphological features such as the beach step, berm and swash cusps are discussed in terms of interacting fluid and sediment dynamic processes.

## 5.2 SWASH KINEMATICS

Wave-induced movement of the water's edge back and forth across the beachface is known as *wave run-up* and consists of two components: a quasi-steady super-elevation of the mean water level, termed *wave set-up*, and oscillations about this level, termed *swash* (Guza and Thornton, 1982). Most studies of wave run-up to date have investigated the nature of wave set-up and swash separately (e.g. Holman and Sallenger, 1985), although there are some notable exceptions (e.g. Nielsen, 1989). While not seeking to over-emphasise this simplified approach, the following discussion will be largely restricted to swash, since wave set-up has already been discussed at length in Section 4.2.7. Some observations of the combined effect of set-up and swash, or total wave run-up, are discussed briefly in Section 5.2.3.

Swash is driven by breaking waves (surf zone bores or plunging breakers), surging waves or standing waves, and fundamentally consists of two flow regimes: the

*Handbook of Beach and Shoreface Morphodynamics*. Edited by A. D. Short.
© 1999 John Wiley & Sons Ltd.

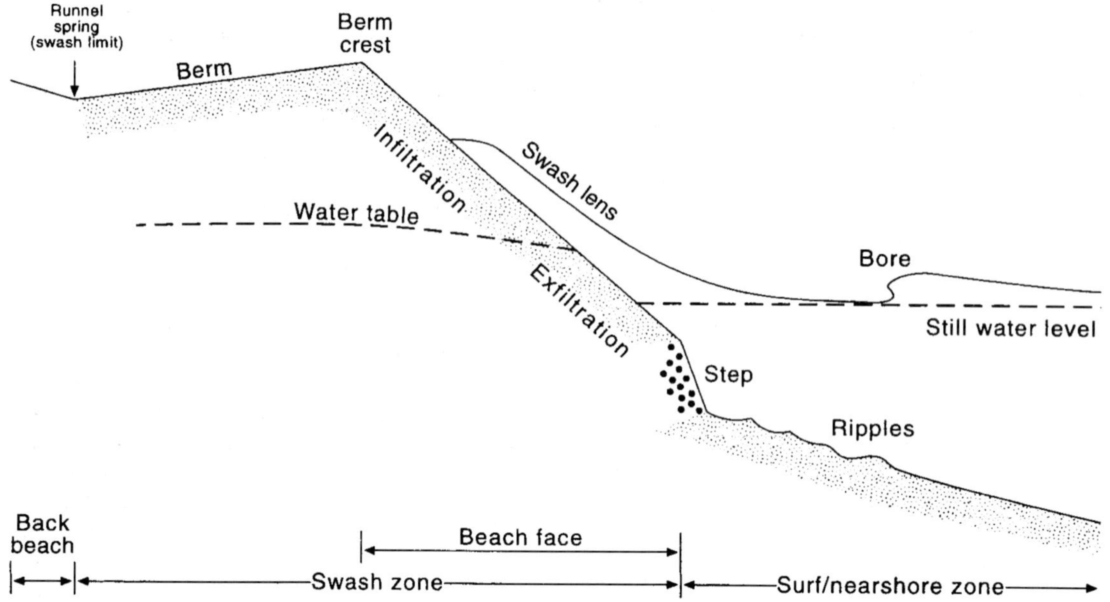


**Figure 5.1** Schematic diagram of beach face morphology showing terminology and principal processes.

landward directed (upslope) flow termed *uprush* and the seaward directed (downslope) flow termed *backwash*. Swash oscillations are generally observed to fall within one or a combination of three spectral bands: incident (typically 0.07–0.2 Hz or 5–15 s period); sub-harmonic (frequencies half the incident wave frequency, typically 0.03–0.07 Hz or 15–30 s period); or infragravity (typically 0.003–0.03 Hz or 0.5–5 min period). As the terminology implies, incident swash is associated with incident waves arriving at the shoreline. Sub-harmonic swash may be associated with sub-harmonic edge waves (e.g. Wright, 1982), or swash–backwash interference at the base of the beachface (e.g. Kemp and Plinston, 1968). Infragravity swash may result from leaky mode standing waves, standing edge waves or progressive edge waves (Sections 4.3.4 and 4.3.5).

Incident and sub-harmonic swash are generally predominant on the *reflective* and steeper *intermediate beach types* (Section 7.3.4). The surf zone width on these beaches is usually insufficient to permit complete dissipation of incident wave energy seaward of the shoreline. In contrast, infragravity swash generally predominates on the *dissipative* and flatter *intermediate beach types* and during storms. On these beaches the surf zone is often sufficiently wide to result in nearly complete dissipation of incident wave energy seaward of the shoreline (Section 7.3.4). The following sections provide an overview of the theory of incident and infragravity swash behaviour, illustrated where available with laboratory and field measurements.

### 5.2.1 Incident swash

#### *5.2.1.1 Linear theory*

Energy associated with non-breaking or surging waves arriving at the shoreline will be almost entirely reflected at the beachface. There will of course be some energy dissipation through bed friction, but energy dissipation associated with turbulent wave breaking will be essentially absent. The incoming incident waves and outgoing reflected waves interact to produce a standing wave that will presumably control swash behaviour on the beachface. A model for incident swash frequently used to interpret field data is that postulated by Miche (1951). This model proposes that the swash amplitude is proportional to the amount of wave reflection and thus the linear standing wave amplitude at the shoreline. Two behavioural regimes for swash are proposed, distinguished by the presence of either non-breaking or breaking waves. In the case of non-breaking waves the swash amplitude is expected to increase with increasing beachface slope, wave height and wave period, since all of these directly affect the amount of wave energy that is reflected at the beachface. In the case of breaking waves, swash amplitude is expected to be proportional to the maximum linear standing wave amplitude permissible for the given wave and beach conditions, while all excess wave energy is presumed to be completely dissipated seaward of the shoreline. This implies that once wave breaking occurs the swash amplitude is independent of wave height and period and is in fact constant, i.e. the energy spectrum in both the inner surf and swash zones is saturated.

There is some support for Miche's model in that Guza and Bowen (1976) have shown standing waves can exist at the landward edge of a turbulent surf zone on a laboratory beach and Guza and Thornton (1982) presented field data showing virtually complete saturation of incident swash on the barless, low gradient Torrey Pines Beach, USA (Figure 4.10). Nevertheless, numerical models demonstrate that incident waves in the inner surf zone can theoretically reach heights several times the water depth (e.g. Peregrine, 1974). Moreover, field data from beaches with moderate to large beach slopes show no evidence of energy saturation in either the inner surf or swash zones (e.g. Holman and Sallenger, 1985). Miche's model cannot be presumed to provide a universal description of incident swash, indeed its relevance seems rather restricted to *dissipative beaches* where the swash is driven predominantly by non-breaking infragravity waves (Section 5.2.2). On steeper *intermediate* and *reflective beaches* where the swash is driven predominantly by breaking incident waves, a non-linear theory is required.

### *5.2.1.2 Non-linear shallow water equations*

Non-linear shallow water theory is emerging as the most useful approach to investigating incident swash, and perhaps infragravity swash as well (Section 5.2.2). The theory involves two differential equations that define conservation of mass and momentum for the flow. Once initial conditions are defined in terms of water velocity and water surface elevation, these equations can be solved to determine the subsequent fluid motion. The *shallow water equations* (SWE) can be stated in their simplest form (first-order, depth-integrated) as:

$$\frac{\partial(u(\eta + h))}{\partial x} + \frac{\partial \eta}{\partial t} = 0 \qquad (5.1)$$

$$\frac{\partial u}{\partial t} + u\frac{\partial u}{\partial x} + g\frac{\partial \eta}{\partial x} = 0 \qquad (5.2)$$

where $u$ is horizontal water velocity directed cross-shore, $\eta$ is water surface elevation relative to still water level, $h$ is still water depth, $x$ is distance in the cross-shore direction, $t$ is time and $g$ is gravitational acceleration. The principal assumptions underpinning these equations are that the slope of the water surface is not much larger than the beach slope, there is no vertical flow (i.e. water pressure is hydrostatic), and the horizontal water velocity does not vary with water depth. For a more complete discussion of the SWE and underlying assumptions see Peregrine (1972).

Two independent, and now considered benchmark, studies used the SWE to investigate separately the behaviour of swash driven by non-breaking (Carrier and Greenspan, 1958) and breaking waves (Shen and Meyer, 1963). Numerous subsequent studies have extended the theoretical aspects of this work, and to a lesser extent field and laboratory experiments have tested the theoretical predictions. The application of the theory to non-breaking waves and incident swash has been limited due to restrictive assumptions in the context of natural beaches. It is therefore discussed only briefly in this context. The remainder of the section will focus on breaking waves and incident swash – a situation most common on natural beaches and one in which the theory has had some measured success in modelling.

### *5.2.1.3 Swash driven by non-breaking waves*

In a similar manner to the standing wave concepts introduced by Miche (1951), Carrier and Greenspan (1958) used standing waves as a starting point for their investigation of swash driven by non-breaking waves (Figure 5.2a). By performing some transformations on the SWE they were able to derive a simpler approximate equation, which they solved for a number of initial wave forms including a non-linear standing wave. Unfortunately some rather restrictive physical conditions must be satisfied for the mathematical approach used by Carrier and Greenspan to remain valid. These conditions are delimited by the surf scaling parameter, $\varepsilon$ (Equation 4.9), which must be less than unity. For a beach slope as large as 0.1 and a wave period as long as 12 s, a solution is not possible unless the swash height is less than 35 cm. For shorter wave periods and/or smaller beach slopes the maximum permissible swash height becomes even smaller. Even a cursory inspection of surging waves at incident frequencies on a natural beach will show that the swash height is usually greater than is admissable. This theoretical restriction explains why the work of Carrier and Greenspan (1958) has not been widely tested in the context of incident swash on natural beaches. It is interesting to note, however, that their theoretical work has been tested against the measured behaviour of a number of wave forms produced on laboratory beaches and found to be quite successful. This suggests that, under appropriate conditions, the mathematical approach is actually consistent with the physics. Some implications of this observation are discussed further in relation to infragravity swash (Section 5.2.2).

### *5.2.1.4 Swash driven by breaking waves*

On most natural beaches incident waves break within the surf zone, and therefore the limited application of the non-linear standing wave solutions to the SWE does not seriously hinder the non-linear shallow water theory as a framework for investigating incident swash. In the case of a progressive wave propagating toward a beach, the SWE predict that the wave front will steepen until the water surface slope becomes much larger than the beach slope (Greenspan, 1958). This situation violates the assumption concerning water surface slope (Equations 5.1 and 5.2), thus the SWE cannot be used to describe fluid motion in the region of the wave front. Mathematicians refer to this region where the SWE are inapplicable as a *bore*, and oceanographers have extended the use of this term to apply to wave breakers in the surf zone (Figure 5.2b). In order to proceed with the theoretical investigation of swash, mass and momentum are assumed to be conserved across the narrow region of the wave front. This allows use of the classic bore equations (Stoker, 1957) to describe the behaviour of the wave front and the SWE can still be applied either side.

Using this approach, it can be shown that the behaviour of progressive waves immediately seaward of the still water shoreline is largely independent of any details of the initial wave shape (Ho and Meyer, 1962). The implication of this rather remarkable theoretical result is that the swash is also independent of initial wave shape. This highlighted the limited scope of previous decades of empirical research, which had tried to predict details of wave run-up from deep water wave conditions seaward of the surf zone. It turns out that the incident swash motion is dependent on only one aspect of the incident wave, its propagation speed upon arrival at the shoreline (Shen and Meyer, 1963). The classic bore equations predict that the wave height decreases as it approaches the beach and experiences a rapid collapse to zero at the shoreline. These equations also predict that the wave velocity and horizontal water velocity immediately behind the wave increase as the beach is approached, ultimately converging to a finite value at the shoreline (Keller *et al.*, 1960). An estimate of the wave velocity at the shoreline can be calculated using an ordinary differential equation derived by Whitham (1958), which subsequent laboratory experiments have shown is very accurate (Figure 5.3).

Once an incoming broken wave (bore) has collapsed at the shoreline a volume of water termed the *swash lens* is driven up the beachface. The water surface slope of the swash lens is similar to the beach slope so the bore equations can now be put aside as the SWE are again capable of describing the entire flow. The motion of the front of the swash lens and some details of the flow within the swash lens can be determined from a set of corollaries of the SWE (Shen and Meyer, 1963). A useful way to visualise the mathematical arguments is to imagine that the swash lens consists of several small fluid elements, each containing the same mass of water at all times. The motion of each fluid element depends only on the pressure exerted by the adjacent elements and gravity. The leading element is always moving faster than the elements behind, thus the pressure acting on it will be negligible. The problem of the swash front climbing the beach therefore simplifies to consideration of the balance of forces acting on this single element only (Ho *et al.*, 1963). One way to picture this is that the balance of forces and resulting motion of the swash front are exactly the same as those that determine the deceleration of a marble rolled up a slope with some initial velocity. The fluid element representing the swash front is set in motion at the still water shoreline by the incoming bore, which instantaneously accelerates it to the terminal bore velocity, $u_o$. The equation of motion for the swash front is therefore:

$$m\frac{d^2X_s}{dt^2} + mg\sin\beta = 0 \tag{5.3}$$

where $m$ is the mass of the fluid element, $X_s$ is its position relative to the initial shoreline position and $\beta$ is beach slope. The first term in this equation refers to the initial acceleration of the swash front upslope due to bore collapse, and the second term describes the effect of downslope gravity acting to slow the swash front and then reverse its direction of travel. From this simple starting point, a number of very useful equations can be derived. Integration yields the swash front velocity, $dX_s/dt = u_s$, with respect to time:

$$u_s = u_0 - gt\sin\beta \tag{5.4}$$

This equation indicates that the rate at which the swash front decelerates is constant. The swash front velocity can also be written in terms of its position on the beachface:

$$u_s = \sqrt{u_o^2 - 2X_sg\sin\beta} \tag{5.5}$$

**Figure 5.2** (a) A non-breaking wave depicted immediately before uprush on Pearl Beach, NSW. (b) Swash lens depicted at the time of maximum uprush, with surf zone bore in the middle ground on Whale Beach, southern NSW (M G Hughes).

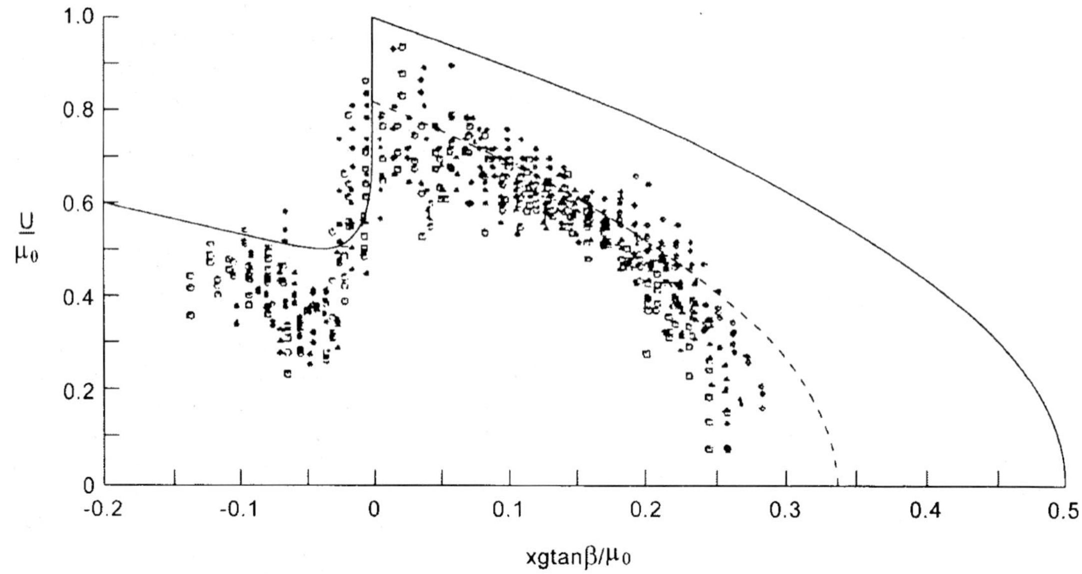


**Figure 5.3** Laboratory measurements of non-dimensional bore velocity (seaward of the still water shoreline) and swash front velocity (landward of the still water shoreline), $U/u_o$, as a function of non-dimensional distance in the shore-normal direction, (after Yeh *et al.*, 1989). Note the still water shoreline is at $xg \tan \beta/u_0 = 0$. The solid line represents the prediction of the non-linear shallow water theory: the differential equation of Whitham (1958) in the case of bore velocity and Equation 5.5 in the case of swash front velocity. The dashed line is the model prediction for an initial velocity reduced by approximately 20% to account for friction effects.

Integration of Equation 5.4 yields the position of the swash front, $X_s$, with respect to time

$$X_s = u_0 t - 0.5gt^2 \sin \beta \qquad (5.6)$$

Substituting into Equation 5.6 the time that maximum uprush occurs ($t = u_o/g\sin\beta$, from Equation 5.4) and employing some trigonometry yields the maximum swash height, $Z_s$

$$Z_s = \frac{u_0^2}{2g} \qquad (5.7)$$

The shallow water theory also provides an approximation for water depth in the swash zone, $h_s$ (Shen and Meyer, 1963)

$$h_s \approx \frac{(X_s - x)^2}{9t^2} \qquad (5.8)$$

This mathematical description of the swash front moving across the beachface is valid for the entire uprush duration, but is valid for only part of the backwash duration. Eventually during the backwash a localised steepening of the water surface slope develops in the interior of the swash lens, i.e. a bore (Figure 5.5), and the SWE are no longer applicable (Shen and Meyer, 1963; Ho *et al.*, 1963).

The first researchers to test the accuracy of these equations used numerical techniques to directly solve the SWE (e.g. Hibberd and Peregrine, 1979). Subsequent researchers have tested their accuracy using both laboratory and field measurements of swash behaviour (e.g. Yeh *et al.*, 1989; Hughes, 1992). In all cases it was found that the general features of bore propagation, collapse and ensuing swash are predicted reasonably well (Figures 5.3, 5.4 and 5.5). It is important to recognise, however, that the equations consistently overpredict the magnitude of the parameters measured, whether they be in the field or laboratory. Laboratory data reported by Miller (1968) also show that the maximum swash height scales positively with beach slope, which is not predicted by Equation 5.7. All these observations are consistent with the fact that bed friction and water loss through infiltration are both absent in the preceding derivations.

To date, the only investigation of infiltration effects on swash is a numerical modelling study by Packwood (1983), which examined the problem of swash on a fine–medium sand beach. The study concluded that infiltration has a minimal effect on uprush, but a significant effect on backwash. If this conclusion is correct then one approach is to presume that the discrepancy between the predictive Equation 5.7 and measured

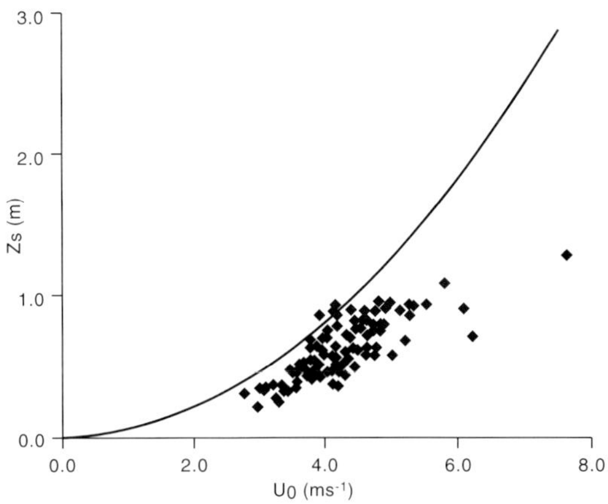


**Figure 5.4** Field measurements of maximum swash height as a function of initial swash velocity (after Hughes, 1992). The solid line represents the prediction of the non-linear shallow water theory (Equation 5.7).

maximum swash height in Figure 5.4 is dominated by the effects of bed friction. By introducing a shear stress term into Equation 5.3 a new set of equations can be derived, consistent with the existing inviscid equations, but including friction effects (Kirkgoz, 1981). The resulting equation for the maximum swash height is rather more complex than the case of no friction

$$Z_s = \frac{-8h_s \sin\beta}{f} \ln\cos\left(\tan^{-1}\frac{u_o\sqrt{f}}{\sqrt{8gh_s \sin\beta}}\right) \tag{5.9}$$

where $f$ is the friction factor and $h_s$ is the water depth a short distance behind the swash front (Hughes, 1995). Now that friction is included in the theory we have a predictive equation indicating that the maximum swash height is positively related to beachface slope. Hughes (1995) showed that Equation 5.9 predicts an initially rapid then somewhat reduced increase in maximum swash height as beach slope increases, which is very similar to the trend evident in the laboratory measurements reported by Miller (1968).

The discussion thus far has been concerned with the

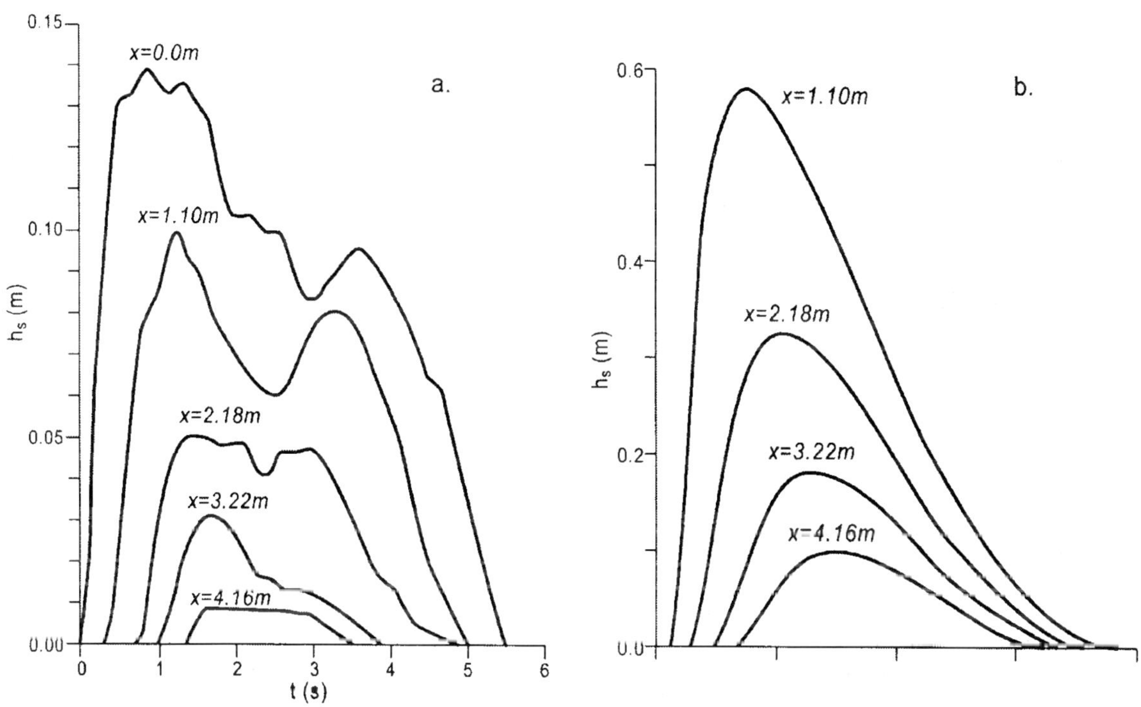


**Figure 5.5** (a) Field measurements of water depth at various locations in the swash zone for a single swash cycle (initial conditions: $u_o = 3.71$ m/s and $\beta = 0.095$). Note the secondary maximum in water depth, which occurred during the backwash and relates to the formation of a retrogressive bore (after Hughes, 1992). (b) The prediction of the non-linear shallow water theory (Equation 5.8).

motion of the swash front, but what of the flow in the interior of the swash lens? Numerical experiments that directly solve the SWE for swash driven by fully developed bores yield the following predictions for any given point on the beachface (Hibberd and Peregrine, 1979). During wave uprush, the onshore flow velocity begins at its maximum value (equal to the swash front velocity) upon arrival of the swash front and then decreases steadily to zero for the remainder of the uprush. During backwash, the offshore flow velocity begins at zero and increases gradually towards its maximum, which occurs just prior to the end of the backwash. As the swash lens recedes past the measurement point an instantaneous drop to zero in backwash velocity occurs. Furthermore, the duration of the uprush flow is predicted to be shorter than the duration of the backwash flow. This general behaviour is confirmed in field measurements obtained on Palm Beach, Australia, where well developed surf zone bores were driving the incident swash (Figure 5.6). The field data do not show an instantaneous response in velocity at the indicated times, but this is probably due mostly to the time constant of the flow meters (Hughes *et al.*, 1997). The combination of flow velocities and swash depths shown in Figure 5.6 indicates that the flow is supercritical during much of the uprush and backwash, due mostly to the shallow water depths. This observation has important implications for the genesis and dynamics of a number of bedforms characteristic of the beachface (Section 12.3.4).

Before leaving the discussion of incident swash, it is appropriate to make one final comment on the equations describing swash driven by bores (Equations 5.3 to 5.9), particularly concerning the range of conditions for which they are valid. To date these equations have only been applied to measurements of swash driven by incident breaking waves. Theoretically a bore exists wherever the wave is sufficiently steep to violate the assumptions of the SWE. The steep, turbulent breaking waves observed in the surf zone are certainly one example of a bore, but the steepness assumption is violated long before breaking occurs. Meyer and Taylor (1972) suggest that on natural beaches bores probably occur in rela-

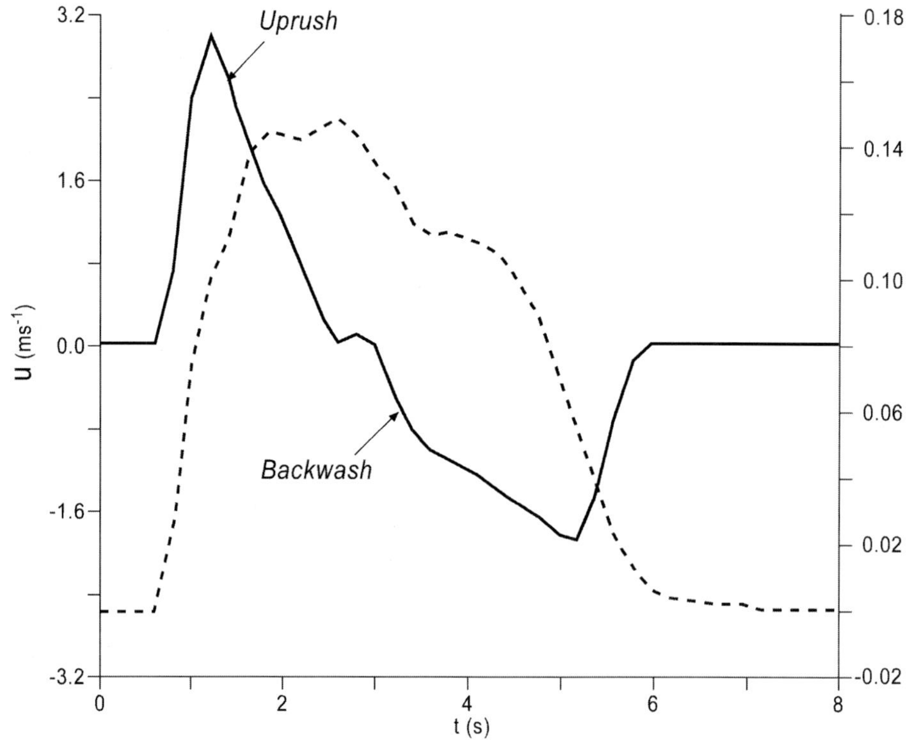


**Figure 5.6** Field measurements of horizontal water velocity (solid line) and water depth (dashed line) for a single swash cycle (after Hughes *et al.*, 1997). The time series were recorded at a point located approximately half-way between the still water shoreline and the position of maximum uprush.

tion to both breaking and non-breaking waves. If this is the case then an important question arises concerning swash driven by surging waves: Is this situation most appropriately modelled by the non-linear standing wave or bore solutions to the SWE? This important question has received little attention to date and remains unresolved.

### 5.2.2 Infragravity swash

The existence and importance of surf zone infragravity waves has only been established relatively recently, and as a result the study of infragravity swash is still in its infancy. Most studies to date have been concerned with collecting swash data to infer the existence of infragravity waves in the surf zone or to simply characterise the nature of the swash spectrum; there have been relatively few attempts to test or refine any theory in relation to infragravity swash. This is somewhat surprising since it has already been noted that infragravity waves are generally standing waves (Section 4.3), thus theoretical predictions of the resulting infragravity swash motion have been available for some time. Both the linear standing wave model (Miche, 1951) and the non-linear standing wave model (Carrier and Greenspan, 1958) discussed in the previous section could be applied.

Guza and Thornton (1982) were the first to report detailed measurements of infragravity swash. Their data were obtained on a barless *dissipative beach* where incident significant wave height varied between 0.5 and 1.5 m (measured in 10 m water depth). Under these conditions the incident ($>0.05$ Hz) swash was saturated; however, the infragravity ( $<0.05$ Hz) swash height displayed a linear increase with incident significant wave height (Figure 4.10). Holman and Sallenger (1985) obtained a more comprehensive data set on a steeper *intermediate beach* displaying a pronounced bar–trough morphology in the surf zone and where incident significant wave height varied between 0.4 and 4.0 m (measured in 20 m water depth, 3 km offshore). Holman and Sallenger also observed a positive relationship between the infragravity swash height and incident significant wave height; however, they found that the Iribarren number provided a more effective scaling and suggested

$$\frac{Z_{is}}{H_s} = 0.53\xi_o + 0.09 \tag{5.10}$$

where $Z_{is}$ is significant infragravity swash height and $\xi_o$ is the Iribarren number (Section 4.6).

Although an empirical relationship between infragravity swash height and incident wave height has been observed, it should not be considered directly causal. The physical connection between the two almost certainly involves some high to low frequency energy transfer mechanism operating at the seaward edge of the surf zone (Section 4.3), since on the basis of either the linear or non-linear shallow water theories we would expect that it is the infragravity standing wave inside the surf zone that is actually driving the infragravity swash.

Holland *et al.* (1995) met with only limited success in using an inviscid linear standing wave theory (i.e. friction is neglected) to model field data from Scripps Beach, USA. They found that measured infragravity ($<0.02$ Hz) swash heights were larger than predicted. This over-amplification reached a maximum of 250% at their lowest resolvable frequency, which was 0.005 Hz. The short cross-shore scale of the over-amplification rules out an edge wave explanation (e.g. Oltman-Shay and Guza, 1987), leaving ignored non-linear effects as the only plausible explanation for the discrepancy between field observations and theory.

The infragravity swash heights and periods typically observed on natural beaches fall within a range sufficient to ensure that the surf similarity parameter is less than unity. It is therefore possible that Carrier and Greenspan's (1958) analytical equations for swash driven by a non-linear standing wave could accurately describe the available infragravity swash data, although as Synolakis (1987) has pointed out, the application of these equations to natural beaches is not trivial. Another alternative is to use numerical schemes to solve the governing equations directly. One such numerical model based on the inviscid SWE, but including a quadratic friction term, has been tested against field measurements of infragravity swash driven by an infragravity standing wave in the surf zone (Raubenheimer *et al.*, 1995). It was found to be accurate in predicting the measured decrease in incident energy and increase in infragravity energy landward across the surf zone, but most importantly, it also correctly predicted both the incident and infragravity energy in the swash zone. These results are encouraging, and should provide the impetus for a great deal more work in this area. The need for a good model of infragravity swash cannot be over-emphasised, since in many cases beach erosion during storms is more attributable to infragravity swash than incident swash.

### 5.2.3 Total wave run-up

In the preceding sections incident and infragravity swash have been described independently of each other and of wave set-up. This approach was necessary in order to simplify the discussion, but it also reflects the way the discipline has approached the study of wave run-up. It is important to keep in mind, however, that the *total wave run-up* on most beaches will consist of a quasi-steady set-up component plus superimposed both incident and infragravity swash – probably interacting in some non-linear fashion. Given present limitations to our understanding of the physics of wave set-up and swash, from a coastal engineering viewpoint it is critical to the management of coastlines that something is known about the statistical distribution of wave run-up, in particular the probability of *extreme run-up* heights.

The two most comprehensive studies of extreme run-up statistics on natural beaches are those of Holman (1986) and Nielsen and Hanslow (1991). Holman's data set was obtained from the CERC Field Research Facility at Duck, USA, and incorporates significant deep water wave heights of 0.4 to 4.0 m, periods of 6 to 16 s and beachface slopes of the order 0.1. Nielsen and Hanslow's data set was obtained from six beaches in NSW, Australia, and incorporates significant deep water wave heights of 0.53 to 3.76 m, periods of 6.4 to 11.5 s and beach slopes of 0.02 to 0.19. Both studies yielded similar results and suggest as a useful rule of thumb

$$\frac{R_{2\%}}{H_s} = 0.9\xi_o \tag{5.11}$$

where $R_{2\%}$ is the wave run-up height that only 2% of the data exceeded. For engineering design purposes, Equation 5.11 offers an easily quantifiable estimate of the height of swash inundation on coastal structures.

## 5.3 BEACHFACE SLOPE

The volume of sediment transported upslope then downslope with every uprush–backwash cycle is often so great that it is difficult to distinguish the overlying granular-fluid layer from the underlying stationary bed. It is therefore perhaps surprising that there exist a number of generally consistent empirical relationships that link beachface gradients with varying sediment and wave characteristics. Though an elusive phenomenon, it is apparent that an 'equilibrium' beachface slope is a meaningful concept. During successive swash cycles, net onshore transport, net offshore transport and zero net transport may all be observed. The beachface is defined as being in a state of *dynamic equilibrium* when the net transport averaged over several swash cycles is zero. Clearly, if offshore transport dominates, the beachface profile will become flatter, and conversely if onshore transport prevails, the profile will steepen.

In nature it is rare to see a beach profile that exhibits a uniform and planar gradient. Typically the beachface profile is concave seaward, and beachface slope may vary along shore between steeper cusp horns and lower gradient cusp swales. To characterise the beachface at any specific location, the gradient observed within the mid-intertidal profile may be chosen, or more frequently a linear interpolation is inferred, extending from the low tide shoreline to the crest of the berm. There is no standard technique for how beachface slope is measured, and this fact alone may well account for some of the scatter that is apparent in the available data.

### 5.3.1 Empirical slope relationships

The slope of the beachface is fundamentally a balance between two factors: the hydrodynamic asymmetry of near-bed velocities in the swash zone tending to move sediment landward, and the opposing force of gravity acting to move sediment seaward. Numerous researchers in the laboratory and field have examined the relationship between beachface slope and a wide range of parameters, including: wave height (e.g. Bascom, 1951), sediment size (e.g. Bagnold, 1940), wave steepness (e.g. Rector, 1954), sediment sorting (e.g. McLean and Kirk, 1969) and swash period (e.g. Emery and Gale, 1951).

Compiled from the results of over 35 published sources, Sunamura (1984b) presents one of the more comprehensive empirical studies of beachface slope. Through the application of dimensional analysis, separate quantitative predictions of beachface slope were derived for the laboratory and field:

$$\beta = \frac{0.013}{(H_b/T\sqrt{gD})^2} + 0.15 \qquad \textit{laboratory} \tag{5.12}$$

$$\beta = \frac{0.12}{(H_b/T\sqrt{gD})^2} \qquad \textit{field} \tag{5.13}$$

where $H_b$ and $T$ are breaking wave height and period, $D$ is grain size and $g$ is gravitational acceleration. The relationship between several key variables and beachface slope (Figure 5.7) are discernible from these equations, most notably a positive correlation with sediment size, and a negative correlation with wave height. The trend that steeper slopes are observed at the laboratory scale is consistent with early research indicating an inverse relationship between foreshore gradient and wave steepness ($H/L$), but when the ratio $H/L$ is held constant, an increase in wave height results in a decrease in beachface slope (King, 1972). Equations 5.12 and 5.13

**Figure 5.7** Beachface slope is partially a function of grain size as indicated by (a) the steep (15°) shingle beach (Cullykhen Bay, Scotland) and (b) the flat fine-grained (1°) dissipative beach (Goolwa, South Australia) (A D Short).

and the data upon which they are based are depicted in Figure 5.8. It is apparent that a great deal of scatter is present, particularly in the field data (Figure 5.8b), indicating that although sediment size and wave characteristics are of fundamental importance in defining beachface slope, the contribution of other factors is not negligible.

The elevation of the water table within the beachface is one example of an additional variable that can influence equilibrium beachface slope (Section 5.3.3), particularly on macro-tidal beaches where two distinct regions of the intertidal profile may be distinguished by the outcrop of groundwater (Section 8.3). To extend the results of Sunamura to include this effect, Turner (1994) undertook a survey of 15 beaches of the central Queensland macro-tidal coast, where tide ranges in excess of 6 m result in a lower region of the beachface at many sites remaining in a permanently saturated state. An empirical relationship for saturated beachface gradients was derived, based on a convenient reanalysis of Sunamura's results by Kriebel *et al.* (1991) in which the dimensionless fall velocity parameter $\Omega$ (Section 4.6.1) is used to characterise sediment and wave parameters

$$\beta = 0.05\left(\frac{W_s T}{H_b}\right)^{0.5} = 0.05\Omega^{-0.5} \quad \textit{saturated beachface} \qquad (5.14)$$

In Figure 5.9 this equation for saturated beachface slope is contrasted to unsaturated beachface gradients predicted by Equation 5.13. The available data are rather limited and Equation 5.14 should be taken only as indicative; however, it is apparent that elevated coastal groundwater in the presence of large tides can significantly reduce beachface gradients.

### 5.3.2 Uprush–backwash transport asymmetry

To move beyond empirical relationships for beachface slope towards a more physically based model, a clear understanding of sediment transport mechanisms within the swash zone is required. It was noted earlier that the beachface is a complex and difficult environment in which to undertake morphodynamic research, and this is reflected in the relative paucity of sediment transport data from the swash zone. A particular difficulty inherent to the prediction of sediment transport by swash (and therefore equilibrium beachface slopes) is that a large amount of sediment is typically carried both upslope and downslope with each swash cycle, while the resulting net transport on an equilibrium beachface must by continuity be zero. This means that any proposed sediment transport model that integrates over many swash cycles must be very accurate, since small errors in the uprush or backwash transport rate will translate into relatively large errors in the predicted net transport and equilibrium beachface gradient.

A simple but enduring theoretical model for beachface slope was developed by Inman and Bagnold (1963) based on the premise that an equilibrium gradient represents the balance between upslope flow asymmetry and the downslope component of gravity. For simplicity, only sediment transported as bedload is considered here. According to the energetics model of Bagnold (1963), uprush transport per unit width of shoreline ($i_{up}$) is

$$i_{up} = \frac{e_{up}\kappa_{up}}{\tan\phi + \tan\beta} \qquad (5.15)$$

where $\kappa_{up}$ is uprush fluid power, $\tan\beta$ the beachface gradient, $\phi$ the internal friction angle of the sediment and $e_{up}$ an uprush efficiency factor representing the ratio of energy expended in transporting sediment relative to total uprush energy. Similarly, downslope transport is simply

$$i_{back} = \frac{e_{back}\kappa_{back}}{\tan\phi - \tan\beta} \qquad (5.16)$$

with $e_{back}$ and $\kappa_{back}$ now representing the backwash efficiency factor and fluid power, and the change of sign in the denominator indicating that gravity assists downslope sediment transport. Inman and Bagnold (1963) reasoned that if the beachface slope is in a state of dynamic equilibrium then upslope and downslope transport must be equal, so by equating Equations 5.15 and 5.16 and rearranging for the *equilibrium beachface gradient* $\tan\beta_{eqm}$

$$\tan\beta_{eqm} = \tan\phi\frac{1-c}{1+c} \qquad (5.17)$$

where

$$c = \frac{e_{back}\kappa_{back}}{e_{up}\kappa_{up}} = \frac{\text{backwash energy dissipation}}{\text{uprush energy dissipation}} \qquad (5.18)$$

This formulation for equilibrium beachface slope is conceptually appealing because a condition of zero uprush–backwash asymmetry ($c = 1$) will result in a horizontal bed, while with increasing onshore asymmetry ($c < 1$) the beachface slope will increase and ultimately approach the angle of sediment repose. Field measurements of sediment transported within the swash zone on a natural beach have been used to investigate the trans-

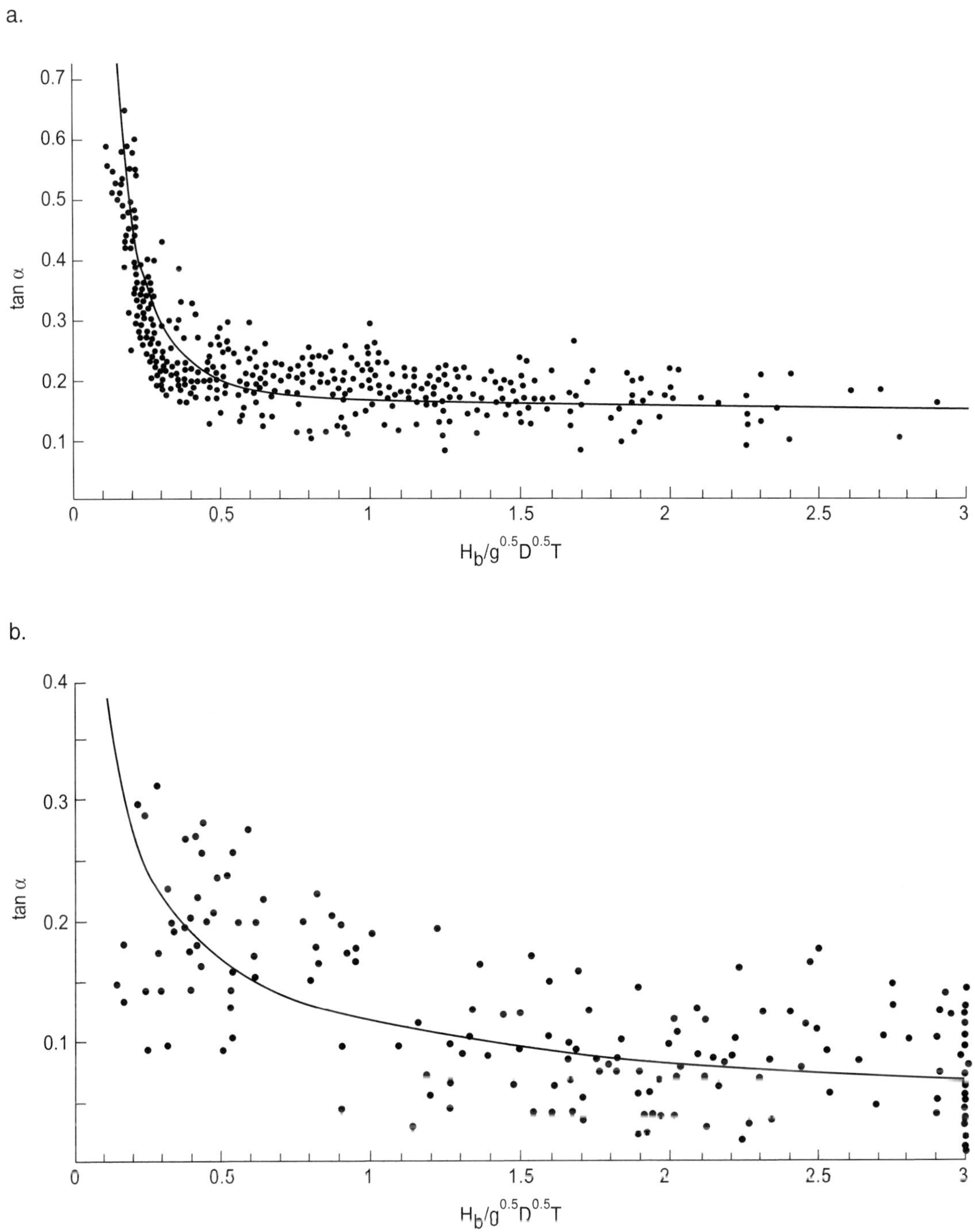


**Figure 5.8** (a) Laboratory and (b) field measurements of beachface slope as a function of wave and sediment conditions (after Sunamura, 1984b). The solid lines represent the empirical relationships (Equations 5.12 and 5.13).

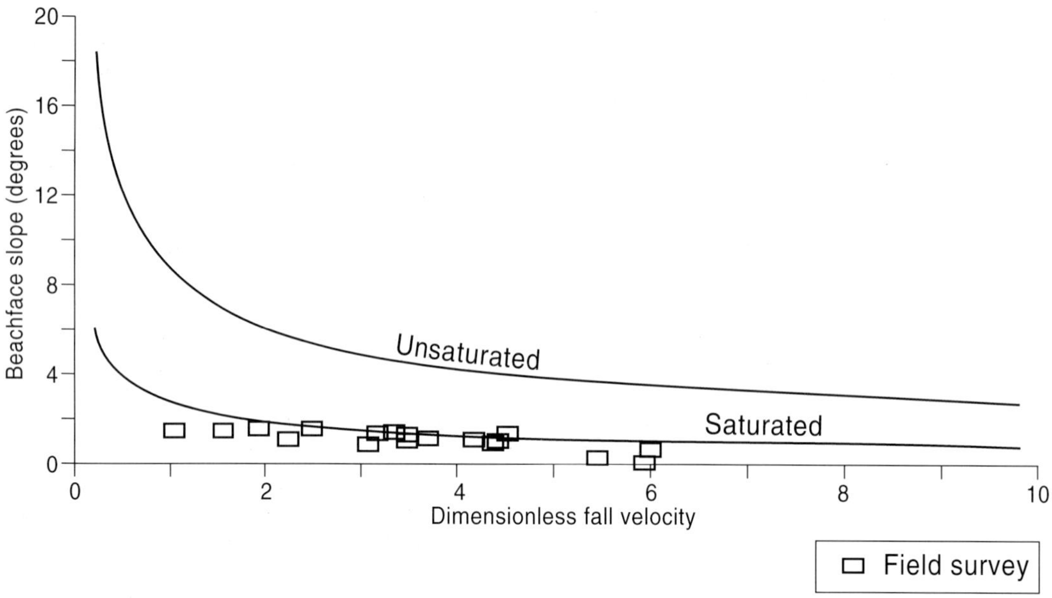


**Figure 5.9** Field measurements of saturated beachface slope as a function of the dimensionless fall velocity parameter (after Turner, 1994). The lines represent the empirical relationships (Equations 5.13 and 5.14).

port relationships given by Equations 5.15 and 5.16 (e.g. Hardisty *et al.*, 1984). Using sediment traps to measure the total load carried upslope then downslope by individual swashes across the beachface at Palm Beach, Australia, Hughes *et al.* (1997) confirmed that the amount of sediment transported by a single uprush was several orders of magnitude greater than net transport per swash cycle. Empirical estimates were obtained for the product $e_{up}.K_{up}$ for the particular prevailing beach and wave conditions, but a sufficiently accurate method to predict sediment transport rates and ultimately beachface slope still does not exist.

### 5.3.3 Swash infiltration/exfiltration

It was noted earlier (Section 5.2.1) that bed friction and infiltration both influence swash kinematics, and therefore both these processes affect hydrodynamic uprush–backwash asymmetry and equilibrium beachface gradient. The tendency for some portion of the swash lens to infiltrate into an unsaturated beachface during uprush, resulting in a reduction in energetics and hence sediment transport during the backwash, has been demonstrated in both the laboratory (e.g. Bagnold, 1940) and the field (e.g. Longuet-Higgins and Parkin, 1962). This asymmetry between uprush and backwash volumes is the simplest way in which infiltration can affect equilibrium beachface morphology, but there are two additional infiltration/exfiltration mechanisms (that are generally less well recognised and understood) that may further contribute to sediment transport asymmetry in the swash zone.

*Swash infiltration* (inflow to the bed) increases the effective normal stress on the bed, while *groundwater exfiltration* (outflow from the bed) decreases normal stresses. Vertical flow exerts a force termed *seepage force* within the beachface, that has the effect of increasing (infiltration) or decreasing (exfiltration) the effective weight of the surficial sediments (Nielsen, 1992). This change in the effective weight of sediment on the beachface may act to stabilise (infiltration) or destabilise (exfiltration) the bed, resulting in the beachface tending to respectively increase or decrease in gradient. The second infiltration/exfiltration mechanism that may influence sediment mobility at the beachface is a change in horizontal shear stresses at the bed due to altered boundary layer flow within the swash lens. There are a number of reasons why flow through a permeable bed will change the characteristics of horizontal flow above it, but in general terms infiltration can be regarded as having the effect of 'thinning' the boundary layer and thereby increasing shear stresses at the bed, while seepage results in boundary layer 'thickening' and hence a reduction in bed shear stress (e.g. Lofquist, 1975). The interesting point to note here is that while infiltration results in a stabilising seepage force, simultaneous boundary layer thinning has the opposite effect of enhancing sediment mobility, and vice versa for exfiltration. It is therefore critical to determine the net effect on beachface stability. At present the literature is contradictory; experiments reported by Oldenziel and Brink (1974) where water was sucked and blown through sand beds in the presence of a horizontal flow above the bed clearly show that infiltration reduces the net sediment transport rate, but Willetts and Drossos (1975) found the opposite effect. A different result again was reported by Martin (1970), who concluded that the two effects nearly balance out and therefore have little effect on the initial motion of sediment. The message from all these studies is that the role of infiltration/seepage in determining sediment transport asymmetry on the beachface requires extensive further investigation (Turner and Masselink, 1998).

## 5.4 BERMS

*Berms* are an accretionary feature common to many beaches, resulting from the accumulation of sediment at the landward extreme of wave influence (Figure 5.10). Berms may be thought of as the most landward morphological features resulting from wave action, and their location marks the transition zone where hydrodynamic sediment transport mechanisms cease and aeolian processes take over (Chapter 6). The shore-parallel berm crest tends to be most distinct on beaches composed of coarser sediment, as the beachface is typically much steeper than the near-horizontal backbeach region. On fine-grained beaches the berm may be indistinct due to similar backbeach and beachface gradients. Berm width can be uniform alongshore, or vary significantly due to rhythmic features at the shoreline including swash cusps (Section 5.6) and local scouring associated with rip feeder channels (Section 4.4.2).

### 5.4.1 Height, evolution and erosion

*Berm height* is equal to the maximum elevation to which sediment is transported during wave uprush. Early descriptions of berm development and the relationship between crest height and maximum wave run-up can be found in the laboratory studies of Bagnold (1940) and the field observations of Bascom (1954). Komar (1976) provides a useful overview of this early work. More recent researchers including Takeda and Sunamura (1982) and Larson (1988) have also derived empirical relationships for the height of berm growth, based on both field and laboratory data (Figure 5.11).

Clearly, higher berm crests are associated with wave run-up extending a greater distance up the beachface. It is therefore not surprising that both berm height and the distance of the berm crest from the shoreline are not only dependent on wave height, but are also a function of wave characteristics at the base of the beachface. In small scale laboratory experiments, Okazaki and Sunamura (1991) showed that collapsing bores result in a higher and further landward berm crest, relative to surging waves of comparable height. It is interesting to note that the dependence of berm crest height on run-up height introduces an intriguing paradox: higher run-up will result in berm crest development at a higher elevation; however, above some critical erosion–accretion threshold, storm waves erode the beachface leading to rapid and complete removal of berm morphology.

Interesting insight into differing mechanisms of berm development is provided by Hine (1979) who observed three distinct patterns of berm formation and widening along a single sand spit. The critical distinguishing factor between these mechanisms is the rate of sediment supply to the lower beachface. Along a straight section of beach exposed to significant longshore transport, the difference between neap and spring tide ranges was observed to be fundamental to berm development. During neap tides, sediment accumulated on the beachface at the elevation of the high tide swash line, to then be redistributed across the beachface with increasing tide range. Over-topping of the berm crest on spring high tides resulted in berm growth both laterally and horizontally. On a second section of the spit where decreasing rates of longshore transport resulted in a surfeit of sediment, berms were observed to develop from the onshore migration and welding of swash bars. On a third and most active section of the spit located at the lateral extent of recurve spit formation, swash bars built vertically at such a rapid rate that over-topping was inhibited during neap tides. Instead, the seaward face of the swash bar steepened to develop a wide berm ridge, followed by subsequent infilling of the landward swale. In this manner the beach was observed during the 16 week study period to translate 115 m seaward in a succession of shore-parallel (berm) ridges.

Berms are observed to erode in a differing manner depending on the prevailing morphodynamic conditions in the surf zone and at the beachface (Section 7.3). Wright (1980) distinguished three quasi-distinct modes of berm cut, and identified relative energy levels associated with each of these mechanisms. On steep, *reflective beaches*, berm erosion occurs under moderate swell conditions by enhanced run-up and subsequent berm over-topping resulting from the resonance of sub-harmonic water level fluctuations in the surf zone. On flatter, *dissipative beaches*, berm cut is associated with

**Figure 5.10** A well developed berm at Warriewood Beach, NSW, showing a relatively straight berm crest and backing runnel (A D Short).

significantly higher incident storm energy producing water level oscillations in the surf zone at infragravity frequencies and with enhanced set-up and bore penetration to the beachface and backshore. Berm erosion on *intermediate beach types* can result from both of the mechanisms described above, but in addition scour adjacent to the embayment of topographically arrested rips can cause localised destruction of the berm even when the net trend alongshore may be accretionary. On both *reflective* and *dissipative beaches*, berm scarping is characteristically relatively uniform along shore; on intermediate beaches berm growth at one location may contrast with berm erosion a few tens of metres to either side. The elevation of the groundwater table at the beachface may also be an important factor determining berm erosion during storms. By field calibrating a steady-state groundwater seepage model, Katoh and Yanagishima (1992) found a convincing correlation between the estimated super-elevation of groundwater effluence at the beachface and the observed critical elevation of berm erosion during approximately 30 storms.

### 5.4.2 Cyclic patterns of berm–bar sediment exchange

The exchange of sediment between berms and the surf zone has attracted the attention of coastal scientists from the earliest days of coastal morphodynamic research. Shepard (1950) is the most frequently referenced early example of such a study, and documents that the offshore movement of sediment during storms to form bars occurs at the expense of the berm and beachface, in contrast to onshore transport and berm development at the expense of bars during calmer wave conditions. On the west coast of the United States where much of this early research was undertaken, winter is generally associated with storm erosion and summer with calmer periods of beach recovery, which has led to the term *summer–winter beach cycle* becoming widely entrenched in the literature. This is unfortunate, because on many of the world's coastlines the prevailing wave climate, and hence beach response, is non-seasonal (Section 2.6).

Useful insight into the episodic exchange of sediment between the berm and nearshore can be achieved through the use of a common statistical technique called

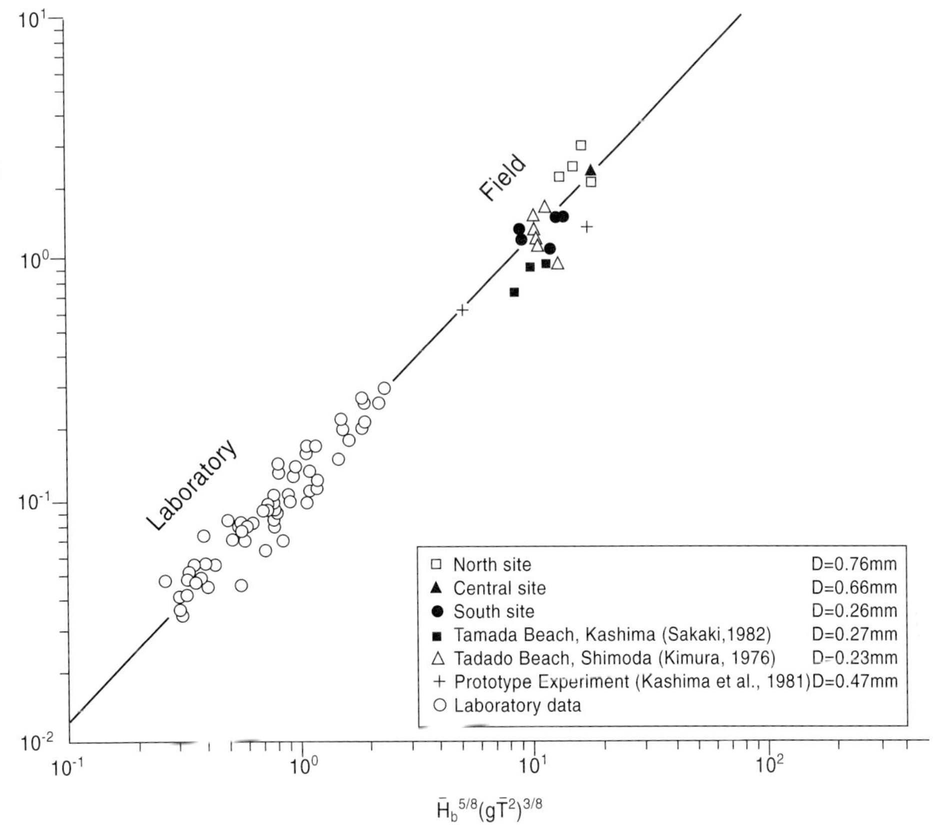


**Figure 5.11** Field and laboratory measurements of berm height as a function of wave conditions. The solid line represents the least squares regression equation, $B_h = 0.125H_b^{5/8}(gT^2)^{3/8}$, where $B_h$ is the berm height, $H_b$ is the wave breaker height, $g$ is the gravitational acceleration and $T$ is the wave period (after Takeda and Sunamura, 1982).

*Principal Components Analysis* (PCA), often alternatively referred to in the coastal literature as *Empirical Orthogonal Function Analysis* (EOF) (e.g. Winant *et al.*, 1975; Aubrey, 1979). The primary advantage of using this method is that it can often identify underlying patterns within beach survey data that are typically highly variable and otherwise difficult to interpret. Conceptually, when applied to a long-term data set of repeated beach profile surveys, PCA separates the spatial and temporal scales of changing profile morphology, and highlights underlying erosion–accretion trends. In Figure 5.12a it is difficult to identify any recurring trends in over five years of approximately monthly surveys from the macro-tidal North Harbour Beach, Queensland, Australia. Figure 5.12b depicts the first two 'spatial eigenfunctions' or underlying patterns to profile variability derived by the PCA technique. The first function is analogous to the mean beach profile, and the second function reveals that the dominant pattern of erosion and accretion involves sediment exchange between the upper beachface (berm) and the lower intertidal and nearshore. This is as anticipated, but it is hard to distinguish this trend directly from the raw survey data. The 'pivotal point' coincides with the mid-beachface and indicates that erosion and accretion of the berm are 180° out of phase with erosion and accretion across the lower part of the profile. For this reason the second spatial function derived from PCA of micro-tidal beaches is frequently referred to as a 'bar–berm' function. The presence of a bar is not apparent in Figure 5.12 because bar formation is typically suppressed on macro-tidal beaches (Section 8.4.1).

## 5.5 BEACH STEP

The *beach step* is located at the base of the beachface (Figure 5.1). Its morphology is characterised by a local steepening in the beachface gradient and it is often, but

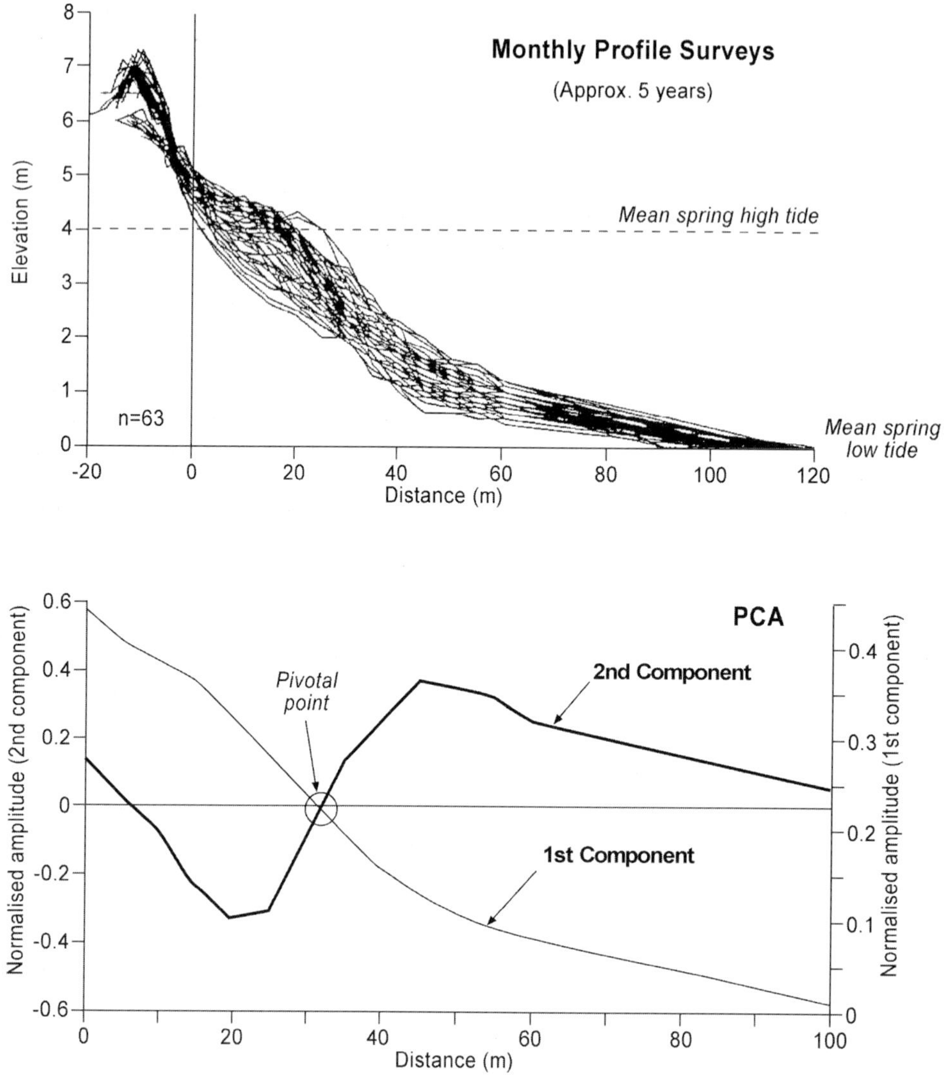


**Figure 5.12** (a) Five years of monthly profile data from the macro-tidal North Harbour Beach, Queensland, Australia. (b) Results of PCA: the first component (thick line) identifies the mean beachface profile and the second component (thin line) identifies an underlying pattern of recurring sediment exchange between the berm and nearshore (after Turner, 1994).

not always, composed of sediment that is significantly coarser than the sediment found both seaward and landward (Bauer and Allen, 1995). The origin of this feature is primarily related to swash zone processes, and hence it is commonly regarded as part of the beachface, even though permanent submergence should technically identify it as part of the surf zone.

The *height* of the step on sandy beaches can range from a few centimetres to over a metre (Hughes and Cowell, 1987). Most observations support the notion that the beach step consists of three units: (1) a steep step face with dip angles slightly less than the repose angle of the sediment; (2) a step crest that separates the step face from the beachface; and (3) a nearly horizontal step base that extends seaward from the step face (Bauer and Allen, 1995). Beach steps are ubiquitous on *reflective beaches* and steep bar–trough type *intermediate beaches*, but they occasionally occur on gentle beach slopes as well (Short, 1984b).

### 5.5.1 Beach step formation

#### *5.5.1.1 Early hypotheses*

There is general agreement that it is a combination of both surf and swash zone processes that develop and maintain the beach step, but the actual mechanism involved has been the subject of debate for some 40 years. Early researchers agreed that the development of the beach step was related to the kinematics of plunging breakers, although there was disagreement on whether the beach step was a depositional or erosional feature. The argument for the former was based on the understanding that onshore sediment transport across the nearshore is driven by shoaling waves, whereas offshore sediment transport across the beachface is driven by backwash. The meeting point for these opposing transport processes is the base of the beach, where the incident waves plunge and collide with the backwash. The result is a highly turbulent region at the base of the beach within which strong sorting occurs – fine sediment is winnowed and swept up the beach by the uprush and coarse sediment is deposited to produce a step (Miller and Ziegler, 1958). More recently, the laboratory experiments performed by Miller (1976) showed that in shallow water depths the vortex enclosed by a plunging breaker can scour a trough in the seabed. Bauer and Allen (1995) suggested that this scour trough, if located close to the beachface, could erode a step-like feature. The difficulty with both of these explanations linking beach step development to plunging breakers is that beach steps have also been widely observed in the presence of surging waves (Figure 5.2a).

#### *5.5.1.2 Backwash vortex*

The most recent explanation for the development of a beach step emphasises the importance of the *backwash* flow. A causal link between the beach step and an unsteady, separated *vortex* formed when the backwash turns back into an oncoming wave was first proposed by Matsunaga and Honji (1980). The vortex forms immediately seaward of the still water shoreline during the backwash phase of the swash cycle. The rotation of the vortex is such that flow is directed landward near the bed and seaward near the surface, in contrast to the plunging breaker vortices previously described by Miller (1976). The laboratory experiments performed by Matsunaga and Honji (1980) clearly showed the backwash vortex developing a scour at the base of an initially stepless beach which, once developed, was self-perpetuating through morphodynamic feedback. Once the step was formed the flow then separated at the lower edge of the step and another vortex was induced, which generated a ripple mark seaward of the step. This process continued and produced a field of ripples that decreased in amplitude seaward of the step, consistent with field observations of the seabed on a number of reflective beaches (e.g. Nielsen, 1983; Short, 1984).

The backwash vortex occurs in the presence of surging, collapsing and plunging breakers and therefore provides the most complete explanation for beach step formation that is currently available (Matsunaga and Honji, 1983). Furthermore, laboratory experiments performed by Larson and Sunamura (1993) show that the backwash vortex also provides the hydrodynamic regime necessary to sort sediment and concentrate the coarsest material in the step. The phases in beach step development and sediment sorting elucidated by their experiments are shown in Figure 5.13.

The size and angular velocity of the backwash vortex

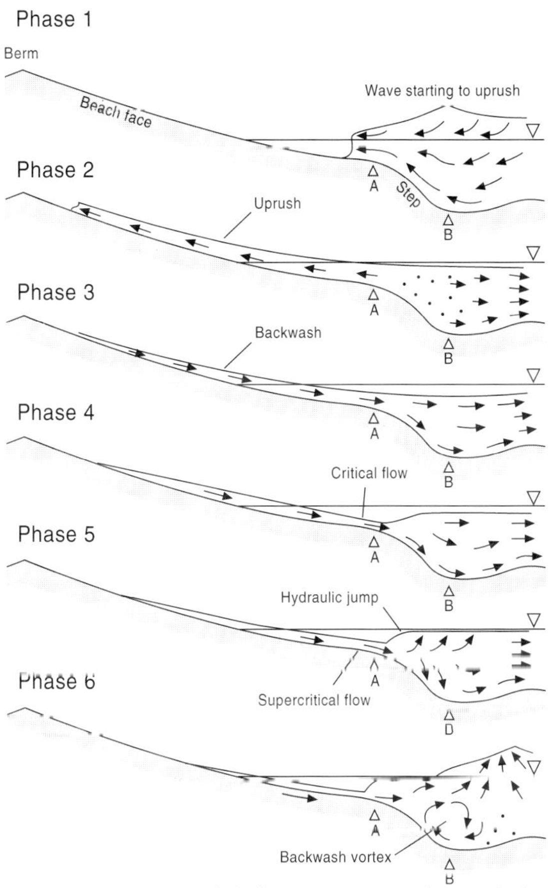


**Figure 5.13** Characteristic flow patterns over the step during uprush (phases 1 and 2) and backwash (phases 3 to 6) (after Larson and Sunamura, 1993).

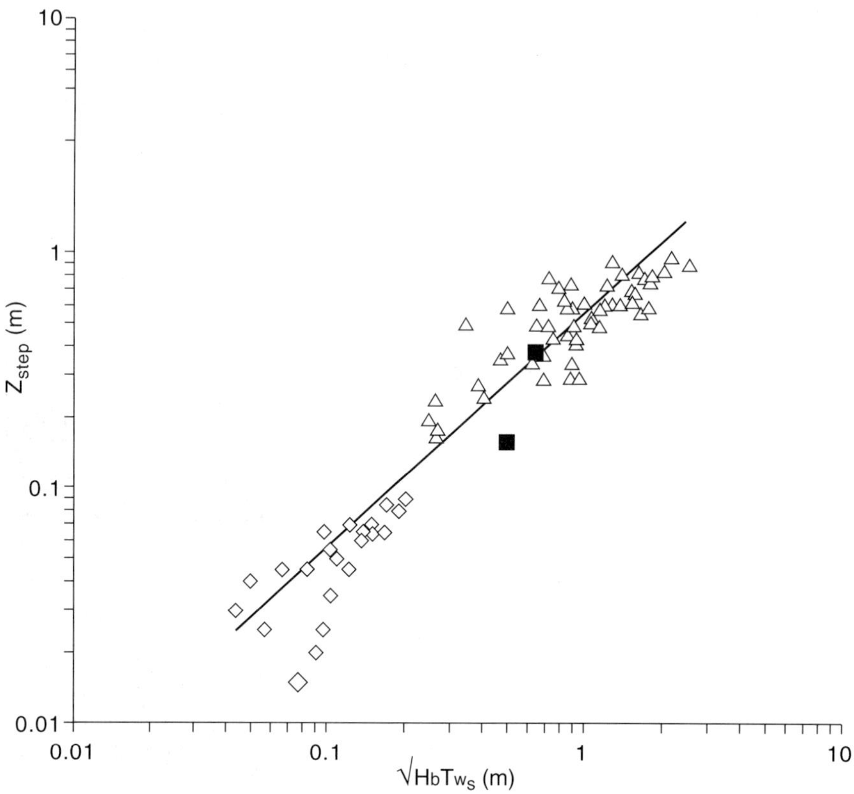


**Figure 5.14** Field (open triangles, Hughes and Cowell, 1987; closed squares, Bauer and Allen, 1995) and laboratory (open diamonds, Takeda and Sunamura, 1983) measurements of step height as a function of wave and sediment conditions. The solid line represents the least squares regression equation, $Z_{step} = 0.55\sqrt{H_b T w_s}$, where $Z_{step}$ is step height, $H_b$ is wave breaker height, $T$ is wave period and $w_s$ is the sediment fall velocity.

is positively related to swash length and swash period, which in turn are positively related to wave height and period immediately seaward of the shoreline. The anticipated positive correlation between wave height and beach step height has been confirmed by field observations obtained on the perennially reflective Pearl Beach, Australia (Hughes and Cowell, 1987). The range of wave periods that this beach experiences is limited, but laboratory experiments confirm that the step height does indeed also scale positively with wave period (Takeda and Sunamura, 1983). In addition to wave parameters, Hughes and Cowell (1987) found that coarse sediment in the step was indicative of a large step height, probably because larger sediment sizes have larger angles of repose and should therefore permit development of larger steps. The available field and laboratory measurements of beach step height are well described by a simple non-dimensional, morphodynamic parameter incorporating wave height, period and grain size (Figure 5.14). These measurements are indicative of beaches where morphology and wave conditions are characteristic of the *reflective beach type*. More measurements are required in order to establish whether the morphodynamic parameter shown in Figure 5.14 is useful on *intermediate beach types*. On these more dynamic beach types the response time of the morphology compared with the hydrodynamics means the two are rarely in equilibrium or linearly related (e.g. Bauer and Allen, 1995; Wright and Short, 1984).

## 5.6 SWASH CUSPS

There is perhaps no single morphological feature of sandy beaches that has sparked as much scientific interest and intrigue as beach cusps. The emergence of regularly spaced crenulations of the shoreline (Figure 5.15)

**Figure 5.15** Swash circulation in relation to cusps at Pearl Beach, NSW. Note that the cusp spacing increases towards the higher energy end of the beach (bottom of the photograph), as indicated by the larger swash length (A D Short).

immediately landward of a highly complex and apparently chaotic surf zone is a puzzle that has fascinated and occupied the attention of many researchers. It is not surprising that the volume of literature examining morphological, sedimentological, hydrodynamic and theoretical aspects of cusp formation is extensive. Space does not permit much of this work to be included here; instead the objective of this section is to provide an overview of past and current ideas of cusp morphology and formation.

### 5.6.1 Definition and occurrence

Beach *cusps* are crescentic, concave-seaward features at the shoreline, their most striking feature being their regular or quasi-regular spacing alongshore. The term cusp has been used to describe many cuspate features that exhibit a wide range of longshore wave lengths. Dolan and Ferm (1968) developed a classification based on spacing, which distinguished between 'typical beach cusps' (8–25 m), 'storm cusps' (70–120 m) and 'giant cusps' (70–1500 m). On lake shores (e.g. Johnson, 1919; Komar, 1973b) and in the laboratory (e.g. Longuet-Higgins and Parkin, 1962; Guza and Inman, 1975) cusp spacing ranging from a few centimetres to 1 or 2 m is commonly observed. With such a wide range of shoreline features exhibiting very different length scales a classification system based on spacing is unconvincing. Instead, it is more useful to distinguish between *swash cusps*, typically displaying a longshore spacing of a few tens of metres, and *rhythmic topography*, exhibiting spacing typically an order of magnitude greater (50 m to hundreds of metres). This terminology was first proposed by Guza and Inman (1975). Swash cusp morphology is generally associated with *reflective beach* systems in contrast to rhythmic shoreline topography which is more characteristic of *intermediate beach* systems (Sections 7.3.4 and 7.3.3). In this chapter discussion is strictly limited to swash cusps, which are linked to the action of wave uprush and backwash acting directly on the beachface (Inman and Guza, 1982). Rhythmic beach and surf zone morphology associated with the quasi-regular alongshore distribution of nearshore cell circulation is presented in Section 4.4.2.

*Swash cusps* are typically composed of medium to coarse sand. They have, however, been observed on beaches composed of all types of sediment, including gravel, cobbles and boulders (e.g. Russell and McIntire, 1965a; Komar, 1976, p.264; Carter, 1988, p.127). The selective sorting of grain sizes and grain shapes between cusp swales and intervening horns has been examined both in the laboratory (Fleming, 1964) and on natural beaches (Longuet-Higgins and Parkin, 1962). Once cusp development commences, it appears that the increased permeability of coarser material that accumulates on the steeper cusp horns may contribute to continued cusp growth. On micro-tidal beaches a single set of cusps is typically observed, which is periodically removed and then re-emerges in response to storm erosion and subsequent recovery of the beachface. On larger tide range beaches a *succession of cusps* may be stepped up the beachface, the landward-most cusps being inactive features at all times other than high spring tide. Cusp spacing may vary with tidal stage; on a typical reflective macro-tidal beach it is likely that cusp spacing will increase landward due to decreased dissipation in the steeper nearshore at high stages of the tide.

### 5.6.2 Swash cusp formation

#### *5.6.2.1 Early models*

A large number of imaginative, and frequently contradictory, mechanisms for swash cusp evolution can be found in the earlier scientific literature. The formation of cuspate shoreline morphology has been attributed to erosional processes (e.g. Johnson, 1919; Kuenen, 1948), accretionary processes (e.g. Branner, 1900; Komar, 1971b) or a combination of both (e.g. Otvos, 1964; Guza and Inman, 1975). A general consensus in the earlier work (e.g. Johnson, 1919; Kuenen, 1948) was that the uniform growth and spacing of cusps results from an unclear mechanism whereby an equilibrium morphology emerges in response to initial *foreshore irregularities*. It was proposed that any slight depression in the beachface would be enhanced by uprush and backwash, with refraction of swash in enlarged hollows on the beachface resulting in erosion of the swale and deposition on an adjacent horn. The regular spacing of cusps was proposed to emerge from an initially random distribution of irregularities alongshore, as smaller hollows are subsumed by further developed features that cease growing when some optimum size is reached. The principal problem with this hypothesis is the fact that cusps are often observed to be uniformly spaced as soon as they form (e.g. Komar, 1973b).

A very different mechanism to explain the rhythmic spacing of swash intensity that is required for cusp formation was proposed by Cloud (1966) who developed a hypothesis based on the *instability of breaking waves*. Cloud suggested that a wave at the instant of breaking approximates a solid fluid cylinder, which will theoretically separate into sections that have a length some 15 to 17 times the diameter of the cylinder (i.e. wave height at breaking). There is some merit to this theory since it can account for the observed positive correlation between cusp spacing and wave height, but there are no suitable data available to test the theory. One problem with this breaking wave instability hypothesis is that cusps have been reported where swash motion results from wave surge as opposed to the run-up of broken waves (e.g. Komar, 1973b).

*Further hypotheses* for the formation of cuspate features on beaches include 'sand waves' (analogous to dunes and antidunes in fluvial systems) originating from bottom irregularities in the presence of a longshore current (e.g. Schwartz, 1972a), 'crossed waves' in the surf zone (Dalrymple and Lanan, 1976) and regular swash 'salients' separating turbulent erosive zones of retarded flow on the beachface (Gorycki, 1973). However, the oblique wave approach necessary for the sand wave and crossed wave hypotheses is not always a precursor to observed swash cusp development (e.g. Longuet-Higgins and Parkin, 1962), and the swash salient hypothesis fails to account for the emergence on a planar beachface of regularly spaced cusps in the presence of irregularly spaced swash excursions.

#### *5.6.2.2 Standing edge waves*

The first inference of the role of standing edge waves (Section 4.3.5) in swash cusp development may be attributed to Escher (1937) who reproduced swash cusps in the laboratory and observed that standing waves at right angles to the incoming waves were present in the experimental wave basin. These 'extra' waves had the same period as the generated incoming waves. Erosion resulted where the height of incoming wave run-up was enhanced, and deposition prevailed where opposing standing wave crests and incoming wave troughs reduced run-up excursion across the beachface. Although these experimental observations were ignored or discounted (e.g. Kuenen, 1948) for several decades, the interaction of incident waves with shore-normal standing waves is currently the most widely accepted mechanism for swash cusp development. Guza and Inman (1975) were the first to test the standing edge wave hypothesis for beach cusp formation, and numerous other researchers have attempted to identify edge waves on natural beaches with varying degrees of success. Holman (1983) provides an interesting overview of several methods that have been pursued to 'catch' an edge wave. A remaining challenge is to distinguish between the various types and modes of wave motion that often coexist in the surf zone (Section 4.3.7).

To recap, *edge waves* are standing or progressive waves trapped at the shoreline by refraction; the crests are normal to the beachface and therefore normal or near-normal to the alignment of incident wave crests. Edge wave amplitudes vary sinusoidally alongshore and are a maximum at the beachface, decaying exponentially through the surf zone (Figure 4.9). *Standing edge waves* result from the interaction of two progressive edge waves of equal frequency, amplitude and mode travelling in opposite directions. The presence of a headland, groyne or jetty will clearly increase the likelihood that standing edge waves are present, due to reflection of progressive longshore wave energy.

By the standing edge wave mechanism the cusp wave length (or the spacing between adjacent cusp swales), $\lambda_c$, is equal to the longshore spacing of run-up maxima arising from incident–edge wave interaction. On beaches exhibiting a steep, reflective beachface, Guza and Davis (1974) demonstrated that mode 0 *sub-harmonic edge waves* predominate, where edge wave period $T_e$ is twice the incident wave period $T_i$ ($T_e = T_i/2$). Though smaller in amplitude, synchronous edge waves ($T_e = T_i$) are also frequently observed. Referring to Figure 5.16, it follows that cusps formed by *synchronous edge waves* are spaced at a distance equal to the standing wave length, and conversely cusps formed by sub-harmonic edge waves are spaced at a distance equal to one-half the standing wave length. There is a contradiction in the literature as to whether cusp horns in this latter case will be situated at antinodes where run-up is a maximum (Sallenger, 1979a; Inman and Guza, 1982) or the intervening position of nodes on the beachface where run-up is a minimum (Huntley and Bowen 1975; Guza and Bowen, 1981).

The principal attraction of the standing edge wave hypothesis for swash cusp development is its ability to explain the quasi-regular spacing alongshore. By Equation 4.22 the spacing of beach cusps produced by a mode 0 sub-harmonic or synchronous edge wave is respectively

$$\lambda_c = \frac{g}{\pi} T_i^2 \tan\beta \qquad \textit{sub-harmonic} \qquad (5.19)$$

$$\lambda_c = \frac{g}{2\pi} T_i^2 \tan\beta \qquad \textit{synchronous} \qquad (5.20)$$

where $\lambda_c$ is distance between two adjacent horns or swales, $T_i$ is incident wave period, $\tan\beta$ is beach slope and $g$ is gravitational acceleration. Using Equations 5.19 and 5.20 it is interesting to check whether cusp spacing observed in the field matches predicted spacing for the prevailing wave conditions. For example, on a beachface of slope 0.1 subjected to waves of period 10 s, the predicted (sub-harmonic) cusp spacing is approximately 31 m. This length scale is indeed consistent with cusp morphology often observed on steep, *reflective beaches*.

In Figure 5.17 a compilation of data by Komar and Holman (1986) shows a plot of observed versus predicted swash cusp spacing reported by a number of researchers from the laboratory and field. Though some outliers exist, these results are generally consistent with the hypothesis of swash cusp formation by sub-harmonic or, in a more limited number of cases, synchronous standing edge waves. In assessing the scatter of these data, it must be remembered that in practice it is often difficult to assign a single wave period from the spectra of incident wave periods observed, and on a concave beachface it is sometimes difficult to identify a characteristic planar beachface slope required for the theoretical prediction.

### 5.6.3 Recent developments: standing edge waves versus self-organisation

A new hypothesis for the generation of swash cusps, which contradicts the widely accepted standing edge wave mechanism, has had quite an impact on the coastal research community. The model of Werner and Fink (1993) in fact marks a return to earlier ideas that cusp formation results from feedback between swash flow and beachface morphology. This hypothesis had previously been discounted as it was hard to envisage how the regular spacing of cusps could result. However, with recent developments in the understanding of non-linear dynamics and complex systems, techniques for understanding patterns in apparently random behaviour are beginning to emerge.

In the so-called 'self-organisation' model, incipient topographic depressions in the beachface are amplified by attracting and accelerating water flow, with the result that erosion of these depressions is enhanced to form cusp swales. The remarkable fact about the elegant simulations that Werner and Fink present to support

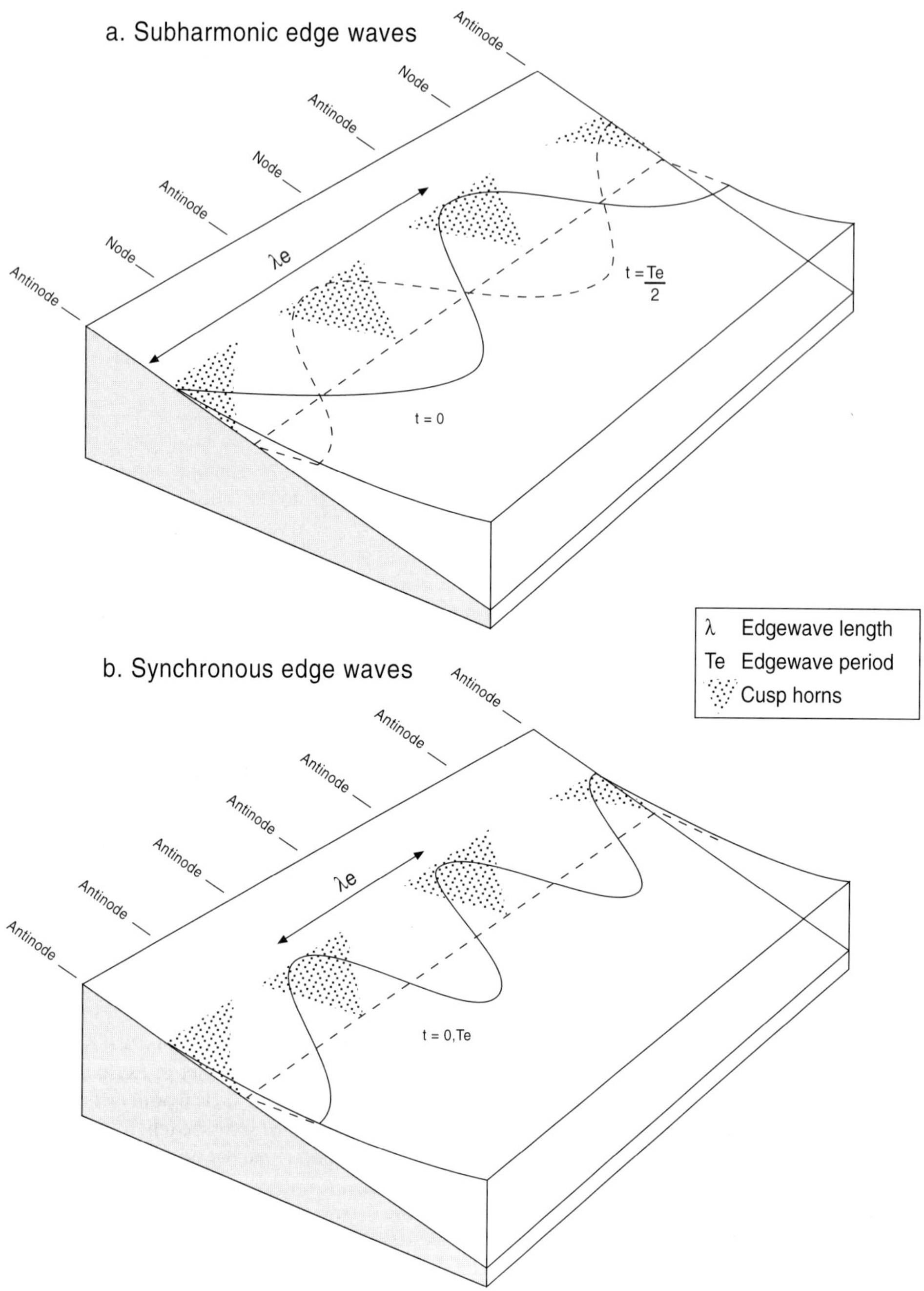


**Figure 5.16** Schematic diagram showing theoretical beach cusp spacing for (a) sub-harmonic and (b) synchronous edge waves (after Komar and Holman, 1986).

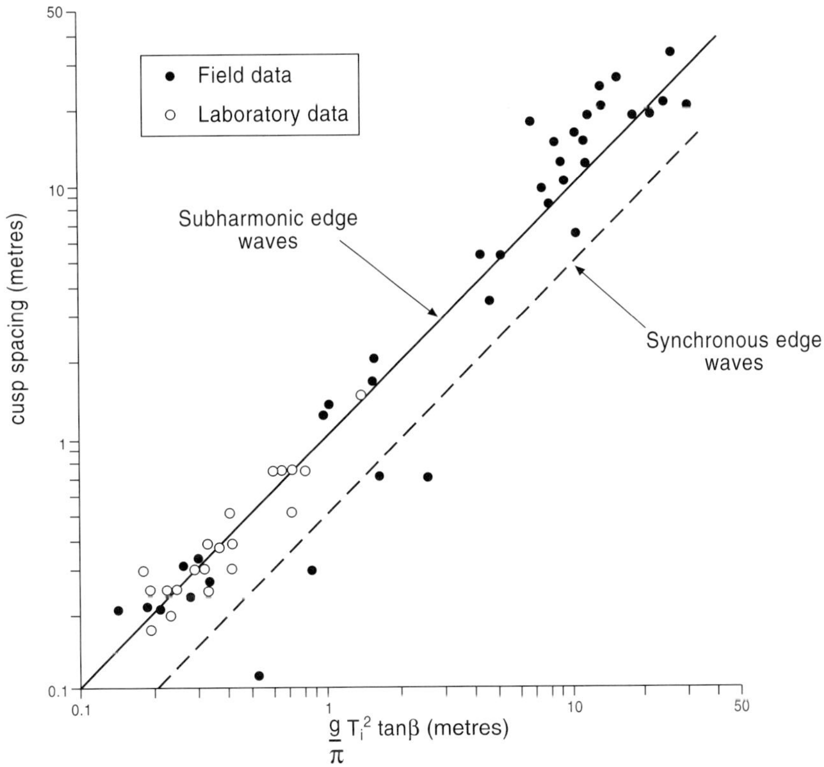


**Figure 5.17** Field (closed circles) and laboratory (open circles) measurements of cusp spacing as a function of theoretical edge wave length (after Komar and Holman, 1986).

their hypothesis is that predicted cusp wave lengths are indistinguishable from cusp spacing predicted by the standing sub-harmonic edge wave model. It would be truly remarkable if there are two contrasting mechanisms that produce identical morphology, particularly if the two hypotheses are mutually exclusive as Werner and Fink suggest.

Field tests to distinguish between the standing edge wave and self-organisation mechanisms have recently been undertaken, but so far the results are inconclusive (EOS, 1994; Burnet *et al.*, 1994). Nicknamed 'Dozer Duck', the work was carried out at the US Army Corps Field Research Facility at Duck, North Carolina, and involved the reshaping of the beachface with a bulldozer. Two sets of experiments were performed: in the first, existing beach cusps were flattened; and in the second, cross-shore channels were cut in the beachface. The first experiment was undertaken to test the idea of Inman and Guza (1982) that standing edge waves provide only the initial beachface perturbations necessary for cusp development; once initiated they continue to develop due to morphodynamic feedback between swash and morphology. Indeed, there are strong theoretical arguments to suggest that the development of a non-planar beachface will have the effect of 'de-tuning' or dampening edge wave generation (Guza and Bowen, 1981). By removing three cusps from a beach with otherwise well developed cusps, by the standing edge wave hypothesis the reformation of cusps was not anticipated to occur. In the second set of experiments, by the edge wave theory the channels would also not be expected to spawn regularly spaced beach cusps because morphology does not initiate the initial longshore pattern. In the majority of 'Dozer Duck' experiments beach cusp formation or reformation was observed; however, the mechanism responsible is still uncertain. It is even uncertain as to whether standing edge waves were present or absent during some of these experiments (EOS, 1994; Burnet *et al.*, 1994). In one of the beach flattening experiments cusps emerged with an irregular spacing varying by 30%; neither the edge wave nor self-organisation model could predict such morphology (EOS, 1994). In five years' time it will be interesting to reflect back on where this present controversy has led.

## 5.7 CONCLUDING REMARKS

In the introduction to this chapter it was noted that the beachface is the most familiar region of a beach to most beach users, but to scientists and engineers it poses some of the greatest questions in beach morphodynamic research yet to be answered. The practical difficulties in undertaking field research in this highly energetic region of the beach undoubtedly pose one of the major challenges. On a more fundamental level, however, the swash zone is a very complex hydrodynamic region and a basic description of many of the pertinent processes is still lacking. This is particularly evident in the discussion of swash kinematics and swash zone sediment transport, where more questions were raised than answers provided. In summary, a recent evaluation of current state-of-the-art coastal sediment transport models concluded that, despite several decades of quantitative beach research, the inclusion of the swash zone within the cross-shore model domain is presently non-existent (Schoonees and Theron, 1995). Before this serious shortcoming can be redressed a great deal more research on nearly all aspects of sediment transport by swash is required.

# 6

# The Beach Backshore and Beyond

*Patrick A. Hesp*

## 6.1 INTRODUCTION

The *backshore* is the zone between the swash limit (sometimes the berm crest) and the seaward edge of a dune or other landform (e.g. a cliff). It is a zone that is episodically impacted by both swash and aeolian sand transport. It is therefore not only a major beach feature, often occupying the bulk of the *sub-aerial beach* (and is considered 'the beach' by most beach users), but it is also the source of most dune sand, and is therefore critical to aeolian sand transport and dune formation. Surprisingly, however, it is both one of the most intensively used, yet scientifically neglected parts of beach systems. It is a critical dynamic boundary between the highly dynamic wave and tide dominated surf and swash zones, and the wind dominated dune system.

Davis (1982) states that the backshore zone is a nearly flat, terrace-like area, slopes slightly landward, is commonly much broader than the foreshore, and in normal conditions is a sub-aerial feature (Figure 6.1; cf. Fairbridge, 1968). Implicit within this definition is an acceptance that the backshore is generally, in fact, the berm terrace tread. Some dissipative, and many meso- to macro-tidal beaches do not display berms, and the high tide swash may extend to the base of the dunes or to the back of the beach (Figure 6.1).

In this chapter, the major processes occurring on the backshore zone of predominantly sandy beaches are reviewed. In addition, the major types of coastal landforms that exist at the landward margin of the backshore (and beyond) are described, and the interrelationships between backshore processes and the evolution and geomorphology of the landward terrains are briefly examined.

## 6.2 BACKSHORE BEACH MORPHOLOGY

Backshore topographic profiles are intimately connected to their lower shorefaces. Chapters 7 and 8 present the range of beach types and associated beach profiles and modal topographies of micro- to macro-tidal beaches. They range from steep, coarse grained, narrow reflective beaches, to wide, low gradient, fine sand, dissipative and ultra-dissipative systems.

The presence and extent of the backshore are dependent on a number of factors including beach type, modal beach width, medium to long term beach status (stable, eroding or accreting), and sand supply. In essence, the backshore portions of the micro-tidal beach types range from wide, flat to slightly concave dissipative backshores (without berms), through moderate width intermediate beaches with more variable berm morphologies, to narrow, relatively steep reflective backshores. As tide range/dominance increases, the intertidal portion widens and under lower waves may consist of an upper, steeper foreshore that dries at low tide, and a lower, flatter saturated low tide beach and terrace. Under higher waves and tides the non-barred to ultra-dissipative beach types prevail with a wider, curvilinear beach. In all cases decreasing grain size and increasing wave height and tide range favour lower, flatter, wetter backshores.

Temporal variations in backshore width and morphology are related to temporal variations in wave height, including storm cycles (e.g. Thom and Hall, 1991), daily and monthly tidal inundation levels, beach type (e.g. the presence, size and intensity of rips; Short and Hesp, 1982), and temporal alongshore and across-shore variations in sediment supply (Davidson-Arnott and Law, 1990).

*Handbook of Beach and Shoreface Morphodynamics*. Edited by A. D. Short.
© 1999 John Wiley & Sons Ltd.

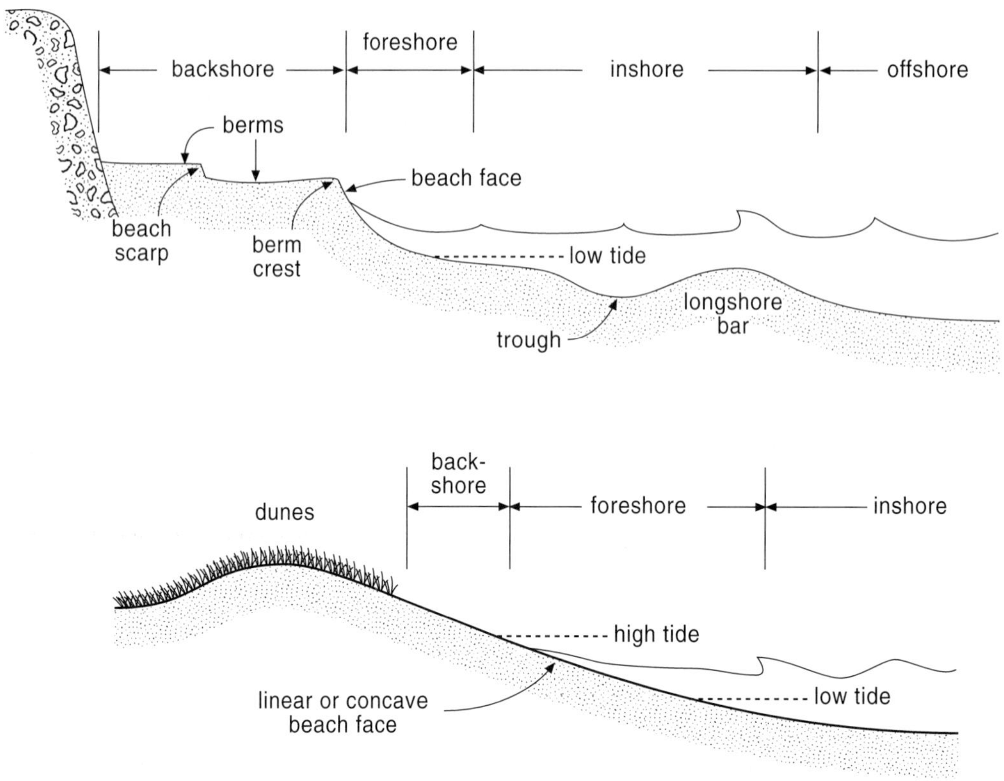


**Figure 6.1** Beach profile terminology for (a) a beach with berm (Komar, 1976), and (b) one without a berm. Note how the backshore may vary considerably in width and morphology depending on beach type.

## 6.3 BACKSHORE SAND TRANSPORT

The most common process occurring in the *sub-aerial* environment on backshores is aeolian sand transport. There have been numerous attempts to model this transport (Bagnold, 1941; Kawamura, 1951; White, 1979). In a review of aeolian sand transport models, Sherman and Lyons (1994, p.383) noted that all basic transport models share the following assumptions: (i) the wind field is steady and uniform; (ii) the surface is flat, horizontal and unobstructed; (iii) the sediments are sands which are non-cohesive, clean and dry; and (iv) transport predominantly takes place via saltation in a layer that is in equilibrium with the local wind field.

However, when it comes to aeolian sand transport on beaches, these assumptions are rarely met. This is because the wind is often highly turbulent coming off the surf zone and onto the beach, and one or more internal boundary layers commonly develop. The morphologies of beaches are seldom flat and planar, rather they are often undulatory both across-shore and alongshore. Beach sands often contain a variety of size ranges, are seldom clean (the presence of organic materials is common), range from nearly completely dry to saturated, and in some circumstances, creep may be more important than saltation. Nevertheless, there are many situations where an attempt to calculate sand transport rates is important.

Sand transport rates may be calculated utilising a variety of sediment transport models (see reviews in Horikawa *et al.*, 1986; Hotta, 1988; Sarre, 1988; Nickling and Davidson-Arnott, 1990; Sherman and Hotta, 1990; Sherman and Lyons, 1994). Generally, there is little variation in calculated potential rates between many (but not all) of the models (Figure 6.2a and b).

In order to calculate potential sand transport off a beach, Bagnold's (1941, p.105) equation may be utilised

$$q = C(d/D)^{1/2} U_*^{\,3} \rho/g \qquad (6.1)$$

where $q$ is the sand transport rate (in kg/m/s), the coeffi-

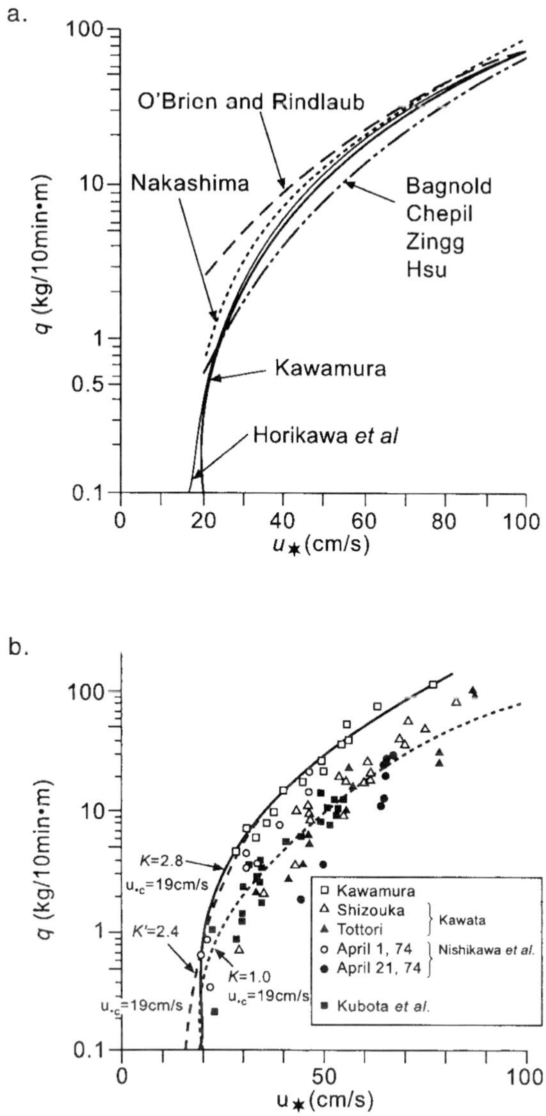


**Figure 6.2** (a) Comparison of aeolian sand transport formulas (from Hotta, 1988). (b) Aeolian sand transport rates measured on various beaches (from Hotta, 1988).

cient $C$ is 1.8, $d$ is the grain diameter, $D$ is a standard grain size diameter (0.025 cm), $U_*$ is the shear velocity (in m/s), $\rho$ is air density (1.22 kg/m$^3$) and $g$ is gravitational force (9.81 m/s$^2$).

The shear velocity is a measure of the velocity gradient of the wind, that is, the rate of increase of velocity with the log height. It is related to the shear stress, $\Omega$, at the bed and the air density by

$$U_* = (\Omega\rho)^{1/2} \tag{6.2}$$

Shear stress is a mean stress per unit horizontal area of ground surface. It is the change of velocity with height that causes the shearing stress in the air and transmits it to the ground as the surface drag. It is the surface drag which moves sediment.

The shear velocity, $U_*$, can be determined from meteorological tower data by utilising the logarithmic relationship describing the wind profile as modified by Howard *et al.* (1978)

$$u_z = 2.5U_* \ln(z/z_{oa}) \tag{6.3}$$

where $u_z$ is the wind velocity at height $z$, $z$ is the height at which wind speed is measured, and $z_{oa}$ is the apparent roughness height when saltation occurs, where

$$z_{oa} = k \exp(-K'U_t/U_*) \tag{6.4}$$

in which $U_t$ is the wind speed at height $K'$ at the threshold of sediment movement, $K'$ is taken as 3 mm following Bagnold (1941, p.104), and $k$ is von Karmans constant (0.41).

In order to obtain $U_*$ in Equation 6.1, Equation 6.3 can be rearranged in the form

$$U_* = u_z - 2.5K'U_{*t}/2.5\ln(z/k) \tag{6.5}$$

The threshold velocity, $U_{*t}$ in Equation 6.5 can be determined from

$$U_{*t} = 680\sqrt{d}\log 30/d \tag{6.6}$$

following Bagnold (1941, p.105), and ignoring the slope term. In the case where velocities at a lower level are required, data obtained from a meteorological station (from which wind speeds at 10 m height are usually obtained), a wind velocity at, say, 2 m can be determined via

$$U(z_1) = u(z_2)/[\ln(z_2/z_0)/\ln(z_1/z_0)] \tag{6.7}$$

in which $z_1$ is wind speed at 2 m, $z_2$ wind speed at 10 m and $z_0$ is the roughness length.

If wind velocity profiles are measured in the field, shear velocity may be determined directly from the lowest two wind profile measurements utilising

$$U2 - U1 = (U_*/K')\ln(z_2/z_1) \tag{6.8}$$

which may be rearranged to

$$U_* = (U2 - U1)K'/[\ln(z_2/z_1)] \tag{6.9}$$

In addition, conditions such as moisture content and a wet sand surface (increasingly important on meso-tidal and macro-tidal beaches), sediment size distribution, beach slope, and three-dimensional beach morphology

**Table 6.1** Factors affecting aeolian sand transport on beaches.

| Factor | Expression | Impact |
|---|---|---|
| Moisture (Section 6.3.1) | Surface water, pore water | Grain cohesion and increased threshold velocity |
| | Rain splash | Impact ejection, then cohesion |
| Temperature (Section 6.3.2) | High temperature | Rapid drying, increased transport |
| | Low temperature: pore ice, frozen grains, frozen swash, snow | Increased grain size, increased threshold velocities, transport reduced and eliminated |
| Surface crusts (Section 6.3.3) | Salcrete, calcrete/carbonate cement, ferrous cements, clay, algae and fungi | Increased threshold velocities, reduced transport |
| Surface roughness (Section 6.3.3) | Shell lags, coarse/heavy sands and gravels, wood and seaweeds | Reduced and restricted transport, increased local turbulence |
| Beach morphology (Sections 6.3.4 and 5) | Cross-shore and alongshore profile variations | Velocity accelerations and decelerations, increased and decreased transport, lateral variations in transport |

will affect actual transport rates (Table 6.1). The impact of each is explored in the following sections.

### 6.3.1 Wet and moist beach and backshore surfaces

*Wet surfaces* and *pore water* act to increase the threshold shear velocity required to initiate aeolian sediment movement (Belly, 1964; Hotta, 1988; Sherman and Lyons, 1994; Namikas and Sherman, 1995). Beach salcrete and salt crystals covering the surface may have a similar effect (Pye, 1980). Aeolian sand transport from a wet surface depends on the evaporation rate, the amount of water content, wind speed (Hotta, 1988), and the vertical variation of moisture content with depth. In the latter case, pronounced sand transport can be observed on beaches where there is a thin (a few millimetres to ~ 1 cm) wet sand layer overlying a relatively drier layer. Once the wet layer is deflated on the upwind edge, it is peeled back by the wind exposing the drier layer underneath. The transport rate then increases rapidly. Hotta (1988), in reviewing other studies, shows that up to a water content, $w$, of 10%, the threshold wind velocity increases linearly with increasing water content of the sand surface, and also with increasing sediment size. In the range 3% to 8% water content, there is little difference in the rate of sand transport, $q_a$, above a shear velocity of 100 cm/s (Hotta, 1988; Figure 6.3). On average, the shear velocity required to initiate transport on a surface with a moisture content of 5% is about twice that when the sand is dry (Sherman and Lyons, 1994).

### 6.3.2 Frozen beaches

Many high latitude and seasonally cold coastal environments have beach and dune systems (Davies, 1980). The presence of pore ice within beach and dune sediments profoundly affects sand transport rates. Pore ice binds the grains and markedly increases the surface resistance to wind action (McKenna Neuman, 1993; van Dijk and Law, 1995). However, it does not necessarily completely restrict sand transport. Sublimation (the reversible change of state where a solid may become a vapour or vice versa without passing through the liquid stage) can occur leading to the release of sand particles for transport.

### 6.3.3 Surface crusts and surface roughness

*Surface crusts* inhibit sand transport by increasing the threshold velocity required to initiate sand movement. Crusts may be formed by carbonate cements, salt aerosol cements (salcrete), ferrous oxide cements, the presence of clay particles, and the presence of algae and fungi (Nickling and Davidson-Arnott, 1990).

### 6.3.4 Backshore slope and morphology

Surface slope affects wind or shear velocity because changes in slope can produce wind velocity acceleration or deceleration, promote turbulence, and potentially act to create the development of internal boundary layers and even flow separation. In addition, increasing slope angles tend to enhance the effect of gravity, potentially reducing sand transport rates.

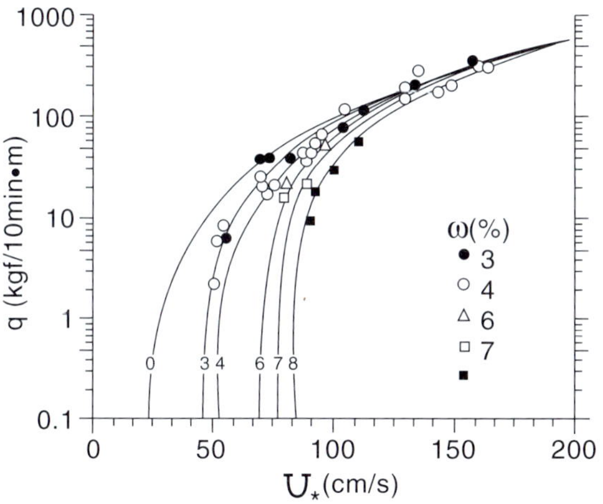


**Figure 6.3** Aeolian sand transport rates ($q$) under high shear velocities ($U_*$) on sand surfaces with varying amounts of moisture ($w$) (modified from Hotta, 1988).

Hardisty and Whitehouse (1988) provide a transport equation which incorporates a slope correction coefficient such that the potential transport decreases as the slope increases. However, the importance of slope in transport rates is still difficult to assess. Field observations of transport on sometimes very steep slopes indicate that ripples remain aligned normal to the slope and appear to be minimally affected by gravity.

### 6.3.5 Longshore morphologies

The calculation and measurement of potential and actual aeolian sand transport on a beach are made more difficult by subtle longshore morphological changes. For example, many dissipative beaches display very subtle mega-cusp horn and bay forms which may only rise millimetres to a few centimetres above the general beach level. These slightly higher areas dry out more rapidly than the surrounding beach, and can provide a localised source of sediment for aeolian transport, when most of the beach sand is wet or moist, and unable to be transported.

Such variations may become more important with a trend from micro-tidal to macro-tidal beaches. The tops of ridges, for instance on ridge and runnel forms exposed at low tide, often dry out and become an important source of aeolian sediment. To date there has been little research on the relationships between tidal range and aeolian sediment transport on sandy beaches. Thus, the validity of the micro-tidal wave–beach–dune model interrelationships of Short and Hesp (1982) and Hesp (1988a) in meso- to macro-tidal environments remains unknown. Large dune systems do, however, exist in many relatively high wave–wind energy, meso-tidal and macro-tidal environments (e.g. Pye, 1983).

## 6.4 BACKSHORE AERODYNAMICS

Beach and backshore slope and morphology have a pronounced effect on the potential and actual rate of landward aeolian sand transport. In very general terms the flatter the surface and the more linear the morphology, the less the flow disturbance. Hence, the less wind velocities fluctuate and the greater the potential for maximising sand transport off, and across, that surface (Bradley, 1968; Hsu, 1977; Mulhearn, 1978; Hesp, 1982; Short and Hesp, 1982; Kaimal and Finnigan, 1994).

Hesp (1982) and Short and Hesp (1982) measured wind fields across three typical micro-tidal beaches (Figure 6.4). As noted, *dissipative* beaches are characterised by wide, flat profiles which may be virtually linear in micro-tidal environments and more curvlinear in meso- and macro-tidal environments (Figure 6.2). Minimum disturbance of the wind structure occurs across these beaches, and with a gentle landward incline, there is often a gradual landward increase in surface–near surface flow speed (Figure 6.4a).

The sub-aerial/intertidal morphology of *intermediate* beaches varies through a number of morphologic states. Consequently, the beachfaces are relatively steep toward the reflective end of the spectrum, and relatively flat on the dissipative end. The backshores may be relatively flat, or troughs and topographic lows may exist. Figure 6.4b illustrates lines of equal mean speed derived from wind profiles measured across a low energy intermediate beach profile. Even though the berm crest is low, and the berm trough subtle, there is a reduction in wind speed across the beach and backshore.

In the case of *reflective* beaches, steep cusps and berms are common on the higher energy reflective beaches and linear, relatively steep beachfaces are common on the low energy beaches. Figure 6.4c illustrates wind flow structure across a relatively wide reflective beach with a steep beachface and a morphologically complex backshore. Marked acceleration occurs across the berm crest, and rapid deceleration occurs landward of the crest with significant disturbance of the flow structure.

Although more research needs to be done on wind flow over backshores, these limited experiments indicate that, all other factors being equal, the minimal disturbance of the flow structure on dissipative backshores by beach morphology would lead to maximum

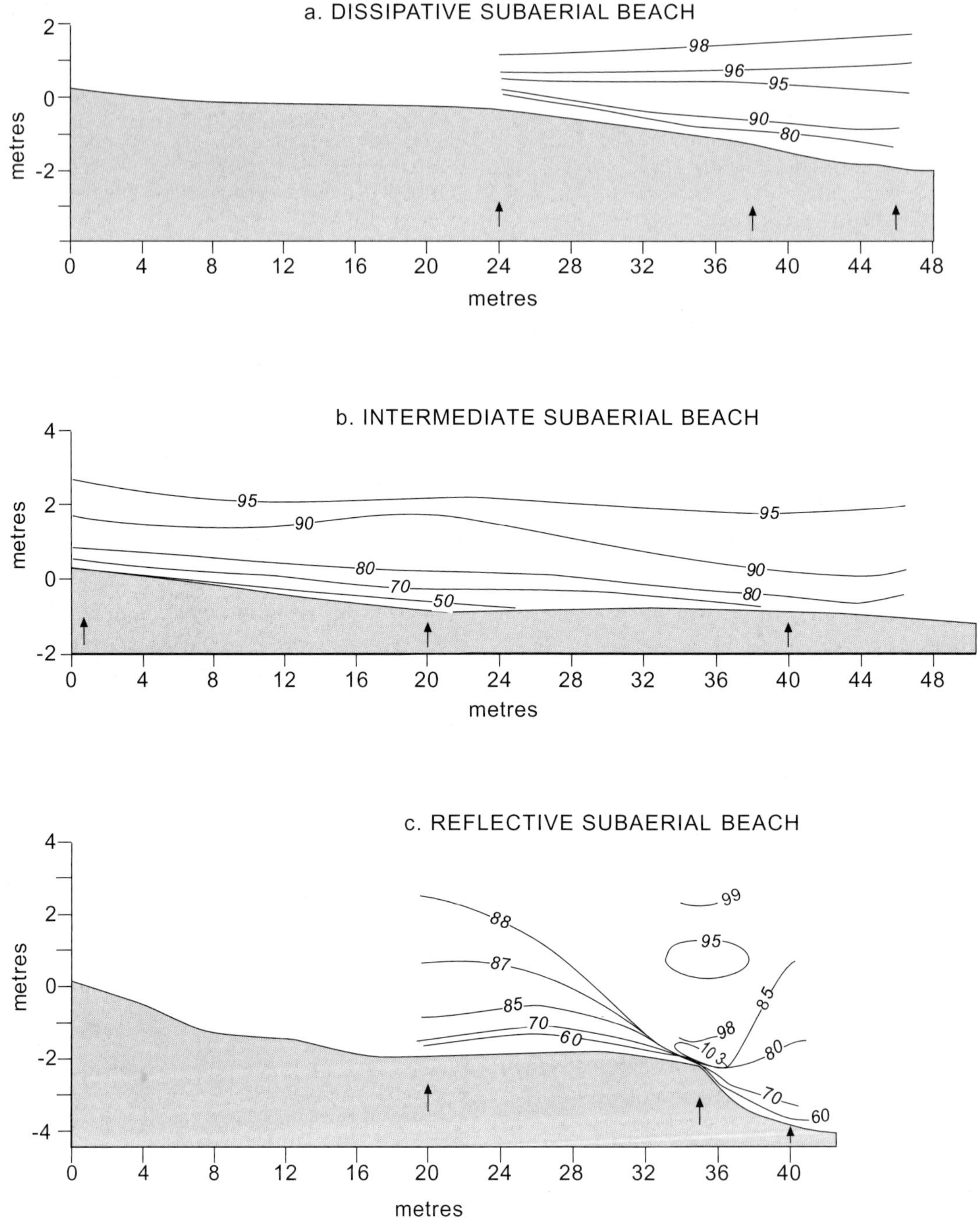


**Figure 6.4** Percentage isolines of equal mean wind speed measured over (a) dissipative, (b) intermediate and (c) reflective beaches. Flow disturbance is minimal and wind speed accelerations occur across flatter, more dissipative backshores. The arrows indicate wind profiling locations.

potential sand transport off dissipative beaches and minimal potential sand transport off reflective beaches (Short and Hesp, 1982; Hesp, 1988a).

Sherman and Lyons (1994) have utilised a numerical model to estimate sand transport off different beach types and, in part, to validate (or not) the Short and Hesp model. Using three modal beach morphologies (reflective, intermediate and dissipative), all 100 m wide and with different grain sizes (finest on dissipative beaches), they found that if just the effects of slope and grain size were taken into account, there was only a 20% difference between transport off the reflective beach (the lowest rate) compared with the dissipative beach. Where moisture content was included, transport rates were nearly two orders of magnitude higher off the dissipative beach compared to the reflective beach.

Research by Davidson-Arnott (1988a) and Davidson-Arnott and Law (1990) further demonstrates the control that beach width (and indirectly beach type) has on wind fetch and sand transport. They found higher rates of aeolian sand transport and deposition on wide beaches compared to adjacent narrow beaches. Proxy measurements (e.g. calculation of dune volumes along barriers) of long term rates of sand transport off different beach types indicate similar results (Hesp, 1988a).

## 6.5 BEACH – BACKSHORE – DUNE SEDIMENT TRANSPORT

Sediment exchange between the beach, backshore and dunes varies according to wind energy, seasonal variations in wind direction and magnitude, wind resultants, tidal range, sediment supply, and surf zone–beach type.

During beach *accretion* phases, swash transports sediment to the backshore via beachface progradation, accretion of swash bars, berms, and plane bed deposition on the intertidal and supratidal beach (Carter, 1977; Rosen, 1979; Short and Hesp, 1982; Hesp, 1983, 1984a,b; Carter and Wilson, 1990; Davidson-Arnott and Law, 1990; Thom and Hall, 1991; Wal and McManus, 1993; Arens, 1994; Arens *et al.*, 1995).

The influence of *beach type* is very important in determining transport. Low energy, micro-tidal *reflective* beaches are characterised by steep, narrow linear slopes. Aeolian accretion is fetch limited. In comparison, on *intermediate* micro-tidal beaches, low tide terraces attach to the beachface and accrete as berms, the steepness of which decreases with decreasing sand size and wave energy. The berms may be well developed and tens of metres wide. *Dissipative* micro-tidal beaches are wide and flat (often bermless). Wave-driven sediment accretion and erosion are accomplished by broad deposition or removal of swash bore laminae. At mid- to low tide a significant beach width is available for aeolian transfer of sediments to the backshore.

With increasing *tidal range*, berms become less prominent as intertidal and backshore profiles become more concave in shape. Low energy macro-tidal beaches have narrower, steeper morphologies than the relatively moderate to higher energy macro-tidal beaches (Figure 6.2) which restricts medium to long term sediment transfer and dune development.

*Oblique and longshore aeolian sediment transport* is important on many beaches. Firstly, it results in an increase in fetch and effective beach width (Arens *et al.*, 1995). Secondly, it often results in the temporary development of barchans, transverse dunes and shadow dunes and hummocks on the backshore (Figure 6.5) (e.g. Walker and Matsukura, 1979; Bourman, 1986; Marques and Julia, 1988; Hesp *et al.*, 1989; Psuty, 1993; Wal and McManus, 1993). This sediment may be eventually redistributed onshore and alongshore into the foredune or other dune types or blown offshore into the surf zone. The impact and importance of longshore wind-driven sediment transport is however difficult to gauge on many beaches, especially where local winds are topographically steered by foredunes (Svasek and Terwindt, 1974; Hsu, 1977; Wilson *et al.*, 1984; Rasmussen, 1989; Arens *et al.*, 1995).

## 6.6 IMPACT OF SALT AEROSOLS

Salt spray or aerosols are ubiquitous to shorelines. *Salt spray* transport is strongly related to wind velocity, and to wave energy or surfzone/width type. The spray tends to be locally produced by transport off the crests of breaking waves, and through the process of water particle or droplet release (or ejection) as waves break.

Seawater contains every element essential to plant growth, and the composition of seawater remains remarkably constant over time (Rozema *et al.*, 1985; Hesp, 1991). Salt spray is semi-continuously supplied to dunes in most coastal environments, and while a proportion of the spray is intercepted by plants, much of it is also taken into the sand surface. Plants have developed a variety of ways to cope with, or adapt to, saline water droplets on their parts and to sand salinity (Doing, 1985); see Hesp (1991) for a full review.

Salt spray aerosols are considered a primary determinant in plant species succession and growth habit in the immediate beach–dune environment (Oosting and Billings, 1942). Salt spray loads and intensities have been correlated with distance inland from the beach (e.g.

**Figure 6.5** Crescentic dune migrating across a frozen beach, Schiermonnikoog, The Netherlands (P A Hesp).

Cartica and Quinn, 1980; Avis and Lubke, 1985; Harwood, 1993), wind speed, height of plant above the ground (Barbour, 1978; Barbour *et al.*, 1985), site microtopography (Barbour, 1978; Salisbury, 1952), and beach–surf zone type (Short and Hesp, 1982; Hesp, 1988a, 1991).

## 6.7 OVERWASH PROCESSES AND DEPOSITS

*Overwash* is the process of waves washing over a berm, backshore, dune, barrier or other surface (e.g. marsh terrain). Typically, overwash occurs during storms and may be associated with storm surge, but may also occur during high tide events. Overwash is most prevalent on transgressive barriers, that is barriers that are generally receding landwards often in environments characterised by rising sea level, sediment deficiency, low dunes and occasional hurricane or cyclone activity (e.g. Pierce, 1970, Godfrey and Godfrey, 1976; Hosier and Cleary, 1977; Godfrey *et al.*, 1979; Leatherman, 1979a,b,c; Nummedal, 1983; Ritchie and Penland, 1988, 1990; Davidson-Arnott and Fisher, 1992; Dubois, 1995; Sexton, 1995; Section 14.4.1). However, minor overwash events may also occur on barriers with larger dunes in stable sea level environments, during storm events, and during periods of erosion or barrier retreat.

Overwash results in the transfer of sediment from beaches, foredunes and other dune types onto the barrier itself, to various positions landward, and into lagoons, estuaries and marshes on the landward side of a barrier (often the backbarrier or interbarrier depression). Overwash can be the single most important process in barrier retreat or transgression, although the relative importance of overwash versus aeolian processes in barrier retreat will vary from place to place, and in particular with dune height and volume (Armon, 1979; Cleary and Hosier, 1979; Rosen, 1979).

The *frequency* of overwash events depends on the exposure of the barrier, frequency of major storms, wave energy, tidal range and dune type(s), height and volume (Leatherman, 1979a). Tidal range is important according to Leatherman (1979a), since overwash is more common on barriers with low tide ranges and high storm frequencies.

A range of *landforms* occurs on barriers and backshores where overwash processes are important. Wash-

**Table 6.2** Terminology of some swash built and aeolian landforms.

| Term | Mode of formation | Also termed | Erroneously termed |
|---|---|---|---|
| Berm | Swash-deposited terrace, sand and/or gravel/shingle | | Beach ridge |
| Beach ridge | Swash and storm wave deposit/ridge, sand and/or gravel/shingle | Storm ridge<br>Shingle ridge | Foredune |
| Storm ridge | Storm/high energy wave-built ridge | Shingle ridge<br>Berm | |
| Shingle ridge | Shingle/cobble/gravel (wave-built) ridge | Storm ridge<br>Beach ridge<br>Gravel ridge (plain) | |
| Chenier | Storm wave-built ridge (sand, shell), often formed in mud-dominated environments | Shell mounds<br>Shell berms | |
| Beach-foredune ridge | Swash-deposited terrace or ridge with *aeolian* capping | | Beach ridge |
| Foredune | *Aeolian* sand deposited in vegetation at rear of backshore | | Beach ridge<br>Transverse ridge<br>Parallel dune |

over fans and sheets, washover terraces, barrier and backbarrier flats, hummocky dunes, dissected foredunes, and foredune remnants may all be common features (see figure 14.10).

*Washover fans* tend to be more discrete, tongue-shaped or linear forms, often with narrow throats, and landward widening fan margins. *Washover sheets* are continuous sheets of sand produced by single or multiple washover events (e.g. Hosier and Cleary, 1977; Cleary and Hosier, 1979; Nummedal, 1983). Coarse sediments including significant quantities of shell material may cover the fans and sheets to various degrees. The landward extent of these landforms varies considerably, and depends on the severity of storm overwash, water levels, acceleration of washover waters through fan throats and foredune breaches among others. Ritchie and Penland (1990) distinguish between washover sheets and terraces, the latter being higher landforms which are overwashed far less frequently, and often only on the upper backshore area and dune toe.

The role *vegetation* plays in washover fan and sheet stabilisation, and in the development of various amorphous, irregular and hummocky dunes across the barriers, can be very important (Hosier and Cleary, 1977). Certain vegetation species can survive some burial and may rapidly revegetate washover surfaces. Deeper burial results in various vegetation responses ranging from death, rapid colonisation by new species, or slow growth through the overwash sediments and gradual recolonisation (Godfrey and Godfrey, 1976). Other species are resilient and survive saline intrusions and flooding (Young *et al.*, 1995). Aeolian erosion on bare over-wash surfaces is common, and blowouts and parabolic dunes may form as a result (see Section 6.10).

## 6.8 SWASH-DEPOSITED BACKSHORE LANDFORMS

A variety of swash constructional landforms are formed on beach backshores. Forms include berms, beach ridges (*sensu strictu*) and storm ridges. There is considerable confusion in the literature in the use (and abuse) of these terms.

### 6.8.1 Berms

The term berm has come to be synonymous with the intertidal 'beach' and backshore, as an asymmetric landform comprising a seaward swash slope (usually the steepest component; the 'riser' in terrace morphological terminology), a crest, and a low, often linear, sometimes concave, beach top slope (the 'tread', see Table 6.2). The berm has often been taken to mean the nearly horizontal portion of the beach (i.e. the tread; see Komar, 1976; Davis, 1982). A *berm* is here defined as an inter- to supratidal deposited shore-parallel terrace or ridge, consisting of a steeper seaward sloping beachface, topped by a shore-parallel crest, the 'riser', usually backed by a flat to slightly landward dipping berm terrace (the 'tread'), which may terminate in a linear runnel, where

**Figure 6.6** Modern gravel beach ridge (with wood debris) forming at Otaki Beach, New Zealand (P A Hesp).

swash collects at (spring) high tide (Figure 5.10). Deposition cusps commonly make up part of the berm and on some beaches are a precursor to full berm development.

Berms may be very pronounced in some cases (e.g. lower energy micro-tidal and meso-tidal intermediate beaches), subtle forms on some beaches (e.g. high energy intermediate beaches), and always absent on others (e.g. modal dissipative beaches). Berms may occur as a single well defined landform, or as a sequence of obliquely stacked features. For example, beaches which experience marked seasonal variations in wave energy (e.g. Hawaii), often display remnant or abandoned berms on the upper backshore (formed during high wave energy seasons or spring tide), fronted by an active berm on the lower beach (formed during lower wave energy seasons or periods, or neap tide).

### 6.8.2 Beach ridges

*Beach ridges* are wave-deposited, linear, alongshore, symmetric or asymmetric, triangular to convex ridges, formed of sand, gravel or shell sediments (or combinations of these). Although the American, Australian and New Zealand literature, in particular, has confused the use of the term (see Hesp (1984b) for partial review) by calling sand ridges that are clearly foredunes, beach ridges (e.g. Bird, 1976; Price, 1982), we restrict the use of the term to those ridges formed only, or primarily, by wave processes, principally swash and overwash processes (Table 6.2).

Beach ridges have been defined by Stapor (1975, 1982) as 'linear mound-shaped ridges roughly paralleling the coast'. Such a broad definition has it problems; for example, Carter (1988) states that substantial high-water sand bodies form berms or beach ridges, that there is no real morphological distinction between them, and that it may be prudent to reserve the term 'berm' for artificial or gravel beach slopes. This cannot be accepted since the term 'berm' has a relatively strong meaning in the micro-tidal literature (see Figure 6.1) as discussed above.

Beach ridges occur in a range of environments and are said to be formed by fairweather swash processes during beach progradation or accretion phases (e.g. Nossin, 1965; Carter, 1986, 1988), by storm wave and swash processes (e.g. Psuty, 1965; Anthony, 1995, Hesp, personal observation), or by both (Taylor and Stone, 1996). In southern Australia Short *et al.* (1989) describe beach ridges and beach ridge plains fronted by extensive intertidal sand flats, in protected low wave environments, and assumed the ridges were produced by episodic high waves/storm surges.

The term 'storm ridges', has often been utilised to reflect the role storm waves play in the formation of discrete or overlapping beach ridges at the limit of storm wave run-up. This process of beach ridge formation is typical in gravel and mixed sand/gravel environments (Kirk, 1980; Carter, 1988; Figure 6.6). All gravel/shingle ridges are beach ridges. However, some may be capped by aeolian sand at a later stage (if such finer sediment is available) to form beach-foredune ridges.

### 6.8.3 Beach ridge plains

A series of beach ridges may be formed in areas experiencing coastal progradation and/or a relative fall in sea level. In many cases wide plains are formed comprising ridge and swale sequences (e.g. Carter, 1988; Tanner, 1988; Short *et al.*, 1989; Stapor *et al.*, 1991; Anthony, 1995). Depending on the rate of formation and progradation, ridges and swales may be closely spaced or widely separated. Most gravel ridges tend to be separated by low amplitude, high elevation swales.

## 6.9 BACKSHORE DUNES

A variety of coastal dunes may be initiated and formed on the backshores of beaches. Where plants play a role, coppice dunes (or nebkha), shadow dunes, foredunes, blowouts, parabolic dunes and some transgressive dunefields may be formed or influenced by vegetation. Where vegetation is unimportant or lacking, a variety of more typically desert dune forms may be present including dome dunes, barchans and transverse dunes, as well as transgressive dunefields (when active).

### 6.9.1 Foredunes

*Foredunes* are shore-parallel, convex, symmetric to asymmetric dune ridges formed on the top of the backshore by aeolian sand deposition within vegetation. Actively forming foredunes occupy a foremost seaward position; however, not all foremost dunes are foredunes. They have been classified into a wide variety of types (see Hesp (1984a, 1988a) for reviews) but generally fall into two main types, incipient and established foredunes, within which there can be wide morphological and ecological variations.

Foredunes have also been called 'frontal dunes' (Guilcher, 1958), 'retention ridges' and a host of other names (Bigarella *et al.*, 1969; Bigarella, 1972; McKee, 1979), 'beach ridges' and 'parallel dune ridges' (Bird, 1965, 1990), and 'transverse' dunes (Shulmeister and Kirk, 1996).

#### *6.9.1.1 Incipient foredunes*

*Incipient foredunes* are newly developing foredunes forming within pioneer plant communities. They may be formed by sand deposition within discrete or relatively discrete clumps of vegetation or individual plants, forming shadow dunes, coppice mounds and nebkha. In toto, such development comprises an incipient foredune zone. Such foredunes may be seasonal if formed in annual plants, and require invasion by perennial plants in order to survive. Incipient foredunes may also form on the backshore by relatively laterally continuous alongshore growth of pioneer plant seedlings in the wrack line or spring high tide region.

Morphological development principally depends on plant density, height and cover, wind velocity and rates of sand transport. Wind velocities experience rapid deceleration on reaching the plants, local acceleration around the plants, and flow separation behind the plants (Bressolier and Thomas, 1977; Hesp, 1981; Greeley and Iversen, 1985; Wasson and Nannings, 1986; Wolfe and Nickling, 1993). Plant species type is important in determining morphological development; species such as the tall, dense *Ammophila* tend to produce higher, more hummocky peaked dune forms than lower, more spreading, rhizomatous plants such as *Spinifex* or *Ipomoea* which produce lower, less hummocky dune forms (Hesp, 1988b). Shadow dunes, hummocks, embyro dunes and nebkha all form due to high localised drag within and behind individual plants and clumps of plants. Secondary factors such as the rate of occurrence of swash inundation, storm wave erosion, overwash incidence and wind direction can also be important in determining subsequent dune evolution (Cowles, 1899; Ranwell, 1972; Davies, 1980).

Relatively continuous plant canopies variously impact the wind/sand flow depending on plant density, distribution and height (e.g. Aylor *et al.*, 1993). High, dense canopies (e.g. marram, *Ammophila arenaria*) act to reduce flow velocities very rapidly, and sand transport (saltation and traction) is markedly reduced from the leading edge (Figure 6.7). Incipient foredunes tend to be asymmetric with the short slope to seawards. Lower, less dense canopies (e.g. *Spinifex* sp., *Uniola* sp.) act to reduce the flow and transport more slowly so that there is often a gradual downwind reduction in transport and

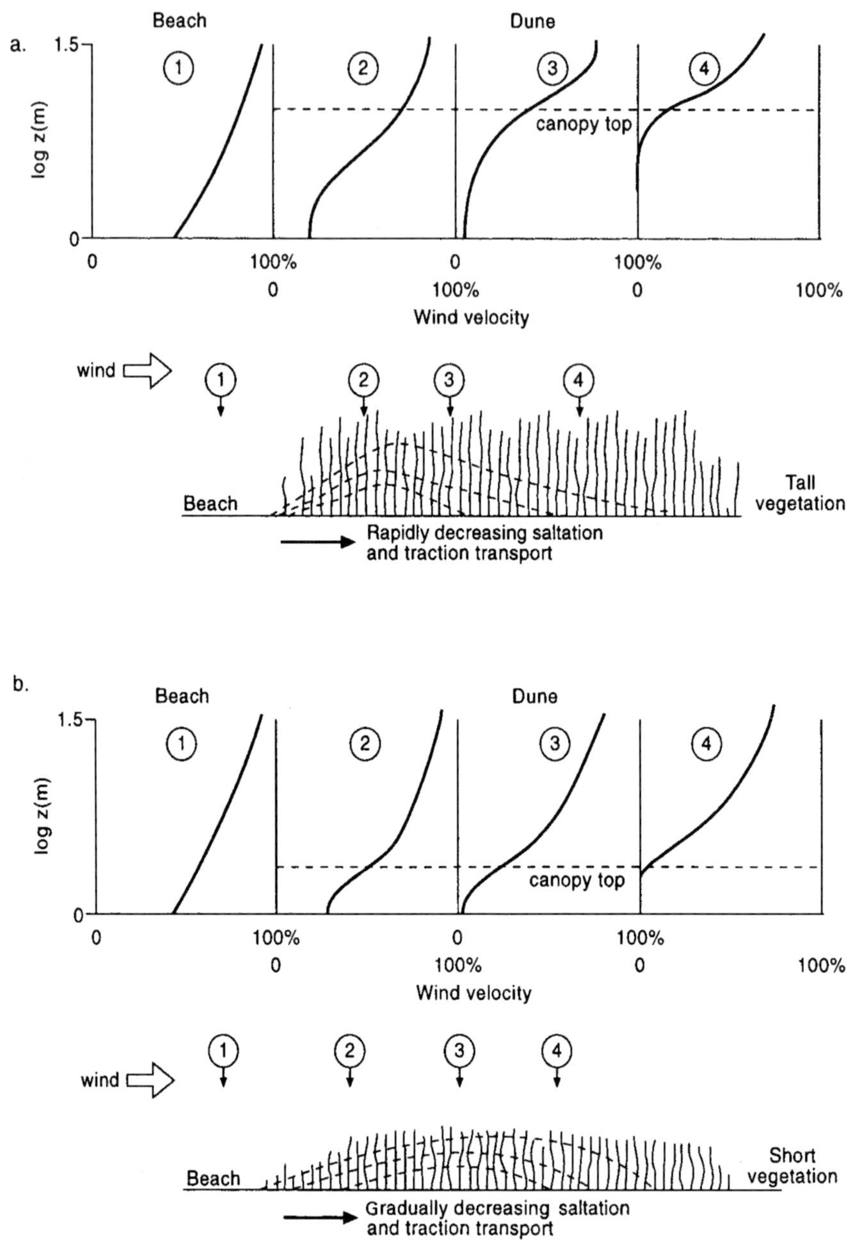


**Figure 6.7** Schematic diagrams of sedimentation in (a) tall (e.g. marram) and (b) short (e.g. spinifex, ipomoea) plants and the sequential development (dashed lines) of incipient foredunes. Typical schematic velocity profiles are indicated. Note that the vegetation generally maintains upward growth and height (not shown) as the dunes evolve.

asymmetric dunes are formed with the short slope on the downwind side. In canopies which vary alongshore in density or distribution (as most do), or in the presence or absence of species, the morphology of the foredune varies alongshore accordingly (Figure 6.8). The greater the variations in density or distribution and sand supply, the greater the morphological variation (Carter, 1988; Carter and Wilson, 1988, 1990; Hesp, 1989). Figure 6.7 summarises some of the basic processes and forms.

**Figure 6.8** Incipient foredune (beach to immediate right) formed in short spinifex grass, Dark Point, NSW, Australia (P A Hesp).

On stable shores, incipient foredunes have an episodic life, tending to be eroded or completely removed by severe storm events, commonly scarped by minor to moderate erosion, and growing to the high tide swash limit during accretionary periods. In southern Australia a well developed incipient foredune may reach 1 to 2 m in height and 50 m in width after a period of 10 years (Hesp, 1983).

### 6.9.1.2 *Established foredunes*

*Established foredunes* develop from incipient foredunes and are distinguished by the growth of intermediate, often woody plant species, and commonly by their greater morphological complexity, height and width.

The morphological development and evolution of established foredunes depend on a number of factors including sand supply, the degree of vegetation cover, plant species present (a function of climate and biogeographical region), the rate of aeolian sand accretion and erosion, the frequency and magnitude of wave and wind forces, the occurrence and magnitude of storm erosion, dune scarping and overwash processes, the medium to long term beach or barrier state (stable, accreting or eroding), and increasingly, the extent of human interference and use.

Foredunes range from very low dunes 1 m or so in height on some barrier islands dominated by overwash and in areas of limited sediment supply, up to 30–35 m in height in some instances (Short and Hesp, 1982). Dunes over this height are rare and tend to occur on erosional coasts where human interference has acted to somewhat artificially induce greater foredune heights by, for example, maintaining absolute lee slope stability (e.g. parts of the Dutch coast). Foredune height and volume are related to beach surf zone type, with larger foredunes occurring on dissipative beaches and the smallest foredunes occurring on reflective beaches, all other factors being equal (e.g. equal potential sediment supply) (Short and Hesp, 1982).

6.9.1.2.1 *Foredune types.* Hesp (1982, 1988b) classified foredunes into five morpho-ecological types. Carter (1988) used the original type descriptions in Short and Hesp (1982) to create a similar, somewhat broader classification. Arens (1994) and Arens and Wiersma (1994) classified foredunes according to their state of accretion, stability or erosion.

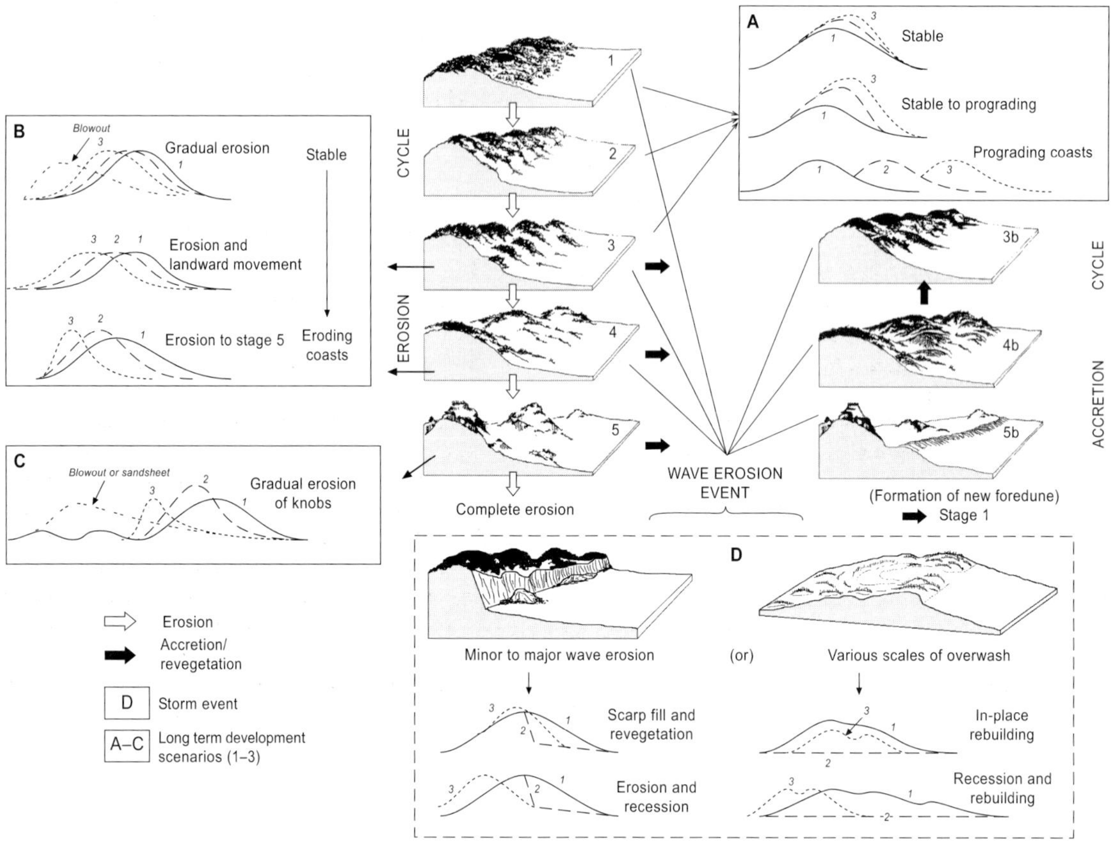


**Figure 6.9** A model of established foredune morphology, dynamics and evolutionary trends on stable, prograding and eroding coasts. A foredune may remain in a particular stage, or it may evolve by erosion, accretion or vegetation processes to another stage. The long term evolution will be strongly influenced by beach surf zone state, sediment supply and coastal trend (prograding, stable or eroding) (based on Carter, 1988; Hesp, 1988a; Arens, 1994).

The former classifications of Hesp (1988b) and Carter (1988) comprise elements of Arens (1994) types. Figure 6.9 is an attempt to meld the classifications. The Hesp foredune types (1 to 5) indicate both the type a foredune may stay in for most of its existence, and the types a foredune may progress through (arrows) in an evolutionary sequence as aeolian erosion and reduction in vegetation cover take place. This evolutionary progression may be reversed (5 to 1) if certain conditions, such as revegetation and stabilisation, and reduced aeolian erosion (due, for example, to changes in climatic conditions) take place. It is unlikely that a type 5 foredune would ever eventually revert to a type 1 or 2 foredune, as it is difficult to envisage a situation where all topographic lows and erosion hollows are filled in by sand accretion. However, it is reasonable to expect that a type 5 foredune might at least revert to a type 4 foredune if the right conditions were met. Figure 6.9 shows that a type 5 foredune could continue to erode while a new incipient foredune forms to seaward, and eventually becomes a type 1 foredune.

6.9.1.2.2 *Short term changes.* At any time, mild change (e.g. minor wave scarping) or catastrophic change may be introduced by moderate to severe storm events. Such events shift the foredune to a more potentially erosional form (e.g. major wave-formed scarps, slumping and dieback of vegetation), or to partial or complete removal by overwash processes. The subsequent development of the foredune (and its type) depends on the degree of revegetation and re-establishment (Figure 6.9).

6.9.1.2.3 *Medium to long term changes.* Figure 6.9 also provides a general model of the medium to long term evolution of a coastal region dominated by foredunes experiencing beach stability (no long term erosion or accretion/progradation), erosion, or accretion/progradation. It attempts to synthesise partly the findings of Hosier and Cleary (1977), Cleary and Hosier (1979), Godfrey *et al.* (1979), Hesp (1988b), Psuty (1988, 1989, 1993), McCann and Byrne (1989), Ritchie and Penland (1990), Saunders and Davidson-Arnott (1991), Arens (1994), Arens and Wiersma (1994), Ruz and Allard (1994), and others. The coastal, regional evolutionary paths (based principally on sediment budgets, but also on sea level rise, fall or stability), are indicated for each foredune type where that morpho-ecological type is *maintained*, at least until the new foredune is formed to seaward of it, or it is removed by landward erosion. With decreasing vegetation cover (types 2 → 5), aeolian sand may be transported to higher parts of the dunes and over the leeward slopes giving the range of depositional sites indicated in Figure 6.9B and C. Type 1, 2 and 3 foredunes may continue to slowly build or new foredunes may form to seawards as beach progradation take place, resulting in a series of foredune ridges (Figure 6.9A). The type 5 foredune, being highly erosional, may erode landwards as blowouts or sandsheets develop. In the case of large scale wave erosion and scarping, or overwash, foredunes of any type may lose a significant quantity of sediment and then gradually retreat as a steep, scarped foredune (Figure 6.9D). Thus, a relatively wide range of medium to long term responses is possible for a relatively small range of initial foredune types.

### *6.9.1.3 Plant succession and species richness of foredunes*

As discussed in Section 6.6, salt spray can be a primary determinant in plant species richness and zonation. Reflective beaches (low waves, narrow surf zones, few breaking waves) tend to have low salt aerosol levels, while dissipative beaches (high waves, many waves, wide surf zones) have high levels (Hesp, 1988a). The impact of these varying levels of salt aerosols operating over hundreds of years on beaches and dunes, in conjunction with varying levels of sand transport rates, is to produce marked variations in plant successional trends and richness on foredunes. There is a tendency to find a higher species richness and well defined laterally compressed zonation on reflective beach foredunes (outside the tropics) while the reverse is true for dissipative

**Figure 6.10** Type 1 established foredune (laterally continuous, stable, well vegetated) on a low energy dissipative beach, Kingston, S. Australia (P. A. Hesp).

beaches. Within the tropics species richness on the backshore–foredune environment is fairly high as non-dune species grow in areas quite close to the beach due to the low sand transport, low aerosol levels, low wind strengths and limited dune development (Hesp, 1991).

### 6.9.2 Foredune plains

Foredunes may become isolated from accretion and erosion processes by the seaward development of a new incipient foredune which itself may evolve into an established foredune. The original foredune then becomes a relict foredune as it is largely or wholly removed from a foremost beach position.

Systematic beach progradation over time frames of tens to thousands of years may lead to the development of wide *foredune plains* (Figure 6.11). Progressive soil development and plant community succession are common, particularly in regions with a temperate climate and siliceous sediment (Hesp, 1991). Examples may be found around the world including South Africa (Weisser and Backer, 1983), Ireland (Carter and Wilson, 1990), Australia and New Zealand (Hesp, 1984a; Shepherd, 1990), England (Salisbury, 1952; Ranwell, 1972), European coasts (e.g. Arens, 1994), Russia (e.g. Zenkovich, 1967), the USA south and east coasts (e.g. Otvos, 1995) and Brazil (Suhayda *et al.*, 1977). See Chapter 14 for a further discussion of barrier types.

As some foredune plains resemble beach ridge plains in plan form and morphology, and vice versa, it is clearly important to establish the genesis of the landform by stratigraphic, stratification and process studies prior to ascribing an origin or term to the feature.

## 6.10 EROSIONAL AND TRANSGRESSIVE DUNES

There are a variety of dune forms that may develop on the backshore (and beyond) that are either purely erosional in origin (e.g. blowouts, parabolic dunes) and/or transgressive in the sense that they transgress a prior landform or landform assemblage (e.g. parabolic dunes and transgressive dunefields).

### 6.10.1 Blowouts

*Blowouts* are commonly found in coastal dune environments. Most authors define a *blowout* as a saucer-, cup-

**Figure 6.11** A 3–5 km wide Holocene foredune ridge plain, Warnboro Sound, Western Australia (S. Chape).

or trough-shaped depression or hollow formed by wind erosion on a pre-existing sand deposit. The adjoining accumulation of sand, the depositional lobe, derived from the depression and possibly other sources, is normally considered part of the blowout (Glenn, 1979; Carter *et al.*, 1990).

Blowout morphology may be highly variable. Although there exists a wide range of types in aeolian environments, it seems that many blowouts can be classified into one of two types defined by Cooper (1958, 1967) (Figure 6.12). *Saucer blowouts* are semi-circular or saucer shaped and often appear as shallow dishes. *Trough blowouts* are generally more elongate, with deeper deflation floors and basins, and with steeper, longer erosional lateral walls or slopes.

However, in nature there is a large degree of spatial and temporal variability in blowout morphologies (Figure 6.13). The initial shape, size and location of blowouts and their subsequent development may depend on several factors. For example, Smith (1960) notes that blowouts initiated on the broad crests of foredunes are generally shallow saucer types, while those initiated on steep stoss faces are generally elongate trough types. Observations in south-east Australia

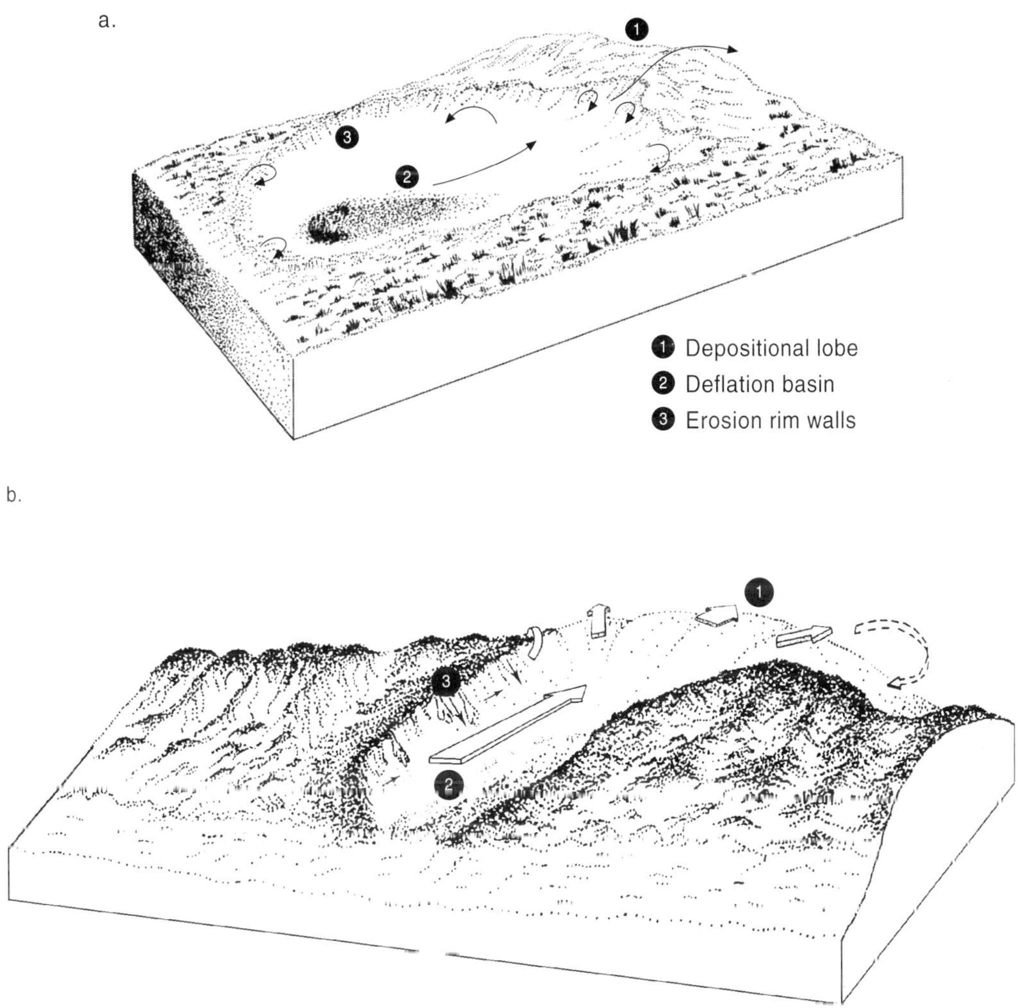


**Figure 6.12** Saucer (a) and trough (b) blowouts. The typical morphologies and flow conditions are indicated for each type (adapted from Hesp and Hyde, 1996).

**Figure 6.13** Multiple blowouts of varying sizes and morphologies in Warnboro Sound, Western Australia (S Chape).

confirm this view (Carter *et al.*, 1990). Once initiated, the subsequent morphologic development may depend on:

- the size of the initial constriction (Borichansky and Mikhailov, 1966);
- the height and width of the dune in which the blowout is developing;
- the degree and species type of vegetation cover (Melton, 1940; Esler, 1970; de Castro Lopo, 1979; Rutin, 1983);
- the magnitude of regional winds (Jennings, 1957; Cooper, 1958; Marta, 1958; Davies, 1980); and
- the degree of exposure to winds from various directions (Hogbom, 1923; Jennings, 1957).

The potentially large variation in one, or a combination, of these factors accounts for the large variety of blowout morphologies described in the literature and observed in the field.

*6.10.1.1 Evolution*

Blowouts may be initiated in a variety of ways.

1. Wave erosion and subsequent airflow acceleration and deflation (Andrews, 1916; Laing, 1954; Ritchie, 1972; Hesp, 1982; Psuty, 1989; Carter, 1990), and overwash development and subsequent aeolian deflation of the fan (Leatherman, 1976, 1979a,b; Godfrey *et al.*, 1979; Ritchie and Penland, 1990).
2. Topographic acceleration of air flow. In the cases cited above topographic acceleration of wind flow over scarps and cliffs and through arcuate bowl-shaped topography is an integral process in the development of the blowouts. Blowouts, however, may be initiated purely by topographic acceleration of flow, e.g. through dune topographic lows (Esler, 1970; Ranwell, 1972; Goldsmith, 1978; Carter, 1988). This may occur in conditions where winds blow onshore or offshore (e.g. Gares and Nordstrom, 1995).
3. Climate change and drought. Blowouts may be initiated where the vegetation cover is weakened, reduced or dies due to a prolonged dry or arid period (Smith, 1964; Jelgersma and van Regteren Altena, 1969; Ahlbrandt *et al.*, 1983; Pluis, 1992; Thom *et al.*, 1994).
4. Vegetation variation or change (Cholnoky, 1902; Onno, 1933; van Dieren, 1934; Melton, 1940; Jennings, 1957; Guilcher, 1958; Steers, 1964; King, 1972; Ritchie, 1972; Bird, 1976; Quinn, 1977; Martini, 1981; Hesp, 1982; Tinley, 1985; Jungerius and van der

Meulen, 1988; van der Muelen and Jungerius, 1989; Carter *et al.*, 1990; Thom *et al.*, 1992).

5. Water erosion (Thompson and Bowman, 1984; Jungerius and van der Meulen, 1988).
6. High velocity wind erosion (Bird, 1974; Hesp and Hyde, 1996).
7. Sand inundation and burial (Marta, 1958).
8. Human activities such as pedestrian track creation (Parker, 1975; Quinn, 1977; Mather and Ritchie, 1978; Carter, 1988), forest felling (Jelgersma and van Retcren Altena, 1969), offroad four-wheel-drive activity (Brodhead and Godfrey, 1977; Godfrey *et al.*, 1978; van der Merwe, 1988), housing and resort development (Nordstrom and McCluskey, 1984), sand extraction and military training (Marston,1986), and fires (Tinley, 1985).

This list demonstrates that blowouts (and parabolic dunes) do not only develop on coasts which are eroding or in negative sediment supply situations as Psuty (1992) indicates. They commonly develop on coasts experiencing significant wind energy and with either a balanced (stable) sediment supply (e.g. parts of the southern Australian coast), or a positive beach and dune sediment supply (e.g. the west coast dunefields on lower North Island, New Zealand).

#### *6.10.1.2 Aerodynamics*

Wind flow and sand transport in blowouts have been examined by a number of authors including Landsberg and Riley (1943), Olson (1958), Harris (1974), Hails and Bennett (1981), Jungerius *et al.* (1981), Jungerius (1984), Lancaster (1986), Gares and Nordstrom (1987, 1995), Finnigan *et al.* (1989), Jungerius and van der Meulen (1989), Robertson-Rintoul (1990), Pluis (1992) and Hesp and Hyde (1996).

In general these studies found that in conditions where the wind approaches a trough blowout directly or parallel to the throat orientation, the flow within the blowout is highly turbulent, characterised by: (i) jets both in the deflation basin and along the erosional walls; (ii) relative flow acceleration and expansion up the depositional lobe; (iii) jets over the depositional lobe crest; (iv) downwind flow separation on the leeward side of the lobe; and (v) flow separation and the formation of corkscrew vortices over the crests of the erosional walls. Saucer blowouts are also characterised by turbulent flow, although the net flow direction appears to be directed less downwind than in the case of trough blowouts. There is a greater degree of flow separation over the upwind rim for any given wind direction. Maximum erosion and transport occur up the deflation basin and onto the depositional lobe. In trough blowouts, the deflation basin usually erodes down to a maximum base level; this occurs to a far lesser extent in saucer blowouts.

### 6.10.2 Parabolic dunes

*Parabolic dunes* are U-shaped or upsiloidal dunes characterised by elongate, trailing ridges which terminate downwind in U-shaped ridges, sand sheets or dunefields. Deflation basins and slacks occupy the area between the trailing ridges. Deflation basins tend to continue to erode until a base level is reached such as the seasonally lowest water table level, a calcrete (or other cemented/indurated) layer, or an armoured surface such as a pebble, shell, pumice or artefact surface (Carter, 1976; Hesp and Thom, 1990).

#### *6.10.2.1 Evolution*

Parabolic dunes typically evolve from blowouts, and they may therefore form under any of the conditions listed above (blowout evolution). In many cases, the blowout depositional lobe continues to advance downwind regardless of what is happening in the rest of the blowout. For example, it is common for foredunes to reform across the blowout throat or entrance area, and for the deflation basin to revegetate. Trailing ridges develop due to deposition of sand in marginal vegetation and the subsequent trapping of that sand while the inside portion of the ridge is eroded (Figure 6.14). Although it is difficult to distinguish accurately the difference between a blowout and a parabolic dune, parabolics may usually be differentiated from blowouts by the existence of trailing ridges.

#### *6.10.2.2 Long walled and imbricate types*

Two principal sub-types are long walled types (e.g. Pye, 1982; Tinley, 1985), and imbricate types (Hesp *et al.*, 1989) (Figure 6.15). *Long walled parabolic dunes* display long trailing ridges and extensive deflation basins. The trailing ridges may range from hundreds of metres to several kilometres long. They are particularly well developed on relatively flat terrain, in regions of low heath or shrubland, high sand supply and strong winds, e.g. the Manawatu, New Zealand dunefields (Cowie, 1963; Esler, 1970), the central and north-west coast of Western Australia (Hesp and Chape, 1984; Hesp and Pelham, 1984), the north Queensland coast (Pye, 1982, 1983), the Callender area, California, USA (Cooper, 1967). *Imbri-*

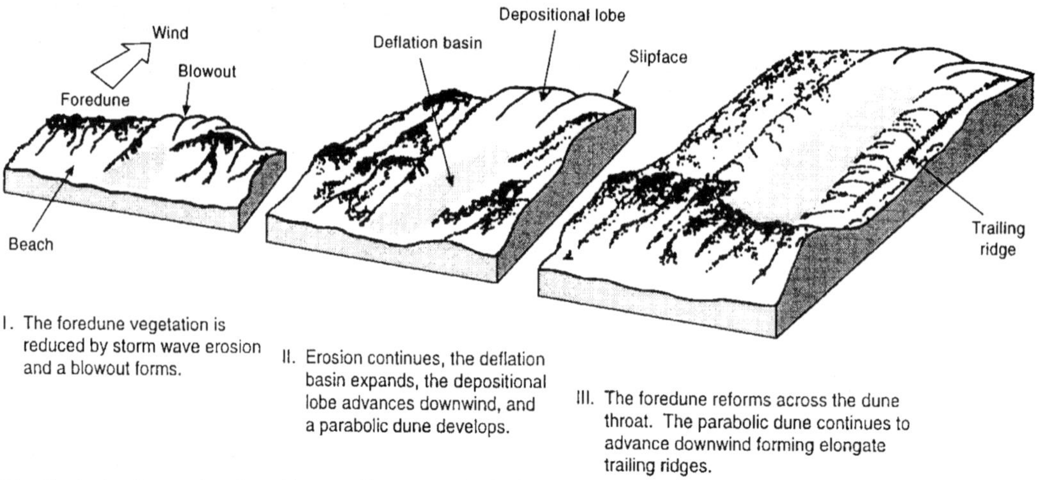


**Figure 6.14** Evolution from a foredune blowout (I) to a parabolic dune with trailing ridges (III) (source: P A Hesp).

*cate or nested parabolic dunes* display a squat, shorter form, often with more semi-circular deflation basins. Multiple development results in the dunes overlapping each other in an imbricate fashion. They commonly develop in wetter areas, on hummocky or relatively steeper terrain, and/or in dense, tall vegetation where the rate of advance is low and/or migration is impeded, e.g. the Monterey area, California (Cooper, 1967), southern Queensland, Australia (Coaldrake, 1962; Thompson, 1983), Britain and Denmark (Landsberg, 1956), Padre Island, Texas, USA (Weise and White, 1991). In some cases complex forms may develop (Filion and Morisset, 1983; Pye and Tsoar, 1990; Robertson-Rintoul and Ritchie, 1990).

### 6.10.3 Transgressive dunefields

*Transgressive dunefields* are aeolian sand deposits formed by the downwind or alongshore movement of sand over vegetated to semi-vegetated terrain. Such dunefields may range from quite small (hundreds of metres in alongshore and landward extent) to draa size fields (many kilometres in extent). They may be completely vegetated (post-formation), partially vegetated or largely unvegetated (Hesp and Thom, 1990). They occur on many of the world's coasts but are best developed on west and south coasts in high wind and wave energy environments, e.g. Coos Bay, Oregon, Californian coast, South African south coast, and more particularly so where these factors combine with a moderate to high sand supply and/or semi-arid to arid conditions and limited pioneer plant growth, e.g. southern and western Australian coasts, some Mediterranean coasts, Peru, Namibia, north-west New Zealand coast (e.g. Brothers, 1954; Bigarella, 1975; Hesp and Chape, 1984; Shepherd, 1987; Illenberger and Rust, 1988; Short, 1988b) (Figure 6.16).

#### 6.10.3.1 *Origin*

Transgressive dunefields develop for a variety of reasons.

1. As a response to rising sea level and/or climatic change, particularly in the period 10 000 to 7000 years BP (e.g. Pye and Bowman, 1984, Hesp, 1993; Shulmeister *et al.*, 1993; Young *et al.*, 1993). Some transgressive dunefields (and parabolic dunefields) have also developed during and at the closing stages of the Holocene Marine Transgression in high wave energy window locations, e.g. Cape York, Queensland (Figure 6.17; Pye, 1983; Pye and Tsoar, 1990) and central west Java. The 'windows' are present because Holocene coral reefs have not kept up with sea level rise and wave energy is higher at these coasts for some time prior to maximum vertical reef growth (e.g. Hopley, 1983).
2. In regions of high alongshore and onshore sediment supply, and in sediment traps often in regions of high wind and wave energy (e.g. Illenberger and Rust, 1988; Fryberger *et al.*, 1990).
3. On coasts experiencing erosion, both natural and human induced (e.g. Enright *et al.*, 1988; Bressolier *et al.*, 1990; Pye, 1990).
4. As continental shelves were exposed and/or climate

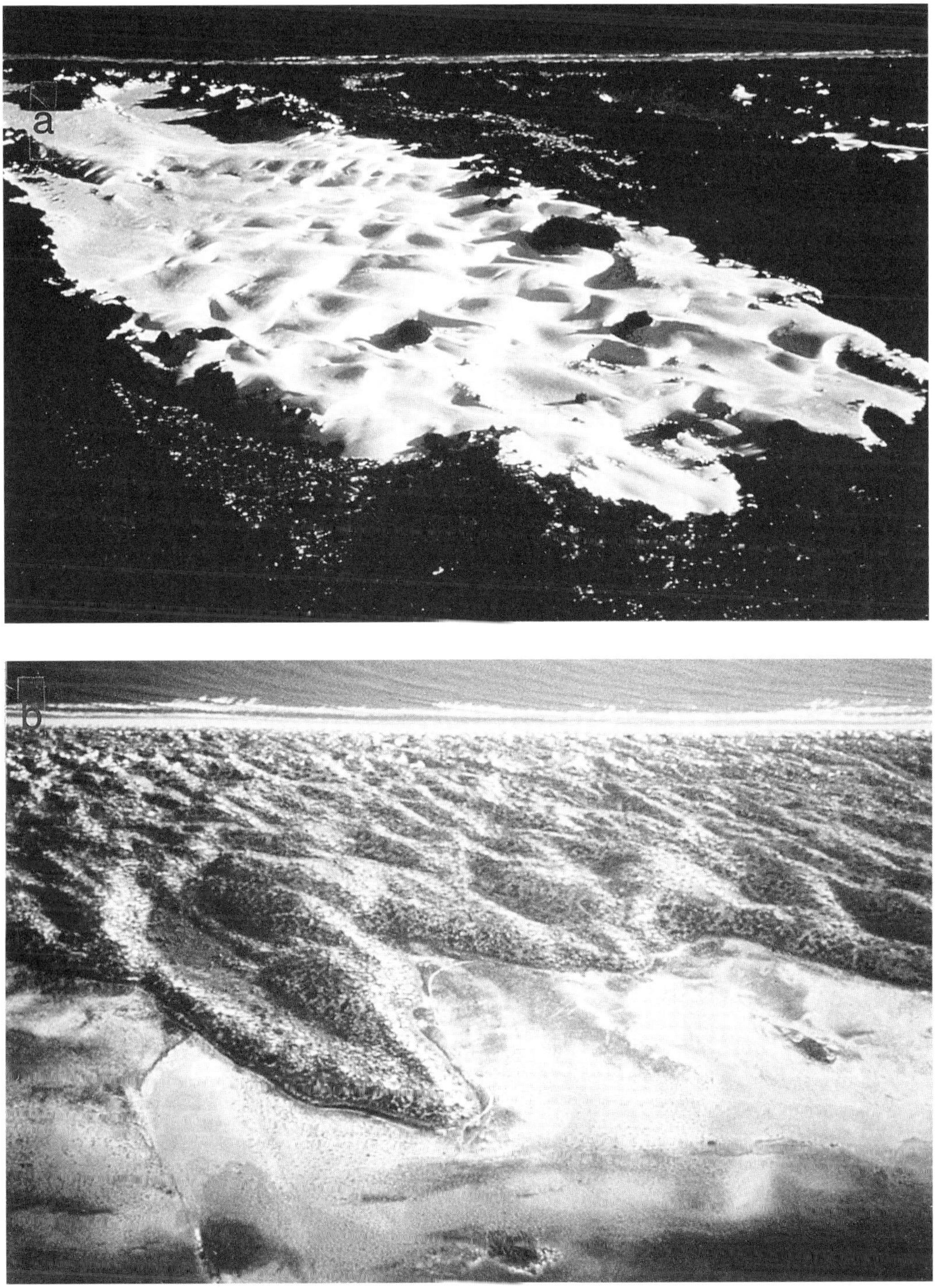


**Figure 6.15** (a) Long walled parabolic dunes near Dongara, Western Australia (S Chape); (b) imbricate nested parabolic dunes near Lake Gregory, Western Australia (S Chape).

a

b

**Figure 6.16** (a) Small developing transgressive dunefield, south of Perth, Western Australia (S Chape); (b) large, complex transgressive dunefield, Alexandria dunefield, South Africa (W Illenberger).

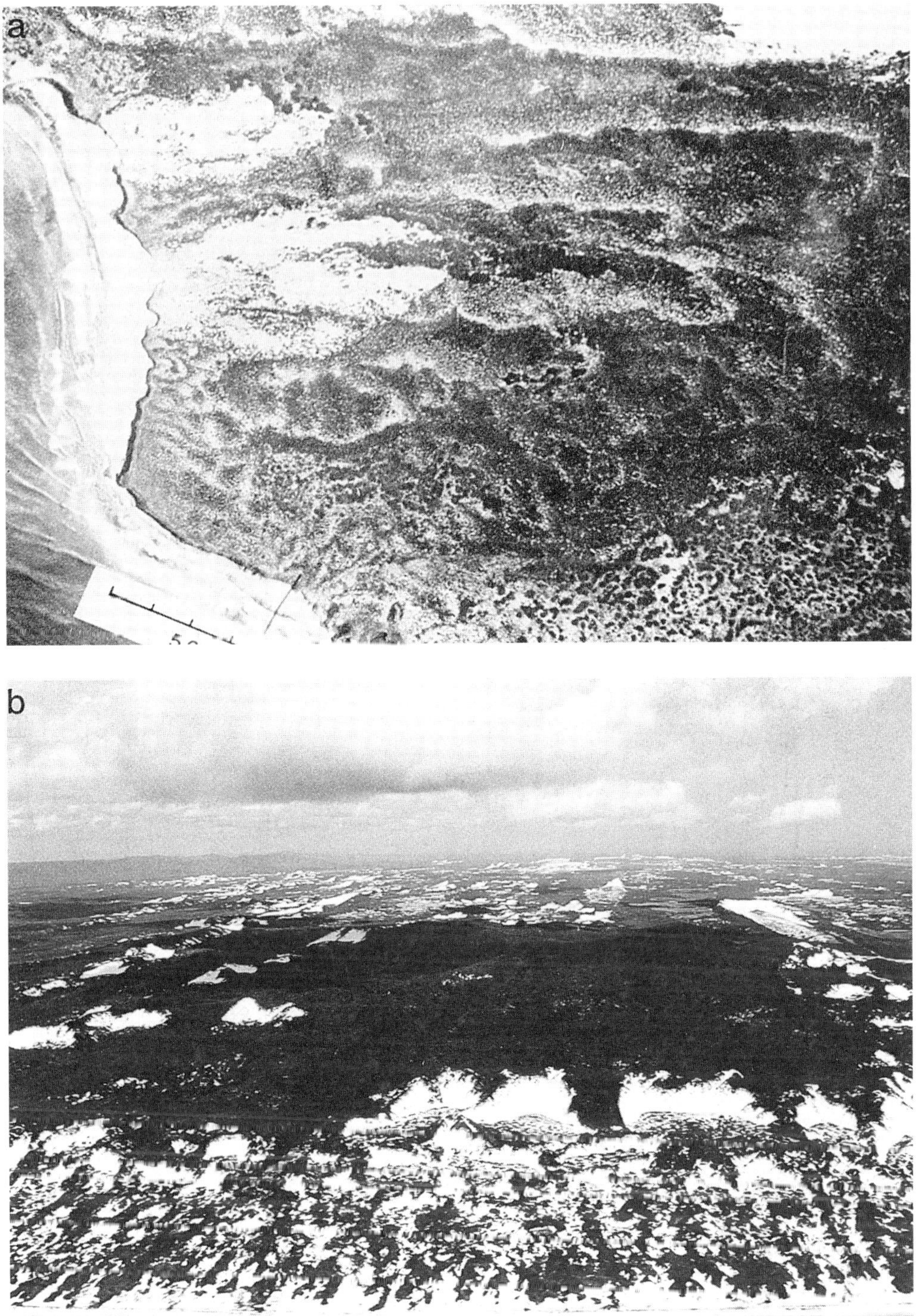


**Figure 6.17** (a) Nested Pleistocene–Holocene transgressive dune fields, Wanna Well, Eyre Pensisula, South Australia; and (b) scores of parabolics and long walled parabolics, Cape York, Queensland (beach to extreme bottom of photo) (A D Short).

changed during the Last Glacial (e.g. Lees *et al.*, 1990, 1995; Bryant *et al.*, 1994; Thom *et al.*, 1994).

5. As a response to periods of regional sea level fall during the mid- to late Holocene (e.g. Christiansen *et al.*, 1990).

In arid and desert regions, coastal transgressive dunefields are virtually indistinguishable from marginal continental 'desert' dunefields, and often merge imperceptibly into the deserts. Examples include the Namib Desert (Ward, 1987; Lancaster, 1989), sections of the Peruvian coast (Finkel, 1959), Chile (Araya-Vergara, 1986), the Baja California, Mexico region (Inman *et al.*, 1966; Fryberger *et al.*, 1990) and various north-western African coastal states (Fryberger and Dean, 1979; McKee, 1979; Seely, 1994).

#### *6.10.3.2 Morphology*

Active transgressive dunefields are usually characterised by a small to extensive, laterally continuous alongshore deflation basin or series of slacks on the seaward side, a small to extensive, mobile to partially vegetated sand sheet or dunefield, and a long walled, commonly sinuous slipface or precipitation ridge on the landward side (Cooper, 1958, 1967; Davidson-Arnott and Pyskir, 1988; Hesp *et al.*, 1989; Hesp and Thom, 1990; Thom *et al.*, 1992). In toto they may take tabular forms, buttress forms, or they may be headland bypass dunefields or climbing and falling dunefields (e.g. Hellstrom and Lubke, 1993).

#### *6.10.3.3 Dune types*

The surfaces of active transgressive dunefields are commonly covered with a variety of dune types ranging from simple transverse dunes and barchans, to barchanoidal and sinuous transverse and oblique dunes, to complex aklé forms (Cooper, 1958, 1967; Rosengren, 1981; Hunter *et al.*, 1983; Tinley, 1985; Carson and McLean, 1986; Orme and Tchakerian, 1986; Hesp *et al.*, 1989; Borowka, 1990; Orme, 1990; Kocurek *et al.*, 1992; Burkinshaw *et al.*, 1993). Some simple transverse and oblique dunes, domes and barchans may be ephemeral or seasonal forms; many are permanent and migrate downwind and landwards at various rates. Star dunes are very rare in coastal dunefields, one of the few examples being the Bilbunya dunes west of Cocklebiddy in Western Australia (Figure 6.18).

#### *6.10.3.4 Stabilisation*

It is common to find fully or partially vegetated transgressive dunefields, and it is not uncommon for a dunefield to be characterised by an active mobile portion and vegetated portion. As the dunefields begin to revegetate or to be colonised by vegetation, a variety of depositional dune forms may occur depending on the subenvironment (e.g. McLachlan *et al.*, 1987). For example, Hesp and Thom (1990) record a temporal sequence of plant colonisation of a deflation plain where isolated hummocks initially appear in the early stages, and eventually give way to a plain covered by hummocky dunes and flats. In some spectacular cases, the whole dunefield may be largely vegetated intact, and even large transverse dunes may be stabilised with little variation in form (e.g. Thom *et al.*, 1992). As transgressive dunefields revegetate, the downwind edge may break up into parabolic forms as individual lobes advance into the surrounding vegetation. Erosional landforms may also be widespread in some revegetated dunefields.

#### *6.10.3.5 Dunefield phases*

Transgressive dunefields may be formed either semicontinuously or in phases or episodically. Thus, it is common to find sequences of deflation basin–dunefield–precipitation ridges/slipfaces where large scale development of dunefields has taken place over thousands of years, e.g. Brazilian dunefields, South African dunefields (Illenberger and Rust, 1988), Australian dunefields (Thom, 1984; Hesp and Thom, 1990; Thom *et al.*, 1992), New Zealand dunefields (Brothers, 1954; Muckersie and Shepherd, 1994), west coast USA dunefields (Cooper, 1958, 1967; Orme, 1990). The dunefields may comprise either one major coherent phase, or several discrete or superimposed, active large scale phases. These may develop as phases, but may not, and at present our knowledge of the forcing factors driving periods or phases of dune development is very limited.

### 6.10.4 Machair and other dune terrains

*Machair* refers to a vegetated to semi-vegetated plain of low ridges and hummocks, which may also include peatlands, deflation plains and basins and blowout topography (Angus and Elliott, 1992). Some machair could clearly be classified as blowouts, parabolic dunes or transgressive dunefields. In other cases, the machair landscape is a somewhat confused, low hummocky topography, much influenced by hundreds of years of human and animal activity (Ranwell, 1980; Randall, 1983).

A variety of shadow dunes, hummocks and irregular 'rim dunes' (around the margins of washover fans) and amorphous dune terrain also develop on barrier islands

**Figure 6.18** A 100 m high star dune, Bilbunya, south-east Western Australia (A D Short).

undergoing frequent overwash and retreat (e.g. Hosier and Cleary, 1977; Leatherman, 1979a; Ritchie and Penland, 1990).

## 6.11 SUMMARY

The backshore links the wave-dominated foreshore and surf zone to the landward wind-dominated dunefields, cliffs or other landforms. It acts both as a storage and transfer zone for marine sediments. Sand can be transferred beyond the backshore by both wave accretion and erosion and, while in storage and during beach accretion, can be transported landward by the wind (aeolian transport). Rates and scales of wave and wind transfer depend on sediment supply, beach types, morphology and mobility, wave and wind energy, and various factors affecting transport rates such as moisture content, grain size and seasonal climates. Storage usually takes place via coastal progradation or stability, and occasionally as storm deposition.

A variety of coastal landforms may be found at the edge of and beyond the backshore. Several are formed there as a response primarily to wave-induced beach construction, namely, berms and storm ridges, cheniers, overwash terraces and fans and beach-foredune ridges. Aeolian processes dominate the formation and evolution of landforms such as incipient and established foredunes, blowouts, parabolic dunes and transgressive dunefields.

# PART C

# BEACH TYPES AND APPLICATIONS



In Part B the beach was divided into sub-systems with little attention to the total beach system, and no consideration of the role of tides or structural impacts. In Part C the role of the essential beach components (waves, tides and sediment) in contributing to beach morphodynamics is considered, first in Chapter 7 which deals with wave-dominated beaches, then in Chapter 8 which looks at the effects of tides on beach morphodynamics. Finally as many beaches are located between or adjacent to headlands and structures, Chapter 9 considers embayed and structurally controlled beaches.

# 7

# Wave-Dominated Beaches

*Andrew D. Short*

## 7.1 INTRODUCTION

All beaches are by definition dominated by waves and composed of sediments. Wave dominance ranges from total on straight, sandy tideless shores, decreasing as other processes, particularly tide, increasingly exert their influence on the shoreface topography. Likewise beach sediments can be dominated and ultimately replaced by rocks, reefs and structures. Beaches are therefore part of a highly variable natural system that can range from wave- to tide-dominated, and from sedimentary to rocky shores. This chapter is concerned solely with sandy beaches in micro-tidal environments, removed from the influence of tides, embayments and other parameters. The three beach zones of shoaling, surf and swash are assumed to be stationary, with minimal horizontal and vertical (tidal) movement. The chapter will first review the development of our knowledge of wave-dominated beach systems, followed by a description of the three wave-dominanted beach types – dissipative, intermediate and reflective. The impacts of tides are presented in Chapter 8 and topography (headlands and structures) in Chapter 9.

## 7.2 HISTORICAL PERSPECTIVE

*Beach type* refers to the morphodynamic character of a beach system, including its beach and surf zone morphology and associated wave shoaling, surf and swash characteristics. Early beach studies observed a wide range of beaches, but often had no satisfactory explanation of the factors controlling the range (e.g. King, 1972). This ultimately led to the desire to develop a more general and uniform classification of beach types, at the same time developing a greater understanding of beach morphodynamics.

### 7.2.1 Two-dimensional beach change

The first attempts to classify beaches and understand beach change were based on the results of sequential *beach profiles*, that recorded two-dimensional beach profiles and beach change. Research groups in the United Kingdom and on either side of the USA led the way in classification and prediction of beach change.

#### *7.2.1.1 Beach profile*

The simplest expression of a beach is its cross-sectional profile. The beach profile records the height, width, slope and volume of a beach, while sequential profiles record changes either through time, or alongshore. Early researchers identified the relationship between beach slope and grain size (BEB, 1933; Meyer, 1933; Bagnold, 1940; Bascom, 1951; Johnson, 1956). King (1972) expressed these results in a sensitivity graph which showed how beach slope decreases with decreasing grain size and increasing wave energy. King (1972) reported on numerous beach profiles from a variety of beaches in Britain, France and the USA as well as laboratory experiments. The beaches included those with a wide, low gradient and smooth profile which she attributed to large tide range and high waves (see Section 8.6.2); those with 'submarine' bars which were found in enclosed tideless seas exposed to steep waves; those with crescentic bars for which she had no satisfactory explanation; and those with ridges and runnels which occurred on some fetch-limited, high tide range beaches. King also reported on the temporal variation on many of these profiles, which depending on seasonal wave action could either be regular (seasonal) or erratic. Finally, she attributed beach cut and offshore bar migration to the occurrence of higher, steeper waves, and

*Handbook of Beach and Shoreface Morphodynamics*. Edited by A. D. Short.
© 1999 John Wiley & Sons Ltd.

beach accretion to lower, flatter waves. While King correctly linked certain profile types with salient environmental conditions, and beach change with dominant processes, there was no clear relationship between the seemingly wide range of beach profiles and the nature and scale of their response.

#### 7.2.1.2 *Sequential beach profiles (cut and fill)*

The first classification of beaches developed out of beach profile observations was initiated in 1936 at the Scripps Pier in southern California. Shepard and LaFonda (1940) observed a distinct seasonal pattern with cut during higher winter storm waves, which formed an offshore bar, and summer beach fill with a steep, barless beach, produced by lower swell waves (Figure 4.39). Whereas they accurately described seasonal beach conditions on southern Californian beaches, they did not show how the two profile forms were linked, nor how they might be applied outside of California. Unfortunately many people have adopted the distinctly seasonal California beach behaviour as a universal norm. As has been discussed in Section 2.6, beach behaviour is a function of wave climate, and as wave climates vary regionally and globally so too will beach behaviour. Consequently, the winter cut and summer fill behaviour is only applicable to beaches receiving distinct winter storms and summer calms, usually on northern hemisphere west coast locations.

In a review of North American beach research Johnson (1956) extended the classification of beaches to include steep, barless beaches produced by calm waves (low steepness, summer); moderate gradient beaches with an offshore bar, which can become discontinuous during low waves; and flat, multi-bar beaches on exposed high energy coasts (high steepness, winter waves). At the same time he reported on regional beach types, including 'Lake Michigan', 'Atlantic coast' and 'Pacific coast' beaches (US Army, 1954).

#### 7.2.1.3 *Sequential beach change*

In order to link 'winter and summer' or any disparate beach profile end members, observations of the transition from one type of beach profile to the other are required. The first study to report on sequential sub-aerial beach profile changes was presented by Sonu and van Beek (1971). Using 1341 sequential beach profiles collected at Nags Head, North Carolina, between 1963 and 1964 by R. Dolan (Russell, 1967a) they found that beach change was driven entirely by waves and consisted of an erosion and accretion sequence (Figure 7.1).

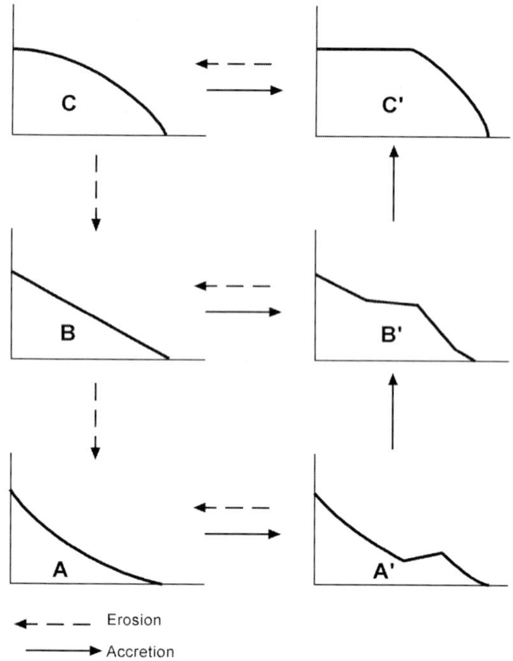


**Figure 7.1** Two-dimensional sequence of sub-aerial beach profile change (from Sonu and van Beek, 'Systematic beach changes on the Outer Banks, North Carolina', *Journal of Geology*, 1971, Vol. 79, pp. 416–425, by permission of the University of Chicago Press).

Sonu and James (1973) demonstrated that the cycles are Markovian with each step in the sequence being dependent on the previous beach profile as well as wave conditions. This sequential change was however limited to the sub-aerial beach, and neglected the role of bars.

### 7.2.2 Three-dimensional beach change

While two-dimensional beach change is a useful indicator of gross beach behaviour, it does not acknowledge the inherent three-dimensional behaviour of beaches and their surf zones. The development of a univeral beach change model would require the linking of two- and three-dimensional beach changes.

#### 7.2.2.1 *Rhythmic topography*

Researchers had been well aware of the three-dimensional nature of beaches, with Evans (1938, 1940) reporting on beach cusps and megacusps in the Great Lakes, and Shepard *et al.* (1941) monitoring rip currents, including their dynamics and relation to 'beach indentations'. The first paper to specifically address this problem was by Homma and Sonu (1962) who observed well developed rhythmic shoreline and bar topography in Japan. They noted that such topography required a gentle bed slope; that the rhythmic forms could be normal or skewed downdrift, depending on angle of wave approach; and the phase coupling of the bars and shoreline rhythms. They were however unable to propose a satisfactory explanation for the rhythms. Sonu (1972) took this further in his work on rip currents and associated rhythmic topography on the Florida coast, and in Sonu (1973) presented the first three-dimensional sequence of beach change (Figure 7.2). This model accounted for both spatial (longshore) and temporal changes in beach profile and plan. He tentatively attributed the crescentic bar formation to standing (edge)

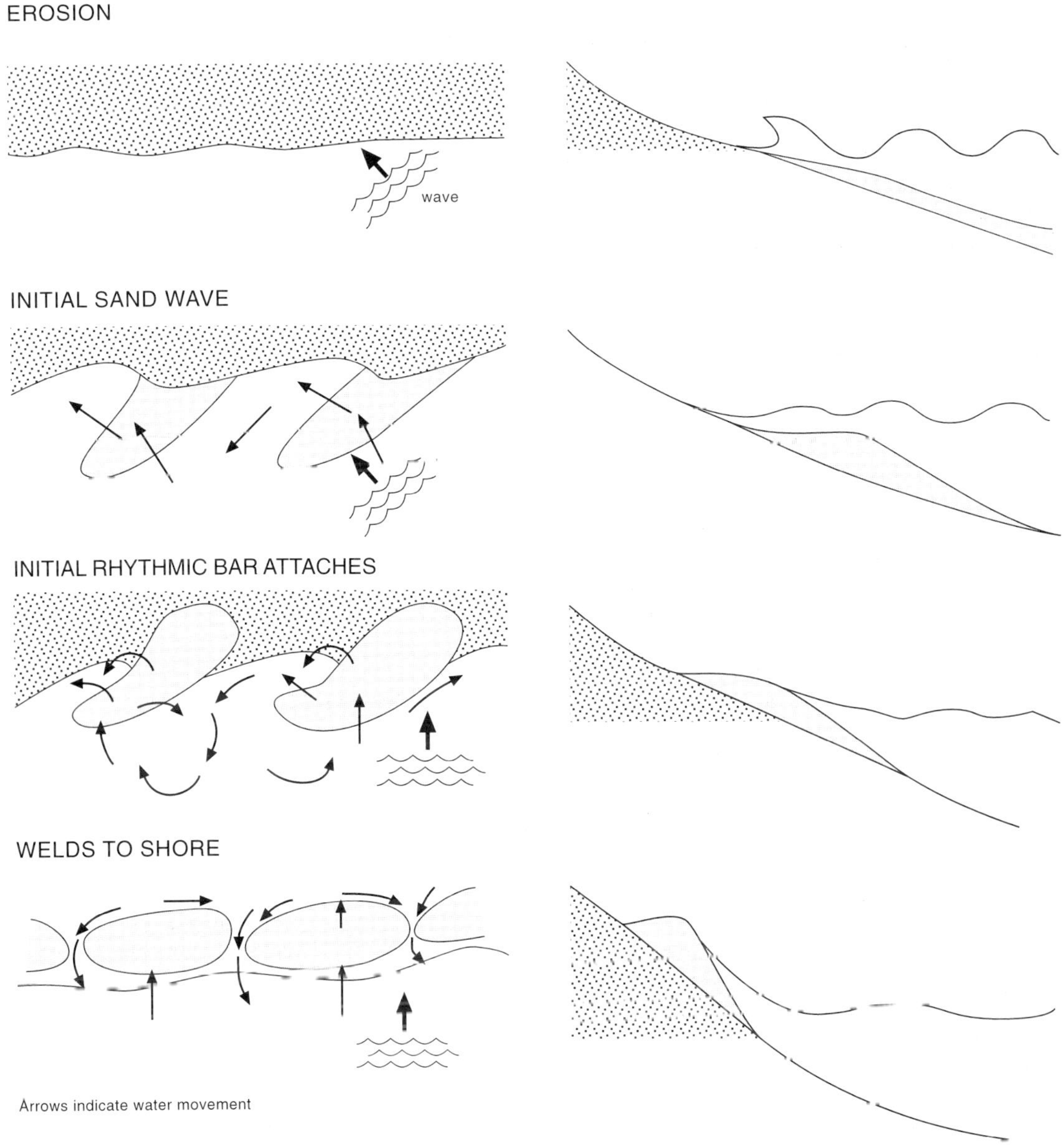


**Figure 7.2** Sequence of three-dimensional beach cycle involving the formation and shoreward migration of rhythmic bars (modified from: Sonu, 'Three-dimensional beach changes', *Journal of Geology*, 1973, Vol. 81, pp. 42–64, by permission of the University of Chicago Press).

waves, and to movement through the beach cycle from storm erosion followed by beach recovery.

### 7.2.2.2 *Surf zone dynamics*

Much early work on beaches and beach change was based on measurements of beach morphology. While waves were observed and assumed to form and drive beach change there was little data on wave and surf zone processes. The first scientific observations of the surf zone were made by the Scripps group. Shepard and LaFonda (1940) recorded waves, tides and beach change. These observations were continued through the 1940s with attempts to monitor surf zone current velocity and circulation patterns. These results and particularly the role of rips in the surf zone were reported by Shepard *et al.* (1941), Shepard and Inman (1950, 1951) and Inman and Quinn (1952). They found that while the seabed topography at Scripps did influence the occurrence and location of rips, rips were also observed on long straight beaches. In 1958 McKenzie made independent observation of rips on several Australian beaches, noting the recurrence of rip patterns on certain beaches and the increase in rip size with wave height.

While the general nature of the beach and surf zone, its circulation and change were becoming better understood, there was still no satisfactory explanation for the overall beach morphology and dynamics being observed. The solution to this problem commenced with Munk (1949) who first observed the periodic rise and fall of the shoreline at periods of 2 to 3 min, a phenomenon he called '*surf beat*'. Tucker (1950) found that the rise in water level or set-up was associated with groups of smaller waves and the set-down with groups of larger waves, a phenomenon now known as *wave groupiness*. Longuet-Higgins and Stewart (1962, 1964) in attempting to explain Tucker's observations introduced the concept of *radiation stress*, that is, the excess momentum flux due to the presence of waves, which produces changes in mean sea level when waves encounter a sloping beach. This phenomenon is addresssed in Section 4.2.6.

Bowen (1969a) and Bowen and Inman (1969) applied Longuet-Higgins' and Stewart's results in wave tank experiments. They provided the first theoretical and laboratory verification of the relationship between (standing) edge waves and surf zone circulation, including rip currents, which was further elaborated on by Bowen and Inman (1971) in relation to crescentic bars (Figure 7.3) and Short (1975b) in relation to parallel longshore bars.

In Japan field experiments in rip-dominated surf zones enabled Sasaki and Horikawa (1975) and Sasaki *et al.* (1976) to identify the full range of surf zone dynamics labelled the *infragravity*, *instability* and *edge wave* domains.

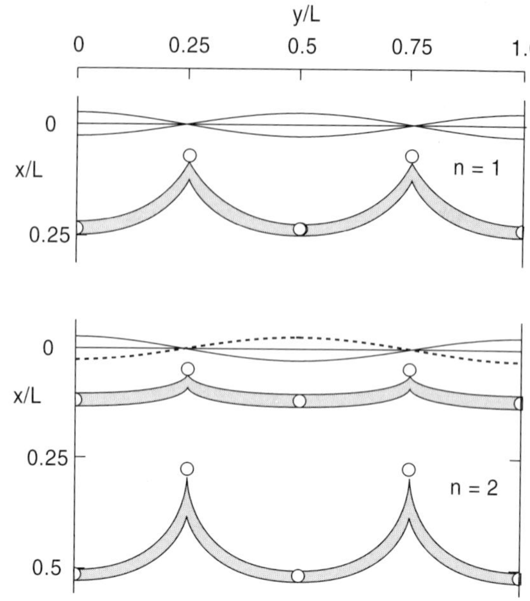


**Figure 7.3** Schematic illustration of the crescentic bars formed by edge waves of modes $n = 1$ and $n = 2$. The horizontal motion of the edge waves at the shoreline (upper lines) suggests a possible mechanism for forming cuspate features (from: Bowen, A J and Inman, D L, *Journal of Geophysical Research*, Vol. 76, pp. 8662–8670, 1971, copyright by the American Geophysical Union).

The results of Bowen and Inman (1971), Short (1975b) and Sasaki (1977) provided the basis on which to develop a more comprehensive explanation of beach behaviour, and in particular the coupling of surf zone process to the resultant morphology.

### 7.2.2.3 *Morphodynamic beach types*

Sonu's work showed that beaches do change sequentially, and Sasaki and colleagues showed that there is a sequential range of surf zone dynamics. Meanwhile Owens (1977) conducted detailed beach and surf zone experiments on either side of Magdalen Island in Gulf St Lawrence, and observed two dominant beach end members. The exposed west coast side had a seasonal storm-dominated double bar system, compared to more sheltered east coast beaches with a more variable morphology. He went on to compare these to the high energy 'west' and lower energy 'east' coast north American beach systems. However the two systems were still not linked.

It took Short (1978, 1979b) 18 months of daily observations of a very dynamic and changeable Narrabeen beach, Australia, to record the full range of micro-tidal beach types and present them in a sequential three-dimensional model. At the same time Wright *et al.* (1979), working on the same coast, had been undertaking field experiments on a select range of beach types aimed at characterising the morphodynamic signatures of each beach type. Wright and Short (1984) combined their results with additional field experiments to form the basis of the present beach type classification for wave-dominated micro-tidal beaches (see Section 7.3).

This model was reviewed by Sunamura (1988) who also incorporated the results of several other workers in Japan, North America and Israel. Finally, Lippman and Holman (1990) used 2 years of daily beach observations from the CERC Facility at Duck, North Carolina, to both verify the model, and with slight modification propose classification criteria. Interestingly, the Duck site is located just 25 km north of the Nags Head site, where Russell (1967a) started the research paradigm that would lead to the morphodynamic classification of beaches.

## 7.3 WAVE-DOMINATED BEACH TYPES

### 7.3.1 Parameterisation

Wave-dominated, micro-tidal beaches can be classified using three parameters – wave height ($H_b$), wave period ($T$) and grain size (as defined by sediment fall velocity, $W_s$). These were first combined by Gourlay (1968) into the *dimensionless fall velocity*

$$\Omega = H_b/W_sT \tag{7.1}$$

Gourlay found that when $\Omega < 1$ laboratory beaches tended to be steep and stable; however when $\Omega > 1$–2 they eroded with the sediment forming a terrace attached to the base of the beach. These same variables and their derivatives have since been used by Battjes (1974, 1988), Guza and Inman (1975), Sunamura (1984a,b), Wright and Short (1984), Balsillie (1985) and Short and Aagaard (1993) to develop a number of parameters for both describing and predicting beach behaviour.

Wright and Short (1984) adapted $\Omega$ for use on natural beaches. Based on both beach experiments from representative beach types, and numerous observations around the south-east Australian coast, they found that when $\Omega < 1$ beaches tended to be *reflective* (steep, barless), when $\Omega > 6$ they tended to be *dissipative* (flat, multi-barred), and in between ($\Omega = 2$–5) they were what they called *intermediate* between the two end members (one to two bars). Table 7.1 lists the three beach types together with the various parameters that have been developed to classify and quantify the extent of each type and their associated wave, current and morphological characteristics. Figure 7.4 illustrates the full range of micro-tidal beach types, and Figure 7.5 plots the sensitivity of each type to wave height, period and sand size. In the following sections each of the three beach types is described, together with their environmental conditions, level of stability and mode of erosion.

The role of the three parameters $H_b$, $T$ and $W_s$ (grain size) in influencing beach type is illustrated in Figure 7.5, which plots the sensitivity of $\Omega$ to each parameter. It is critical to understand that beach type and change are a function of all three parameters, with each contributing to change in $\Omega$ and thereby beach type. Increasing $H_b$ and decreasing $T$ and $W_s$ favours dissipative beaches, while decreasing $H_b$ and increasing $T$ and $W_s$ favours reflective beaches, with intermediate beaches lying in between.

### 7.3.2 Dissipative beaches

*Dissipative beaches* represent the high energy end of the beach spectrum, the beach type to which all beaches begin to head during periodic high waves and accompanying sub-aerial beach erosion. These beaches are characterised by wide, low gradient surf zones across which spilling breakers dissipate their energy, hence the name *dissipative* (Guza and Inman, 1975). While all beaches at some time move in this direction, most never reach the fully dissipative state. Of the beaches that do, many return to lower energy intermediate types once wave height drops. Because of the requirement of persistent high waves only a few, high energy beaches with fine sand are modally dissipative.

#### *7.3.2.1 Occurrence*

Dissipative beaches occur where $\Omega > 6$ and $\xi$ is between 20 and 200 (Table 7.1). They require fine sand (< 0.2 mm), relatively high waves (> 2.5 m), and preferably short wave periods (Figure 7.5). For this reason they are only found on open swell coasts ($T > 10$ s) in areas of fine to very fine sand with persistent high swell (> 2 m), as in storm wave and high energy west coast swell environments (Figure 1.2, Section 2.6). They have been reported on the coasts of southern Australia (Fig-

**Table 7.1** Wave-dominated, micro-tidal, beach types and their characterisation (modified from Sasaki and Horikawa, 1978; Short and Hesp, 1982; Carter 1988).

| | Reflective | Intermediate | Dissipative |
|---|---|---|---|
| *Dimensionless fall velocity*, $\Omega = H_b/TW_s$ (Wright and Short, 1984) | $< 1$ | 2–5 | $> 6$ |
| *Surf scaling index* $\xi = (a\sigma^2)/(g \tan^2 \beta)$ (Guza and Inman, 1975) | 0.1–2.5 | 2.5–20 | 20–200 |
| *Surf similarity index* $\Sigma = \tan \beta/(H_o/L_o)^{0.5}$ (Battjes, 1974) | $> 1$ | 1–0.23 | $< 0.23$ |
| *Beach stability (transport direction)* $K = H_b^2/gT_d^2$ (Sunamura, 1984a) | $< 3.5$ | 10–20 | $> 20$ |
| *Bar parameter* $B^* = (x_s/g \tan \beta T_i^2)$ (Short and Aagaard, 1993) | $< 20$ | 20–50 | $> 50$ |
| *Waves* | | | |
| *Breaker type* | Surging | Spilling--plunging | Spilling |
| Iribarren number $\Sigma_b = \tan \beta/(H_b/L_o)^{0.5}$ (Battjes, 1994) | $> 5$ | 5–0.64 | $< 0.64$ |
| Breaker equivalent (Balsillie, 1985) | Plunging–collapsing | Spilling–collapsing | Spilling |
| *Number of waves in surf zone* | 1 | 1–3 | $> 3$ |
| *Wave reflectivity* | High | Moderate | Low |
| *Relative wave energy* | Low | Medium | High |
| *Edge wave type* | Trapped (in swash) | Trapped (in trough) | Leaky |
| *Edge wave period* | Sub-harmonic ($T_e = 2T$) | $T_e = 2\text{–}4T$ | = wave group period $T_e \gg T$ |
| *Currents* | | | |
| *Plan* | On–offshore | Cellular | On–offshore |
| *Profile* | Uprush, backwash | Onshore over bars, seaward in rips | Surface wave bores and bed return flow |
| *Morphology* | | | |
| *Beach profile* | Steep | Rhythmic (megacusps) | Flat |
| *Bars* | No bar | Rhythmic | Multiple, parallel |
| *Slope* | $> 4°$ | 2–6° | $< 2°$ |
| *Cusps* | Common | On megacusp horns | Rare |
| *Megacusps* | Absent | Common | Rare |
| *Sediment transport* | | | |
| *Net longshore* | Low | Medium | High |
| *Onshore/offshore* | Low | Medium | High |
| *Dominant mode* | Bedload | Mixed | Suspended |
| *Texture* | Medium–coarse | Fine–medium | Fine |
| *Aeolian* | Low | Medium | High |
| *Mobility* (Short and Hesp, 1982) $y^b$ = *beach width* | | | |
| *Shoreline mobility*, $S_m = \sigma_{yb}$ | Low ($< 10$ m) | High ($> 10$ m) | Low ($< 10$ m) |
| *Backshore mobility*, $B_m = CV_{yb} = \sigma_{yb}/y_b$ | Low ($< 0.1$) | Moderate (0.1–0.5) | High ($> 0.5$) |

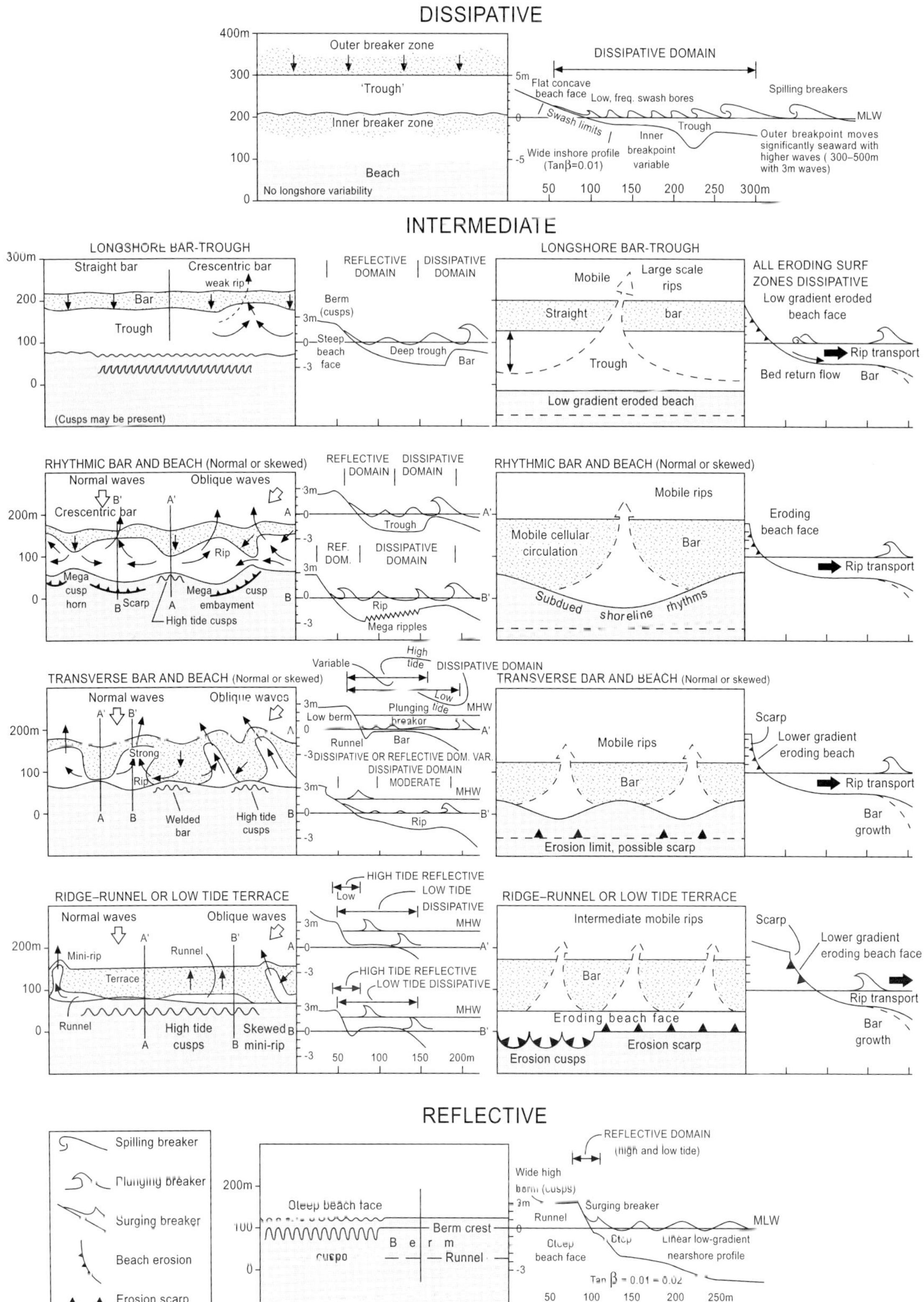


**Figure 7.4** Three-dimensional sequence of wave-dominated beach changes for accretionary (left) and erosional (right) wave conditions (based on Short, 1979b; Wright and Short, 1984; Sunamura, 1988; Lippmann and Holman 1990).

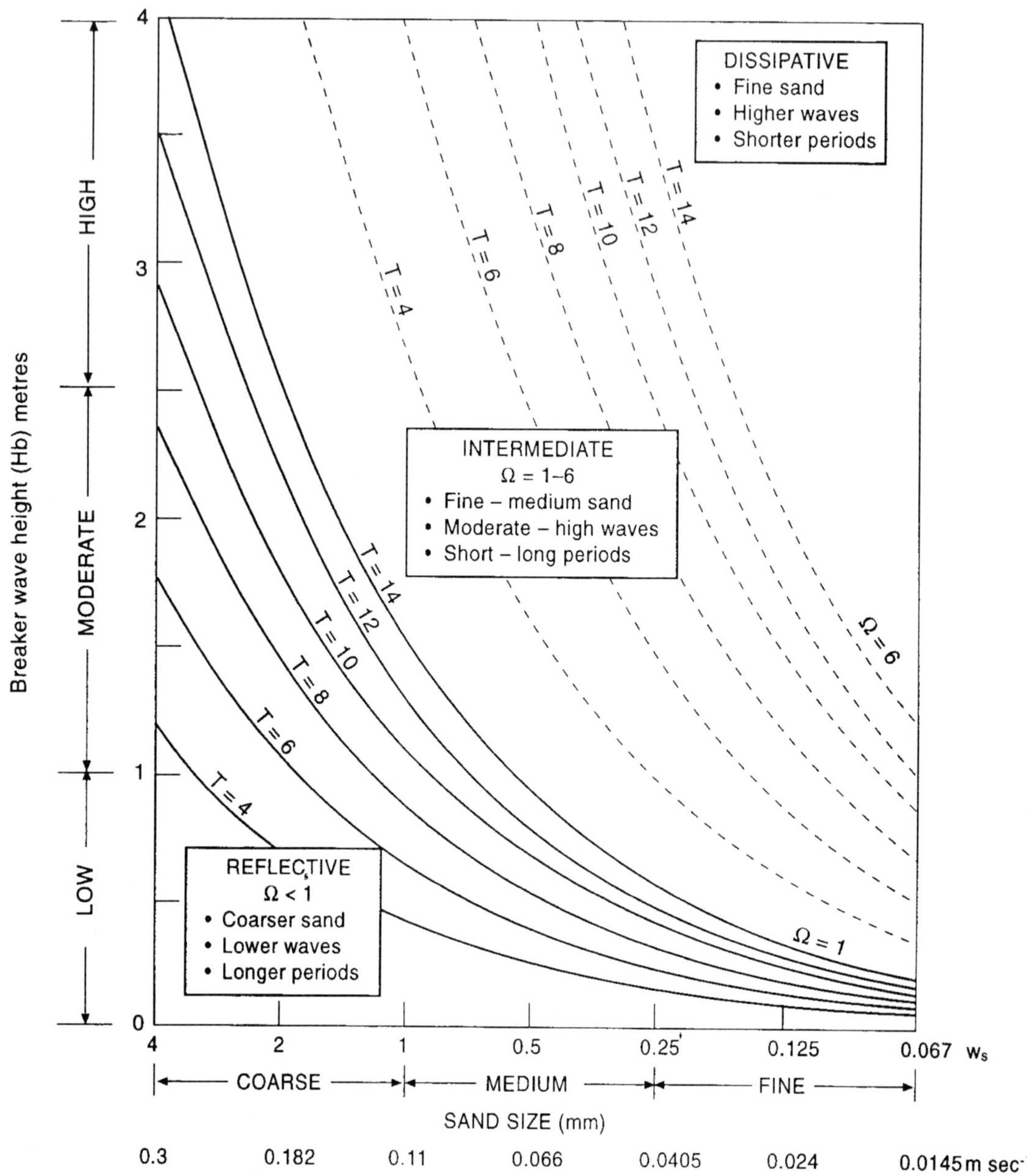


**Figure 7.5** Sensitivity plot of the contribution of wave height, sediment size and wave period to Ω and beach type. To use the chart determine the breaker height, period and grain size/fall velocity (phi or cm/s). Read off the wave height and grain size, then use the period to determine where the boundary of reflective/intermediate, or intermediate/dissipative beaches lies. Ω = 1 along solid *T* lines, and 6 along dashed *T* lines. Below the solid lines (Ω < 1) the beach is reflective; above the dashed lines (Ω > 6) the beach is dissipative; between the solid and dashed lines (Ω = 1–6) the beach is intermediate (modified from Short, 1986).

ure 7.6a) (Short and Hesp, 1982; Wright and Short, 1984; Short, 1996), South Africa (McLachlan *et al.*, 1993), west coast New Zealand (Figure 7.6b), and the coasts of Oregon (Fox and Davis, 1978), Brazil (Figure 7.6c; Suhayda *et al.*, 1975) and Chile (McLachlan *et al.*, 1993). They are however more common in storm-dominated seas and larger lakes and bays, where high, but short period ($T = 2$–$5$ s) storm waves act on fine sand beaches, such as parts of the Arctic Ocean (Short, 1975a,b), eastern Nicaragua (Owens and Roberts, 1978), Great Lakes (Greenwood and Osborne, 1991), Baltic (Aagaard, 1990), Mediterranean (Goldsmith *et al.*, 1982a,b), North Sea (Short, 1992), Gulf St Lawrence (Owens, 1977; Aagaard and Greenwood, 1995b) and Port Phillip Bay (Short, 1996). In between the storms the dissipative morphology usually stagnates as inactive outer bars, with only the inner bar remaining active usually in an intermediate form, under the intervening lower waves or calms (Aagaard, 1990).

#### 7.3.2.2 *Morphology*

Dissipative beaches are characterised by a wide, low gradient swash zone, that extends from the foot of the dune to the low tide zone, and may be more than 100 m wide at low tide, with the fine compacted sand firm enough to drive a car on. The swash zone is usually flat and featureless, apart from occasional widely spaced cusps on the high tide beach, and backwash-induced rhomboidal and jump ripple features across the low tide beach. The surf zone is usually between 200 m and several hundred metres wide and contains at least two (swell coasts) and up to five (storm coasts) subdued shore-parallel bars and troughs. The dune face, beach, shoreline, troughs and bars are parallel longshore with minimal longshore variation in morphology (Figure 7.6, Table 7.1).

#### 7.3.2.3 *Dynamics*

Wave breaking begins as high, steep *spilling breakers* ($\xi_b < 0.23$) breaking over a broad outer surf zone. The waves reform in the trough(s) to break repetitively on the inner bar(s). In doing so they dissipate their energy across the surf zone which may be 300 to 500 m wide. Multiple lines of bores and/or non-saturated spilling breakers are present and the surf zone is partly or entirely saturated with respect to incident wave energy. Currents are driven by both the spilling incident breakers and group bound long waves (Section 4.3.3). Incident waves gradually decay towards the beach, and bed-return currents and standing infragravity waves (edge waves and/or leaky modes) dominate the vertically segregated cross-shore hydrodynamics, particularly in the inner surf zone. During oblique waves, longshore currents may be strong due to the large energy dissipation.

At the shoreline the standing wave is manifest as a periodic ($> 30$ s) rise in the water level (set-up), followed by a rapid fall in the water level (set-down). As a rule of thumb the height of the set-up is 0.3 to 0.5 times the height of the breaking waves (i.e. 1 to 1.5 m for a 3 m wave, Figure 4.10). Because the wave is standing, the water moves with the wave in a seaward direction during set-down, with velocities of 1 to 2 m/s closer to the seabed. As the water continues to set down, the next wave is building up in the inner surf zone often to a substantial wave bore, $> 1$ m high. The bore then flows across the low gradient beachface–swash zone as set-up. This process continuously repeats itself at infragravity frequency.

#### 7.3.2.4 *Sediment transport*

The high amplitude set-up and set-down, coupled with the fine beach sand, produces a planed-down, wide, low gradient beach, with the high tide swash reaching to the back of the beach, often leaving no dry sand at high tide. In the inner surf zone bed-return flows result in offshore-directed sediment transport, but large sediment transport gradients or even reversals may occur within the surf zone if the infragravity waves are sufficiently energetic and/or narrow banded. The morphological response is usually a seaward bar migration with concomitant reduction of the nearshore slope resulting in even more pronounced incident wave dissipation.

#### 7.3.2.5 *Rates of change*

Dissipative beaches tend to have relatively stable morphologies, and exhibit minimal shoreline change (Short and Hesp, 1982; Short, 1983). This can be attributed to a number of factors. First, the fine low gradient profile is less susceptible to change compared to steeper, coarser profiles (Figure 7.7). Second, the beach is suited to high storm waves, and consequently experiences little erosion during storm events. Third, coupled with this is the fact that the beach accommodates higher waves by causing them to break further seaward on the gentle nearshore slope thereby dissipating their energy over a wider surf zone. Fourth, the greater dissipation in turn increases the elevation of wave set-up which accommodates the greater inshore volume of water. In other words the atmosphere is 'eroded' rather than the beach.

**Figure 7.6** Examples of dissipative beaches. (a) Goolwa Beach on the southern Australian coast is a typical high energy dissipative beach and surf zone. The spilling waves in this view are 3 to 4 m high and breaking across a 500 m wide surf zone. (b) Muriwai beach, New Zealand, receives waves averaging over 2 m and its wide two to three bar surf zone is typical of the high energy west facing New Zealand beaches. (c) Aracaqu beach in north-east Brazil is exposed to persistent moderate trade wind waves of moderate period, which combine with the fine sand to produce a modally dissipative beach. (d) A low to moderate energy three to four bar dissipative beach near Israelite Bay, Western Australia (A D Short).

c

d

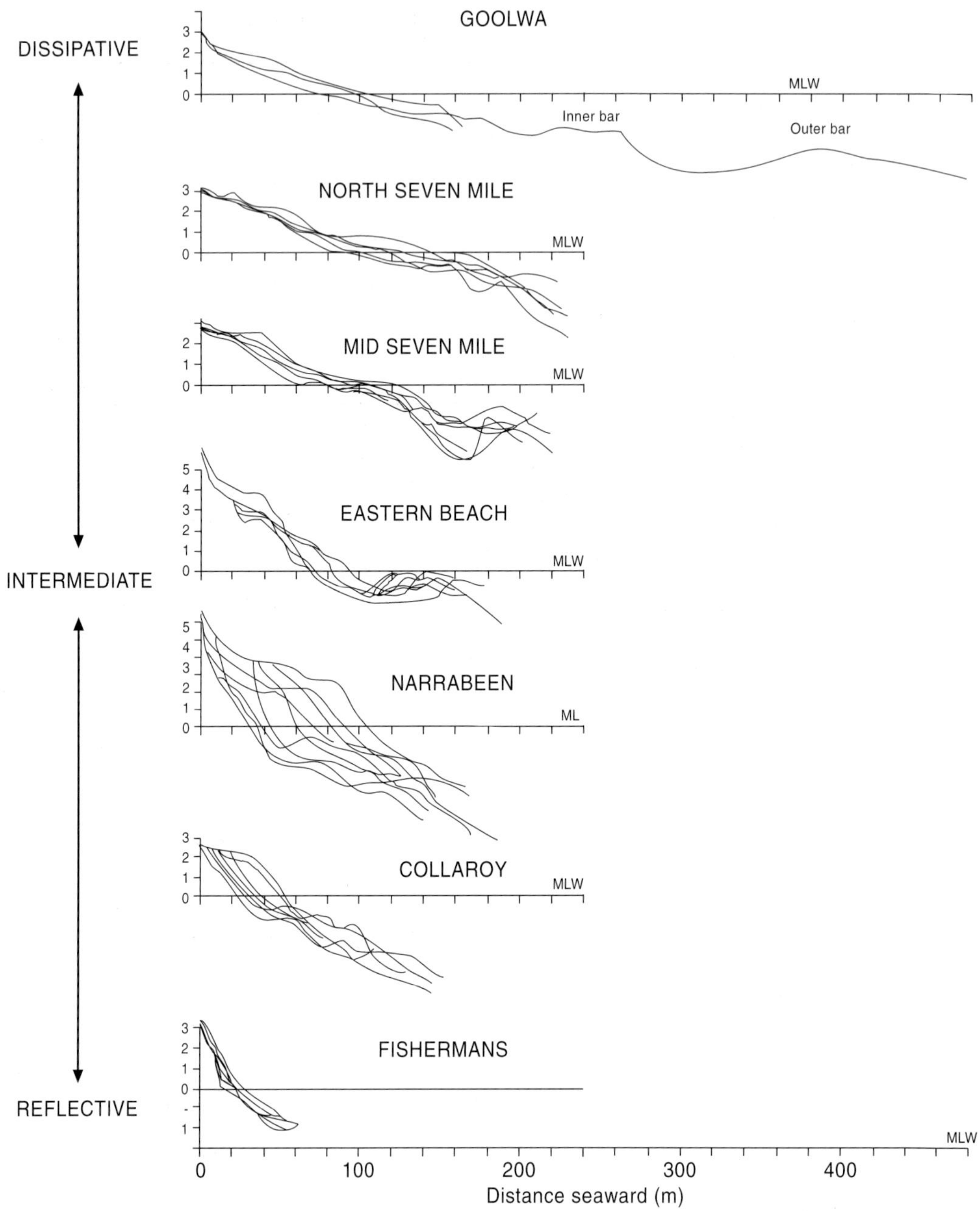


**Figure 7.7** Cross-sectional profile and beach change on dissipative, intermediate and reflective beaches (reproduced with permission from: Wright, A D and Short, A D, in Komar, P D (ed.), *CRC Handbook of Coastal Processes and Erosion*, 1983, pp. 35–64. Copyright CRC Press, Boca Raton, Florida).

Finally, there is minimal longshore variation, which minimises the need for spatial or temporal readjustment. As cross-shore wave dissipation gradients decrease following the above adjustments, set-up gradients and bed-return currents weaken. Such conditions may result in a situation where the topography approaches a state of equilibrium with the hydrodynamics, and the morphology will become relatively stable.

*7.3.2.6 Mode of erosion*

Higher waves lead to a growth in infragravity wave energy, in the size of the standing wave and accompanying wave set-up and set-down, and in the strength and seaward extent of the bed-return flow (Section 4.4.1). As a consequence this may lead to a net offshore transport from the inner surf zone and growth of bars in the outer surf zone (Lee *et al.*, 1995). However the most noticeable and measurable erosional impact is at the backbeach. The higher swash waves and infragravity bores, riding on a higher level of wave set-up, cause the backbeach to be severely attacked (Wright, 1980). This leads to dune erosion and parallel scarping, where backed by dunes, and massive, shore-continuous overwashing on low barriers (Short and Hesp, 1982).

### 7.3.3 Intermediate beaches

*7.3.3.1 Occurrence*

*Intermediate beaches* refer to beaches that are intermediate between the high energy dissipative and the lower energy reflective beaches (Wright and Short, 1984). In intermediate systems, spatial wave transformation gradients are larger, nearshore slopes are steeper and incident waves tend to break through plunging, particularly when they have a relatively low steepness. The most obvious characteristic of intermediate beaches is the presence of a horizontally segregated surf zone with bars and rips. They occur when $\Omega$ is between 2 and 5 and $\xi$ between 2.5 and 20 (Table 7.1). These conditions are favoured by a wide range of environmental conditions, ranging from moderate to high waves (0.5 to 2.5 m), fine to medium sand, and longer wave periods (Table 7.2 and Figure 7.5).

However, between those beaches produced by 0.5 m waves and by 2.5 m waves is a substantial range in the morphodynamic character of these beaches. For this reason intermediate beaches are classified into four beach states. Immediately below the high energy dissipative is the *longshore bar and trough*, then the *rhythmic bar and beach*, next the *transverse bar and rip* and finally the *low tide terrace* (Wright and Short, 1984). Each of these beach states is described below.

**Table 7.2** Beach type, Ω, and wave height on Narrabeen Beach, Australia.

| Beach *type*/state | Ω mean* | Ω sd | Wave height (m) |
|---|---|---|---|
| *Dissipative* | > 5.5 | — | > 2.5 |
| Longshore bar and trough | 4.70 | 0.93 | 2.0–2.5 |
| Rhythmic bar and beach | 3.50 | 0.76 | 1.5–2.0 |
| Transverse bar and rip | 3.15 | 0.64 | 1.0–1.5 |
| Low tide terrace | 2.40 | 0.19 | 0.5–1.0 |
| *Reflective* | < 1.5 | — | < 0.5 |

*Mean $W_s = 0.04$ m/s; $T = 10$ s.

Intermediate beaches are the most common beach type and occur both as single barred intermediate beaches, particularly on open swell coasts, and commonly as the lower energy inner bar or bars of multi-barred dissipative beaches, particularly on short period sea coasts. In the following description a single bar is assumed, while multi-barred beaches are discussed in Section 7.5.

*7.3.3.2 Longshore bar and trough*

7.3.3.2.1 *Morphology. Longshore bar and trough* (LBT) *beaches* are characterised by a near-continuous longshore bar located up to 100 and 150 m seaward of the beach (or inner bar), and separated by a 50 to 100 m wide, 2 to 3 m deep longshore (Figure 7.8). The beachface, like a dissipative beach, is straight alongshore; however depending on grain size, the beachface gradient can range from low to moderate and may contain cusps or a berm, with a step at their base. The shoreline and trough are relatively straight, the bar can range from straight to slightly rhythmic and is usually crossed by rip currents every 250 to 500 m on swell coasts (Short, 1985).

7.3.3.2.2 *Dynamics.* Higher waves tend to break, by spilling to plunging, continuously along the bar, with lower waves not breaking in the vicinity of the deeper rip channels. The wave bores cross the bar and enter the deeper longshore trough where they quickly reform and continue shoreward as a lower wave to break or surge up the beachface.

**Figure 7.8** Longshore bar and trough beaches at (a) Eastern Beach, Victoria, and (b) Wanganui, New Zealand. Note the deep trough and continuous longshore bars (A D Short).

At the shoreline the waves lowered by breaking over the bar, coupled with net onshore transport in the trough, may be sufficiently low to produce a reflective, cusped beachface (single bar), or lower energy beach type on an inner bar (multi-bar). The water moving shoreward into the trough returns seaward using two mechanisms. First, the water piles up along the beachface as wave set-up. As the water sets down it moves both seaward as a standing wave (bed-return flow), and longshore close to the shore, as a rip feeder current. Second, as adjacent feeder currents converge they cause additional set-up in lee of the rip. As this set-up sets down, the rip pulses seaward across the bar as a relatively narrow, strong rip current.

As incident wave dissipation decreases, infragravity waves weaken and bed-return flows become spatially limited. Sediment transport becomes increasingly dominated by incident waves which tend to transport sediment onshore. Therefore, the bar(s) tend to move onshore. To complicate matters, spatial gradients also exist in the longshore direction. In fact, the presence of longshore gradients in processes and morphology is characteristic of intermediate beaches. The complex morphology contributes to similarly complicated sediment transport patterns. Onshore directed oscillatory transport (mainly due to incident waves), and in some cases onshore-directed mean flows due to bore-induced mass transport or infragravity wave drift velocities (Aagaard and Greenwood, 1994b), prevail on the bar crest, while offshore-directed mean transport takes place in rip channels.

### *7.3.3.3 Rhythmic bar and beach*

7.3.3.3.1 *Morphology. Rhythmic bar and beaches* (RBB) consist of a distinctly rhythmic (crescentic) longshore bar with phase-coupled rhythmic shoreline (Figure 7.9). The bar crest alternates longshore between deeper seaward-protruding sections (rip channels), *separated* by shoaler landward-protruding 'crescentic bar' sections. The entire bar is separated from the shore by a continuous, though rhythmic trough. The beachface is usually rhythmic with seaward-protruding megacusp horns in lee of the shoaler bar sections, and deeper megacusp embayments behind the rips. The embayments may have an eroded and scarped beachface, while the accreting horns may have beach cusps. The surf zone may be up to 100 to 150 m wide and the rhythmic bar and rips spaced every 150 to 500 m alongshore on swell coasts, but closer (50 to 150 m) on sea coasts.

7.3.3.3.2 *Dynamics.* Waves break by plunging over the shallower bar sections, while they spill or pass unbroken, and move more quickly, across the deeper channel sections, producing a distinct longshore gradient in wave breaking and direction in the surf. In lee of the bar sections wave bores move into the trough and at the shoreline are deflected longshore toward the adjacent embayments, as rip feeder currents. The feeder currents are weakest where they diverge behind the centre of the bar, but increase in velocity, intensity and depth toward the adjacent embayment (Wright and Short, 1983).

In the *embayment* waves break less or possibly not at all. They may move unbroken into the deeper channel section to finally break or surge up the steeper swash zone. The strong uprush often causes slight erosion of the beachface and cuts a high tide erosion scarp. The backwash returning down the embayment beachface combines with flow from the adjacent rip feeder channels. This water sets up, then pulses seaward as a strong, narrow rip current. On swell coasts, the currents pulse every 30 to 90 s, depending on wave conditions. The rip current accelerates with each pulse and persists with lower velocities between pulses. Rip velocities are usually less than 1 m/s (3.5 km/h), but will increase up to 2 m/s under higher waves and toward low tide (Short and Hogan, 1994; Brander, 1996)

The *rhythmic shoreline* is produced *after* the formation of the rhythmic bar. The longshore variation in wave breaking and refraction acts firstly to cause nodes of lower (in lee of bars) and higher (in lee of rips) waves at the shoreline. Secondly, wave refraction around the shoaler bars bends the waves so they alternately converge in lee of the bar, and diverge in lee of the rip. The resulting longshore variation in wave/swash height and direction at the shoreline produces slight erosion in lee of the rips, and slight deposition in lee of the bars. This produces a rhythmic shoreline, building a few metres seaward behind the attached bars as deposition occurs, and being scoured out and usually scarped in lee of the rips. The rhythmic undulations are called megacusp horns (behind the bars) and embayments (behind the rips). They have an identical spacing to the bar and rips, 150 to 500 m on swell coasts.

### *7.3.3.4 Transverse bar and rip*

7.3.3.4.1 *Morphology. Transverse bar and rip* (TBR) *beaches* receive their name from the fact that the bars are transverse or perpendicular to shore. They are similar to the rhythmic bar and beach, except that the bars are *attached* to the beach, which in turn segregates the individual rip systems. The bars and rips are usually

**Figure 7.9** Rhythmic bar and beach systems at (a) Evans Head, Australia; (b) Torrey Pines, USA; (c) Egmond, The Netherlands; and (d) Panama City, Florida (A D Short).

regularly spaced every 150 to 250 m, but can reach spacings of 700 m on high energy swell coasts (Short and Brander, in press; Figure 7.10a). This produces a beach with alternating shallow bars and deeper rip channels, with protruding megacusp horns attaching to the bars, and often scarped embayments in lee of the rip channels. Their bars and surf zones range from 50 to 150 m in width.

7.3.3.4.2 *Dynamics.* Waves plunge heavily on the shallow bars (particularly at low tide) dissipating much of their energy, while in the deeper rip channels they break less by spilling or possibly not at all, leaving more energy to be expended as a plunging or surging shorebreak at the beachface. Consequently across the inner surf zone and at the beachface there is an alternation of lower energy converging swash in lee of the bars, and higher energy diverging swash/shorebreak in lee of the rips.

The surf zone has a cellular circulation pattern (Figure 4.20). Waves tend to break more on the bars and move shoreward as wave bores. This water flows both directly into the adjacent rip channel and closer to the beach to the rip feeder channels located at the base of the beach. The water in the rip feeder and rip channel then returns seaward in two stages. First, water collects in the rip feeder channels and the inner part of the rip embayment and sets up against the lower beachface, as part of a standing (edge) wave. It then sets down and pulses seaward as a relatively narrow ($\geqslant$ 10 m) accelerated flow, the rip current. The velocities range up to 1.5 m/s, with higher flows occurring toward low tide when the current is more constricted by the topography (Short and Hogan, 1994; Brander, 1996). The water usually moves through the rip channel, out through the breakers, and seaward for a distance usually less than the width of the surf zone, that is, several tens of metres. The pulsing is at infragravity frequency ($>$ 30 s).

#### 7.3.3.5 *Low tide terrace*

7.3.3.5.1 *Morphology.* *Low tide terrace* (LTT) *beaches* (also known as 'ridge and runnel') are the lowest energy intermediate beach type. They are characterised by a moderately steep (grain size dependent) beachface, which is joined at the low tide level to an attached bar or terrace, hence the name low tide terrace. The bar usually extends between 20 and 50 m seaward (though it may reach 100 m in fine sand) and continues alongshore attached to the beach (Figure 7.11). It may be flat and featureless, have a slight central crest called a ridge (ridge and runnel), and may be cut every several tens of metres by small, shallow rip channels, called *mini-rips* (Short, 1979b).

Low tide terrace beaches differ from transverse bar and rip beaches in that the rips are essentially infilled, and the beach straightened, resulting in no rhythmic topography. If rips are present they are small with shallow to negligible channels.

7.3.3.5.2 *Dynamics.* Low tide terrace beaches are favoured by waves less than 1 m, which at high tide may pass right over the bar and not break until the beachface, which will then behave like a *reflective* beach. At spring low tide the entire bar is usually exposed as a terrace or ridge running parallel to the beach. At this time waves break by plunging heavily on the outer edge of the bar and *dissipate* across the terrace. At mid tide waves usually break right across the shallow bar and may induce weak, ephemeral rip circulation.

#### 7.3.3.6 *Rates of change*

Intermediate beaches are the most dynamic beach type both in space (longshore) and over time. As Figure 7.4 illustrates, the alternate bars and rips, coupled with cellular rip circulation, provide for a continual longshore variation in cross-sectional beach morphology (shallow bars and deep trough and/or channels) and dynamics (onshore wave bores and seaward-pulsating rip currents), as well as a potentially rhythmic shoreline. Consequently shoreline mobility reaches its maximum on intermediate beaches (Table 7.1 and Figure 7.7).

#### 7.3.3.7 *Bar migration*

The high level of shoreline mobility is also apparent in the surf zone, where it is manifest in potentially rapid rates of onshore bar migration. Observed rates range from 1 to 30 m/day (Orme, 1985), while most average a few metres per day. The order of magnitude range in migration rates can be attributed in part to the prevailing level of wave energy. Rates can range from zero with no waves to 20–30 m/day under moderate waves ($H_b$ = 1–1.5 m). The migration rates infer two additional facets of these beaches: first, intermediate beaches are inherently unstable, as manifest in the dynamic bars; and secondly, as the bar migrates and reshapes the surf zone and shore, the beach type will change.

#### 7.3.3.8 *Mode of erosion*

The erosion of intermediate beaches is dominated by the presence of the seaward-moving rip currents, with maximum erosion occurring in and in lee of arrested rips (Wright, 1980), in what Short and Hesp (1982) termed *rip embayment erosion*. As wave height increases on an intermediate beach, Short (1985) has shown that the rip size (and intensity) will increase. This means that while erosion is maximum in lee of the rip, as the rip spacing increases, the rip locations migrate longshore causing a rip-induced planing down of the beach. As rip spacing continues to increase with wave height (see next section) a point may be reached where the spacing becomes infinite, and dissipative conditions prevail. Consequently intermediate beach erosion begins with relatively closely spaced, through segregated rip-dominated erosion, but may end with continuous standing wave-dominated *dissipative erosion* dominated by bed-return flow (see Section 4.4.1). On embayed beaches, this condition is prevented from occurring by the headlands, as will be discussed in Chapter 9.

Intermediate beaches also undergo what Short (1979b) called *second phase erosion* during general beach accretion. This is produced by the longshore variation in surf zone and swash dynamics, discussed above, which induce rip embayment scarping (and horn progradation) during general beach accretion. The shore-

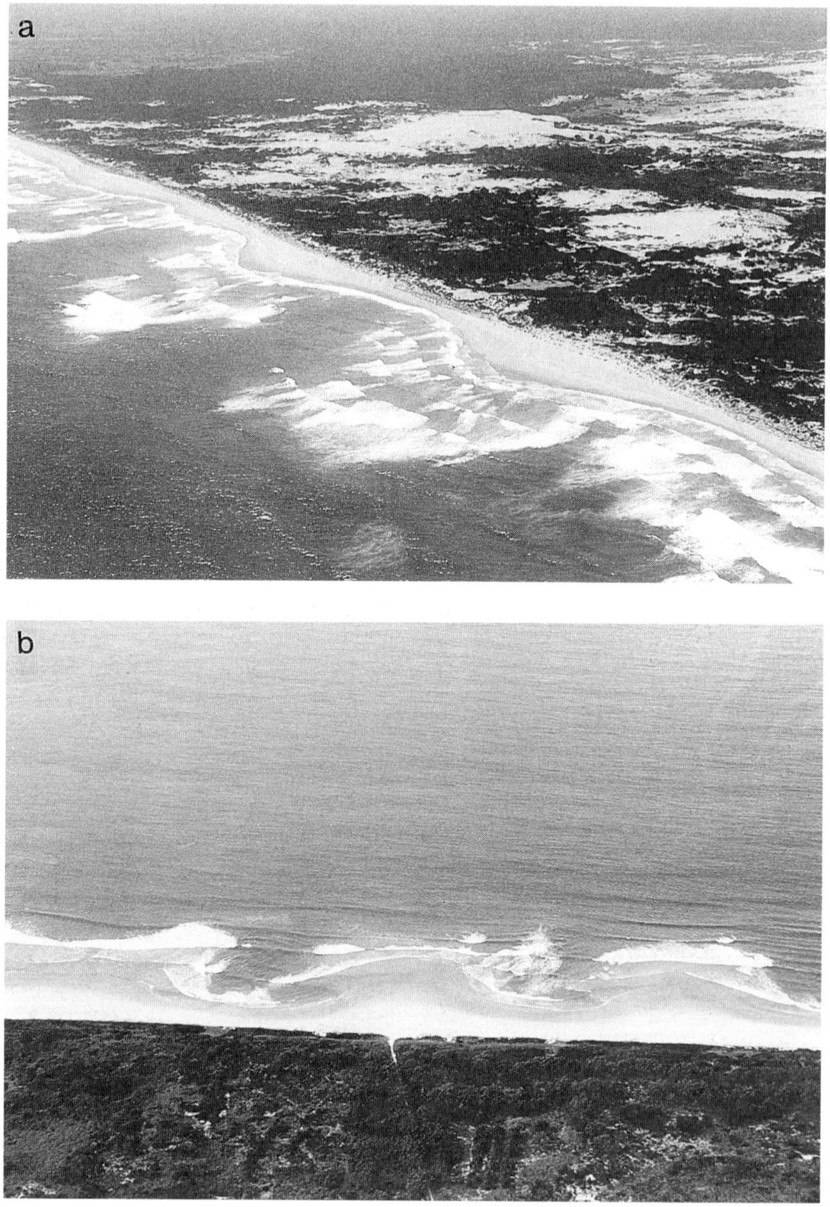


**Figure 7.10** (a) A high energy transverse bar and rip beach system with rips spaced at 400 m, Discovery Bay, southern Australia. (b) A moderate energy system with rips spaced 200 m at Stuart Point, eastern Australia. (c) A highly rhythmic series of bars and rips, Forster Beach, Eastern Australia (A D Short).

line can therefore alternate from eroding embayments to prograding horns. Consequently even though net accretion or fill is occurring, those parts of the beach in lee of the rips will go through a second phase of wave-induced erosion, which may be more severe than preceding general beach erosion. This erosion can occur even under low waves. It will cease as the adjacent horns and bars prograde and eventually infill the rip.

Intermediate beach erosion is therefore characterised by discrete erosion maxima in lee of rips. When the rips are stationary, as in an accreting phase, pronounced (rhythmic) embayments and beach scarping may occur. When the rips are mobile or increasing in spacing, the rips plane down the entire beach which, in conjunction with the dissipative standing wave erosion, produces a flatter, more gently undulating to ultimately straight eroded beach.

#### *7.3.3.9 Rip types*

There are four types of rips (Short, 1985). Two types, erosion and accretion rips, may occur on all beaches, while the third and fourth, topo-rips and megarips, require topographic influence and are discussed in Section 9.3. *Erosion rips* occur when wave height is increasing and the beach is eroding. They are hydrodynamically driven and 'free' to choose their spacing and location, within the prevailing edge wave controls (Section 4.4.1). As wave height increases rip spacing increases, until a point is reached where the rips stabilise or, having become infinitely long, the dissipative, ripless state is reached. These are also called 'overfit' rips in that the rip circulation is too large for the prevailing morphology, which has to be enlarged (eroded) in order to fit the rip currents. Brander (1996) introduced the rip-fit parameter Rf, where

$$Rf = M - O/Ar \tag{7.2}$$

where $M$ is the morphological rip area, $O$ the overbank rip area and $Ar = M + O$. Rips are underfit when $Rf > 1$ and overfit when $Rf < 1$.

*Accretion rips* occur when waves are low or falling; they tend to be underfit, particularly at low tide. These rips are 'fixed' in that they are arrested by surf zone topography (they are located between the rhythmic bars). They are interlinked in the longshore bar and trough and rhythmic bar and beach state, but become segregated in the transverse bar and rip state. They are also 'underfit' in that the rip current is too small or weak compared to the size of the inherited channel, $Rf > 1$ (similar to an underfit stream; Dury, 1968), and unable to excavate or even maintain the infilling channel. Under uniform falling waves these rips remain fixed in location until they infill. This may take up to several weeks (Goldsmith *et al.*, 1977; Short, 1979b). However if wave conditions change thay may erode or under oblique waves migrate longshore (e.g. Wright *et al.*, 1982b). Brander (1996) developed a morphodynamic model of low energy rip evolution, between the LBT and LTT beach states, that highlights both the morphological evolution of the rips and the changing dynamic response.

#### *7.3.3.10 Rip current velocity*

The *velocity* of the rip currents varies tremendously. On a typical beach with waves less than 1.5 m they peak at about 1 m/s, or 3.5 km/h, about walking pace (Short and Hogan, 1994). However under high waves they may double that speed. They also increase with falling tide as the rip currents become constricted by the adjacent bar topography, thereby leading to flow accelerations (Figure 7.12; Brander, 1996). What this means is that under average conditions a rip may carry someone out from

**Figure 7.11** (a) A transverse bar and rip beach grading into a low tide terrace at Crowdy Bay, eastern Australia. (b) A wide low tide terrace at Cape Patterson, southern Australia (A D Short).

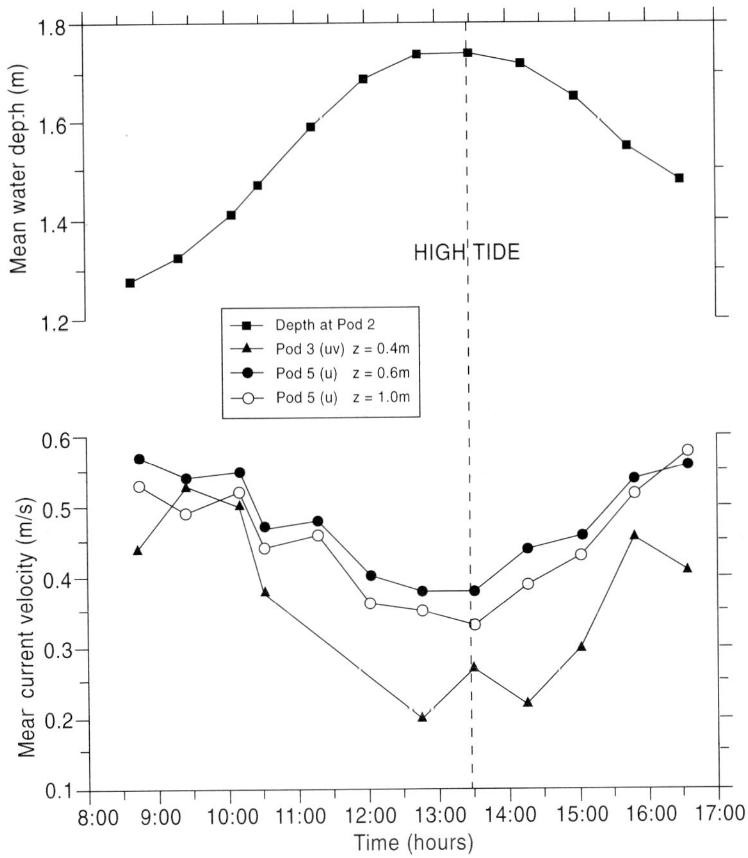


**Figure 7.12** Time series of rip current velocity versus tide height. Note the increase in velocity at all three rip located sensors during low tide (source: Modified from Brander, 1996).

the shore to beyond the breakers in 20 to 30 s, at speeds exceeding the velocity of an Olympic swimmer. The consequent implication of rips in beach safety is also discussed in Chapter 13.

### 7.3.4 Reflective beaches

#### *7.3.4.1 Occurrence*

*Reflective* sandy beaches lie at the lower energy end of the beach spectrum, though not necessarily in areas of low waves. They occur when $\Omega < 1$ and $\xi < 2.5$, which are favoured by combinations of lower waves and/or longer periods or coarse sediments (Figure 7.5, Table 7.1). Where beaches are composed of fine to medium sand, waves must be less than 1 m to produce reflective beaches. For this reason on swell coasts they are usually found at the lower energy end of embayed beaches and inside the entrance to estuaries, bays and harbours, away from higher energy ocean swell (Figure 7.13a). On sea coasts they are less prevalent, despite the usually lower waves, because of the shorter wave periods and morphological inertia following episodic high waves. On beaches composed of coarse sediments, from coarse sand to boulders, they can exist with significantly higher waves; for this reason coarse beaches tend to be always reflective, no matter what the wave height (Figure 7.13b).

#### *7.3.4.2 Morphology*

Reflective beaches always have a relatively narrow beach and swash zone, with beach cusps commonly present. The swash zone is relatively steep, with actual gradient depending on grain size (Figure 7.1). If there is a range of sand size, the coarsest sand will accumulate at the base of the swash zone as a *step* (Hughes and Cowell, 1987; Bauer and Allen, 1995; Section 5.5). Immediately

**Figure 7.13** Four reflective beaches illustrating their relatively steep beachface, common cusps, absence of a surf zone and surging waves. (a) Round Head, Queensland; (b) Napier, New Zealand; (c) Pearl Beach, eastern Australia (all A D Short); and (d) a well developed beach step at Seaford, Victoria (M Hughes).

seaward of the step the nearshore zone is low and featureless, apart from wave ripples (Short, 1985). They have no bar or surf zone and waves move unbroken to the shore where they collapse or surge up the beachface ($\Sigma_b > 5$).

#### 7.3.4.3 Dynamics

Reflective beach morphology is a product of two factors. First, seaward of the breakpoint, shoaling waves induce onshore sediment, as well as a preferential shoreward movement of coarse sand. As there is no surf zone this sediment accumulates in the swash zone. Over time this results in a relatively deep, low gradient nearshore zone, terminating at a steeper swash. Second, because the waves break at the base of the beachface they must expend all their remaining energy in the swash zone. Much of the energy goes into the wave uprush which builds a relatively steep beachface (aided by coarser sand) and strong backwash which is reflected back out to sea as a reflected wave, hence the name reflective (Guza and Inman, 1975).

The *step* is always located at the base of the beachface around the low water mark (Figure 7.13d). It consists of a band containing the coarsest material available including rocks, cobbles, even boulders and often numerous shells. Because it is coarse its slope is steeper than the swash zone, hence the step-like shape. Steps are usually a few decimetres in height, reaching a maximum of perhaps a metre (Hughes and Cowell, 1987). Immediately seaward of the step the sediments usually fine markedly and assume a lower slope.

The reason for the step is twofold. The unbroken waves sweep the coarsest sediment continuously toward the beach and the step. The same waves break by surging over the step and up the beachface. However the uprush deposits the coarsest, heaviest material first, only carrying finer sand up onto the beach, then the backwash rolls any coarse material back down the beach. The coarsest material therefore converges on and is 'trapped' at the base of the beachface by both the incoming waves and the uprush and backwash. On modally reflective beaches this can produce a longer term sorting of the sediment, which is manifest as a very sharp, distinct break between the coarser, steep step, and the lower finer nearshore (Figure 7.13d).

The strong swash, in conjunction with the usually coarser sediment, builds a high, steep beachface. The *cusps* which often reside on the upper part of the beachface are a product of sub-harmonic edge waves, meaning that the waves have a period twice that of the incoming wave (Section 5.6.3; Wright *et al.*, 1979). The edge wave period and the beach slope determine the edge wave length, which in turn determines the cusp spacing, which on swell coasts can range from 20 to 40 m (Figure 7.13).

#### 7.3.4.4 *Rates of change*

Reflective beaches are relatively *stable*. This is a product of three factors. First, they represent the lower energy end of the beach change spectrum, and can therefore only move in one direction, that is when eroded by higher waves. Second, waves are usually low and induce small changes. Third, when the beachface is eroded, the sediment is usually deposited as a bar or terrace at the base of the beach, from where it can rapidly return onshore. Short (1979b) found that modally reflective beaches are the most stable of the three beach types (Figure 7.7), while Short and Hesp (1982) also found that they have stable backshores (Table 7.1)

#### 7.3.4.5 *Mode of erosion*

Reflective beaches are eroded through the growth of both incident wave swash and sub-harmonic edge waves. First, increasing swash volume saturates the beach which ultimately leads to failure of the beachface and the formation of an erosion scarp, particularly during periods of short waves and/or neap or low tide. During long waves and/or high or spring tide, it can lead to overwash and slumping of the beachface (Short, 1979b). The net result is erosion of the sub-aerial beach and its deposition usually as an attached bar or (low tide) terrace at the base of the beach. Secondly, if the erosion is accompanied by the growth of sub-harmonic edge waves they will induce maximum erosion at length scales determined by the edge waves (tens of metres) leading to the formation of '*erosion*' cusps (Short, 1979).

## 7.4 SEQUENTIAL BEACH CHANGES

Section 7.3 presented the three beach types including the four intermediate beach states. In nature beaches actually exist in a continuum or spectrum of types, hence the types and states presented in Figure 7.4 merely illustrate idealised stages along this spectrum. Where a particular beach usually resides in the spectrum is dependent on its modal wave and sediment parameters. How much change occurs on the beach, the direction (erosion/accretion), speed and duration of change are all dependent on the changes in wave height, period and grain size. If any of these change so too must the beach. The changes may be short term in response to changing wave conditions, taking hours, days or weeks, which is discussed below; or they may be long term in response to changes in sediment supply or climate which can result in net beach progradation or erosion, which is discussed in Chapter 14.

The sensitivity of a beach to changes in wave height, period and grain size is illustrated in Figure 7.5. Increasing wave height and decreasing period favour more dissipative conditions, while increasing grain size favours more reflective conditions. Most beaches have a set sediment size and so this will not be considered as a source of temporal change in this chapter; however its spatial impact will be discussed in Chapter 10. In this section, beach change induced by changing wave conditions, primarily wave height, is presented, based on Short (1979b) (Figure 7.4). Sequences of a similar pattern have also been described by Sonu (1973), Goldsmith *et al.* (1977), Sunamura (1984a), Wright and Short (1984), Lippman and Holman (1990) and Brander (1996). To briefly explain how and why beaches change shape, either by eroding or accreting, two sequences of beach change will be described. First, the accretionary sequence of beach change is presented beginning with the high energy dissipative beach moving down through the six beach types/states to low wave reflective beach. Then beginning with a low wave reflective beach, the changes that occur as waves continue to increase in height and erode the beach, will be described. It must be understood, however, that in nature most beaches never, and none regularly, go through this entire sequence; rather most oscillate between two or three beach states.

### 7.4.1 Accretionary sequence: dissipative to reflective

A fully dissipative, fine sand beach on a swell coast exposed to gradually lowering waves, for a period of several weeks, can slowly move through the entire sequence from intermediate to the reflective beach state (Figure 7.4).

*Fully dissipative beaches* (Section 7.3.2) occur in areas of fine sand exposed to high waves ( > 2.5 m) for periods of weeks to months. The high waves acting across the wide surf zone produce shore-parallel standing waves, which are in turn responsible for the location of the shore-parallel bars and troughs (Section 4.6.3; Figure 7.6).

If waves decrease to less than 2 m in height (swell coast) and stay low (< 1–1.5 m) for a period of months, the following sequence of beach change will occur. First, the standing wave dynamics of the fully dissipative surf zone is replaced by standing (edge) waves. Edge waves (Section 4.3.5) stand perpendicular to shore and generate residual surf zone currents that initiate cellular surf zone flow patterns. Once this circulation pattern is established, with the assistance of incident waves, it begins to move more water and sand shoreward under the shoreward cell flows, and water (and some sand) seaward under the offshore cell flows. The result is the formation of large scale rip circulation (alternating cellular on- and offshore currents) and the initiation of spatially segregated bars (onshore flows and sand transport producing shoreward bar growth) and rip channels (offshore flows and deeper bar). While the bar is increasingly segregated longshore becoming 'crescentic' in shape, the beach remains straight (Figure 7.4). This is called the *longshore bar trough* beach type, distinguished by its continuous, straight to crescentic bar, continuous trough and the straight beach (Section 7.3.3).

As the waves drop to about 1.5 m, increasing incident wave dominance continues to move the bar shoreward. The most rapid bar movement is where the flow is shoreward, the least or most retarded where the flow is seaward in the rips. Once the hydrodynamics has expressed itself in the morphology (bars and rip channels) a positive feedback occurs such that the morphology helps to reinforce and extenuate the dynamics. This is called *morphodynamic coupling*. Over the bars, shoreward-moving wave bores accelerate bar migration leading to the formation of crescentic bars, while between the bars increasingly intense seaward-moving rip currents maintain a deep rip channel and seaward-protuding rip head bar.

The bar and rip topography accentuates the longshore variation in the degree of wave breaking and the direction of wave approach across the inner surf zone. This produces a reworking of the beachface with deposition occurring in lee of the bars (lower converging waves) initiating the growth of megacusp horns, and erosion commencing in lee of the rips (higher diverging waves) scouring megacusp embayments, as described in Section 7.3.3. At this stage the beach is called *rhythmic bar and beach* after the crescentic bar and rips, and the rhythmic or undulating shoreline owing to the megacusp horns and embayments (Figure 7.4).

Continued low waves ( < 1.5 m) will maintain the morphodynamic coupling, extenuating the topographically induced cellular rip circulation. The bars will rapidly move shoreward (metres/day), separated by strong rip currents and well developed rip channels. At the same time there is a slight ( < 10 m) seaward growth of the megacusp horn, while beachface scarping may continue in the rip embayment. Eventually, after a few days to a week or more, the bars will begin to reach the beach and attach to the base of the megacusp horns. At this stage the beach becomes a *transverse bar and rip beach* type (Section 7.3.3). Both the bars and the rips are *transverse* or perpendicular to shore, the bars are attached to the megacusp horns and the segregated rips front the megacusp embayments (Figure 7.10). The prominent megacusp horns and embayments ensure the beachface remains rhythmic.

As waves continue to fall ( < 1 m) the whole bar moves shoreward. In the process the surf zone decreases in width and the bar expands laterally as the rips infill. The rips and megacusp embayments stay in the same location but infill and shoal, decreasing in size to narrow, shallow mini-rips, separated by a shallow bar. At the same time the backing megacusp embayments are infilled leading to a straightening of the beachface. At this stage the *low tide terrace* beach type is reached. It is characterised by a continuous attached bar, a relatively straight beachface with the mini-rips which eventually fill in and disappear (Figure 7.11). This beach state is described in more detail in Section 7.3.3.

Finally, continued low waves ( < 1 m) enable the low tide terrace to be washed up onto the beach by the waves. This process usually takes several days and is usually manifest as a series of low, regularly spaced (20 to 40 m) beach cusps (Short, 1979b). As more sand moves onto the beach the stationary cusps increase in height and volume. The cusp embayments may then completely infill to form a long, straight berm crest, backed by a slightly landward-dipping berm, at the rear of which is a runnel, where swash collects at high tide. By this time the bar has moved completely onto the beachface and a relatively deep, barless nearshore zone fronts the beach (Figure 7.13). This is the fully *reflective* beach type described in Section 7.3.4.

In nature movement through the complete accretionary sequence is rare. On swell coasts waves rarely stay low enough for long enough to achieve the full transition. The southern Californian winter cut (dissipative) and summer fill (reflective) is an example of the type of transition that should occur in a full sequence. On sea coasts, the calms between the occasional storms result in the arrest and preservation of the dissipative morphology (but not dynamics) until the next storm. Consequently a higher energy outer surf zone morphology may prevail as part of the nearshore wave shoaling zone, while only the inner bar adjusts dynamically, and morphologically, to the lower wave conditions (Section 7.3). Sequential movement through a substantial part of the sequence has been reported by Short (1979b), Goldsmith *et al.* (1982a) and Brander (1996) while Lippman and Holman (1990) observed all but the two end members at Duck, North Carolina. Section 7.6 elaborates on the capacity for beaches to move through the sequence.

### 7.4.2 Erosional sequence: reflective to dissipative

The erosional sequence (Figure 7.4) is based on Short (1979b, 1985) and has also been recorded in part by Goldsmith *et al.* (1977) and Sunamura (1988). Essentially the erosion sequence consists of two complementary mechanisms designed to accommodate the higher waves and greater volume of water moving in and across the surf zone. First, beach erosion lowers and widens the beach–surf zone by moving sand from the beach–inner surf zone and depositing it in the outer surf zone. Second, this process causes waves to break further seaward and across a wider surf zone thereby lowering incident wave height and energy at the shore, and providing a wider surf zone for both energy dissipation and surf zone circulation.

When a swell coast *reflective* beach is exposed to an increase in wave height ( $> 0.5$ m) the beach begins to rearrange itself to accommodate the larger waves. The higher waves give rise to large wave set-up, leading to a rise in the beach water table and eventual saturation of the beachface. The combined action of berm overwashing, berm breaching and strong swash action results in slumping and collapse of the lower beachface and possibly the formation of an erosion scarp and storm cusps. The sediment removed from the berm and beachface is deposited immediately seaward of the beachface where it forms an attached incipient bar. Depending on wave height and state of the tide, this process may scarp or scour the beachface. The eroded sand shoals and widens the surf zone, causing waves to break further seaward, and thereby expending more energy crossing the wider surf zone.

The sand eroded from the reflective beachface is initially deposited immediately seaward of, and attached to, the lower beachface, forming an *erosive low tide terrace* (LTT′). Small transient ephemeral rips usually cross the terrace, helping to move more sand seaward.

When waves exceed 1–1.5 m the rips increase in size and intensity, moving more sand seaward, thereby extending the bar and scouring deeper rip channels. While some parts of the bar are still attached to the shore this is called the *erosional transverse bar and rip* beach type (TBR′).

When the waves exceed 2 m the rips not only grow in size, intensity and spacing, but they begin to link up, forming a continuous longshore trough, occupied by widely separated rips (300 to 600 m), joined by rip feeder channels lying behind the still rhythmic bar. This is the *erosional rhythmic bar and beach* type (RBB′).

Increasing wave height ( $> 2.5$ m) causes larger rips to form where the longshore flow converges, returning water seaward as a strong rip current, whereas between the rips in zones of divergence, water is increasingly returned seaward by means of bed-return flow, driven by standing wave set-down. The bar lies 100 m or more seaward of the beach, with a continuous longshore trough and widely spaced rips (500–1000 m). At this stage the beach enters the *erosional longshore bar and trough* beach type (LBT′).

Finally, if waves reach 2–3 m and maintain that height for several days the fully *dissipative* beach type is reached. The bar is moved to its most seaward limit, separated from the beach by a broad, moderately deep trough. The rips increase in spacing to more than 2 to 3 km on long beaches and eventually disappear, being replaced by the more efficient standing wave mechanism for seaward return of the water, described in Sections 4.4 and 7.3.3. On many of the longer, higher energy swell beaches these conditions also produce a second bar seaward of the first, the first lying up to 100 m offshore and the second up to 250 m offshore.

On shorter embayed beaches, or whenever beaches are less than about 4 km in length, the beach cannot reach this fully dissipative state, because the presence of headlands, reefs and/or the curvature of the beach will produce a longshore gradient in wave height. The impact of embayments on surf zone circulation and beach dynamics is presented in Section 9.3.

It is important to recognise that the evolutionary sequences presented above are based on observations made on a modally high energy, single bar, micro-tidal beach, and are fairly specific to such beaches. On multi-bar beaches, more complex responses of the morphology to changing wave conditions occur, with the inner bar responding relatively rapidly and exhibiting frequent fluctuations and the outer bar responding more slowly and over longer periods (Birkemeier, 1985; Lippman and Holman, 1990, Short and Aagaard, 1993). In addition, during an accretionary sequence a certain bar morphology may remain 'arrested', when the wave energy level becomes too low to transport sand shoreward and to move the bar any further in this sequence (Aagaard, 1988). The presence of antecedent bar morphology is especially apparent on multi-barred beaches in storm-dominated lacustrine (Greenwood, 1987) and sea environments (Short, 1992). Multi-bar beaches are discussed more fully in Section 7.5.

#### *7.4.2.1 Rates of bar migration*

A quantification of the bar sequences is provided by the determination of bar migration rates. Average onshore bar migration rates vary between 1 and 30 m/day (Orme, 1985) and small bars move faster than large bars because less sediment is involved in their migration (Sunamura and Takeda, 1984). Davis *et al.* (1972) further note that the bar migration rate decreases with increasing tide range due to the fact that at some stage the shoreward-moving bar becomes emerged during low tide. In addition, sediment transport patterns are expected to be unstable on tidal beaches. On the seaward slope of the bar, for example, onshore transport (due to incident waves) may occur at high tide, and offshore transport (due to mean flows) at low tide. Onshore bar migration during storm recovery is generally slower than offshore bar migration during storms (Birkemeier, 1985) as energy levels and sediment transport rates decrease. Few measurements of offshore bar migration rates are made due to the difficulties associated with measuring beach morphology under storm conditions. Over a 3 day storm period with 3.5 m waves, Birkemeier (1985) monitored an offshore migration of 90 m. Sallenger and Holman (1985) reported a seaward bar migration of 57 m in 5 days and Aagaard and Greenwood (1995a) measured a nearshore bar that shifted 25 m offshore within 10 hours.

Cross-shore changes in bar morphology may also be the result of longshore migration of rhythmic bar features. Dolan (1971) reported on the longshore movement of sand waves associated with crescentic bars and noted that alternating periods of shoreline erosion and accretion may occur as the rhythmic forms migrate past the reference point. Kroon (1994) monitored the beach morphology on the multi-barred Dutch coast and demonstrated that the offshore migration of a bar (in the cross-shore profile) was due to a longshore migrating crescentic bar system, a phenomenon also observed at Duck, North Carolina by Lee *et al.* (1995). In addition, cross-shore bar change may be accomplished by longshore growth of bar systems (Guillen and Palanques, 1993).

#### *7.4.2.2 Rate of beach change*

The rate of change between beach states and types is dependent on wave conditions. A study of beach changes on Narrabeen beach by Wright *et al.* (1984) showed that the accretionary changes under relatively low waves ( $<$ 1.5 m) average less than one state per day, while the erosional sequence under high waves ( $>$ 2 m) can move as rapidly as one state per day. However if waves decrease sufficiently, the entire morphology can be arrested, and essentially no change occurs.

This implies several important factors relating to beach change: that the prevailing conditions are a poor indicator of beach type; that antecedent morphology is critical; that there is an inherent lag in beach change of up to a few weeks; that the importance of antecedent morphology decreases with time; that each beach possesses its own unique relationship, which will be dependent on both sediment and prevailing and antecedent wave conditions; and finally that accurate prediction of beach change will require sequential information for individual beaches.

## 7.5 MULTI-BAR MICRO-TIDAL BEACHES

The occurrence of double and multiple bars along a beach has long been the subject of general interest and scientific investigation. Johnson (1956) noted that beaches with flat gradients exposed to wave action usually have several bars. King (1972) provides a brief review of the early studies of 'submarine bars' including multiple bars and crescentic bars. King's major conclusions were that multiple bars are more prevalent in tideless seas, may be attributed to waves of different heights, and that the inner bar is always the most mobile.

Many micro-tidal beaches have more than one bar. Two and occasionally three bars are common on swell coasts, while in seas three to fives bars are more common. In order to accommodate multi-bar beaches in a general sequence of beach types, two additions to the single bar model are required. First, the environmental conditions that favour multiple bars, which are not covered in Figure 7.5, and second, the impact on beach/bar type and change in multi-bar systems.

### 7.5.1 Bar occurrence

Short and Aagaard (1993) developed the bar parameter ($B^*$, Table 7.1) to help explain the presence and number of bars in micro-tidal beach environments, with bar number increasing as gradient and/or $T$ decrease. This relationship is a product of the surf zone dynamics that control bar formation and spacing. Infragravity standing waves will induce residual flows such that bars tend to form under the standing wave antinodes, while troughs will occur under the nodes (Short, 1975b; Section 4.6). The spacing of the standing waves is negatively related to wave period. In other words the bars are more closely spaced where period is short and gradient low. Their occurrence is also favoured by low surf zone gradients which not only place the bars closer to the shore, but also place the seabed closer to the formative standing wave currents than deeper seabeds. They are least likely to occur where the period is long and gradient steep. Longer periods and steeper gradients produce more widely spaced standing waves which puts the potential bar-forming dynamics into deeper water, where it is less likely to impact the deeper seabed. In other words for a given beach gradient, short period sea environments will produce more, more closely spaced bars, whereas in long swell environments fewer, more widely spaced bars can form.

### 7.5.2 Bar hierarchy (spatial change)

In multi-bar systems the outer bar will always receive more wave energy than inner bars as a result of wave breaking and decreasing breaker height across the bar system. As a consequence there is always a *hierarchy of bar types*, with the outer bar being the highest energy type, and incident breaker wave energy decreasing shoreward. Consequently unlike single bar beaches, most multi-bar beach types have a shoreward transition in bar type from an outer, more dissipative, high energy end member, to intermediate bars and even reflective bar/beach at the shoreline. Finally, because the outer bar(s) are produced and maintained by episodic storm/high swell waves, they commonly become inactive and stagnate, once the formative waves cease, particularly in sea environments (Aagaard, 1988; Short, 1992). Figure 7.14 illustrates an idealised single, double and multi-bar beach sequence.

### 7.5.3 Temporal bar change

Changes in bar type are driven locally by changing wave conditions. The outer bar(s) usually remain dissipative, or at best move through the higher energy beach types (dissipative–LBT–RBB). Outer bar movement is more likely to occur in swell environments, while stagnation is more likely in storm-dominated sea environments, where calms may prevail between storms in multi-bar systems. On these coasts the inner bar(s) can be assumed to behave like a lower energy single bar beach, moving through a range of usually lower energy beach types (reflective–LTT–TBR).

Figure 7.15 illustrates the nature and frequency of occurrence of double bar systems along the open, swell-dominated New South Wales coast. Note that the outer bars are dissipative only during low frequency high waves ($<1\%$) and shift to LBT (36%), RBB (58%) and reach TBR attaching to the inner bar only 5% of the time. In contrast the inner bars are usually LTT (56%) or TBR (30%) rarely RBB (3%) and occasionally reflective (10%).

Figure 7.15 also indicates that bar change can occur in two modes. Horizontal movement across the figure indicates an inactive outer, but active inner bar. This is likely to occur when formative conditions are followed by waves too low to activate the outer bar and/or the bar is in water that is too deep. Movement down or diagonally across the chart indicates that bars are both active and changing, which is more likely to occur on energetic swell coasts, as suggested here. The modal double bar situation for the NSW beaches is accommodated in the combination of outer LBT or RBB, and inner TBR or LTT, which account for 83% of the observations. Care must be taken is applying Figure 7.15 to other regions and environments, as changing wave and sediment combinations may produce additional bar combinations and certainly different frequencies of occurrence.

## 7.6 BEACH 'STATE' CURVES: BEACHES IN TIME AND SPACE

In order to quantify the nature and degree of variability for any beach, Short (1979b) developed the beach state curve, which is simply a frequency plot of the occurrence

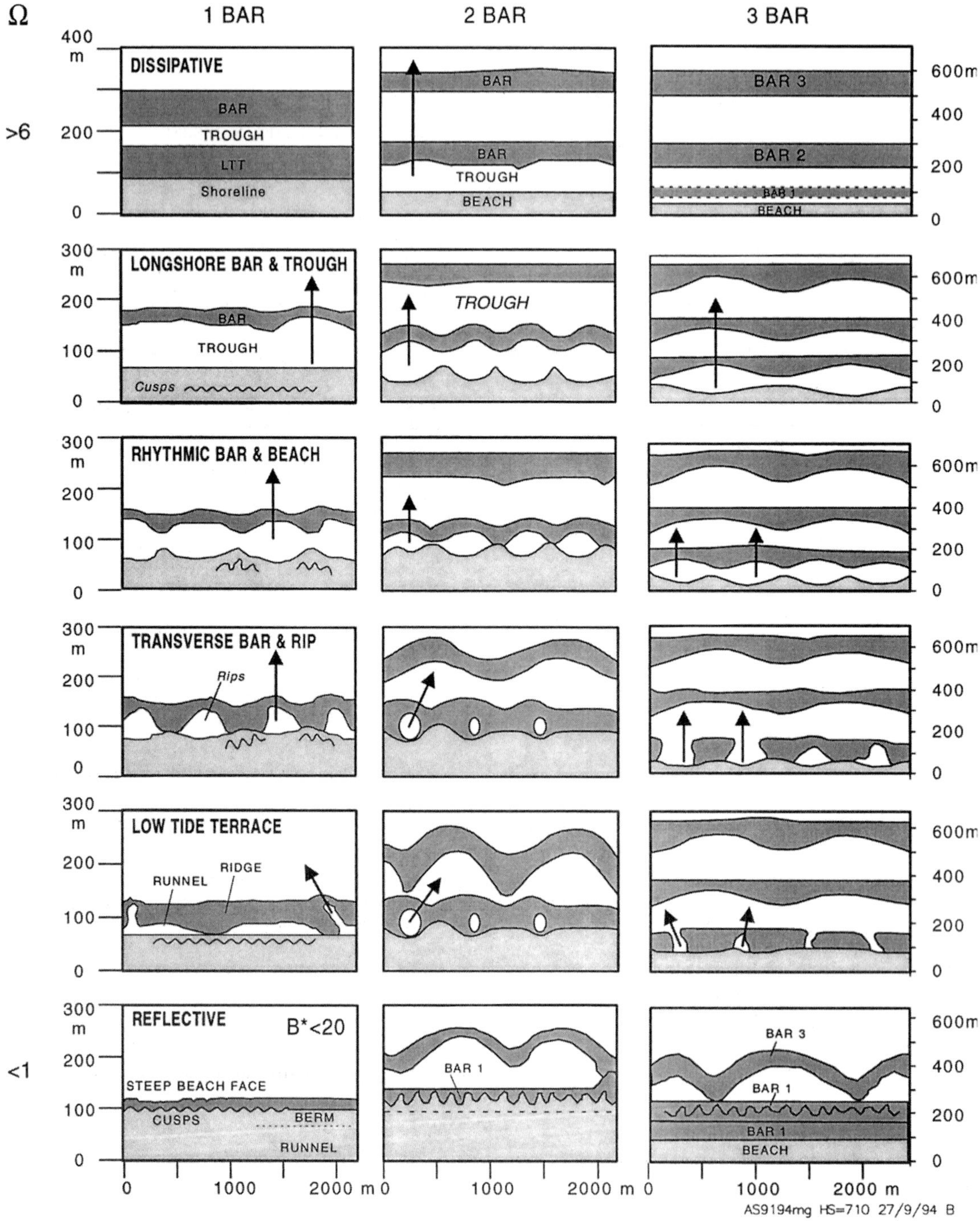


**Figure 7.14** A generalised model of one, two and three bar beach systems. The two and three bar systems have several important characteristics not present on the single bar model (source: Short and Aagaard, 1993; reproduced from the *Journal of Coastal Research* by permission of the Coastal Education and Research Foundation).

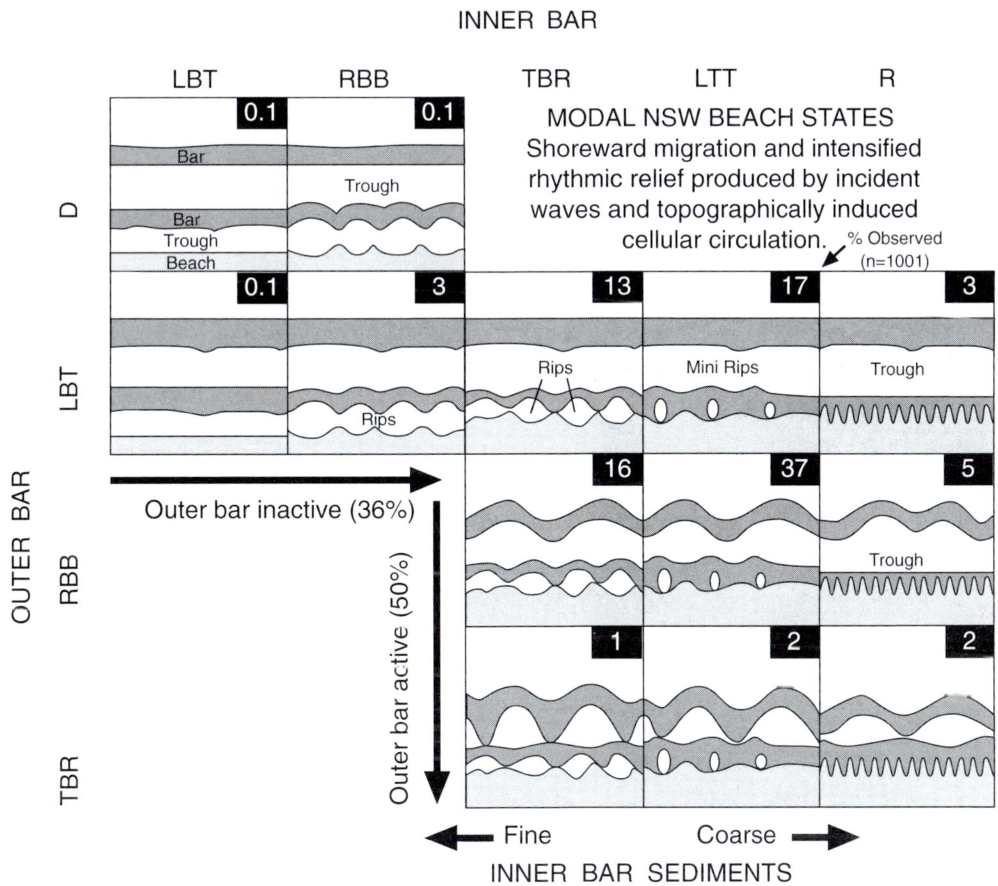


**Figure 7.15** Observed frequency of outer and inner bar combinations on double bar beaches, New South Wales coast, Australia (modified from: Short and Aagaard, 1993; reproduced from the *Journal of Coastal Research* by permission of the Coastal Education and Research Foundation).

of a particular beach type over a given period of time. If a sufficiently long (at least 12 months) record of preferably daily observations is available, the character of the beach can be accurately presented as a frequency curve (Figure 7.16). Using this method the morphodynamic character and variability of a single beach or group of beaches can be readily presented and compared.

### 7.6.1 Temporal beach change

Temporal change in beach state is driven by changing wave conditions, particularly $H_b$, but also $T$ and direction, i.e. horizontal movement in Figure 7.16. Changing prevailing wave height and/or period modify $\Omega$, which in turn leads to a change in beach state (Figure 7.5). Likewise changing wave direction while not necessarily changing beach state will require a reorientation of wave approach, surf zone currents and surf zone and beach morphology. Finally, constant wave conditions with an oblique wave approach can also lead to a longshore migration of beach morphology, as observed by Wright *et al.* (1982b). Sediment ($W_s$) is unlikely to cause a temporal change in beach type because it remains relatively constant on any particular beach. However if natural or induced (e.g. nourishment) changes in sediment size occur then they can lead to a temporal change in beach type under identical modal wave conditions, as observed by Short (1984a).

### 7.6.2 Spatial beach changes

Spatial changes in beach type along or between beaches, as indicated by vertical movement in Figure 7.16, are driven by both changes in modal wave conditions

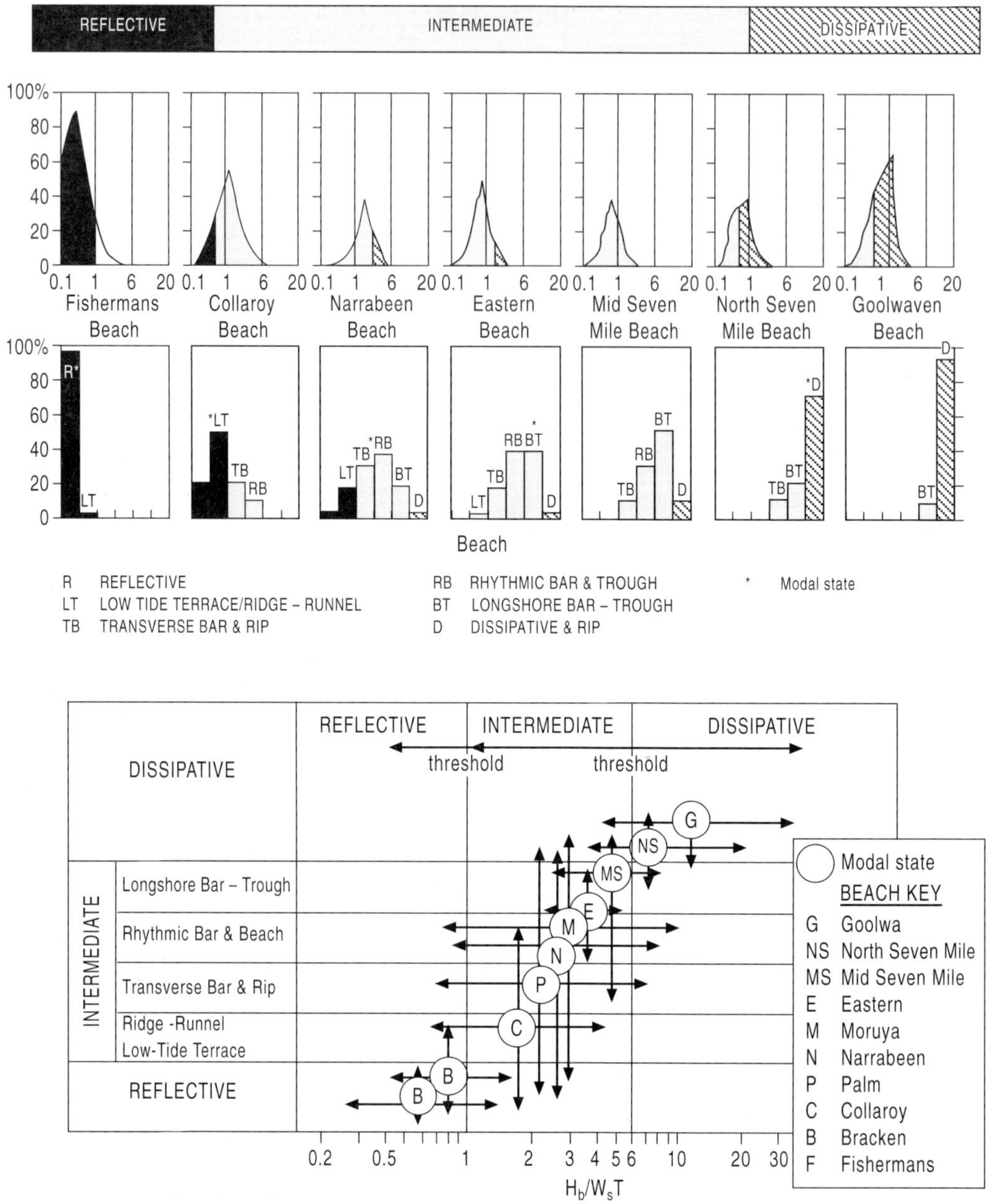


**Figure 7.16** A plot of seven contrasting (reflective to dissipative) south-eastern Australian beaches showing their range of dimensionless fall velocity ($\Omega$) based on $H_b$, $T$ and $W_s$ (top). The bar graphs (centre) show the corresponding occurrence of each beach type. The lower panel plots the modal beach type and $\Omega$ (circled), together with the temporal range in each (modified from Wright and Short, 1984).

**Table 7.3** Relationship of (a) mean grain size to beach state; and (b) mean grain size to breaker height, based on analysis of 430 New South Wales beaches (source: Short and Ni, 1997).

| (a) Beach state | Mean *D* (mm) | (b) $H_b$ (m) | Mean *D* (mm) |
|---|---|---|---|
| R | 0.45 | 0–0.5 | 0.32 |
| LTT | 0.25 | 0.6–1.0 | 0.26 |
| TBR | 0.25 | 1.1–1.5 | 0.28 |
| RBB | 0.23 | 1.6–2.0 | 0.27 |

(breaker wave climate), as well as changing sediment characteristics. Sediment through $W_s$ also modifies Ω and influences beach type. Whereas individual beaches tend to have constant sediment characteristics, different beaches and different parts of longer beaches can have markedly different sediment characteristics, leading to different beach types. Short and Ni (1997) plotted beach type against grain size for beaches with equivalent wave climates and found that coarser sand favoured more lower intermediate and reflective beaches, while finer sand favoured more higher intermediate beaches (Table 7.3a).

### 7.6.3 Wave height and sand size

Traditionally wave height has been directly and positively correlated with beach sediment size (King, 1972, p.308–311). However the foregoing indicates that higher energy beaches can be composed of fine sand through coarse sediments. Furthermore a regional survey of 430 beaches by Short and Ni (1997) found that there was no correlation between wave height and sediment size, if anything the higher energy beaches having finer sand (Table 7.3b). What this indicates, at least for the south-east Australian coast, is that sand size is inherited and is not selected by the prevailing waves. This is not to say that on drift coasts, in particular, there cannot be a longshore sorting of beach material, which may correlate positively with wave height (e.g. Bascom, 1951; Komar, 1976, pp 351–355: Bird, 1996a). Finally, in summary, Box 7.1 presents a qualitative overview of some of the major factors affecting beach type, change and bar number.

**Box 7.1** Summary

*Beach type* is a product of wave height, period and sediment size.

- *Dissipative* beaches (fine sand and high waves, and/or short periods) have wide, low gradient swash and surf zone and infragravity-dominated horizontal flows.
- *Intermediate* beaches (fine to medium sand, moderate to high waves) are dominated by cellular rip circulation.
- *Reflective* beaches (coarse sand, low waves, and/or long periods) have steep, narrow beaches and no surf zone.

*Beach change* is driven by changes in waves and sediment.

- *Temporal changes* in beach type are usually driven by changing prevailing wave conditions.
- *Spatial changes* in beach type are driven by changes in both wave climate and sediment characteristics.

*Bar number* increases with decreasing wave period and beach gradient (sand size), and requires high waves.

*Multi-bar beaches* have the following features:

- a *hierarchy* of bar types with a shoreward shift to lower energy beach/bar types owing to wave breaking;
- the outer bar/s are less active and commonly *inactive* during lower non-formative wave conditions, particularly in sea environments.

# 8

# The Effect of Tides on Beach Morphodynamics

*Gerhard Masselink and Ian Turner*

## 8.1 INTRODUCTION

The most important processes that control the morphology and hydrodynamics of beach environments are associated with the dissipation of incident wave energy. These processes include wave breaking and attenuation, nearshore cell circulation, bed return flow, infragravity waves, wave set-up and swash (Chapters 4 and 5). A secondary role – and on beaches with large tidal ranges an important one – is played by the ocean tides. The tide-generating forces and the modification of tidal waves in shallow water are not discussed within this chapter. A comprehensive discussion of tidal theory can be found in many other coastal texts, including Komar (1976), Pugh (1987) and Carter (1988). Instead, the focus here is on the role of tides in beach morphodynamics. The diurnal or semi-diurnal rise and fall of the ocean surface has the *direct* effects of modulating the surf zone water level and generating tidal currents in the nearshore zone, and the *indirect* effect of causing the beach groundwater table to fluctuate. These two mechanisms significantly modify the morphodynamic behaviour of a beach.

The degree to which tide-related processes have an effect on coastal morphodynamics is expected to increase progressively with tidal range. This notion has resulted in a geographical classification of coastal environments based on the absolute tide range. It is popular to divide coasts into three categories: *micro-tidal* ranges of less than 2 m, *meso-tidal* ranges between 2 and 4 m, and *macro-tidal* ranges in excess of 4 m (Davies, 1964; Chapter 2). Distinct coastal 'morpho-types' and 'process regimes' may be identified that are characteristic of each of these tide range classes. The common assumption is that micro-tidal coasts are by definition wave-dominated and typified by beaches and barriers, in contrast to macro-tidal environments which are tide-dominated and characterised by tidal flats. By this simple scheme, meso-tidal coasts experience a mixed wave/tide influence. Despite the widespread use of Davies' (1964) classification in the coastal literature, the approach is somewhat misleading when applied to shoreline morphology. As will become apparent in this chapter, the concept of a *relative* rather than *absolute* tide range is a more useful morphodynamic concept to pursue (Davis and Hayes, 1984).

The majority of beach research has been carried out in micro-tidal environments and as a result tidal effects have received comparatively little attention in the literature. Notable exceptions are the studies concerning tide-induced beach profile changes (e.g. Duncan, 1964; Strahler, 1966; Clarke *et al.*, 1984; Eliot and Clarke, 1988), the work on British ridge and runnel morphology (e.g. King and Williams, 1949; King, 1972; Parker, 1975; Mulrennan, 1992), intertidal bedform studies (Hunter *et al.*, 1979; Hawley, 1982), morphodynamic investigations (Wright *et al.*, 1982b; Jago and Hardisty, 1984; Horn, 1993), the hydrodynamic and sediment transport studies of the British B-Band programme (Russell *et al.*, 1991; Davidson *et al.*, 1993), and the more recent work on the macro-tidal coast of central Queensland, Australia (e.g. Short, 1991b; Masselink, 1993; Masselink and Short, 1993; Turner, 1993a, 1995a,b; Masselink and Hegge, 1995). This chapter will primarily draw from the results and conclusions of this latter Australian research.

*Handbook of Beach and Shoreface Morphodynamics*. Edited by A. D. Short.
© 1999 John Wiley & Sons Ltd.

## 8.2 TIDE-INDUCED MIGRATION OF HYDRODYNAMIC PROCESSES

### 8.2.1 Tidal translation rate

Tide-induced changes in nearshore water levels cause the shoreline to periodically oscillate across the intertidal beach profile. The lateral extent to which the shoreline migrates is simply a function of the tide range $TR$ and the beach gradient $\tan\beta$, and in turn defines the width of the intertidal zone $x_{IT}$

$$x_{IT} = TR/\tan\beta \tag{8.1}$$

The most extensive intertidal zones are found on gently sloping beaches with large tidal ranges, and it is not unusual for this region to be over 500 m wide on macro-tidal beaches.

The rate with which the shoreline shifts across the intertidal profile under the influence of the tide is referred to as the *tidal translation rate*. The time-varying vertical fluctuation of the mean water level $\eta_t$ can be approximated by a simple sinusoid with amplitude $A_{tide}$ (equal to half the tide range) and period $T_{tide}$

$$\eta_t = A_{tide}\sin\left(\frac{2\pi}{T_{tide}}t\right) \tag{8.2}$$

where $t$ represents time. If the instantaneous shoreline position is defined as $S_t$, then the tidal translation rate $\partial S_t/\partial t$ is given by

$$\frac{\partial S_t}{\partial t} = \frac{\partial \eta_t}{\partial t}\frac{1}{\tan\beta} = \frac{TR\pi}{T_{tide}\tan\beta}\cos\left(\frac{2\pi}{T_{tide}}t\right) \tag{8.3}$$

Maximum translation rates over a tidal cycle are therefore attained at mid-stages of the tide, when $t = 1/4T_{tide}$ or $3/4T_{tide}$. The largest tidal translation rates are found on gently sloping beaches that experience macro-tidal, semi-diurnal tides. For example, on a beach with a gradient of 0.02, a tide range of 8 m and a tidal period of 12 h, the tide will translate horizontally across the intertidal profile at the mid-ebbing tide ($t =$ h) with a rate of around 3 cm/s or 9 m every 5 min. In extremely low gradient coastal environments, such as tidal flats (Hale and McCann, 1982; Short, 1991b), the tidal translation rate may exceed walking pace.

### 8.2.2 Migration of morphodynamic zones

Tides do not directly affect wave driven hydrodynamic processes because the associated time scales are very different (hours as opposed to seconds). However, the tidal stage determines *where* certain nearshore processes operate and the tidal translation rate prescribes for *how long* these processes can act upon the bed and affect the beach morphology. The tide represents an additional level of complexity to beach morphodynamics by introducing a temporal (over the tidal cycle) and spatial (across the intertidal profile) variation in morphodynamic processes.

In Chapter 1, the beach profile was sub-divided into four morphodynamic regions, each of which can be characterised by its typical morphology and hydrodynamic processes (Figure 1.1). Wave shoaling and refraction occur on the low gradient, often fine grained shoreface. Nearshore currents and wave-breaking processes govern the medium gradient and sometimes barred surf zone. Swash processes and their interaction with beach groundwater take place on the steep gradient, relatively coarse grained beachface. Finally, the backshore is primarily affected by aeolian and/or overwash processes.

On tideless beaches, the location and spatial extent of each of these morphodynamic zones are relatively fixed and the hydrodynamic processes all operate in their 'own' region. On tidal beaches, however, aeolian, swash, surf zone and shoaling wave processes shift both vertically and horizontally with the rise and fall of the tide, and the intertidal beach profile is influenced to varying degrees by a combination of these processes. If the tide range is sufficiently large, as on macro-tidal beaches, certain parts of the beach profile may be subjected to the entire suite of processes through a single tidal cycle (Figure 8.1). Consequently, the equilibrium conditions on the intertidal profile are defined by a mixture of different hydrodynamic processes that depend on location-specific wave/tide/sediment characteristics (Masselink, 1993).

Wright *et al.* (1982b) conducted experiments on Cable Beach, Broome (Western Australia; mean spring tide range (MSR) = 9.5 m) and over a half lunar tidal cycle calculated the rate of wave energy dissipation across the intertidal profile. Only energy dissipation by bed friction was considered, since this is equivalent to the rate of work done to the bed, and hence fundamental to moulding the beach morphology. To investigate the relative contribution of differing hydrodynamic processes to total energy dissipation, surf zone processes (breaking waves) were distinguished from nearshore processes (unbroken shoaling waves). The results of this research are shown in Figure 8.2, indicating the proportion of total wave energy dissipation accounted for by the breaking surf zone waves for different deep water wave heights $H_0$. Under all wave conditions, even during relatively rare 2 m waves, only a minor part of the morphological work done on the lower part of the

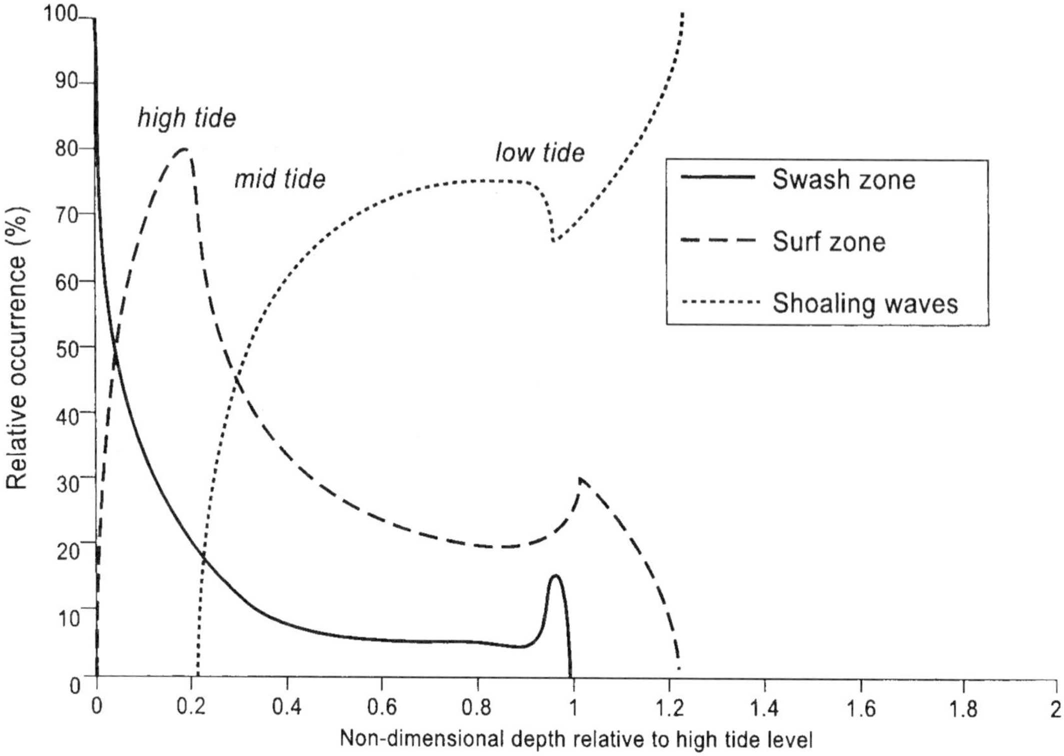


**Figure 8.1** The relative occurrence of swash, surf zone and shoaling wave processes across the beach profile calculated over one tidal cycle. The input parameters are wave height $H = 1$ m, wave period $T = 8$ s, sediment fall velocity $w_s = 0.03$ m/s, breaker criterion $\gamma = 0.8$ and tide range $TR = 6$ m. The location on the beach profile has been normalised by the tide range, such that non-dimensional depths of 0, 0.5 and 1 represent high tide, mid-tide and low tide, respectively (after Masselink, 1993).

intertidal profile is by surf zone waves. Under modal wave height conditions ($H_0 < 1.0$ m) this is also true of the middle part of the intertidal zone, and only the upper part of the intertidal profile is dominated by surf zone processes. Wright *et al.* (1982b) concluded that, dynamically, the lower part of the intertidal profile of macro-tidal beaches is similar to the nearshore/offshore zones fronting micro-tidal beaches, since these parts of the beach profile are dominated by shoaling unbroken waves and, to a lesser degree, by tidal currents.

Similar conclusions were reached by Jago and Hardisty (1984) and Horn (1993) who investigated the sedimentology and morphodynamics of two British macro-tidal beaches (respectively, Pendine Sands in south-west Wales with MSR = 7.5 m, and Ballakinnag Cronk on the Isle of Man with MSR = 5.5 m). On the basis of the sequence of sedimentary structures, Jago and Hardisty (1984) inferred that swash zone processes establish the upper intertidal facies, surf zone waves during storms the mid-intertidal facies, and shoaling waves the lower intertidal and sub-tidal facies. According to Horn (1993), the lower part of the beach profile is in equilibrium with near-bed flow velocities produced by shoaling waves, whereas other processes must be invoked, such as the effect of velocity asymmetry due to percolation losses (Section 8.4.2), to explain morphology and sediment texture on the upper beach.

### 8.2.3 Tidal currents

Tidal currents typically operate seaward of the surf zone at considerable water depths. Generally, they have little influence on beach morphodynamics. However, as the tide range increases, and nearshore processes are introduced to the intertidal zone, tidal currents can play an increasing role across the lower part of the intertidal zone and in the shallow sub-tidal zone. Wright *et al.* (1982b) measured synchronous time series of water surface elevation and longshore current velocity, collected approximately 3 m below mean low water spring level

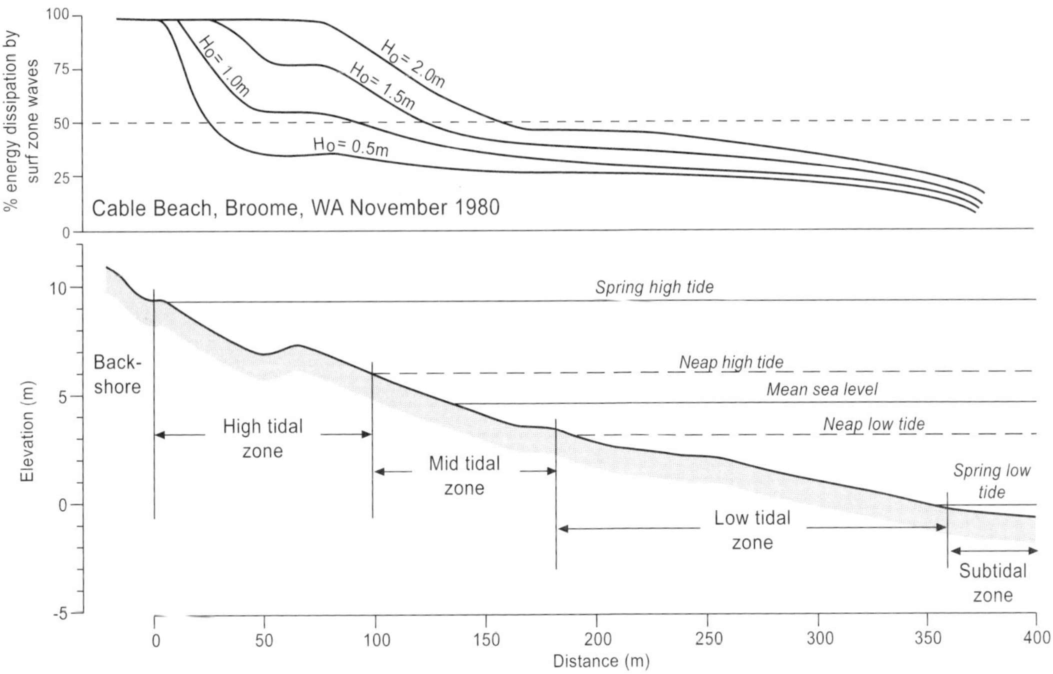


**Figure 8.2** Percentage of surf zone dissipation rate relative to the total dissipation rate calculated for a range of incident wave heights $H_0$ across the profile of Cable Beach, Australia. The dissipation rates have been calculated over a half lunar tidal cycle, and indicate that under modal wave conditions ($H_0 < 1$ m), over the larger part of the beach profile most work is performed by non-breaking shoaling waves, rather than surf zone processes (after Wright *et al.*, 1982b).

on Cable Beach. They found that the semi-diurnal periodicity of the current record and the phase correspondence between the longshore current and the water level clearly demonstrated that the shore-parallel flow is tide-driven. The most important feature of their current record was the pronounced asymmetry, characterised by relatively strong northerly currents with maxima of 0.4 m/s and considerably weaker southerly flows of shorter duration with maxima of only 0.25 m/s. As a consequence, there exists a residual current that flows towards the north with a mean velocity of 0.13 m/s.

Residual tidal currents are a product of the direction of tidal progradation and nearshore topographic effects and are common along tidal coastlines (Wright *et al.*, 1982b; Masselink and Hegge, 1995). Although these currents may not directly affect the beach morphology – the associated current velocities are often too small they may significantly influence longshore sediment transport rates. For this reason, the presence of residual currents, and the direction and magnitude of the flow, has potentially major implications for the large-scale sediment budget of tidal coastlines.

### 8.2.4 Temporal variation in beach morphodynamics over a tidal cycle

#### *8.2.4.1 Wave height*

Most beaches are characterised by a concave-upward intertidal/sub-tidal. As a consequence, the shoreface (i.e. that part of the seabed affected by surface ocean waves) is wider at low tide than at high tide, especially on larger tide range coasts. During low water the greater width of the shoreface during low water results in increased rates of wave energy dissipation due to bed friction, and reduced nearshore wave energy levels (Jago and Hardisty, 1984). It is therefore common to observe larger wave heights in the surf zone at high tide relative to low tide.

#### *8.2.4.2 Beach morphodynamics*

The beach gradient that is encountered by shoreward propagating incident waves is also dependent on the tidal stage, again due to the concave character of tidal beach profiles. As a result, beach morphodynamic state changes with the tidal water level, such that surf zone

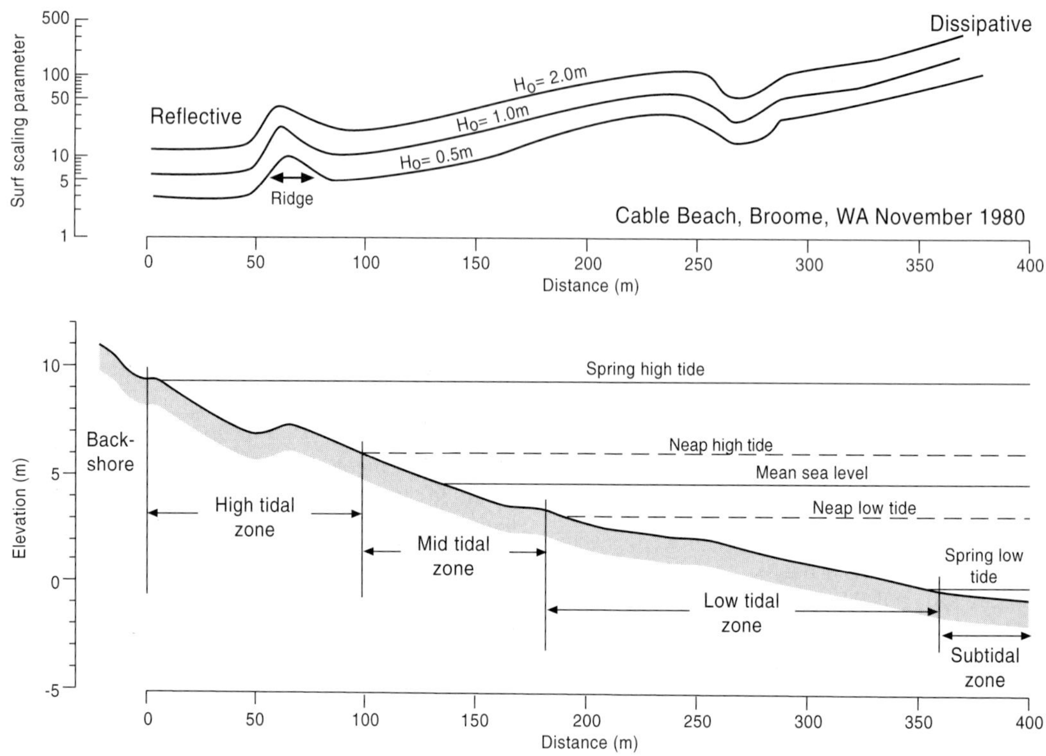


**Figure 8.3** Tidal zonation and cross-shore variation of the surf scaling parameter ε for Cable Beach, Australia (after Wright *et al.*, 1982b).

conditions on the (steeper) upper part of the profile may be morphodynamically different from those occurring on the (lower gradient) lower part. A useful indicator of surf zone morphodynamics is the surf scaling parameter ε, which characterises the type of surf zone (Table 7.1). Figure 8.3 plots the surf scaling parameter as a function of cross-shore position on the intertidal profile for Cable Beach, Western Australia (Wright *et al.*, 1982b). On Cable Beach, and on many other larger tidal range beaches, surf zone conditions around high tide are reflective to intermediate with plunging/surging breakers and a narrow surf zone, whereas dissipative to ultra-dissipative conditions prevail during low tides characterised by spilling breakers and a wide surf zone.

### *8.2.4.3 Wave breaking*

On micro-tidal beaches, surf zone morphodynamics may also depend on the tidal stage with waves breaking over the nearshore bar morphology during low tide and breaking directly on the beachface under high water conditions. Alternatively, on multi-barred beaches, waves may be breaking on the outer bar(s) under low water conditions and on the inner bar(s) during high tide. The degree of breaking-induced wave energy dissipation over the nearshore morphology, and hence the amount of incident wave energy that traverses the surf zone to reach the shoreline, depends primarily on the tidal water level and the associated water depth over the bar(s) (Tucker *et al.*, 1983). Due to the filtering effect of nearshore bar morphology, high tide swash processes are often more energetic than those prevailing at low tide.

### *8.2.4.4 Rip current velocities*

Early observations by Shepard *et al.* (1941) and McKenzie (1958) have suggested that the nearshore cell circulation system intensifies towards low tide. This is confirmed by measurements (Sonu, 1972; Masselink and Hegge, 1995) and documented by the increase in rip velocities and rip-related rescues towards low tide (Short and Hogan, 1994). The larger flow velocities in the feeder and rip channels during lower water levels are primarily the result of an increase in flow concentration. As the nearshore water level falls, the nearshore bar morphology becomes increasingly shallow. This inhibits offshore flows across the bar and confines the flow to the channels, resulting in increased current velocities. It has also been suggested that the seaward drainage of beach groundwater during the falling tide may contribute to increased cell circulation during low water conditions (Komar, 1976). However, the volume of beachface seepage (Turner and Nielsen, 1997) is probably insignificant compared to that associated with wave-driven cell circulation.

#### 8.2.4.5 *Suspended sediment*

Temporal variation in suspended sediment transport induced by the tide was observed by Russell *et al.* (1991) and Davidson *et al.* (1993), who conducted field experiments on two macro-tidal beaches in Great Britain. On both beaches, a distinct asymmetry in the cross-shore sediment transport was monitored, with little net transport occurring on the flooding tide, and strong net offshore transport on the ebbing tide. Davidson *et al.* (1993) ascribe the asymmetry to the destruction of the high-tide wave ripple field during the falling tide, resulting in increased levels of sediment resuspension. The effects of beach dewatering during the falling tide may also have been important (Russell *et al.*, 1991).

### 8.2.5 Tidal zonation of the beach profile

The varying degrees to which swash, surf and shoaling wave processes influence the different parts of the profile of tidal beaches have resulted in the recognition of a typical zonation found on meso- and macro-tidal beaches (Figure 8.3). Each of the zones is distinguished by different morphologic and sediment characteristics, frequencies of inundation, and dynamic regimes (Wright *et al.*, 1982c). Unfortunately, to date most researchers have used differing terminology – for example, high tidal zone (Wright *et al.*, 1982c), upper foreshore (Jago and Hardisty, 1984), high tide zone (Horn, 1993), upper intertidal (Masselink and Hegge, 1995) – and this can lead to some confusion. For convenience, the terminology of the most recent study is adopted here.

The *upper intertidal zone* extends from spring high tide to neap high tide level and has the steepest beach gradient. The morphodynamics of the upper intertidal zone are primarily governed by swash and, to a lesser extent, aeolian processes. Relatively stationary tide conditions are experienced in the upper intertidal zone at high tide. The *mid-intertidal zone* is centred on the position of mean sea level, between neap high tide and neap low tide level, and is dominated by a mixture of swash and surf zone processes. The greatest tidal translation rates, and hence the least stationary tide conditions, are encountered in the mid-intertidal zone. The *lower intertidal zone* is located between neap low tide and spring low tide level and the most important morphodynamic processes are those driven by surf zone and shoaling waves. During low tide, the tide level remains quasi-stationary over the lower intertidal zone for a significant period of time. Therefore bar morphology may be found in the lower intertidal zone. The *sub-tidal zone* extends seaward of spring low tide level and is characterised by the lowest gradient and the finest sediments. Shoaling wave processes are most important in the sub-tidal zone.

The identified zones will merge as the tide range decreases, with each of the zones progressively losing its identity. On most micro-tidal beaches a division of the beach profile into four regions is impractical. On such beaches it is convenient to divide the beach into three zones: a swash zone, surf zone and a sub-tidal zone.

## 8.3 TIDE-INDUCED FLUCTUATIONS OF THE BEACH GROUNDWATER TABLE

In addition to the direct effect of tides in modulating the occurrence of swash, surf and shoaling wave processes across the intertidal profile, tide-induced beach groundwater table fluctuations can play an important part in shaping the morphology of tidal beaches. The role of beach groundwater in beach morphodynamics has received only limited attention in the past; however a resurgence of interest since the 1980s is beginning to redress this. In Chapter 5 the physical mechanisms were presented that support the field and laboratory observation that an unsaturated beachface promotes sediment deposition, whereas a saturated beachface encourages sediment erosion. Because the elevation of the water table determines the time-varying saturation characteristics of the intertidal profile, an understanding of the dynamics of beach groundwater is required for a full description of beach morphodynamics in tidal environments.

The *water table* is most correctly defined as the mean water surface where porewater pressure is equal to atmospheric pressure (Figure 8.4). The elevation of the water table within beaches is determined by prevailing hydraulic conditions, including wave run-up, tide range and rainfall, and the characteristics of the beach material that define hydraulic conductivity, principally sediment size, sorting and porosity. Defining the water table surface by pore pressure distribution is more precise than an intuitive distinction between saturated and

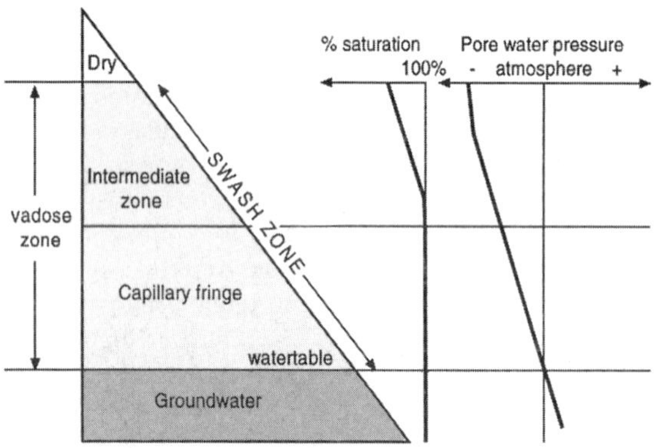


**Figure 8.4** Definition of the water table as the surface where pore pressure is equal to atmospheric pressure. The saturated capillary fringe may extend up to 0.5 m above the water table in fine sand beaches, and hence a perhaps more intuitive definition based on saturation characteristics is imprecise (after Turner, 1993b).

unsaturated zones within the beach, because capillary forces can result in a saturated region extending up to 0.5 m above the water table in sandy beaches (Turner, 1993b). The neglect of this significant region of saturated sand *above* the water table has mistakenly led some studies to associate rapid beach water table fluctuations with high rates of *upward* vertical flow (Chappell *et al.*, 1979). In reality, when the capillary fringe extends to the beachface surface, with each swash cycle a minute *downward* infiltration of water will produce the observed phenomenon of near-instantaneous water table rise (Turner and Nielsen, 1997).

Tide-induced water table fluctuations have been recorded across a wide range of beach morphologies, including pebble/shingle beaches (Kirk, 1966), protected estuarine beaches, tropical carbonate beaches (Dominick *et al.*, 1971), and macro-tidal beaches (Turner, 1993a). Additional mechanisms that can cause the water table to rise and fall include seasonal variations in the coastal groundwater elevation (e.g. Clarke and Eliot, 1983, 1987), barometric pressure changes associated with the passage of weather systems (e.g. Lanyon *et al.*, 1982; Eliot and Clarke, 1986), the propagation of shelf waves (e.g. Lanyon *et al.*, 1982), the occurrence of storm events (e.g. Turner *et al.*, 1995) and infragravity and incident surf zone waves (e.g. Waddell, 1976; Hegge and Masselink, 1991).

### 8.3.1 The 'classic' description of tidal groundwater oscillations

An *aquifer* is defined as a saturated geologic unit through which water can flow with relative ease. Beaches and dunes are termed *unconfined* aquifers because commonly the upper boundary to groundwater flow is defined by the water table itself, rather than by some surface layer of impermeable material. For simplicity, it is common to consider only shore-normal (one-dimensional) groundwater flow, and to assume that the magnitude of tide-induced water table fluctuations is small relative to the depth $d$ to some underlying impermeable layer (e.g. clay or rock).

The fundamental equation describing the rate of horizontal groundwater flow $u$ through unconfined aquifers is Darcy's Law

$$u = -K\frac{\partial h}{\partial x} \tag{8.4}$$

where $K$ is hydraulic conductivity and $h$ is the elevation of the free water surface (water table) at a distance $x$ from the beachface. This simple, yet widely applicable expression indicates that the rate of groundwater flow is proportional to the prevailing hydraulic gradient, or slope of the water table. By combining Darcy's Law with the necessary requirement that total water volume is conserved, the differential equation for unsteady (or time-varying), unconfined groundwater flow can be obtained

$$\frac{\partial h}{\partial t} = \frac{Kd}{S}\frac{\partial^2 h}{\partial x^2} \tag{8.5}$$

In groundwater hydrology texts the product $Kd$ is invariably referred to as aquifer *transmissivity* and the term $S$ is often called *drainable porosity*, indicating that it is equivalent to the potential volume of water released per unit volume of the aquifer. Alternatively, $S$ is frequently referred to as *specific yield*.

When an idealised beach is considered, consisting of a vertical beachface overlying a horizontal impermeable layer (Figure 8.5), Equation 8.5 can be solved to describe fluctuations of the beach water table in response to tides. By defining the rise and fall of the still water level as a simple sinusoidal oscillation around the elevation of mean sea level (Equation 8.2) and assuming that no water table fluctuations occur at a distance far from the ocean (i.e. the landward boundary condition), the solution to Equation 8.5 is given by

$$h(x,t) = \frac{1}{2}TRe^{-x\sqrt{\pi S/T_{tide}Kd}}\sin\left(\frac{2\pi t}{T_{tide}} - x\sqrt{\pi S/T_{tide}Kd}\right) \tag{8.6}$$

Equation 8.6 is the standard solution to tide-induced groundwater fluctuations that can be found in many

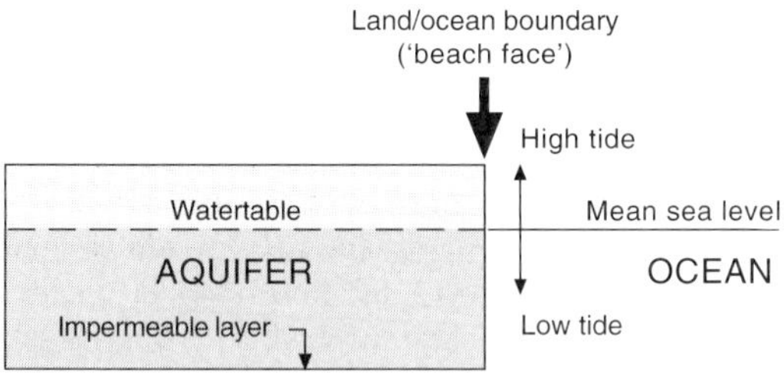


**Figure 8.5** An idealised coastline may be thought of as a vertical beachface, underlain by a horizontal, impermeable layer (e.g. bedrock or clay). This approximation to 'real' beachface morphology greatly simplifies the analysis of tidal groundwater fluctuations. However, the common use of the approach in the past has meant that a number of the more important features of beach water table dynamics are frequently neglected.

popular hydrogeological texts (e.g. Todd, 1980; Fetter, 1988). By maximising the sine-term in this equation, the magnitude of groundwater fluctuations at a distance $x$ from the shoreline is simply

$$TRe^{-x\sqrt{\pi S/T_{tide}Kd}} \tag{8.7}$$

with the tidal wave propagating landward through the aquifer at a velocity

$$\sqrt{4\pi Kd/T_{tide}S} \tag{8.8}$$

Typically, a 1.0 m tide range will result in fluctuations of the water table of the order of 0.1 m at a distance of 10 m inland of the beachface. In addition, water table fluctuations landward of the swash zone usually lag the tide by 1 to 3 h.

In summary, Equation 8.6 describes an exponentially decaying sinusoidal wave propagating inland around the mean elevation of the ocean. Most significantly, according to this classic approach, the time-averaged elevation of the water table at the beachface is equal to mean sea level.

### 8.3.2 Seepage face development

The complete analytical description of tide-induced water table fluctuations is extremely complex. By adopting a perturbation approach, Nielsen (1990) develops a solution that incorporates at least some of the effects of a sloping beachface. However, fundamental to this work is the assumption that the water table follows the movement of the tide across the intertidal profile, and this is a significant limitation. In reality, this condition is rarely met; during the falling tide the water table

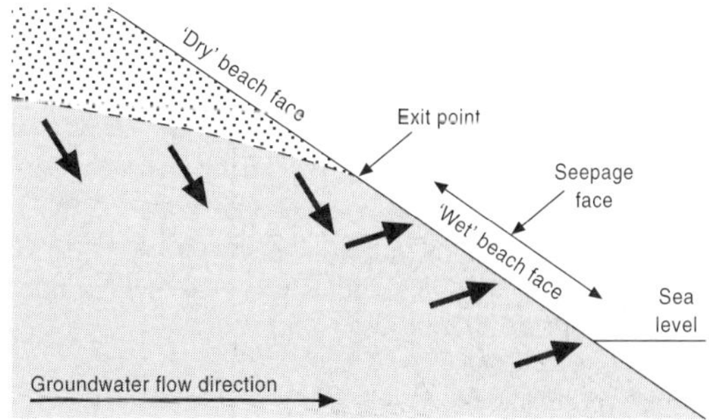


**Figure 8.6** Definition sketch of the *exit point*, distinguishing the upper boundary of the *seepage face* where the water table coincides with the beachface. The general direction of flow within the beach is indicated.

typically decouples from the ocean, resulting in the formation of a *seepage face* where groundwater outcrops on the beachface. Referring to Figure 8.6, the *exit point* defines the intersection of the groundwater surface with the intertidal profile. Following ocean–water table decoupling, the continued motion of the exit point remains independent of the still water level, prior to overtopping by the flooding tide. The importance of seepage face formation is that it distinguishes an upper beachface region of potential swash infiltration, from a lower region of groundwater effluence.

A recent field study of seepage face development on tidal beaches is reported by Turner (1993a, 1995b). By making the simplifying assumption that the dynamics of the exit point are independent of the pressure distribution within the beach and ignoring inertial terms, the maximum rate of vertical exit point fall $V_{ep}$ across a beachface of slope tan $\beta$ can be approximated by

$$V_{ep} = -\frac{K}{S}\sin^2\beta \tag{8.9}$$

where $K$ and $S$ are beachface hydraulic conductivity and specific yield. The presence/absence of seepage face development can now be predicted as a function of tide and beach characteristics. At the instant of water table–tide decoupling, the vertical velocity of tidal fall will, by definition, equal the maximum velocity of exit point fall. Differentiating Equation 8.2 with respect to time and equating to Equation 8.9 gives

$$\frac{\pi TR}{T_{tide}}\cos\left(\frac{2\pi}{T_{tide}}t\right) = \frac{K}{S}\sin^2\beta \tag{8.10}$$

The maximum rate of tidal fall during the ebbing tide ($0 < t \leq T_{tide}/2$) occurs at time $t = T_{tide}/4$. Substitution into Equation 8.10 and rearrangement yields

$$\Sigma = \frac{\pi TR}{T_{tide}} \frac{S}{K \sin^2 \beta} \tag{8.11}$$

where $\Sigma$ is termed the *seepage face parameter* (Turner, 1995a). In Equation 8.11, the variables are grouped to distinguish those defined by characteristics of the tide (range $TR$ and period $T_{tide}$) and the beachface (hydraulic conductivity $K$, specific yield $S$ and beachface angle $\beta$). Water table–tide decoupling is then simply predicted by

$\Sigma \leq 1$ no seepage face development
$\Sigma > 1$ seepage face development

In Figure 8.7 the seepage face parameter $\Sigma$ is plotted as a function of beachface slope and hydraulic conductivity. Micro- to macro-tidal ranges from 1 to 6 m are indicated, and a semi-diurnal tidal period of 12 h assumed. To aid interpretation, the empirical results of Krumbein and Monk (1942) have been used to indicate sediment characteristics of typically well sorted ($\sigma_\phi = 0.35$) sandy beaches. For low profile slopes (1–2°) and/or fine sands it is evident that groundwater will outcrop on micro- to macro-tidal beaches. In contrast, on beaches composed of the coarsest of sand ($\approx$ 1 mm) and steepest of slopes observed in nature (7–8°), a seepage face will not develop. For more typical beachface slopes (3–5°), seepage face development is anticipated for all grades of fine to coarse sediment.

The simplification that within-beach pressure distribution can be neglected to describe the dynamics of the exit point should be approached with some caution, although in reality rates of groundwater seepage to the coast, relative to rates of tidal rise and fall, are typically sufficiently small to limit the error this assumption introduces. The predictive capability of this approach is encouraging when it is used to simulate recorded exit point motion. In Figure 8.8 simulated and observed seepage face development are depicted through a 6.0 m spring tide monitored at North Harbour Beach, on the central Queensland coast, Australia (Turner, 1995b). Close agreement is apparent, distinguishing the time-varying extent of saturated and non-saturated regions of the intertidal profile through a tidal cycle.

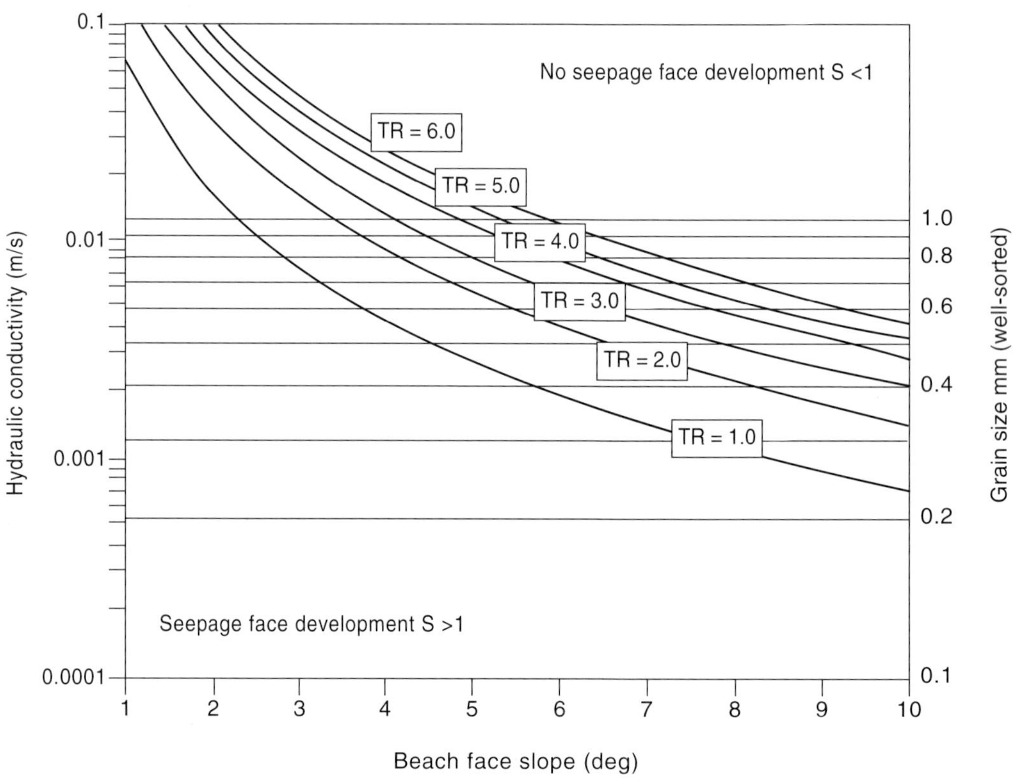


**Figure 8.7** *Seepage face parameter*: this figure can be used to predict the presence ($\Sigma > 1$) or absence ($\Sigma \leq 1$) of seepage face development as a function of beachface sediment characteristics and slope (after Turner, 1995a).

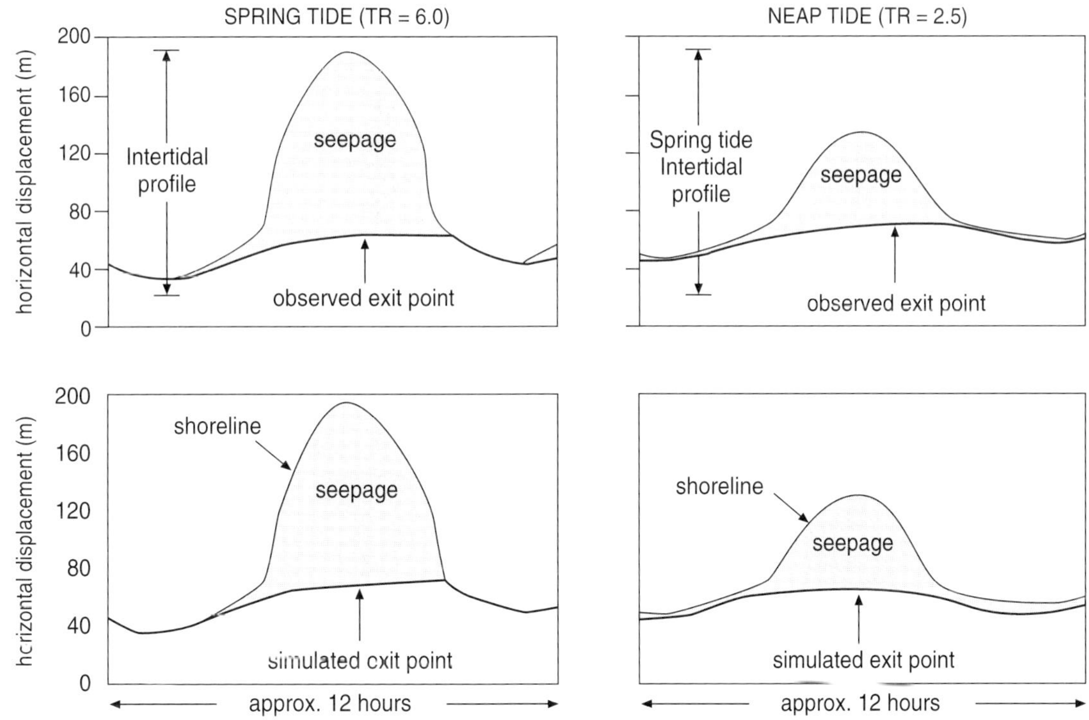


**Figure 8.8** Observed and simulated seepage face development at North Harbour Beach, Queensland, Australia. Two contrasting tide ranges are indicated; a spring tide range of 6.0 m, and a neap tide range of 2.5 m (after Turner, 1995b).

### 8.3.3 Wave–tide–water table interactions

Water table fluctuations due to the rise and fall of the tide are of primary interest here, but for the sake of completeness it is important to note that both wave run-up and set-up are also significant to the dynamics of beach water table fluctuations. Wave set-up at the shoreline results in a raising of the effective elevation of the ocean, and the run-up of waves across the beachface further elevates the potential zone of infiltration (Chapter 5). An extensive field investigation reported by Hanslow and Nielsen (1993) found that, due to the combined effects of wave run-up and set-up, the exit point of the water table on natural beaches is raised approximately 40% of the root mean square (r.m.s.) offshore wave height above the still water elevation of the ocean tide. The presence of tides will regulate the occurrence of potential infiltration in the swash zone. At high tide infiltration will occur as run-up extends above the elevation of the water table outcrop, but at low tide infiltration may be inhibited if the exit point lies above the elevation of the maximum run-up limit.

## 8.4 MORPHOLOGICAL RESPONSE TO TIDES

The periodic shifting of morphodynamic zones over a tide cycle, superimposed on an intertidal profile that is alternately saturated/unsaturated due to tide-induced groundwater fluctuations, are important processes that significantly affect the morphology of beaches located on tidal coasts. In the following sections these mechanisms are discussed separately; however, it is important to remember that this is only for the sake of convenience. The morphological response of beaches to tides results from a combination of both transient morphodynamic zones and the dynamics of beach groundwater; the relative contribution of these to resulting beach morphology is inseparable.

Tidal cycle changes in beach morphology are observed to occur at two distinct time scales: those associated with the variation in ocean water levels through a single semi-diurnal (or diurnal) tide (e.g. Duncan, 1964; Strahler, 1966; Otvos, 1965), and the lower frequency response of the beach profile to the lunar neap–spring cycle (e.g. Inman and Filloux, 1960; Aubrey *et al.*, 1976;

Clarke *et al.*, 1984; Wright *et al.*, 1987; Masselink and Hegge, 1995). In general, on smaller tide range beaches ($TR < 2$ m), the periodic rise and fall of the tide induces cyclic patterns of profile response that result in little net morphological adjustment. But with increasing tide range, the 'equilibrium profile' becomes increasingly influenced by tide-induced processes.

### 8.4.1 Profile adjustment to shifting morphodynamic zones

The ever-transient shifting of contrasting morphodynamic zones across the profile of tidal beaches will influence both the overall profile form, and the location and nature of specific morphological features in the nearshore and intertidal.

#### *8.4.1.1 Profile form*

Early morphodynamic models considered oscillatory flow asymmetry to be of fundamental importance to equilibrium beach gradient (e.g. Bowen, 1980; Bailard and Inman, 1981; Hardisty, 1986). Unbroken shoaling waves are characterised by relatively symmetrical flows and favour gentle gradients, whereas the flow in the swash and surf zone is dominated by large asymmetries and hence promotes the formation of steep bed gradients. The dominant role of shoaling waves and tidal currents on the lower part of beaches with large tide ranges and the increasing importance of swash and surf zone processes in the onshore direction generally result in a low gradient, concave-upward beach profile. As swash and surf zone processes become less important with increasing tide range, the overall beach gradient is expected to decrease. However, apart from the simulation model of Masselink (1993) confirming such a trend, no systematic research has been carried out to explore the relationship between beach gradient and tide range.

#### *8.4.1.2 Nearshore bar morphology*

Many micro-tidal beaches are characterised by nearshore bar morphology (Chapter 7). However, the (twice-)daily sweep of hydrodynamic processes across the beach profile due to the tide inhibits the formation of surf zone bars (Watts and Dearduff, 1954). The laboratory results of Hedegaard *et al.* (1991) suggest that a tide range in excess of three times the wave height is a critical threshold, above which the tidal translation rate does not permit sufficient time for morphodynamic processes to produce bar morphology. Even on micro-tidal beaches the effect of tides on nearshore bars is apparent with bar topography being more subdued during spring tides and more accentuated during neap tides (Wright *et al.*, 1986b, 1987). Thus, an important effect of tides is the tendency towards a smearing of beach morphology. On beaches with large tidal ranges this often results in a 'washed out' and rather featureless profile form.

#### *8.4.1.3 Swash bar (ridge and runnel) morphology*

Swash bar morphology – low amplitude, shore-parallel, single or multiple undulations within the intertidal profile – is a common sight on many tidal beaches. They occur over a wide range of environmental conditions, display different types of morphodynamic behaviour and probably result from a range of genetic origins. There has been considerable debate in the literature regarding their terminology and formation (e.g. Orford and Wright, 1978; Orme and Orme, 1988). A lot of confusion still exists; Orme and Orme's (1988) conclusion that ridges and runnels comprise a range of similar-looking features with dissimilar origins, is an appropriate starting point for discussing these morphological features.

Swash bars found on macro-tidal beaches often belong to the category of 'classic' ridges and runnels that have been described predominantly by British geomorphologists (e.g. King and Williams, 1949; King, 1972; Parker, 1975; Orford and Wright, 1978; Wright, 1984; Mulrennan, 1992). Ridge and runnel beaches exhibit quasi-permanent and quasi-stationary morphological highs and intervening lows across the intertidal profile of low gradient sandy foreshores (Figure 8.9). The ridges tend to be removed during periods of storm waves, whereas calm weather conditions cause the ridges to build in amplitude (King and Williams, 1949; King, 1972). According to King (1972), ridge and runnel beaches are associated with large tide ranges, restricted fetch and abundant sand. However, Mulrennan (1992) suggests that this under-estimates the ranges of possible conditions under which ridges and runnels will develop.

The central question of how and why such ridges form a discrete series across the intertidal zone of larger tide range beaches, and in particular the role of the tides in their genesis, is not yet clearly understood. According to King and Williams (1949) and King (1972), the ridges are the result of an attempt by the waves to produce an equilibrium swash zone gradient, on a beach of which the overall gradient is flatter than the equilibrium gradient. Following this logic, the ridges are expected to develop on the intertidal profile during periods of quasi-stationary water level (high and low tides), but this is not always the case (e.g. Wright, 1984). Carter (1988) con-

**Figure 8.9** Ridge and runnel beach at Skegness, England (A D Short).

cludes that the ridges are in fact multiple nearshore bars that survive tidal immersion, and Short (1991b) infers that infragravity standing waves are responsible for their formation and maintenance, particularly during lower wave post-storm conditions.

### 8.4.2 Profile adjustment to groundwater seepage/infiltration

Swash *infiltration* on an unsaturated beach enhances onshore transport and hence profile steepening, whereas groundwater *seepage* on a saturated beach promotes offshore sediment transport and profile lowering (Chapter 5). The interaction of the tide with an oscillating water table results in some regions of the profile alternating between saturated and unsaturated states through a single or neap spring tide cycle, while other regions remain in a permanently saturated or unsaturated state. This in turn leads to characteristic patterns of both profile adjustment and equilibrium morphology on tidal beaches.

#### *8.4.2.1 Eroson and accretion*

Grant (1946, 1948) was the first coastal scientist to establish that an elevated beach water table promotes beachface erosion, in contrast to a lowered water table promoting deposition. These insights were interpreted by Duncan (1964) to explain the cyclic pattern of beachface cut and fill commonly observed on beaches and associated with the rise and fall of the tide (Figure 8.10). During the rising tide, sediment within the swash zone is removed from the lower saturated region and deposited on the unsaturated beachface above. This deposition is promoted by run-up infiltration as the swash zone extends above the water table exit point. In contrast, seepage face development during the falling tide results in the entire swash zone remaining in a saturated state, with groundwater seepage enhancing the seaward transport of sediment.

A number of subsequent researchers have similarly observed cyclic patterns of intertidal erosion and accretion in response to both daily and longer period (lunar) tidal cycles in a range of beach environments (e g Harrison, 1969; Clarke *et al.*, 1984; Nordstrom and

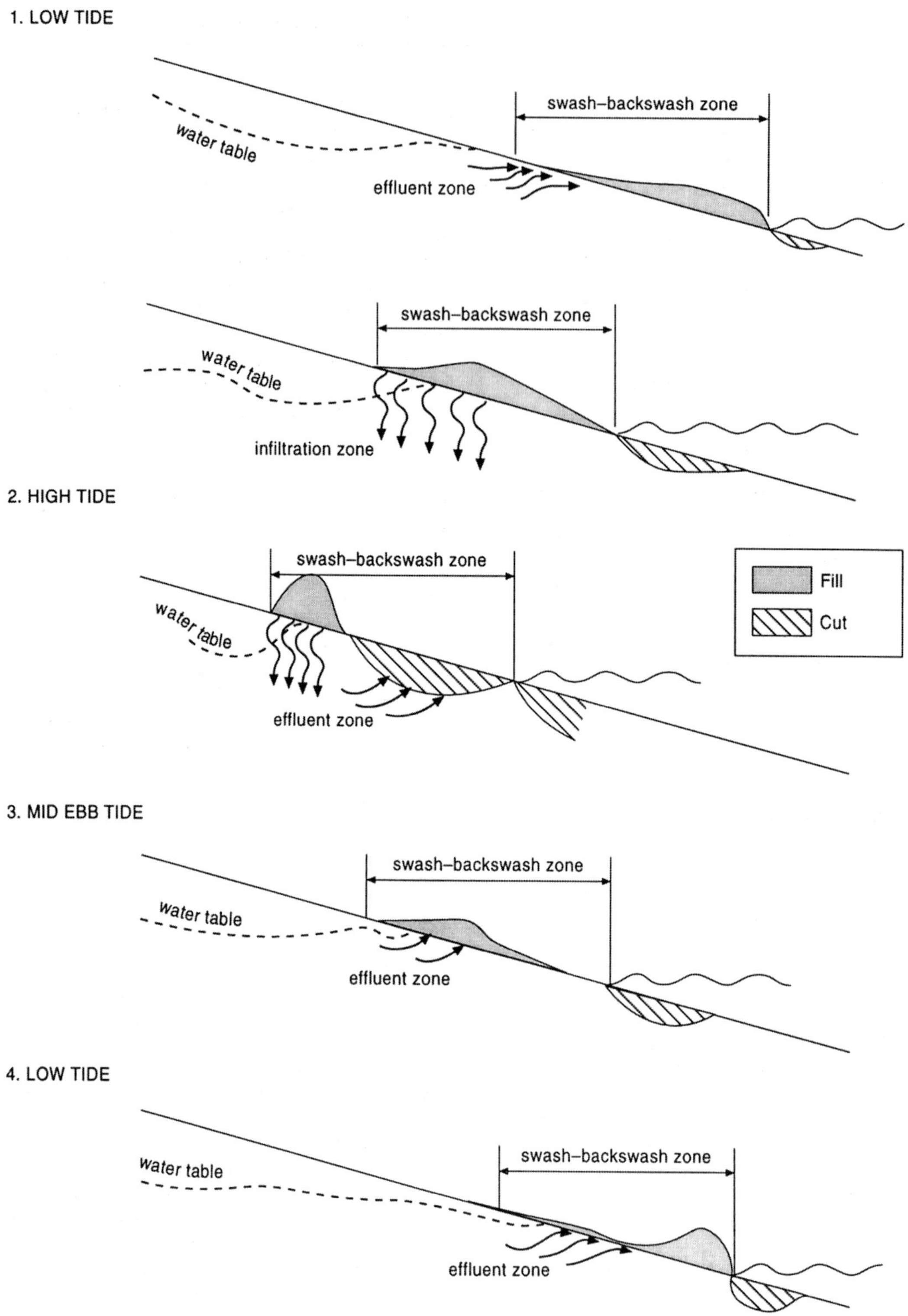


**Figure 8.10** Patterns of beachface erosion and accretion through a tidal cycle as first observed by Duncan (1964). The water table exit point distinguishes zones of deposition and erosion: across the unsaturated beachface *above* the exit point deposition predominates; across the saturated seepage face *below* the exit point, erosion prevails. The movement of the exit point through the tidal cycle translates these contrasting regions across the intertidal profile (after Duncan, 1964).

Jackson, 1990). Probably the most comprehensive empirical study of cyclic profile adjustment in response to the relative elevations of the tide and water table outcropping is detailed in Eliot and Clarke (1988), who applied a moving-axis technique to transform beachface deposition and erosion trends through a semi-diurnal tide cycle. They monitored the tidal migration of the swash zone and seepage face, together with zones of erosion and deposition. Maximum degradation occurred when the beachface was most saturated (1–2 h after high tide), while deposition was associated with unsaturated conditions above the exit point.

#### 8.4.2.2 *Intertidal profile adjustment*

The maximum rate with which the water table exit point can fall is independent of tide range (Equation 8.9). Therefore, on many beaches subject to larger tides ($TR > 2\,m$), the relative elevations of the tide and exit point define two distinct morphodynamic domains: an upper intertidal region that is alternately saturated/unsaturated through the tide cycle, and a lower region within the intertidal profile that remains in a permanently saturated state (Figure 8.11). The intersection of these contrasting intertidal zones marks a point of divergent sediment transport, and provides an explanation for the presence of a distinct break in slope often observed on large tide range beaches (e.g. Komar, 1976; Dyer, 1986; Horn, 1993; Turner, 1993a). On the upper (unsaturated) profile steepening is enhanced by swash infiltration on the rising tide, which in turn promotes profile drainage through subsequent tidal cycles. In contrast, offshore transport is promoted across the lower (saturated) intertidal, maintaining this region in a low gradient form. The selective sorting of coarser sediment upslope and finer sediment downslope reinforces such morphology, with rates of beachface drainage subsequently increasing/decreasing across the upper and lower regions of the intertidal respectively.

A simple numerical model to simulate macro-tidal profile development that incorporates the interaction of the fluctuating tide and outcropping beach water table is described in Turner (1995a). In Figure 8.12 the results of a 360 h simulation are depicted, incorporating a full lunar tide cycle (spring tide range = 4.0 m, neap tide range = 2.0 m). The development of a characteristic break in slope is apparent, resulting from the landward translation of sediment and berm development across the alternately saturated/unsaturated upper intertidal, in contrast to profile lowering and widening across the permanently saturated lower intertidal. These qualitative trends match patterns of equilibrium profile morphology and adjustment determined from the statistical analysis of 5 years of profile observations from the Central Queensland coast of Australia (Turner, 1994), but a quantitative test of the model is yet to be undertaken. It is interesting to note that Masselink (1995a) interprets empirical survey results to speculate that a pronounced intertidal break in slope will only occur on larger tide range beaches composed of sediment coarser than 0.3 mm, with beaches composed of finer sediment exhibiting a more uniform, low gradient intertidal profile. This suggests a critical grain size below which seepage face development precludes the drainage (and hence steepening) of a significant upper region of the intertidal profile.

## 8.5 THE PARAMETERISATION OF TIDAL EFFECTS

The morphodynamic implications of the daily sweep of the tide across the intertidal profile are expected to become more pronounced as the tidal range increases. Similarly, an increase in the relative importance of tide-related processes is also anticipated when wave height decreases. Masselink (1993) developed an empirical model to simulate the tide-induced migration of hydrodynamic processes across a beach profile and found that a useful parameter to quantify tidal effects was the relative tide range $RTR$

$$RTR = \frac{TR}{H} \qquad (8.12)$$

where $H$ is the wave height and $TR$ is the tide range. The larger the relative tide range, the more important become tidal effects relative to wave effects.

The model of Masselink (1993) calculates the relative amount of time that different locations on the beach profile are under the influence of swash, surf and shoaling wave processes. Swash processes operate between the quasi-stationary mean sea level (tide level) and the maximum run-up height. Surf zone processes affect the area between the tide level and the wave breakpoint. Shoaling wave processes only act seaward of the breakpoint. For each hydrodynamic region an empirical relationship between waves, sediment and an equilibrium beach gradient is assumed. The simulation model combines these relationships with the relative occurrences of swash, surf zone and shoaling wave processes for different parts on the beach profile and computes an 'equilibrium' beach profile.

Figure 8.13 shows the relative occurrence of swash, surf and shoaling wave processes and the resulting

**Figure 8.11** (a) Nine Mile Beach and (b) Sarina Beach, Queensland, Australia. Note the effluence of groundwater across the intertidal profile, resulting in a lower region remaining in a permanently saturated state (I L Turner and A D Short).

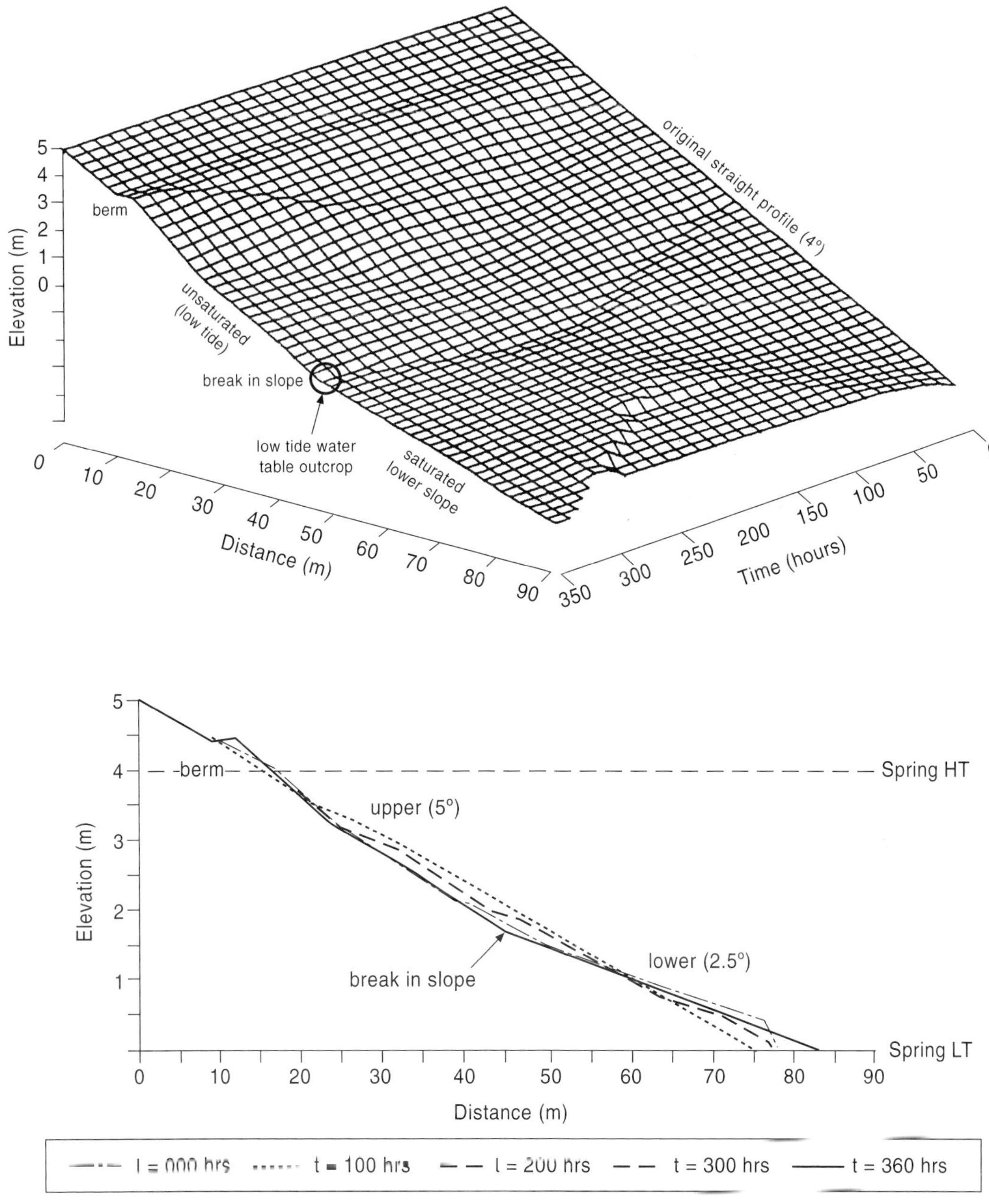


**Figure 8.12** Simulated beach profile evolution incorporating the effects of groundwater seepage/infiltration. Note the development of a break in slope in the intertidal profile, characteristic of many macro-tidal beaches (see Figure 8.14). Over the 14 day (360 h) simulation period, upper beachface berm development is accompanied by lower profile widening and lowering in gradient (after Turner, 1995a).

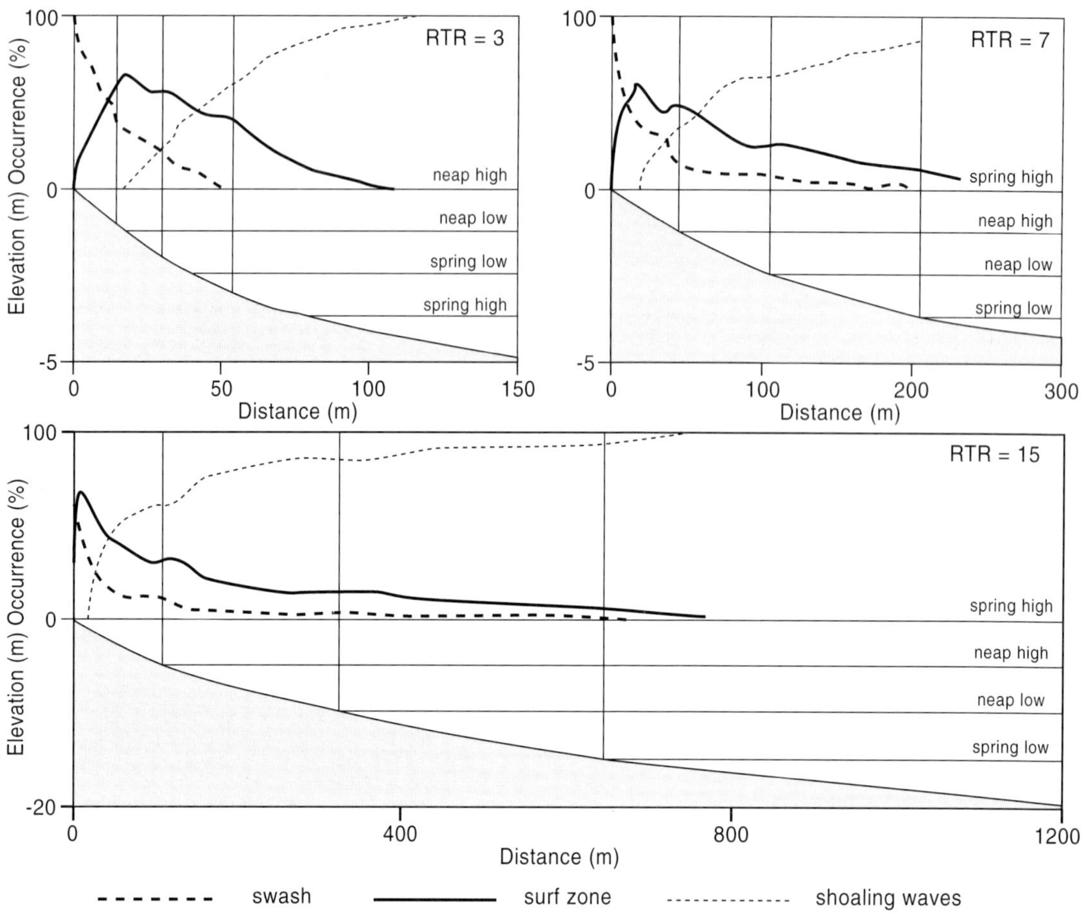


**Figure 8.13** The relative occurrence of swash, surf zone and shoaling wave processes across the beach profile calculated over a half lunar tidal cycle for three different relative tide range values ($RTR = 3$, 7 and 15). In order to generate a considerable neap to spring variation, two tidal constituents were considered, the principal lunar (M2) and the principal solar (S2), whereby S2 is half the amplitude of M2. Spring tide range is then given by 2(M2 + S2) and neap tide range is 2(M2 − S2). Other input parameters to the simulation were $H = 1$ m, $T = 8$ s, $w_s = 0.03$ m/s, $\gamma = 0.8$, and $TR = 3$, 7 and 15 m. It should be noted that running the model with different $H$ and $TR$, but identical $RTR$, would produce similar results with only swash processes slightly increasing in importance with decreasing $H$ and $TR$ (after Masselink and Short, 1993).

beach profile calculated over a half lunar tidal cycle for three different relative tide range values. For $RTR < 3$, the majority of the intertidal zone is dominated by swash and surf zone processes, and only seaward of neap low tide level do shoaling waves play a significant role. At mean sea level, swash and surf zone processes operate 90% of the time. For $RTR = 7$, the swash and surf zone processes displays four maxima, decreasing in amplitude and cross-shore spacing in the seaward direction. These maxima are associated with spring high tide, neap high tide, neap low tide and spring low tide level. At mean sea level, swash and surf processes operate less than 50% of the time. For $RTR = 15$, swash and surf zone occurrences display similar maxima as observed for $RTR = 7$. However, only between neap high tide and spring high tide do swash and surf dominate over shoaling waves. At mean sea level less than 20% of the time over a half lunar tidal cycle do swash and surf zone processes operate.

Obviously, the relative *occurrence* of a certain process is not necessarily an absolute measure of the relative *importance* of this process to resulting beach morphology. Swash and surf zone processes are more energetic than shoaling wave processes and may be more important even when they operate for a shorter time period. However, the results of Wright *et al.* (1982b) suggest that the relative time of occurrence of swash, surf and shoaling waves may serve as a surrogate in assessing its importance. For example, the calculations of Wright *et al.* (1982b) suggest that around mean sea level 45–50% of the total work being done by waves is by surf zone waves for $H_b = 1.5$ m, MSR = 9.5 m ($RTR = 6$) and $T = 11$ s (Figure 8.2). For the same conditions and an *RTR* of 7, the model predicts around 50% of the work is contributed by swash and surf zone processes (Figure 8.13).

The usefulness of the *RTR* parameter lies in the fact that it distinguishes the relative effects of waves and tides. This concept mirrors the earlier work of Davis and Hayes (1984) who first proposed that the geomorphology of coasts is a function of the relative influence of tides and waves, and the absolute values of each are not of major consequence. The classification after Davies (1964) of coastal environments as either micro-, meso- or macro-tidal is therefore an over-simplification, and the terms 'wave-dominated' and 'tide-dominated' relate more to relative tide range values, rather than the absolute tide range. Wave-dominated beaches can be found on macro-tidal coasts when the wave height is large enough, whereas tide-dominated tidal flats occur in micro-tidal environments when the wave height is very small. The concept of a relative rather than absolute tide range provides an effective scaling for the mutual effects of waves and tides. Consequently, the morphodynamics of micro-tidal estuarine beaches are in many aspects similar to those on macro-tidal beaches, since the difference in tidal range is compensated by the variation in wave energy (Inman and Filloux, 1960; Nordstrom and Jackson, 1990).

## 8.6 A MORPHODYNAMIC CLASSIFICATION OF TIDAL BEACHES

### 8.6.1 Early models

Carter (1988) categorised macro-tidal beaches on the basis of wave energy level and the occurrence and types of bedforms. His classification is a continuation of the work by Hawley (1982), who showed that the distribution of intertidal bedforms results from the interaction between swash motions and the action of breaking/shoaling waves. According to Carter (1988), at the high energy end of the spectrum of macro-tidal beaches, the profile is washed out (no ripples), and characterised by a flat, uniform surface. Under slightly lower energy conditions, some wave-lain bedforms may survive tidal emergence, resulting in a rippled beach surface. When wave energy level decreases even more, large intertidal dunes may be preserved as the tide falls. On low energy macro-tidal beaches, wide, low amplitude, parallel or sub-parallel bars (ridge and runnels) are found in the intertidal zone. Finally on exceedingly low energy macro-tidal coasts, ridge and runnel morphology is replaced by transverse bars.

The classification schemes of Carter (1988) and Short (1991b) are essentially based on a gradation from high to low incident wave energy, and therefore beaches exhibiting disparate tide ranges, sediment sizes and morphologies are included within a single category. A further conceptual problem with these models is that, rather than representing part of a continuum of wave/tide deposits, the macro-tidal beach types identified do not grade easily into the micro-tidal beach types previously discussed in Chapter 7.

### 8.6.2 Beach model of Masselink and Short (1993)

By combining tidal effects with more extensively researched ideas on micro-tidal beach morphodynamics (see Chapter 7), Masselink and Short (1993) proposed a grouping of natural beaches on the basis of four physical constraints: modal breaking wave height, modal breaking wave period, sediment characteristics of the upper beachface and mean spring tide range. These variables are simply quantified by two dimensionless parameters: the *dimensionless fall velocity* $\Omega$ (Equation 7.1) and the *relative tide range*, *RTR* (Equation 8.12). The attraction of a classification scheme based on this approach is that both morphological and hydrodynamic factors are incorporated. The dimensionless fall velocity parameter indicates whether reflective, intermediate or dissipative surf zone conditions will prevail. The relative tide range parameter indicates the relative importance of swash, surf zone and shoaling wave processes in determining profile morphology.

The beach model shown in Figure 8.14 is a simplified version of the model proposed by Masselink and Short (1993) and consists of seven major beach types that fall into three broad groups, depending on the value of *RTR*. This classification scheme is summarised in Table 8.1. As both $\Omega$ and *RTR* increase, a logical sequence of characteristic beach morphologies is identified. The three wave-dominated beach types were discussed in Chapter 7 and therefore only the morphodynamics of the mixed wave–tide beaches and the tide-dominated beaches will be outlined below. A summary of the mor-

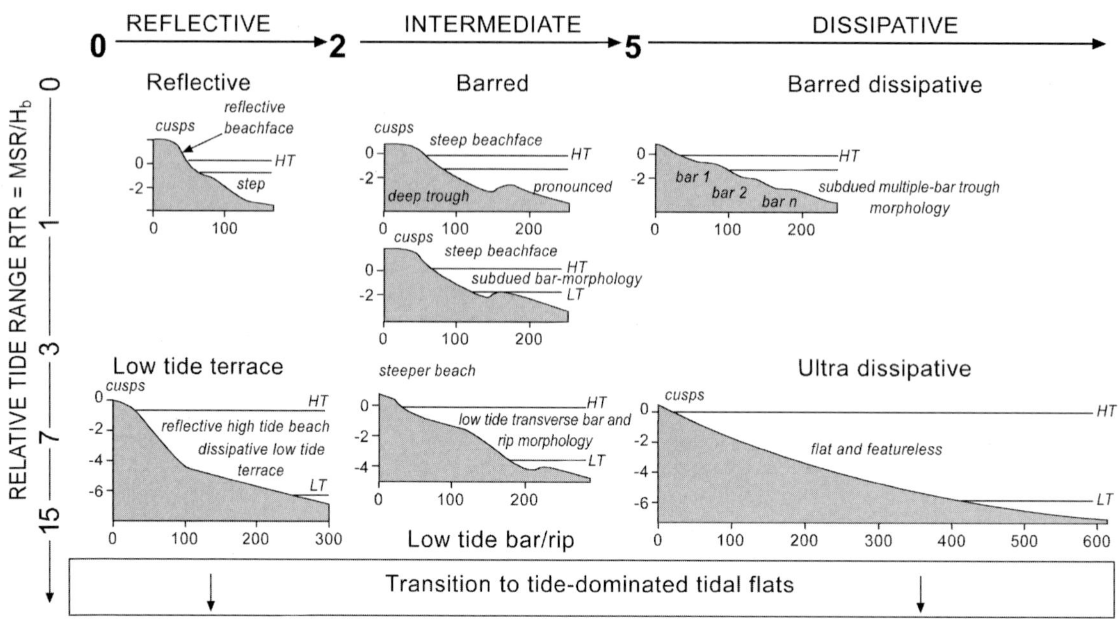


**Figure 8.14** Classification of beaches on the basis of the modal breaking wave height $H_b$ and period $T$, the high tide sediment fall velocity $w_s$ and the mean spring tide range MSR. These four morphodynamic variables are represented by two dimensionless parameters: the dimensionless fall velocity $\Omega = H_b/w_sT$, and the relative tide range $RTR = \text{MSR}/H_b$ (after Masselink and Short, 1993).

phodynamic characteristics of the mixed wave–tide beaches is further presented in Table 8.2 and Figure 8.15.

### *8.6.2.1 Low tide terrace beaches*

Low tide terrace beaches are characterised by a steep upper part of the profile and a gentle gradient low tide terrace (Figure 8.16a) and develop when $RTR = 3$–$15$ and $\Omega < 2$. The sediments on the steep upper part of the beach profile, where the grain size generally exceeds 0.3 mm, are usually significantly coarser than those found on the low tide terrace. The textural discontinuity coincides with a distinct break in slope, and often the low tide water table outcrop (effluent line) is located at this position, saturating the low tide terrace. Typically, beach cusps are found around high tide level and drainage rips form on the low tide terrace during low tide. For $RTR = 3$–$7$ the drainage rips may discharge into subdued surf zone rips, but for $RTR = 7$–$15$ the low tide terrace will be relatively featureless. A difference in high tide and low tide morphodynamics typifies low tide terrace beaches. Reflective conditions prevail at high tide when waves break surging/plunging on the beachface, whereas the low tide surf zone is dissipative with several lines of spilling breakers across the terrace (Horn, 1993; Masselink and Hegge, 1995).

The upper intertidal zone of low tide terrace beaches is dominated by swash processes, whereas a combination of swash and surf zone processes controls the morphodynamics of the mid-intertidal zone. Despite the dominance of swash processes on the upper part of the intertidal profile, pronounced swash bar morphology is not likely to form, since the overall beach slope is already steep and in quasi-equilibrium with swash processes. The lower intertidal zone is governed by a mixture of surf zone and shoaling wave processes and may be characterised by subdued bar/rip morphology for $RTR = 3$–$7$. Under certain circumstances, swash processes may be locally important on the low tide terrace, resulting in the formation of swash bars (King, 1972; Druridge Bay, Great Britain). Shoaling waves dominate the sub-tidal zone.

Only minor changes in beach morphology occur over a lunar tidal cycle on low tide terrace beaches (Masselink and Hegge, 1995). However, during major storms up to 2 m erosion of the steep upper part of the profile may occur with subsequent deposition on the low tide

**Table 8.1** Summary of beach morphodynamic classification scheme incorporating tidal effects (after Masselink and Short, 1993).

| Relative tide range | Group | Beach type |
|---|---|---|
| $RTR < 3$ | Wave-dominated | 1. Reflective<br>2. Barred<br>3. Dissipative |
| $3 < RTR < 15$ | Mixed wave–tide | 4. Low tide terrace |
| $3\ RTR < 7$ | | 5. Low tide bar/rip |
| $3\ RTR < 15$ | | 6. Ultra-dissipative |
| $RTR > 15$ | Tide-dominated | 7. Transitional (beach to tidal flat) |

**Table 8.2** Summary of morphodynamic characteristics of low tide terrace, low tide bar/rip and ultra-dissipative beaches (after Masselink and Hegge, 1995).

| Parameter/process | Low tide terrace | Low tide bar/rip | Ultra-dissipative |
|---|---|---|---|
| $\Omega$ | < 2(3) | 2–5 | > 5(3) |
| $RTR$ | 3–15 | 3–7 | 3–15 |
| *High tide conditions and upper part of the intertidal profile* | | | |
| Sediment size | > 0.3 mm | 0.2–0.4 mm | < 0.3 mm |
| Tan $\beta$ | > 0.05 | 0.03–0.05 | 0.02–0.04 |
| $\varepsilon$ | < 5 | 5–20 | 5–30 |
| Breaker type | Surging/plunging | Plunging | Plunging/spilling |
| Attenuation coefficient | 0.6–0.8 | 0.5–0.7 | 0.4–0.6 |
| Dominant process | Swash | Swash/surf zone | Swash/surf zone |
| Beach cusps | Common | Occasional | Rare |
| Morphological change over lunar tidal cycle | Steepening during neap tides, flattening during spring tides | Swash bar development, destruction and migration | Minor, possible profile changes due to swash bar |
| Storm response | Major erosion of steep upper part | Erosion of upper part and of swash bar morphology | Erosion of upper part |
| *Low tide conditions and lower part of the intertidal profile* | | | |
| Sediment size | < 0.2 mm | < 0.3 mm | < 0.2 mm |
| Tan $\beta$ | < 0.03 | 0.02–0.04 | < 0.03 |
| $\varepsilon$ | > 30 | > 20 | > 30 |
| Breaker type | Spilling | Plunging/spilling | Spilling |
| Attenuation coefficient | 0.3–0.5 | 0.3–0.6 | 0.3–0.5 |
| Dominant process | Surf zone/shoaling | Surf zone | Surf zone/shoaling |
| Swash bars | Occasional | Common | Occasional |
| Rip currents | Drainage rips at low tide | Surf zone rips at low tide | Absent |
| Tidal currents | Important | Minor importance | Important |
| Morphological change over lunar tidal cycle | None | Minor bar changes | None |
| Storm response | Deposition | Deposition and destruction of bar morphology | Deposition |

terrace (Horn, 1993, Turner, 1995a). According to Carter (1988), low tide terrace beaches are commonplace on high latitude coasts where the high tide beach consists of gravel fronted by a fine grained low tide terrace. Many harbour and estuarine beaches may also be classified as low tide terrace beaches (Nordstrom, 1992; Nordstrom and Jackson, 1993).

#### 8.6.2.2 *Low tide bar/rip beaches*

Intertidal swash bar morphology and transverse bar/rip morphology at low tide level are characteristic of low tide bar/rip beaches (Figure 8.16b). The low tide bar/rip beach type forms when $RTR = 3$–$7$ and $\Omega < 2$–$5$. At high tide the bar morphology is under the influence of

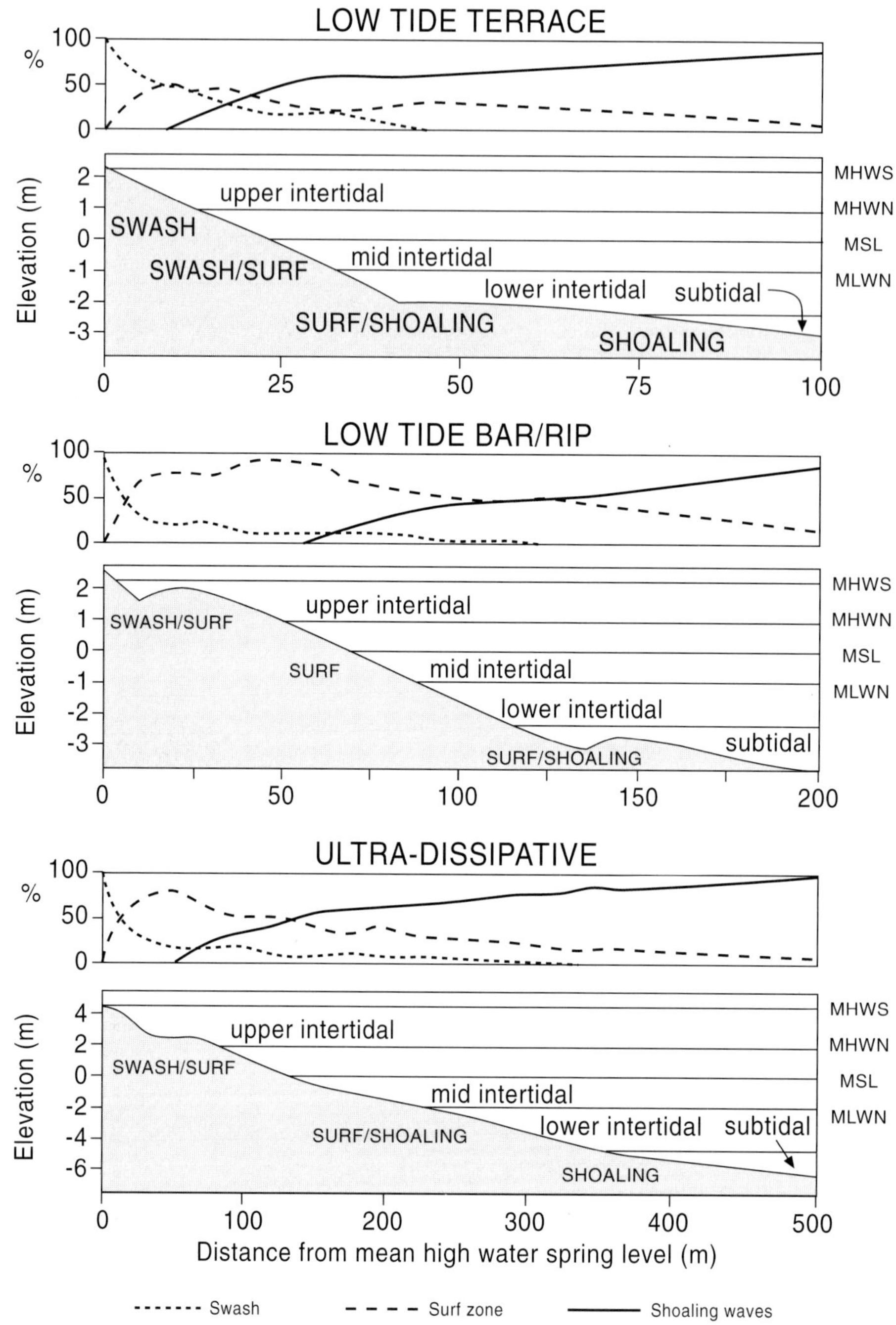


**Figure 8.15** Relative occurrence of swash, surf zone and shoaling wave processes calculated over a half lunar tidal cycle for low tide terrace, low tide bar/rip and ultra-dissipative beaches (after Masselink and Hegge, 1995).

**Figure 8.16** Photos of (a) a low tide terrace beach (Lambert's Beach, Qld); (b) low tide bar/rip beach (rip highlighted by dye, Nine Mile Beach, Qld); (c) and (d) ultra-dissipative beaches (Rossili Bay, Wales and Farnborough Beach, Qld); (e) ridged tidal sand flats (Queens Beach, Qld); and (f) beach and tidal mud flat (Caldwell, Qld) (A D Short).

?

c
d

e

f

shoaling waves and the surf zone is confined to the upper beach where plunging waves break across a relatively narrow intermediate surf zone. During falling tide waves break increasingly on the bar and reform in the trough and approaching low tide cellular rip circulation is enhanced, particularly as feeder and rip currents are increasingly confined to the rip topography (Masselink and Hegge, 1995).

The morphodynamics of the upper intertidal zone on low tide bar/rip beaches are governed by a mixture of swash and surf zone processes and may result in the development of pronounced swash bar morphology. The mid- and lower intertidal zones are dominated by surf zone processes, but typical surf zone morphology (bars and rips) is not present due to the effects of tidal smearing and non-stationarity of the surf zone processes. Surf zone and shoaling wave processes control the sub-tidal zone and have resulted in subdued bar/rip morphology.

Low tide bar/rip beaches exhibit a large temporal variability in beach morphology over a lunar tidal cycle due to modifications to inter- and sub-tidal bar morphology (Masselink and Hegge, 1995). Major storms destroy the bar morphology resulting in a planar, concave-upward profile (Fox and Davis, 1978). Low tide bar/rip beaches are not very common since a relatively narrow hydrodynamic window is required for them to form (Figure 8.14). They are generally found on medium to high wave energy, meso-tidal coastlines (MSR $\approx$ 3–4 m; $H_b \approx 1$ m) including the Oregon coast (Fox and Davis, 1978).

### *8.6.2.3 Ultra-dissipative beaches*

Ultra-dissipative beaches have a low gradient featureless profile and occur when $RTR = 3$–$15$ and $\Omega > 5$ and when $RTR = 7$–$15$ and $\Omega > 2$. The term 'ultra-dissipative' is adopted from McLachlan *et al.* (1993), and refers to both the extreme dissipativeness of the surf zone conditions with multiple lines of breakers, and the extreme width of the low gradient dissipative profile (Figure 8.16c). Surf zone conditions are dissipative with spilling breakers throughout the tidal cycle (Jago and Hardisty, 1984), except for high tide conditions on beaches with intermediate $\Omega$ values when plunging breakers may occur (Wright *et al.*, 1982b). These latter beaches often have somewhat coarser sediments and steeper high tide beach gradients.

Swash and surf zone processes control the morphodynamic processes of the upper intertidal zone of ultra-dissipative beaches and may result in the formation of a spring high tide swash bar. The mid- and lower intertidal zones are primarily governed by a mixture of surf zone and shoaling wave processes. Swash processes, however, may be locally important and result in the formation of swash bars (King, 1972; Blackpool Beach, Great Britain). The sub-tidal zone is dominated by shoaling wave processes.

Dissipative beaches are temporally very stable, and even major storms are not expected to have much impact due to the overall low gradient of the beach profile and the lack of secondary morphological features (Jago and Hardisty, 1984).

### *8.6.2.4 Transition from beach to tidal flats*

When $RTR > 15$, the mixed wave–tide beaches grade into tidal flats, along the lines of Short's (1991b) Group 3. Such environments are characterised by a relatively steep, coarser grained high tide beach which grades abruptly into a finer grained, very low gradient tidal flat (Figure 8.16f). During spring high tide waves may break plunging on the beach, but at low tide extremely dissipative conditions will prevail on the tidal flat. Often, a major part of the incident wave energy is dissipated through bottom friction, rather than wave breaking. Only the upper part of the intertidal profile is controlled by swash and surf zone processes. The morphodynamics of the tidal flat are mainly governed by shoaling wave processes and tidal currents. Little is known about the morphodynamics, nor the temporal stability of these coastal systems.

### *8.6.2.5 Application of the conceptual model*

Several notes of caution regarding the application of the above morphodynamic model are warranted. First, the model is conceptual and primarily based on morphological data collected on the macro-tidal coastline of central Queensland, Australia (Masselink and Short, 1993). Although parametric thresholds are suggested to separate the different beach types, these should not be taken as absolute. Second, the model only considers the wave and tidal energy level in a relative way, neglecting the importance of energy thresholds. For example, the importance of the tidal translation rate (Equation 8.3) is neglected. Thirdly, micro-tidal, low wave energy coastal environments are very sensitive to small changes in environmental parameters (e.g. Davis and Hayes, 1984), and particular care should be exercised when applying the model to these environments. Finally, with decreasing modal wave energy an increasing number of morphological features will be inherited from higher wave events (e.g. storm cusps, ridges and bars). Although these morphological features appear to be part of the modal beach morphology – they are always present – they are not related to the modal hydrodynamic conditions.

## 8.7 SUMMARY

Beach morphology and nearshore/surf zone hydrodynamic processes are primarily determined by the level of wave energy arriving at any particular coast. However, ocean tides modulate the action of waves, as well as driving groundwater fluctuations and tidal currents, all of which can alter the morphodynamic character of tidal beaches. The *direct* effect of tides on beach morphodynamics is the recurring oscillation of hydrodynamic processes across the profile. As a consequence, with increasing tide range, large regions of the intertidal profile may be subjected to the entire range of aeolian, swash, surf and shoaling wave processes within a single tidal cycle. The *indirect* effect of tides on the morphodynamic behaviour of beaches is the introduction of groundwater as an additional morphodynamic variable. Tide-induced groundwater fluctuations induce time-varying saturation characteristics of the intertidal profile that through long-term erosion and accretion trends determine the shape of the equilibrium profile.

More traditional approaches to beach geomorphology have incorporated tidal effects through a simple classification of the world's coastlines by absolute tide range. From recent research it is apparent that the use of a relative measure of tides and waves is a more useful method for incorporating tides, permitting a more process-based assessment of factors shaping sandy coastlines. In particular, this classification scheme brings together the previously disparate characteristics of non-tidal, estuarine, open coast, and macro-tidal beaches within a single continuum, recognising the similarities as well as differences that exist between beaches exposed to varying wave energy and tidal characteristics.

# 9

# Embayed and Structurally Controlled Beaches

*Andrew D. Short and Gerhard Masselink*

## 9.1 INTRODUCTION

While beaches are composed of sediments, they all at some point must have a terminal boundary. On long beaches this may be a sandy foreland or inlet, which is also a product of wave and tidal processes. However on many beaches it is a rocky headland, which is totally independent of the formative beach processes and which in fact can have a major influence on beach planform, sediment transport and morphodynamics. This chapter will address the role of headlands in controlling beach planform (Section 9.2) and morphodynamics (Section 9.3), as well as the embayed phenomena of beach rotation (Section 9.4) and headland sediment bypassing (Section 9.5). Finally structural impacts will be considered in Section 9.6.

Either headlands and/or structures bound embayed beaches. Headland embayed beaches are characteristic of rocky shorelines associated with hilly and mountainous coastal topography (Figure 9.1a). The length and spacing of these beaches in entirely dependent on the pre-existing bedrock topography which has been partially drowned by the Holocene sea level transgression, with the flooded valleys partially filled with sediment, while the valley sides and spurs remain as headlands and rocky coast. Along the hilly coast of New South Wales and Victoria, the 1280 beaches have an average length of 1.4 km ($\sigma = 4.6$), with most of them bounded by headlands at each end (Short, 1996). A total of 51% of the world's coast consists of hilly or mountainous coastline (Inman and Nordstrom, 1971), an indication of the prevalence of embayed beaches. *Structurally embayed beaches* are more common in groyne fields and adjacent to major coastal structures such as training walls and breakwaters. On the central Dutch coast 158 groynes have been placed along three sections of coast totalling 43 km (Short, 1991a). The result has been to divide the once continuous beach into groyned embayments averaging 300 m in length. Whereas headland embayed beaches are considered aesthetically pleasing and represent relatively stable sections of the coastline, structurally embayed beaches are a testament to attempts to stabilise unstable shorelines.

This chapter is primarily concerned with headland embayed beaches and will address these using the term 'embayed' beaches, unless otherwise indicated. Embayed beaches typically have a curved planform and in the literature have attracted a plethora of terms, including zeta bays, crenulate shaped bays, spiral beaches, curved beaches, hooked beaches or headland bay beaches.

## 9.2 BEACH PLANFORM

A typical characteristic of embayed beaches is the close correspondence between the beach planform and the refraction pattern associated with the prevailing waves. Davies (1958) astutely observed that:

> 'Anyone who has looked at air photographs of waves approaching relatively enclosed sandy beaches in particular will have been struck by the way in which they frequently foreshadow not only the general plan of the beach but even the smaller kinks and bends while still many yards offshore'.

On the basis of these observations, Davies correctly concluded that the orientation and planform of em-

*Handbook of Beach and Shoreface Morphodynamics*. Edited by A. D. Short.
© 1999 John Wiley & Sons Ltd.

bayed beaches are controlled by the refraction pattern associated with the prevailing (swell) waves. The important role of refraction and diffraction in determining the shape of embayed beaches has been confirmed by numerical models (Rea and Komar, 1975; LeBlond, 1979).

If the embayment is in equilibrium with the hydrodynamic conditions, refraction is such that no net longshore transport occurs due to obliquely incident waves, because all breaking waves arrive normal to the beach along the entire embayment. Weak longshore currents may be present, however, driven by the alongshore gradient in the wave height (O'Rourke and LeBlond, 1972). These latter currents are expected to be very weak and flow away from areas of maximum wave height to regions of minimum wave energy. Outgoing currents (rips) can be expected at the extremities of the embayment.

Generally, embayed beaches have an *asymmetric planform* characterised by a strongly curved shadow zone, a mildly curved centre of the embayment and a relatively straight downcoast end (Figure 9.2). The shoreline at the *straight section of the embayment* is often parallel to the dominant wave crests. From model tests and field observations certain equilibrium characteristics of the planform of embayed beaches are known. Subsequently, a comparison of observed bay geometry and predicted geometry can reveal the stability of embayed beaches (Silvester, 1960). Figure 9.2 plots the ratio of bay indentation ($a$) to headland spacing ($b$) as a function of the obliquity of the dominant wave crests to the headland alignment ($\beta$). The obliquity of the dominant wave crest to the headland alignment has been approximated by the angle between the shoreline of the downdrift section of the bay and the headland alignment. For this to be correct, the dominant wave crests would have to be approaching parallel to this section of the coast, i.e. the littoral drift would have to be zero. Embayments that plot below the predicted curve are less indented than theoretically expected and littoral drift in these embayments is probably not insignificant. Embayments that plot above the predicted line are more indented than expected and rarely occur in nature.

The planform of the *sheltered and curved section of embayed beaches* can be approximated by logarithmic spirals (Yasso, 1965). However, the relatively straight and more exposed section of the embayment is not accounted for by the logarithmic spiral. In addition, the centre of the spiral does not match the point at which diffraction takes place, rendering it difficult to apply to real embayments. Therefore, Hsu *et al.* (1987) developed a more universal relationship describing the planform of embayed beaches based on the parabolic shape. The parabolic shape proposed by Hsu and co-workers (Hsu *et al.*, 1987, 1989a,b; Hsu and Evans, 1989) is 'tied' to two control points (Figure 9.3). The first control point (upcoast control point) is the diffraction point around which the incident waves diffract as they propagate into the embayment. This point may be the tip of a headland, an island or a coastal structure. The second control point (downcoast control point) may be a second headland or, if the latter is not present, can be taken as the start of the straight section of the coastline where the shoreline orientation is parallel to the incident wave crests. The line connecting the two control points is referred to as the control line and has a length $R_0$. The angle between the incident wave crests and the control line represents the wave obliquity $\beta$. A series of lines with length $R$ can be drawn from the point of diffraction to the beach at an angle $\theta$ to the wave crests (Figure 9.3). The parabolic equation derived by Hsu and co-workers then predicts the ratio $R/R_0$ for any value of $\theta$ for a given bay with known wave obliquity $\beta$

$$R/R_0 = C_0 + C_1(\beta/\theta) + C_2(\beta/\theta)^2 \qquad (9.1)$$

where the values of the constants $C_0$, $C_1$ and $C_2$ depend on $\beta$ (Table 9.1). The polar coordinates $R$ and $\theta$ can easily be converted to Cartesian coordinates $x$ and $y$ using trigonometry.

Figure 9.4 shows the equilibrium bay shape calculated using Equation 9.1 for $\beta = 20°$ and $\beta = 60°$. The control points are formed by headlands. It can be seen that the larger the angle between the crests of the incident waves and the control line, the more deeply indented the embayment is (see also Figure 9.2). It is also apparent that the alignment of the relatively straight section of the embayment is parallel to the approaching wave crests.

The parabolic shape only applies to embayed beaches that are in *static equilibrium*, where the dominant waves arrive at right angles to the entire periphery of the bay with the result that there is no littoral drift within the bay. Embayed beaches can also be in *dynamic equilibrium* with the sediment supply and the wave climate, where there is an active through flow of littoral drift in the embayment. In such embayments the shoreline will usually be seaward of the static equilibrium position. When the sediment supply of a bay in dynamic equilibrium is reduced or cut off, erosion will attempt to reshape the shoreline into the static equilibrium state. Such a condition may be referred to as *disequilibrium*, because the embayment is trying to adjust to changing conditions. Comparison of the predicted shoreline using

**Figure 9.1** (a) View of embayed Elizabeth and Shelly beaches (NSW); note the control on beach length and the 90° change in orientation between the beaches. (b) Brunswick River (NSW); training walls have resulted in an offset in the beach planform. Note the updrift build-out of the beach, while downdrift erosion is evident in the overwashing and has resulted in the total loss of several houses. (c) The training wall at Camden Haven (NSW) has resulted in a beach developing between the southern wall and the headland, a low energy beach inside the walls, and sand accumulation against the northern wall (A D Short).

Equation 9.1 with the actual shoreline can indicate whether a beach is stable or not.

Empirical evidence suggests that embayed beaches in static equilibrium are relatively stable and recede mainly due to recession of the headlands. Hence, it should be possible to stabilise a section of shoreline by constructing artificial headlands within the embayment. If stable artificial headlands are provided, and the dominant wave direction is known, the ultimate shape of the embayed beach can be predicted using Equation 9.1. This philosophy has been utilised in a type of engineering that is known as *headland control* (e.g. Silvester and Ho, 1970; Silvester *et al.*, 1980; Bishop, 1983; Silvester and Hsu, 1993).

An example of headland control to stabilise the coast is the extension of an existing upstream control point. This is illustrated in Figure 9.5, based on a real-life example discussed in Silvester and Hsu (1993). The figure shows an embayment initially in dynamic equilibrium with the prevailing waves and sediment supply. There is an active throughflow of nearshore sediment and the existing shoreline is located well seaward of the static equilibrium shoreline. Now, imagine that as a result of human activities in the upstream embayment, the sediment supply is largely cut off. As a result, the coastline will be receding toward the static equilibrium shoreline. Extension of the upstream control point to position B by means of a breakwater will result in a new static equilibrium shoreline that overlies the existing and eroding shoreline. Such a modification of the upstream control point will shift the static bay shoreline from AC to BC and essentially halt the erosion.

Other applications of headland control are the construction of offshore breakwaters resulting in the formation of salients and the addition of new control points. A large range of examples of the application of headland control to stabilise beaches are outlined in Silvester and Hsu (1993).

## 9.3 BEACH MORPHODYNAMICS

Headlands, rocks, reefs and structures will all impact the beach and surf zone through their influence on wave refraction and attenuation, and by limiting the development of longshore currents, rips and rip feeder currents.

When deep water waves ($H_o$) enter an embayment with a given width ($C_l$) then the wave energy will be redistributed along the embayment shoreline ($S_l$), such that the energy at the shoreline is

$$E_b = H_o^2(C_l/S_l) \tag{9.2}$$

If shoaling and bed friction are ignored then breaker wave height is

$$H_b = \sqrt{(H_o^2(C_l/S_l))} \tag{9.3}$$

Assigning a typical surf zone gradient of 0.01 (1/100), the

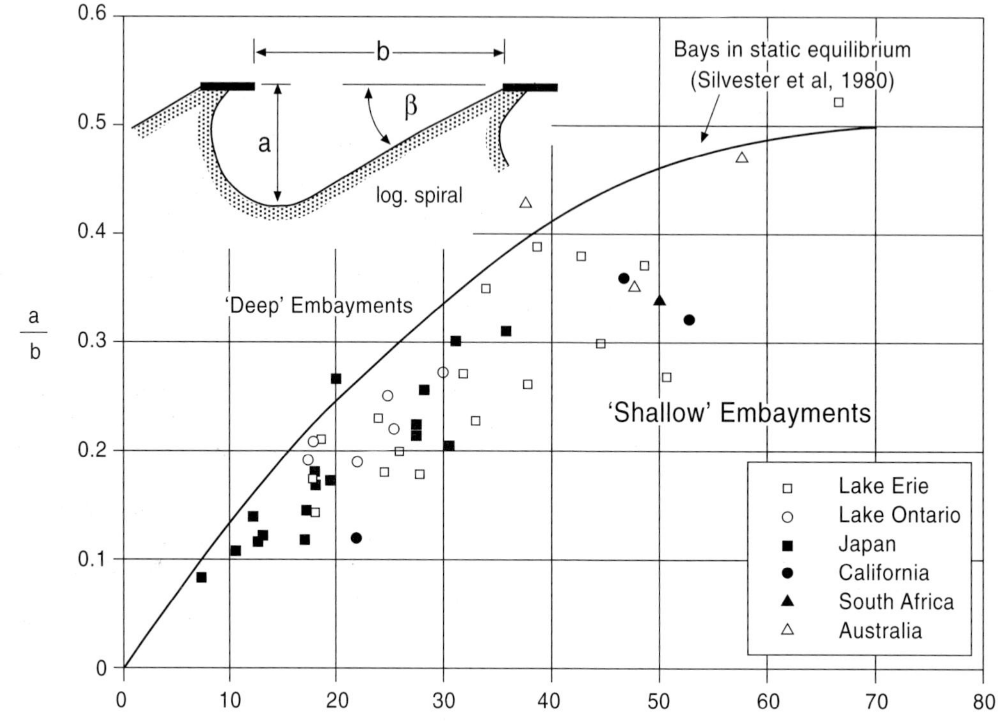


**Figure 9.2** Measured values of the ratio of bay indentation, *a*, to headland spacing, *b*, versus the obliquity of the dominant waves to headland alignment, *β*, for some natural embayments and the theoretical prediction for embayments in static equilibrium (after Bishop, 1983).

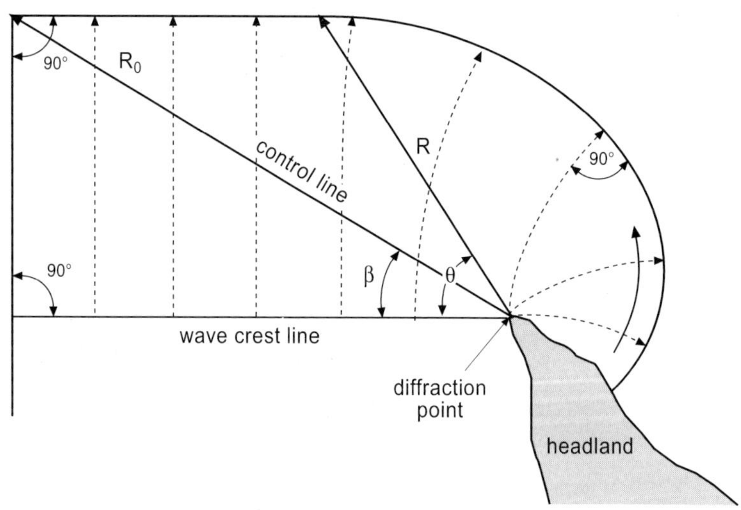


**Figure 9.3** Definition sketch of the parabolic approach to bay shape (after Silvester and Hsu, 1993).

**Table 9.1** Values for the constants $C_0$, $C_1$ and $C_2$ in the parabolic equation (after Silvester and Hsu, 1993).

| $\beta$ | $C_0$ | $C_1$ | $C_2$ |
|---|---|---|---|
| 20° | 0.054 | 1.040 | −0.094 |
| 30° | 0.046 | 1.136 | −0.186 |
| 40° | 0.003 | 1.322 | −0.328 |
| 50° | −0.088 | 1.598 | −0.512 |
| 60° | −0.227 | 1.936 | −0.710 |
| 70° | −0.405 | 2.276 | −0.867 |
| 80° | −0.600 | 2.526 | −0.910 |

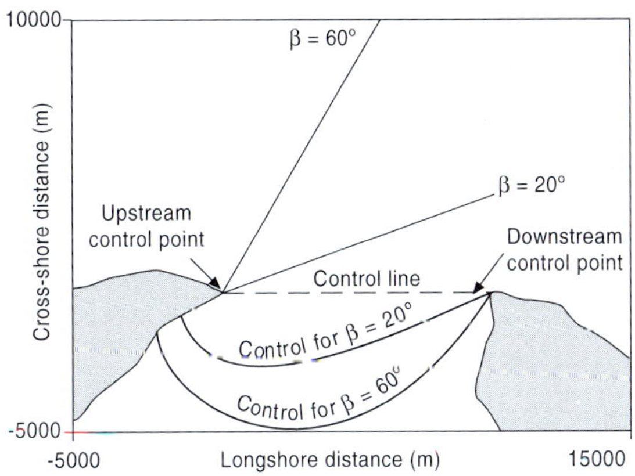


**Figure 9.4** Equilibrium bay shape calculated using Equation 9.1 for $\beta = 20°$ and $\beta = 60°$. The larger the wave obliquity $\beta$, the more indented the resulting equilibrium bay shape.

width of the surf zone is

$$X_s = 100H_b \quad (9.4)$$

which can be expressed as

$$X_s = 100\sqrt{(H_o^2(C_l/S_l))} \quad (9.5)$$

The number of surf zone widths, $W$, that fit into an embayment $S_l$ is then $S_l/W$ or $100H_0\, C_l^{0.5}/S_l^{-1.5}$. This term has been defined by Martens *et al.* (in press) as the non-dimensional embayment scaling parameter $\delta'$ where

$$\delta' = S_l^2/100C_lH_b \quad (9.6)$$

Martens *et al.* (in press) found that in embayed beaches as wave height ($H_b$) increases and shoreline length ($S_l$) decreases, a critical threshold is reached where the wave-dominated beach model (Chapter 7) is increasingly modified by *end effects* (Figure 9.6). On beaches with no headlands, 'normal' surf zone circulation prevails. When headlands/structures are present and are widely spaced and/or the beach receives low waves, the beach becomes sub-intermediate, as the headlands impact surf zone circulation only adjacent to the headlands with normal circulation between. As wave height increases and/or the headlands are closer together, the entire beach circulation may eventually become impacted by the end effects. In extreme conditions with wave height exceeding a few metres, large scale, topographically controlled rip systems, called megarips, prevail (Short,

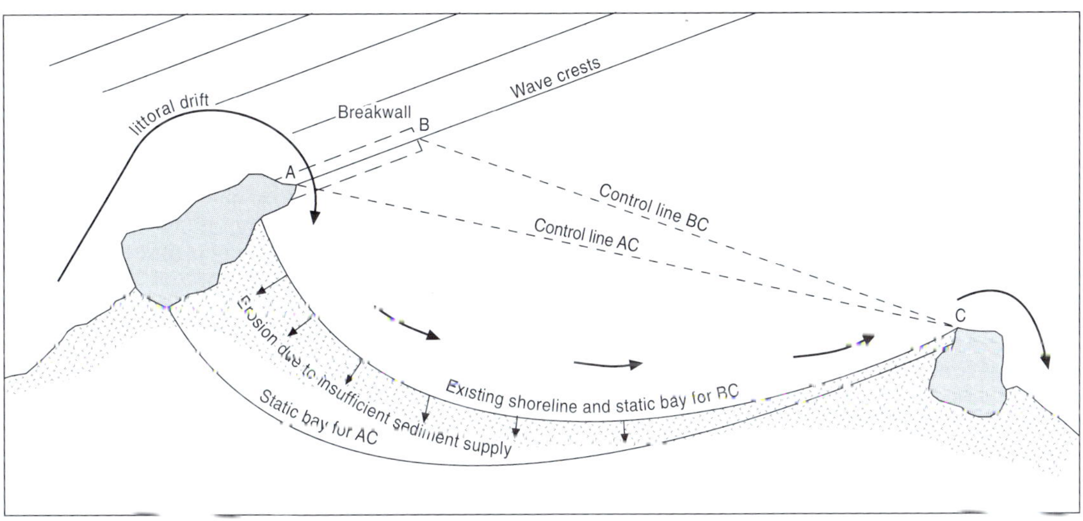


**Figure 9.5** Stabilisation of an eroding shoreline by means of headland control. See text for explanation (after Silvester and Hsu, 1993).

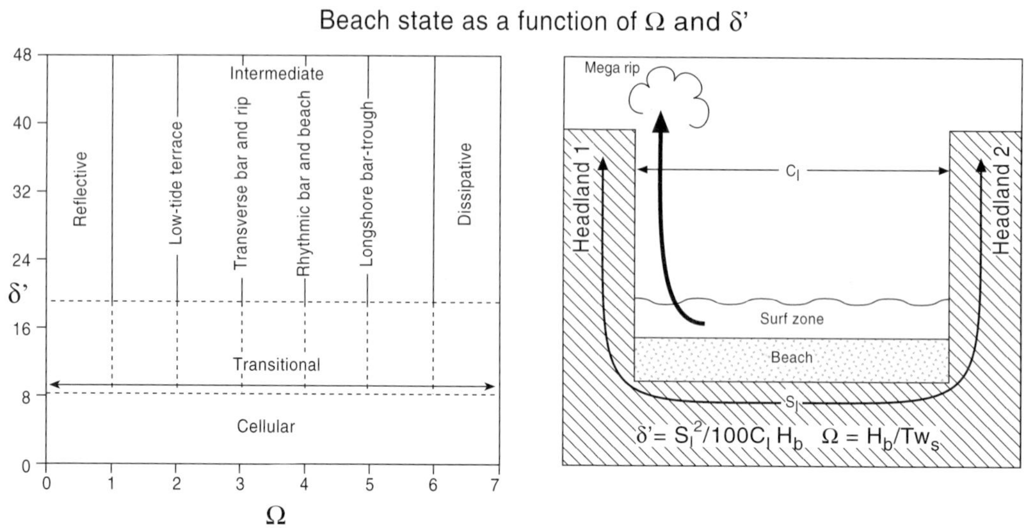


**Figure 9.6** The dimensionless fall velocity (Ω) plotted against the dimensionless embayment scaling parameter ($\delta'$). The micro-tidal beach model applies when $\delta' > 20$. As $\delta'$ moves between 20 and 8, a transitional stage is reached as embayment ends begin to increasingly influence circulation, while when $\delta' < 8$, the embayment size and shape dominate the entire surf zone circulation (modified from Martens *et al.*, in press).

1985). The degree of headland impact or embaymentisation can be predicted using the non-dimensional embayment scaling parameter $\delta'$

| | |
|---|---|
| $\delta' > 19$ | normal beach circulation |
| $\delta' = 8$–$19$ | transitional circulation |
| $\delta' < 8$ | cellular beach circulation |

*Normal* beach circulation is what is described above for micro- and macro-tidal beaches (Figure 7.4).

*Transitional* circulation occurs when the embayment size and shape begin to increasingly influence the surf zone circulation, by initially causing longshore currents to turn and flow seaward against each headland, while still maintaining some normal beach circulation away from the headlands (Figure 9.6).

*Cellular* circulation occurs when the topography (end effects) dominates the circulation within the entire embayment. Longshore flow dominates within the embayment, with strong, seaward-flowing megarips occurring at one or both ends of the embayment, and in longer embayments also away from the headlands.

It has been established that an average rip spacing is between four and eight times the surf zone width (Hino, 1974; Huntley and Short, 1992). Using an average rip spacing of four times the surf zone width, then when $\delta' < 8$ only one or at most two 'natural' rips would fit in the embayment (Figure 9.7C). These would be found at the embayment extremities, hence the surf zone circulation is completely controlled by the end effects. If $\delta'$ is between 8 and 20 about four rips can be expected, two at the extremities and two 'natural' rips in the middle (Figure 9.7B). When $\delta' > 20$, the majority of the circulation is 'natural' and unconstrained by the end effects (9.7A).

The occurrence of large scale rips was first described by Shepard and Inman (1950), and particularly McKenzie (1958), who observed that on embayed beaches increasing wave height strengthens rip intensity while reducing rip number, and that during high wave conditions strong rips tended to occur in similar topographically controlled locations. Short (1985), working on some of the same beaches as McKenzie, called these large scale, topographically controlled rips 'megarips'. Figure 9.8 shows megarips on the Sydney coast, including Dee Why beach, McKenzie's main field site.

In addition to the above impacts, embayments, and particularly the cellular or megarips they may produce, also influence beach erosion and the seaward extent of surf zone circulation. Whereas normal beach rips usually begin to dissipate seaward of the breaker zone, megarips have been observed to flow at high velocity up to 1 km seaward of the breakers (Martens *et al.* in press; Figure 9.8). This phenomenon has important implications for beach erosion and seaward transport of sediments, nutrients and organisms. Because the cellular circulation is constrained by the headlands, in order to

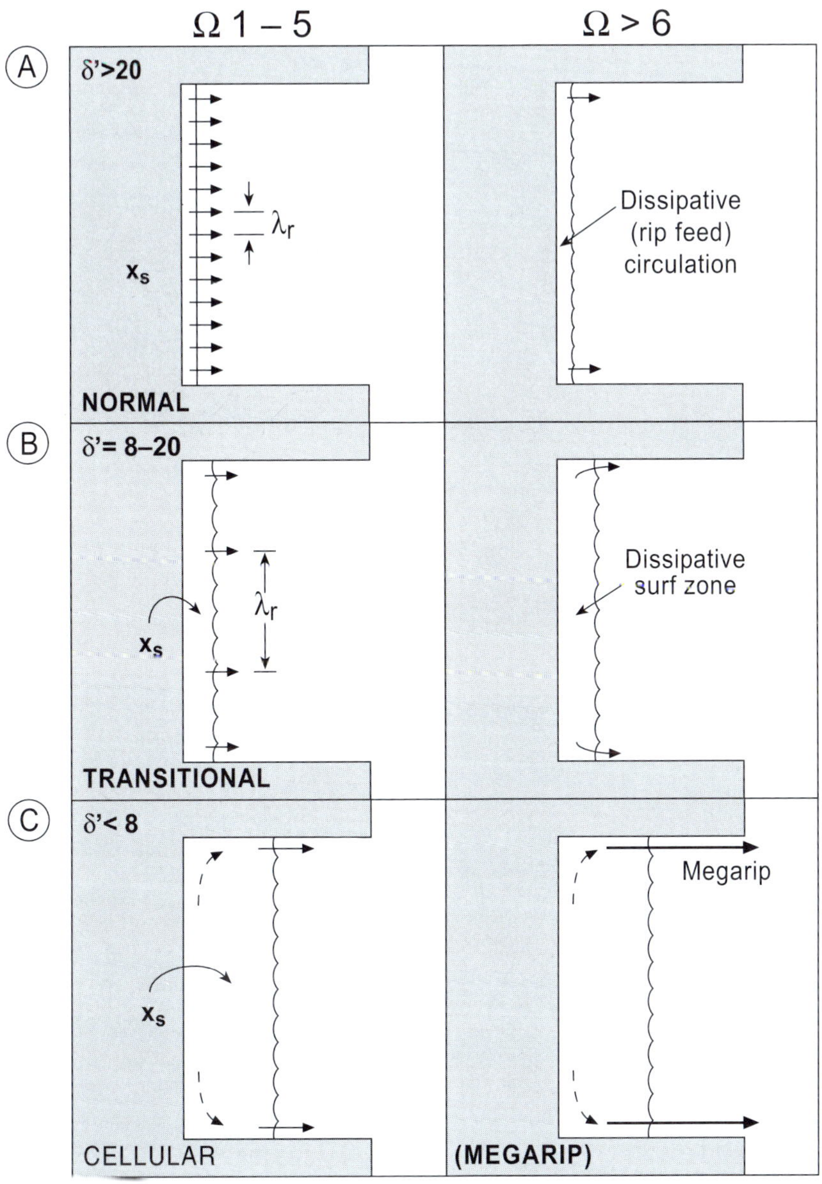


**Figure 9.7** A schematic illustration of the impact of embaymentisation on surf zone circulation based on $\delta'$ and $\Omega$. (A) Normal with end effects ($\delta' > 20$, $\Omega < 6$ and $\Omega > 6$; (B) transitional ($\delta' = 8$–$20$, $\Omega < 6$ and $\Omega > 6$); and (C) cellular ($\delta' < 8$, $\Omega < 6$ and $\Omega > 6$).

discharge water from the surf zone, rip velocities are higher, reaching 2 m/s and more. This in turn leads to greater beach erosion, with erosion most severe in the lee of the rip (Short, 1979b). As the velocities are higher, the rip currents can carry more material and coarser material, and as they penetrate further out to sea, they can carry this material a greater distance from the shore and to greater depths. The net result is more rapid and more severe beach erosion on such beaches, with the eroded sediment being deposited further seaward and consequently taking longer to return to shore, commonly 2 to 5 years following severe erosion (Short *et al.*, 1995). It may even be lost from the system, to longshore or inner shelf deposits (Roy *et al.*, 1994).

**Figure 9.8** Megarips at (a) Whale Beach, (b) Avalon Beach and (c) Dee Why Beach, all Sydney, Australia, and (d) Point Peter, western South Australia. The Whale Beach rip is running out right beside and 300 m seaward of the headland as three distinct pulses; the Avalon rip shows a clearly defined rip head; the Dee Why rip was produced by normal south-east waves with the rip exiting from the centre of the beach (exiting to right centre of photo), while the Point Peter rip is a permanent feature of this small, high energy, embayed beach (A D Short).

c
rip

d
rip

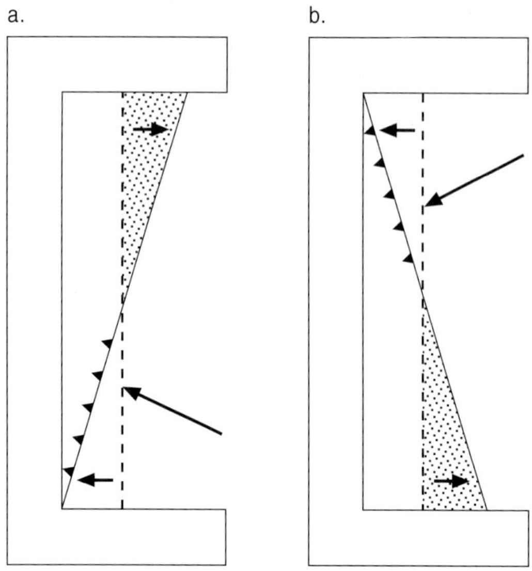


**Figure 9.9** A simplified model of beach rotation. (a) Waves from one direction produce a longshore movement of sand which accumulates against the downdrift headland, with erosion at the updrift end. (b) Waves from another direction produce the reverse. The net result is an apparent 'rotation' of the beach planform.

## 9.4 BEACH ROTATION

*Beach rotation* refers to the periodic lateral movement of sand towards alternating ends of an embayed beach, which is manifest by the apparent 'rotation' of the beach planform (Figure 9.9). Beach rotation has been observed on a number of beaches (Bird, 1993; Shyuer-Ming and Komar, 1994; Short *et al.*, 1995). In most cases it is attributed to a seasonal or periodic shift in wave climate, in particular wave direction. Waves arriving from one direction induce longshore sediment transport which eventually accumulates against the downdrift headland (Figure 9.9a). A shift in wave approach from the other direction leads to the reverse.

The rotation may be seasonal (Bird, 1993; Shyuer-Ming and Komar, 1994; Masselink, personal communication, Figure 9.10), or over longer periods (Short *et al.*, 1995). Net shoreline accretion and erosion may occur at either end, with no net change in the sediment budget, and net erosion may occur during periods of relatively low waves, when adjoining beaches are well accreted. Furthermore, small shifts in wave direction may induce large shoreline movements, particularly in long embayments (Gordon, 1987). Finally, the rotation may be a mechanism for releasing pulses of sand around headlands, known as headland bypassing (Section 9.5).

Short *et al.* (1995) provide a 20 year record of a rotating beach with an aperiodic 3 to 8 year cycle. Under normal conditions dominant south-easterly and sub-dominant east to north-east waves prevail resulting in a normal beach planform. However, during periods of lower, less frequent east to north-east waves, the dominant southerly waves persist permitting a net northern drift resulting in erosion of the southern end and accretion on the northern profile. The result is a 180° phase shift between the northern and southern half of the 3.6 km long beach, with a correlation coefficient of $r = -0.72$ between beach profiles at each end (Figure 9.11).

## 9.5 HEADLAND SAND BYPASSING

Headlands can act as natural obstacles to longshore transport and result is a downdrift accumulation of sediment. If the embayed beach experiences both longshore drift and beach rotation, a mechanism thereby exists to periodically enable sediment to escape around the headland as illustrated in Figures 9.12 and 9.13. Headland bypassing of this nature has been observed on several headlands in northern New South Wales and south Queensland. All are relatively long beaches ($> 5$ km), with net northerly transport and the potential for beach rotation. The bypassing manifests itself in two ways. First is the period of accumulation of sediment on the downdrift side of a headland (Figure 9.12a). Second is the apparent sub-aqueous movement of this sand around the headland, which first visually manifests itself as a large sub-aqueous shoal/bar on the updrift side of the headland (Figure 9.12b). Next the shoal moves shoreward and may result in substantial beach accretion; however most of the sand remains sub-aqueous. The sand waves then migrate updrift, the size and forms ranging from small sand waves to large elongate spits. The migrating wave may then move as an offshore bar (Figure 9.12c), as a wide shallow attached bar, or as an elongate spit which may impound a backing swash lagoon (Figure 9.12c). The latter two forms usually induce downdrift erosion immediately in front of the advancing sand bar/spit (Figure 9.12d). Finally the bar/spit merges with the surf zone and/or beach and is no longer identifiable (Figure 9.12e).

## 9.6 STRUCTURAL IMPACTS

Large coastal structures constructed across a beach and surf zone act as artificial headlands and can have im-

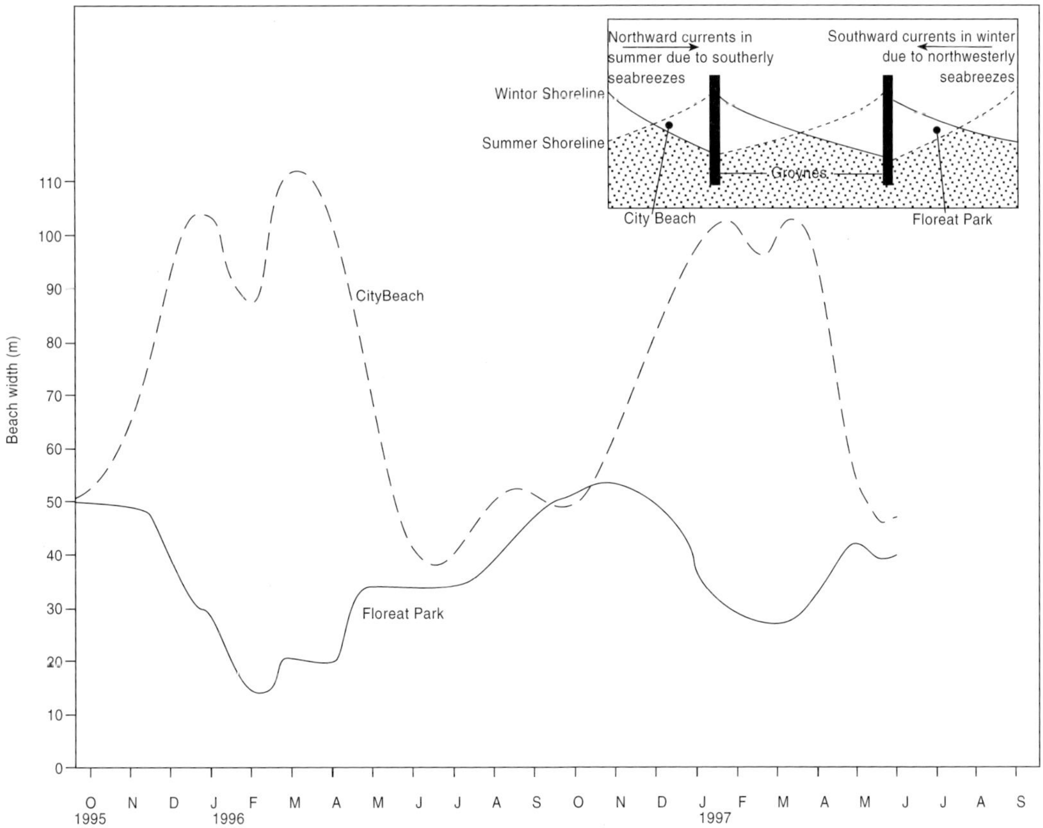


**Figure 9.10** Seasonal 'rotation' of City and Floreat beaches, Western Australia, induced by the distinctly seasonal shift in wave direction from south-west in summer to north-west in winter, and enhanced by the presence of groynes (Masselink, personal communication).

pacts identical to those detailed above. Because we are able to observe the impact of structures on a previously natural beach system, the structures provide an insight into the modification of beach processes and planform that take place following the placement of the structure. Horikawa (1988) provides an excellent overview of 'morphological changes caused by the construction of structures'.

For the majority of this century hard structures have usually been the preferred options for controlling a range of beach problems, including beach erosion and longshore drift. The replacement of sandy beaches and foredunes with rock groynes and seawalls has been referred to as the 'Newjerseyisation' of the shore, based on the substantial amount of rock placed along the small New Jersey shoreline (Pilkey and Dixon, 1996). The end result is to stem the real or perceived problem, but often at the expense of what attracted the development in the first place – the shore and beach. In the latter part of this century a new perception of beaches as a valuable natural system and public amenity, together with a new approach to shoreline management, is resulting in a swing from hard (groyne/seawall) to softer management options, including do nothing, retreat, beach nourishment and dune stabilisation.

The following sections describe the impact of some of the common coastal structures.

### 9.6.1 Groynes

*Groynes* are shore-perpendicular structures that are constructed, usually of rock but also of wood and sand tubes, across the sub-aerial beach and usually some distance into the inner surf zone, with the aim of impeding longshore sediment transport and/or realigning the shoreline (Figure 9.14). Groynes are a traditional, and

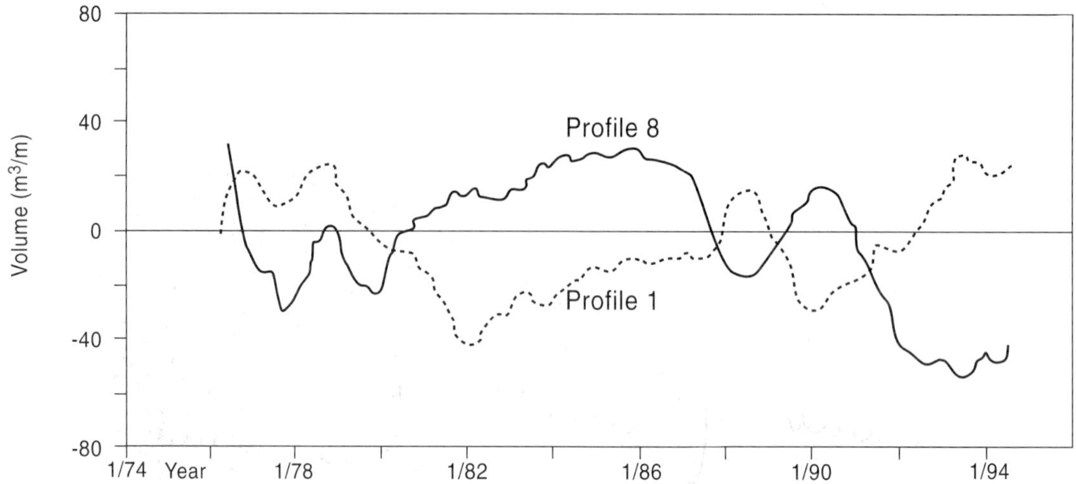


**Figure 9.11** Moving averages of the northern (profile 1) and southern (profile 8) ensemble beach volumes on Narrabeen beach. The two profiles are located toward either end of the 3.6 km long embayment and have a negative ($r = -0.72$) volume correlation, which is coherent with adjacent profiles, while a central profile has weak alternating correlation with either end, and may represent a 'rotation' or point (modified from Short *et al.*, 1995).

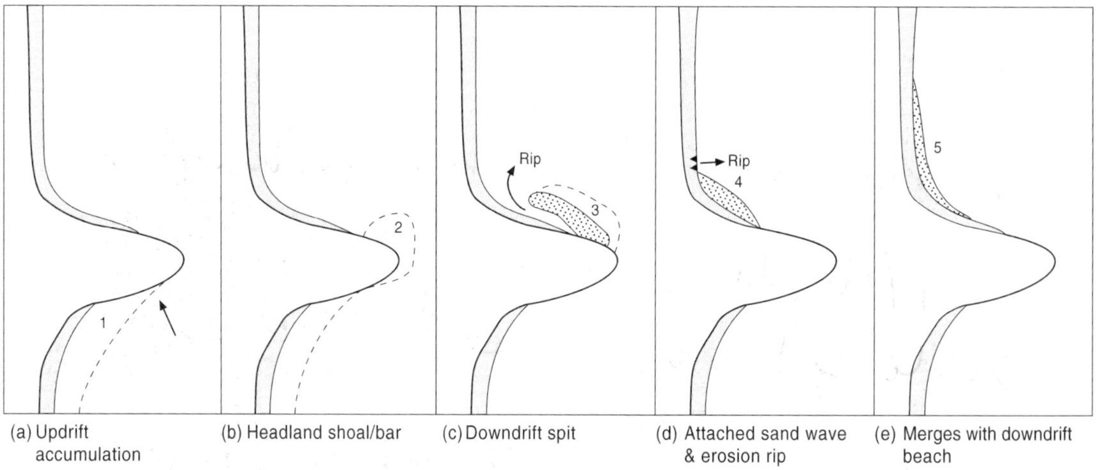


**Figure 9.12** Schematic illustration of headland sand bypassing observed on the northern NSW and southern Queensland coast, eastern Australia. (a) It is assumed that longshore sand transport, perhaps assisted by beach rotation, causes sand to accumulate updrift of the headland. (b) The sand manifests itself as a substantial sub-aqueous sand wave on the tip and immediately downdrift of the headland. (c) The sand wave moves around and along the downdrift side as an elongate spit which often encloses a backing lagoon. (d) When the sand wave/spit is attached to the beach it initiates a topographically controlled rip that migrates in advance of the wave, often causing severe localised beach erosion. (e) The sand finally merges with the beach causing slight accretion.

once commonly used, coastal protection structure, and there are many examples worldwide (e.g. Wiegel, 1964; Silvester, 1974; Komar, 1976; Coastal Engineering Research Center, 1984; Horikawa, 1988). They have in recent years fallen out of favour as it has become increasingly apparent that they do not necessarily achieve their objectives in preventing shoreline retreat, and in fact it has been suggested that they may exacerbate erosion problems (Short, 1991a). There is widespread acknowledgement of their impact on causing downdrift

**Figure 9.13** Examples of headland sand bypassing from eastern Australia. (a) Sand regularly pulses around Cape Byron (the easternmost point of Australia) from south (right) to north, seen here as the sand wave/spit. (b) Sand moving round the training walls at Cudgera Creek mouth as a submerged sand bar. (c) (*see over*) A 1 km long spit moving along the north side of Double Island Point, enclosing a temporary lagoon. (d) An attached sand wave at Cudgen with a topographically controlled migratory rip eroding the beach in advance of the wave (compare to b) (A D Short).

c

d
rip
sand
wave

**Figure 9.14** Groyne field along the central Dutch coast. All groynes have groyne (topographically) controlled rips running out adjacent to one or both groynes (A D Short).

erosion on drift shores, and there are more than enough examples of engineers chasing the erosion problem down the beach, with each new groyne or seawall inducing still further erosion, requiring yet another groyne, and so on. Furthermore they can be an unsightly structure on a natural beach system.

Groynes in the surf zone act as a permanent narrow structure whose presence can lead to:

- beach realignment owing to wave refraction around the groyne, as in Section 9.2,
- updrift accumulation and downdrift erosion of sand on drift coasts, leading to realignment of the shoreline;
- a rip current induced against the downdrift groyne (oblique waves, short groyne spacing; Figure 9.15), or rip currents against both groynes, with possible central rip(s) (shore-normal waves, more widely spaced groynes);
- shift of beach type to more immediate–cellular circulation, including dissipative to cellular, and low tide terrace to transitional (see Figure 9.15);
- owing to groyne-induced rips and megarips the potential for greater offshore and alongshore transport of sediment, hence beach erosion.

### 9.6.2 Training walls and jetties

*Training walls and jetties* are shore-perpendicular structures (usually rock) built on one or both sides of an inlet or river mouth, with the aim of both training, that is channelling and deepening the inlet, and preventing the movement of sand from the adjoining beach(es) into the inlet. Because training walls extend right across the surf zone, and some distance seaward, they have a greater impact on beach processes and morphology than the smaller groynes (see Figures 9.1b and 9.14).

Training walls act as a headland resulting in major changes in beach planform, morphodynamics and shoreline stability. Beach planform is affected by wave refraction round the wall, which, as discussed in Section 9.2, will induce a realignment of the shoreline to fit the refracted wave. Beach morphodynamics will also be affected by any changes in wave direction and breaker height owing to wave refraction which will induce an accompanying change in beach orientation and type. Finally, net shoreline stability will be impacted if the wall modified any longshore sediment transport, which will lead to major updrift sand accumulation and shoreline progradation, and downdrift erosion and shoreline retreat. Such impacts are well documented in many locations (e.g. Komar, 1976, 1984; Griggs, 1990; Murray *et al.*, 1993).

### 9.6.3 Breakwaters

*Breakwaters* are shore-parallel structures (usually rock) built usually some distance seaward of the shore and surf zone, with the aim of lowering wave energy in their lee. They may be detached and are called *breakwaters*, or attached to the shore via a wall, particularly in the case of harbours (*attached breakwaters*), and occasionally are submerged (*submerged breakwaters*). Pilarczyk and Zeidler (1996) provide an excellent text on breakwaters and their impacts.

Breakwaters are commonly built in association with ports located on open coasts and are often part of a series of structures designed to provide safe entrance to and anchorage within the port, as well as space for associated port facilities. As their overall aim is to reduce wave height in their lee, the morphological impact of breakwaters is largely related to the amount of wave

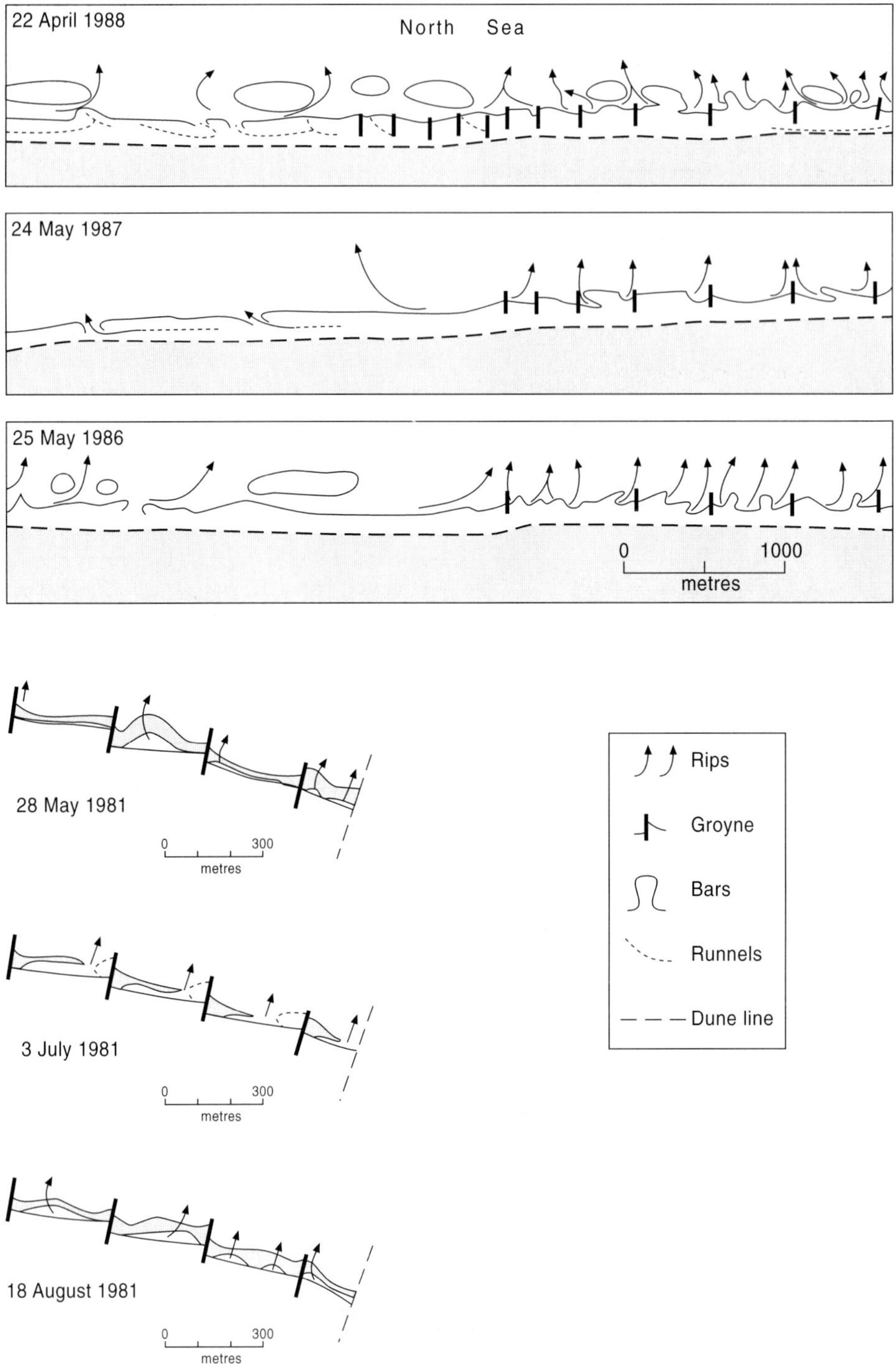


**Figure 9.15** Examples of the influence of groynes on beach morphodynamics and rip formation from the Dutch coast (upper) and Presque Islands, Lake Erie (lower). Note the increase in cellular rip circulation on the Dutch coast once the groyne field is encountered, and the persistent rips adjacent to the groyne along the Presque Island shore (modified from Short, 1992; Nummedal *et al.* 1984).

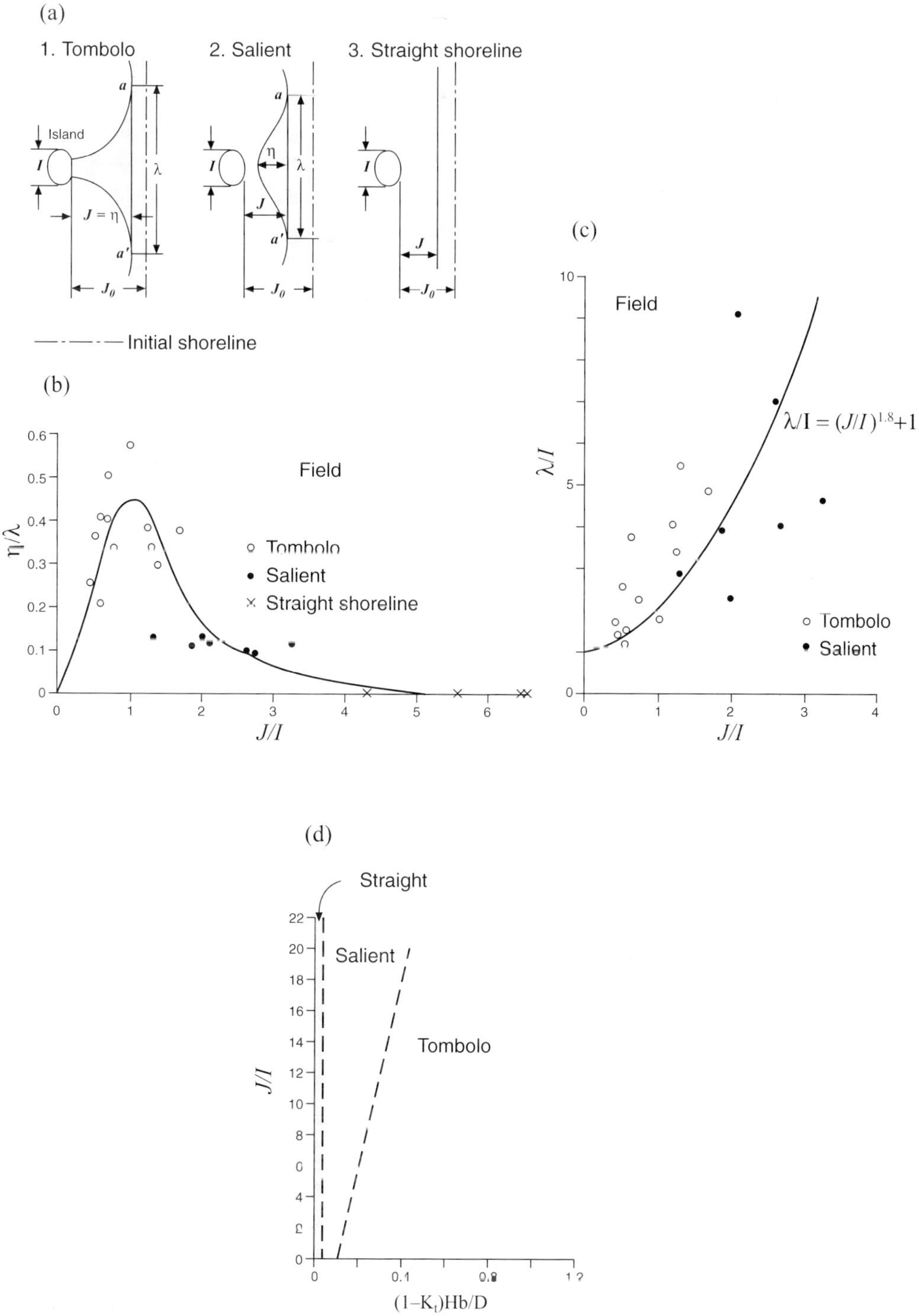


**Figure 9.16** Relationships between shoreline salients and offshore breakwaters (modified from Sunamura and Mizuno, 1987 (a, b) and Dean-Rosati *et al.*, 1992 (c)).

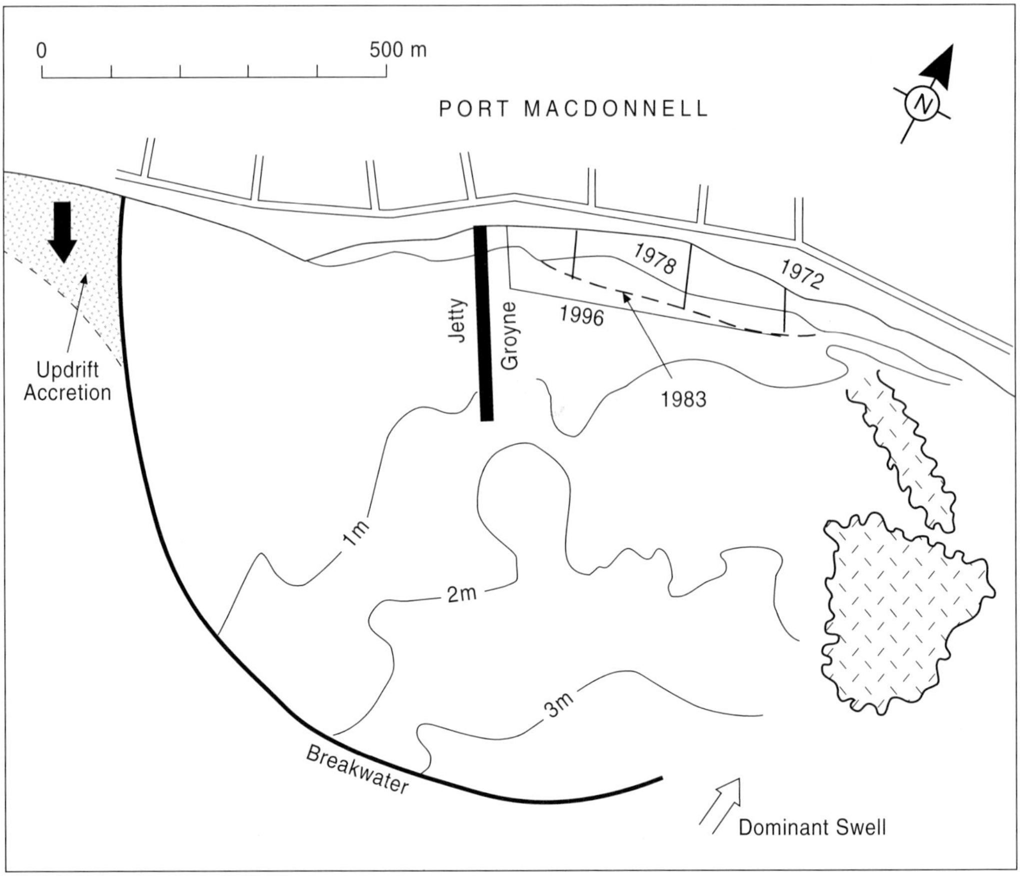


**Figure 9.17** The attached breakwater built in 1978 at Port MacDonnell in southern Australia. It has produced a wave shadow zone in its lee, where the previously stable 500 m long section of shoreline has been prograding at a rate of 5 m per year.

shadowing and the size and distance offshore to the breakwater. Depending on these combinations they may have no shoreline impact, produce shoreline accretion called a salient, or have substantial impact whereby the shoreline links with the breakwater forming a tombolo. Sunamura and Mizuno (1987) quantified this relationship whereby the shape of the shoreline response is a function of structure length and gap distance. Dean-Rosati *et al.* (1992) quantified the same responses based on the same parameters, but also incorporating wave height and period, depth at structure and structure transmission. The two classifications are illustrated in Figure 9.16 and clearly show that the shoreline impact increases with increasing size of the structure, shorter gap distance and higher waves. Equation 9.1 can be used to predict the shape of the equilibrium shoreline in lee of the breakwater.

The major impact is to cause both wave refraction round the end(s), and often a 100% attenuation of waves in their immediate lee. The net result depends on the size of the breakwater, its distance offshore and the level of wave energy. All breakwaters will lower wave energy in their lee; some will induce shoreline progradation in the lee and formation of a salient which in some cases will reach the breakwater forming a tombolo. At Port MacDonnell in South Australia, a harbour constructed in 1972 has resulted in a combination of these effects as shown in Figure 9.17. In addition to lowering the breaker wave height and possible shoreline progradation, the lower wave energy between the shoreline and breakwater will also induce a lower energy, even stagnant, zone, while at the shoreline some beaches, in the absence of waves, transform into tidal flats.

### 9.6.4 Seawalls

Seawalls are shore-parallel structures, usually made of rocks but also timber, that are built on the sub-aerial

**Figure 9.18** A concrete revetment, crude boulder seawall and groynes (background) have all been used, together with beach nourishment, at Folly Beach, South Carolina, in an attempt to prevent beach erosion (A D Short).

beach at some point between the shoreline and the foredune, with the aim of either stopping shoreline retreat and/or protecting backing property.

The degree of seawall impact on the beach is dependent first on the nature and size of the seawall, and second, on its location relative to the shoreline and surf zone. Seawalls will have the greatest impact when they are vertical and impermeable, and located within the active swash or surf zone. They will have least or no impact when they have a low gradient and are porous, and are located landward of the swash zone. All seawalls will annihilate that part of the beach or foredune upon which they are built, which can adversely affect the beach amenity, irrespective of its physical impacts (Figure 9.18). The major physical impacts are however debatable, and clearly depend on the type and location of the seawall, as well as overall beach processes.

A number of studies have documented the adverse impacts of seawall including the overall impacts (Kraus and Pilkey, 1988), consistently narrow beaches (Pilkey and Wright, 1988; Hall and Pilkey, 1991; Fletcher *et al*., 1997), and more rapid beach erosion and downdrift erosion (Griggs and Tait, 1988). The general argument (Griggs *et al*., 1991) is that seawalls have two adverse impacts. First, they lower beach permeability which results in a raising of the water table owing to slower beach drainage. The higher water table will in turn lead to a lowering (i.e. erosion) of the beach below the water table (Section 8.4.2). Secondly, when a seawall is in the wave swash zone it will induce greater wave reflection. The interaction of the reflected waves with incoming waves induces a wave clapotis, which produces increased turbulence at the wave base and consequent potential sand transport. This phenomenon, in conjunction with the raised water table, induces more rapid and more severe beach erosion and slower beach recovery.

However, a more recent literature review of seawalls by Kraus and McDougall (1996) has questioned some of these impacts. They concluded that wave reflection is probably not a significant contributor to beach profile change; that scour does not necessarily occur at seawalls; and that during storms the same amount of sand remains in front of walled and natural beaches. They went on however to recommend continued long-term beach surveying at seawalls, large scale as opposed to small scale physical modelling of seawall impacts, investigations of water table impacts, and numerical modelling of beach profile change.

Finally, if the seawall terminates next to a natural beach, then *end effects* can result in scouring of the adjacent beach (Griggs and Tait, 1988). The end effects are a product of surf zone currents being both deflected longshore by the wall, and accelerated by the constricting influence of the wall. Consequently when the current exits the wall section it can induce severe erosion of the immediately adjacent beach.

## 9.7 SUMMARY

All beaches crossed or bordered by headlands and structures are directly impacted by the nature and scale of the boundary. The degree of impact increases with decreasing spacing between embayment ends, and increasing seaward protrusion of the ends. The major impacts are, firstly, through wave refraction to modify wave direction at the shoreline, which usually produces a beach curved to fit the wave crests. Secondly, through modification of breaker wave height owing to wave refraction and attenuation, it will in turn lead to a longshore variation in breaker height and thereby beach type. Thirdly, because cellular surf zone circulation is induced by headlands and groynes, increasing embaymentisation will progressively shift a beach type from normal to transitional to cellular, whereby the structure totally dominates the circulation. Finally, any structures placed in or across the beach and surf zone will over time have similar impacts to natural headlands and reefs. In order to achieve this impact they will usually realign the original shoreline by processes of wave refraction, increased rip and cellular circulation and sand erosion and trapping. In the process of realignment, short to medium term beach change, including erosion and/or accretion, will impact the beach.

# PART D

# BEACH SYSTEMS AND IMPACTS



The foregoing chapters have dealt with beaches from a primarily superficial point of view. They have been concerned with the contemporary morphodynamic processes that control beach behaviour and type. In Part D a number of beach sub-systems and related issues are addressed. Chapter 10 examines the role of a number of non-essential, though highly influential, environmental parameters that modify beach behaviour and character; Chapter 11 reviews the nature of the biota that inhabit most beach systems; Chapter 12 reviews our knowledge of beach bedforms and related structures; and Chapter 13 assesses the impact of beach dynamics on public risk and safety.

# 10

# Beach Modification: Natural Impacts on Beach Morphodynamics

*Andrew D. Short*

## 10.1 INTRODUCTION

In Chapters 3 to 6 the major parameters that control beach morphodynamics were presented, while in Chapters 7 to 9, the impacts of waves, tides, and headlands and structures were considered. This chapter summarises the contribution of all these parameters to beach morphodynamics, together with the effect of modifications to each parameter, and hence beach morphodynamics, by non-essential environmental parameters, particularly aspects of climate and biota.

In two dimensions there are five primary parameters that control the basic beach characteristics, namely wave height, wave period, sediment size, tide range and nearshore gradient. In three dimensions, beach length and embaymentisation will determine the influence of end effects. Finally, regional climate will influence beach character, as in high latitudes beaches have the additional influence of coastal ice on their beach morphology, while in low latitudes biological impacts of reefs and chemical impacts of beachrock all produce distinctive beach features. In order to assess the role of these parameters in beach morphodynamics, a flow chart is presented for use in predicting beach type, based on the contribution of the primary parameters. Secondly, the role of climate is assessed in terms of its specialised regional influences; and finally, the degree to which each parameter can be controlled and modified by natural and human influences, and can thereby modify beach morphodynamics, is briefly assessed.

## 10.2 DETERMINING BEACH TYPE, CIRCULATION AND BAR NUMBER

The previous chapters have each highlighted the role of various parameters in contributing to beach morphodynamics. In Table 10.1 the five major parameters – tide range, wave height, wave period, grain size and beach length/embaymentisation – are incorporated in seven equations that can be used to predict or describe the major features of beach systems, including beach type (relative tide range *RTR*, $\Omega$ and $\xi$) beach slope ($\tan\beta$), erosion/accretion ($K_*$), number of bars ($B_*$) and degree of impact of end effects (embaymentisation) ($\delta'$).

In order to use Table 10.1 the following sequence is recommended.

1. Obtain *spring tide range* and *modal breaker wave height* to determine *RTR*.
2. Obtain *modal wave period*, *beach slope* and *grain size* (fall velocity) to calculate $\Omega$ and $\xi$ and thereby beach type. Note, for grain size use swash sediment for reflective, inner surf zone for intermediate and outer surf zone for dissipative beaches.
3. Obtain overall beach/surf zone *nearshore gradient* for prediction of bar number ($B_*$).
4. If embayed by headlands or structures measure *embayment shoreline* and *chord length* for calculation of $\delta'$ and prediction of circulation pattern.

When using the equations in Table 10.1 the following issues must also be considered.

- The *relative tide range* (*RTR*) is a predictive parameter that simply locates a coastal environment based on the relative height of the waves and tide.

*Handbook of Beach and Shoreface Morphodynamics*. Edited by A. D. Short.
© 1999 John Wiley & Sons Ltd.

**Table 10.1** Impact of salient environmental parameters on beach type, stability, circulation and bar number. Shaded areas indicate beach type.

| Tide range, $TR$ (m) | Wave height, $H_b$ (m) | Wave period, $T$ (s) | Sediment size, $W_s$ (m/s), $D_{50}$ (mm) | | Embayment geometry $S_1$ (m), $C_1$ (m) | (Gradient, $\tan\beta$) |
|---|---|---|---|---|---|---|
| | | *Beach type* | *Beach stability* | *Slope* | *Embayment impact* | *Bar number* |
| $RTR = TR/H_b$ | $\Omega = H_b/W_sT$<br>$\xi = (a\sigma^2)/(g\tan^2\beta)$ | | $K_* = H_b{}^2/gT^2D$ | $\tan\beta = 0.12K_*{}^{-1/4}$ | $\delta' = S_1^2/100C_1H_b$ | $B_* = X_s/\tan\beta T_i^2$ |
| (Equation 8.12) | (Table 7.1) | | (Equation 10.1) | (Equation 10.2) | (Equation 9.6) | (Table 7.1) |
| | (Section 7.2) | *Wave dominated* | *Erosion/accretion*[1] | *Slope*[2] | *Circulation* (Section 9.3) | Bar number (Section 7.4) |
| $< 3$ | $\Omega < 1$, $\xi < 2.5$ | Reflective | $K_* < 3.5$ | Steep | $\delta' > 19$ normal | $< 20 = 0$ bar |
| | $\Omega = 2$–$5$, $\xi = 2.5$–$20$ | Intermediate | $K_* = 3.5$–$20$ | Moderate | $\delta' = 8$–$19$ sub | 20–50 = 1 bar |
| | $\Omega > 6$, $\xi > 20$ | Dissipative | $K_* = > 20$ | Low | $\delta' < 8$ cellular | 50–100 = 2 bars |
| | | | | | | 100–400 = 3 bars |
| | (Section 8.6) | *Tide dominated* | | | | $> 400 = 4 +$ bars |
| 3–7 | $\Omega < 2$ | Reflective and LTT | | | $\delta' > 19$ normal | |
| | $\Omega = 2$–$5$ | LT bar and rips | | | $\delta' = 8$–$19$ sub | |
| | $\Omega > 5$ | Ultra-dissipative | | | $\delta' < 8$ cellular | |
| 7–15 | $\Omega < 2$ | Beach and LTT | | | $\delta' > 19$ normal | |
| | $\Omega > 2$ | Ultra-dissipative | | | $\delta' = 8$–$19$ sub | |
| | | | | | $\delta' < 8$ cellular | |
| $> 15$ | | Sand (tidal) flats | | | | |

[1]Erosion/accretion = $K_* = H_b^2/gT^2D$ (Sunamura, 1984b).
[2]Beach slope = $\tan\beta = 0.12/(H_b/g^{0.5}D^{0.5}T)^{0.5} = 0.12K_*{}^{-1/4}$ (Sunamura, 1984b).

- The *dimensionless fall velocity* ($\Omega$) is a predictive equation that indicates the beach type that will occur if the wave and sediment parameters are allowed to reach equilibrium.
- The *surf scaling parameter* ($\xi$) on the other hand, is a descriptive equation that actually describes the state of the waves and beach gradient.
- When sand size is also incorporated then Sunamura's (1984b) $K_*$ parameter can be used to predict beach erosion (offshore sediment transport) and beach accretion (onshore sediment transport).
- Furthermore by applying a constant to $K_*$ beach slope is also predicted.
- On embayed beaches the impact of headlands can be predicted using the *embayment parameter*, $\delta'$.
- The number of bars can be predicted using the *bar parameter* ($B_*$).

All the parameters other than $\xi$, are predictive, and require the beach to be able to fully respond to the impact of the salient parameters, particularly $H_b$ and $T$, which may take days to weeks.

Consequently it is important to realise that while Table 10.1 can be used to predict beach type, slope, circulation (embayed beaches), bar number and stability (erosion/deposition), the prediction is based on the beach being fully adjusted to the input parameters. Beaches are however notorious for dependence on antecedent morphology, and beach change may lag processes by days to months. Therefore, predictions based on short term or prevailing processes may not necessarily agree with the observed morphology. Wright *et al.* (1984) found that to predict prevailing, as opposed to equilibrium, beach morphology, required input of the antecedent processes over the previous 30 days on an energetic beach (and probably much longer on less energetic beaches). Therefore Table 10.1 should only be used to give an indication of the modal or equilibrium beach conditions for a particular site, based on modal wave/sediment parameters.

## 10.3 MODIFICATION BY COLD CLIMATES

Climate, through temperature, has major positive and negative impacts on coastal and beach environments. These have been discussed briefly in Section 2.3 and summarised in Table 2.2. In this section the direct impact of freezing temperatures and the indirect effect of warm tropical water on beach morphology are presented.

When a beach surface temperature falls below 0°C and seawater below −4°C they freeze. The seasonal freezing of beaches and the adjacent seas and oceans impacts most beaches above 60–70° latitude, a substantial section of the world's coast. The duration of the freezing ranges from occasional frozen days in marginal areas to several months in polar regions including all of Antarctica and most of the Arctic coast (Figure 2.2). The impact of ice on beaches is both active and passive. Active ice processes that can transport sediment include ice rafting, ice push on the beach and ice gouging seaward of the fast ice (Reimnitz *et al.*, 1990). Passive processes include tides which build an ice foot between the fast and floating ice (e.g. Nichols, 1961; Farrelly, 1972; McCann and Carlisle, 1972; Marsh *et al.*, 1973), and sub-zero temperatures which freeze the beach surface and then the adjacent ocean out to a maximum depth of 2 m and preclude wave activity at the shoreline.

High latitude beaches go through three distinct seasonal phases: open water when normal beach conditions operate, freezeup when the beach and sea surface freeze and the ice may impact the beach, and breakup when the ice melts and beach thaws (Figure 10.1; e.g. Short, 1976). Each is briefly discussed below.

### 10.3.1 Freezeup

As soon as zero degree days (< 0°C) arrive in late spring beach surfaces begin to periodically freeze. Once the beach surface is frozen swash begins to freeze toward the limit of run-up, gradually building an ice berm toward the swash limit. Seawater remains unfrozen until sub-zero temperatures (< −4°C) are reached. As temperatures fall below this threshold the sea surface begins to freeze as a soupy *slush ice* consisting of uncongealed ice crystals. Once the ice freezes into a solid crust it can still be broken into thin pancake-sized pieces called *brash ice*. Both slush and brash ice can be washed onto the beach (Short and Wiseman, 1974). As the ice crust thickens continuous rigid sea ice is formed, which during the winter period may thicken to 1 m or more.

During the freezeup actual wave, wind and tide regimes will dictate which combination of ice-sediment processes occur. The most common sequence is the deposition of frozen swash, called *ice foot*, which may be interlaid with sediment in the swash zone, and capped by frozen swash and snow (Figures 10.2a and 10.3). McCann and Carlisle (1972) found that ice foot tends to increase in size with increasing tide range, while Davies (1980) identified seven types of ice foot based on their location and formative mechanisms. If a storm occurs during freezeup then ice cakes and small ice boulders, derived from thin broken sea ice, may be piled up in the swash zone forming a swash-deposited berm or ridge of ice boulders including slabs of ice a few metres across (Figure 10.3).

The frozen beach itself usually remains dormant until breakup. However seaward of the beach the fast ice slowly thickens to reach a maximum of 2 m in high latitudes. At the boundary of the fast ice the floating ice will rise and fall with the tide. If sea ice is pushed onto the fast ice it can build either an *ice ridge* where the ice slides up unbroken, or if it breaks up an *ice pile* of broken ice up to 20 m high. Also sea ice can ridge and pile up on offshore bars or shoals. Both processes produce ice ridging and piles along the floating/fast ice boundary and over bar crests and tidal shoals. Barnes *et al.* (1993) use the term 'nearshore ice complex' to describe the combination of ice–shore interaction features which can occur along freezing shores (Figure 10.4).

In addition any pack ice that is tilted over by the push will form a sub-aqueous keel, up to 40 m deep, that can be dragged over the bed gouging into the sediment. Reimnitz *et al.* (1990) found that *ice gouging* extends out to at least 40 m water depth, and actively reworks the nearshore sediments.

### 10.3.2 Breakup

Breakup begins as soon as positive degree days (> 0°C) arrive. The melting snow and ice progress through several stages (Short and Wiseman, 1975; Barnes *et al.*, 1993), beginning in the northern hemisphere with meltwater from the hinterland flooding the rivers and creeks. Next the meltwater reaches the coast and either flows directly onto the sea ice or over frozen lagoon surfaces. As the lagoon ice floods some water ultimately reaches the frozen inlets and begins to flow out onto the shorefast and sea ice. At this stage the beaches are still frozen and covered by snow and shorefast ice.

The water flows across the shorefast ice which in front of an inlet may extend further seaward over ebb tidal

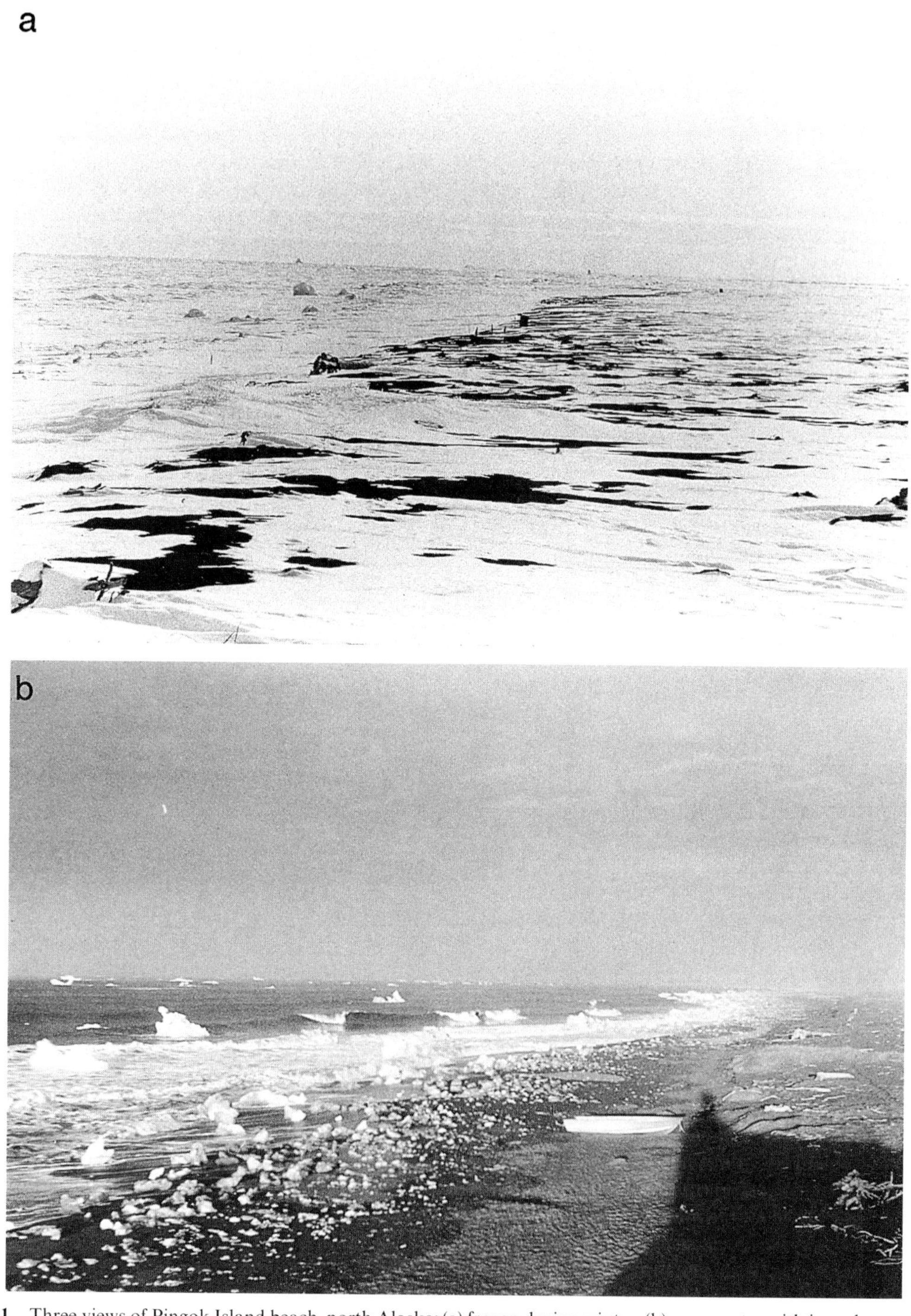


**Figure 10.1** Three views of Pingok Island beach, north Alaska: (a) frozen during winter; (b) open water with ice cakes on beach; and (c) during freezeup with snow on beach (A D Short).

shoals. As soon as the floating sea ice is reached the water flows through tidal cracks and finally into the sea. The water and sediment induce changes in albedo of the ice which accelerate melt of all the water-covered ice which results in an initial breakup of the lagoons and inlets (Short and Wiseman, 1975).

The beaches with no inlets or away from the inlets melt more slowly. First, snow covering the backing beach, which may be a berm or low grassy dunes, melts. Next the floating and cracked sea ice melts along the fracture lines and begins to breakup and both float away and melt in situ. By this stage the beach is bare of surface ice down to the fast ice. The shorefast ice tends to melt in situ and is the last surface ice to melt.

In addition to the sea ice and surficial beach snow and ice, ice may be embedded in and buried beneath the beach sand surface. This occurs when wave activity accompanies beach freezeup. The freezup processes can produce sediment-covered *ice mounds*, which are well insulated by the overlying sediment. They only begin melting once the overlying snow and ice are gone and may take several days to weeks to fully melt. As they melt the overlying sediment collapses and forms a range of thermokarst or melt features in the swash zone (Figure 10.5). The shape and size of the mounds and holes depend on the processes that took place during the previous freezeup.

A final ice feature that can occur during breakup and even during summer open water if sea ice is present inshore, is the deposition of ice cake and blocks in the swash zone with a sand matrix. During summer the ice melts relatively rapidly (a few days) and ice melt potholes are produced in the swash zone (Short, 1976).

The impact of shore ice and frozen beaches is both positive to the degree that ice gouging, ice push and ice foot transport sediments, and negative in that little or no action takes place during the frozen winter period. Over time this restricts the evolution of the coast to open water periods, which in the high Arctic and Antarctica can be as little as a few weeks each year.

### 10.3.3 Open water

The final melt of the sea ice heralds the open water season, which may take several weeks of melt to reach, often not occurring until mid-summer along Arctic coasts. Open water brings normal wave, tide and wind activity with processes identical to non-frozen beaches. The first waves will however rework the melting beach removing thermokarst features. Only the deeper ice gouge features might survive the summer reworking (Ruz and Allard, 1994).

**Figure 10.2** (a) A 2 m high ice foot containing frozen swash and sediment at Point Lay, north Alaska; and (b) the same site 8 weeks later during open water. (c) Brash ice being washed onto Pingok Island beach during freezeup; and (d) frozen sediment, swash and ice boulders, all capped by snow, on the berm crest looking landward at Pingok Island, north Alaska (A D Short).

## 10.4 MODIFICATION BY TROPICAL CLIMATES

Tropical coasts are defined by their low latitudes and high air and water temperatures, both of which indirectly impact beach morphology. This section briefly reviews some of the major impacts of the high water temperatures in particular, which contribute to a range of biological and chemical impacts. Biological impacts are generally related to the growth of structures which modify nearshore processes and topography, while the chemical impacts are related to cementation of the beach sediments, both of which have major impact on affected tropical beaches.

### 10.4.1 Biotic impacts

The major biotic impact on wave-dominated tropical coasts is the prolific growth of coral and algal reefs in suitable locations. Depending on nearshore–inner shelf gradient and coastal river discharge, these may develop as barrier reefs, some distance off the coast, as with the

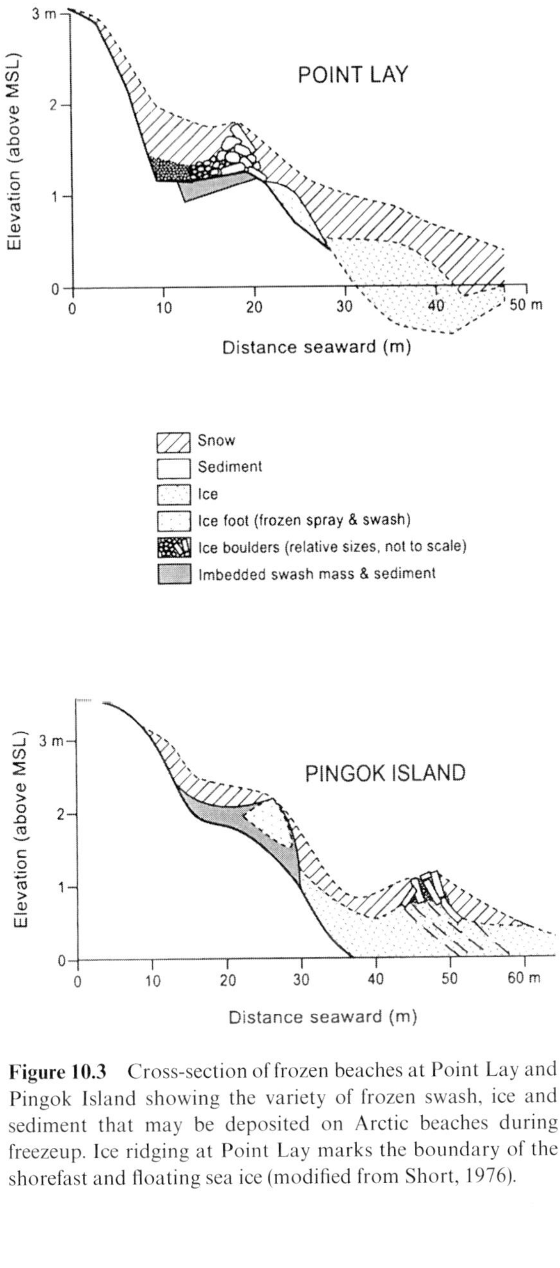


**Figure 10.3** Cross-section of frozen beaches at Point Lay and Pingok Island showing the variety of frozen swash, ice and sediment that may be deposited on Arctic beaches during freezeup. Ice ridging at Point Lay marks the boundary of the shorefast and floating sea ice (modified from Short, 1976).

Great Barrier Reef; fringing reefs against the coast; or as independent atolls containing beach systems as part of their structure.

### *10.4.1.1 Barrier reefs*

Barrier reefs cause wave breaking and dissipation of all or most ocean wave energy. They therefore lead to a substantial lowering of wave height in their lee. Depending on their elevation, size and tide range they may partially or completely eliminate deep water waves. When complete elimination occurs the backing beaches must rely on fetch-limited (reef to shore) winds and waves to drive the beach morphodynamics. Figure 10.6 shows the change in deep water and breaker wave climate caused by the Great Barrier Reef, with modal wave height decreasing from 1.2 m to 0.5 m and period from 10 s to 5 s.

In addition, on the tropical Queensland coast Hopley (1982) suggests that lagged upward growth of large sections of the Great Barrier Reef gradually eliminated a mid-Holocene 'higher energy' wave window. During the mid-Holocene the reef was lower permitting more ocean wave energy to reach the coast, resulting in higher energy beaches, whereas as the reef grew up to low tide level this energy was gradually eliminated, resulting in a temporal decrease in regional breaker wave height and beach processes, and a transition to lower energy beach systems. Analysis of barrier morphostratigraphy has supported this model and also suggested that reef growth may have led to an enhanced tide range (Masselink and Lessa, 1995).

### *10.4.1.2 Fringing reefs*

Fringing reefs by definition lie at the base of the shoreline. On sandy coasts they fringe the base of the beach, forming below mean sea level. They range in width from a few to tens of metres (Figure 10.7). They tend to occur on tropical coasts away from major rivers and sediment discharges, and particularly on arid tropical coasts. Their impact is fourfold. First they truncate the subtidal beach, permitting only the swash zone to form and operate; secondly, they dissipate wave energy, particularly at low tide, so that the beach is usually only active at mid- to high tide. In this way they behave like a *platform beach* in lee of a rocky intertidal zone. Thirdly, the lowering in wave height and duration leads to the formation of predominantly lower energy, steep reflective beaches. Finally, variable reef topography will generate a longshore variation in wave height and direction, which will in turn modify the level of swash energy and shoreline topography.

### *10.4.1.3 Seagrass and kelp beds*

Seagrass and kelp beds form in the sub-tidal zone seaward of beaches predominantly in temperate climates. Kelp beds can form in high wave environments such as along parts of the Californian coast, while seagrasses

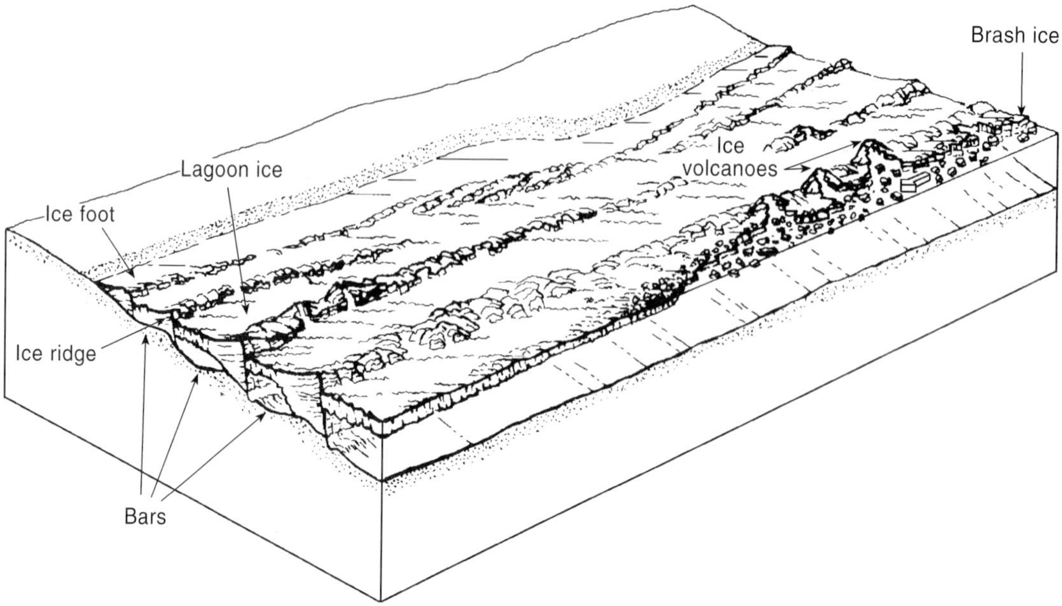


**Figure 10.4** A schematic diagram of an idealised 'nearshore ice complex' showing the ice foot, ice ridges and lagoons, ice push ridges (ice volcanoes), with drifting brash and slush ice offshore (from Barnes *et al.*, 1993; reproduced from the *Journal of Coastal Research* by permission of the Coastal Education and Research Foundation).

favour more protected locations. The impact of both is to increase bed roughness in the nearshore zone, thereby increasing wave attenuation and lowering wave height at the shoreline. This will lead to lower energy beach types than would otherwise occur. Across seagrass-dominated areas of the southern Australian coast, high wave events erode seagrass and transport it onshore where it is deposited in the upper swash zone, forming hummocky seagrass berms which can extend for tens of kilometres alongshore and be up to 2 m high (Figure 10.8).

Seagrass meadows can also be areas of high carbonate production, and in suitable environments can supply carbonate detritus to the adjacent beach systems. In parts of western and southern Australia, beach systems backing seagrass meadows may derive up to 100% of their sediment from the meadows, e.g. Shark Bay (Logan, 1970) and Spencer Gulf (Belperio *et al.*, 1988).

#### *10.4.1.4 Mangroves*

In the transition from low wave energy beaches to tidal flats, tide-dominated beaches can occur in lee of mangrove woodlands. As mangroves grow between neap high tide and mean sea level beach development is restricted to the neap to spring high tide level and swash limit. The beaches associated with mangroves are characteristically low energy and usually inactive, particularly during low tide and calm conditions. They are activated only during strong onshore wind and wave conditions toward high tide (Tedesco *et al.*, 1995). The mangroves themselves also act to filter wave energy and thereby further lower high tide wave energy at the beach.

### 10.4.2 Chemical impacts

The major chemical impact on beaches is the formation of beachrock. Beachrock is an indurated intertidal beach, where the sediments are cemented by calcite crystals. Beachrock has been the subject of considerable attention since the work of Guilcher (1961) and Russell (1962) and has been found to occur on all tropical coasts (e.g. Russell, 1959; Russell and McIntire, 1965b; Miller and Mason, 1994) as well as on some temperate coasts (Semeniuk and Searle, 1987; Davies, 1980). In areas of warm coastal waters, seawater becomes supersaturated with calcium carbonate. In the intertidal zone of beaches (sand through cobbles) the calcium carbonate precipitates on the sand grains as crystals of low magnesium calcite (about 1 $\mu$m in size) cementing the gains

**Figure 10.5** Sand and gravel beach with sediment-covered ice mounds (a) and pothole formed by ice melt (b), Pingok Island, north Alaska (A D Short).

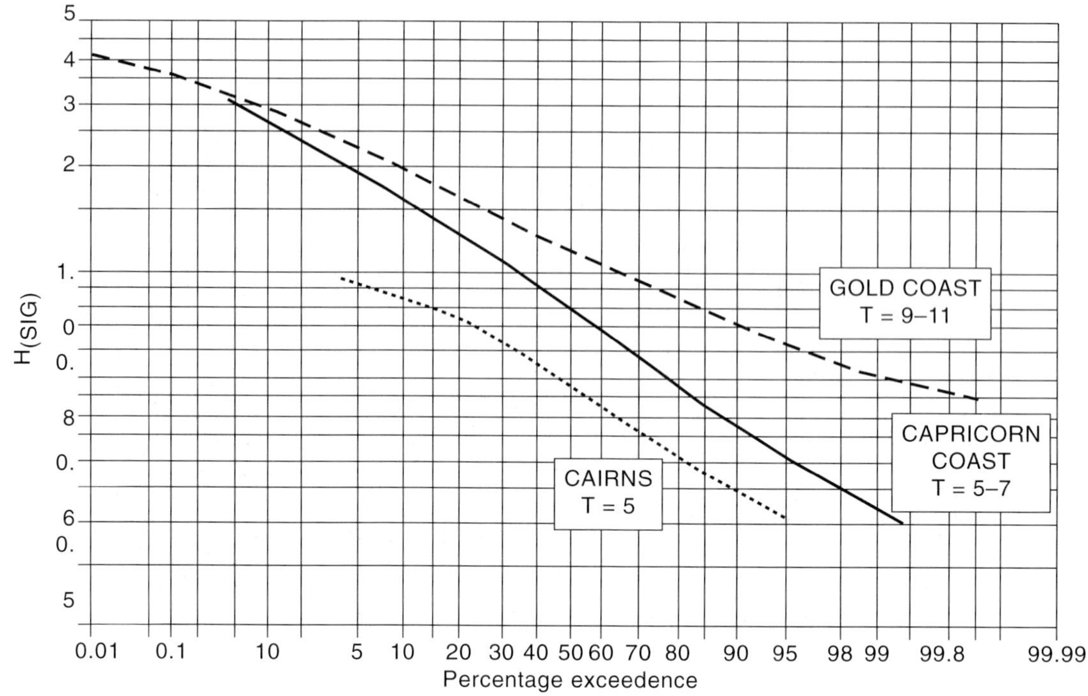


**Figure 10.6** Wave exceedence curves for the exposed Queensland Sunshine coast, and two locations in lee of the Great Barrier Reef, which receive no ocean swell. Modal wave height decreases from 1.5 m to 0.5, period from 11 s to 5 s and overall wave energy by an order of magnitude (source: Beach Protection Authority, Queensland).

(Semeniuk and Searle, 1987; Gishler and Lomando, 1997). Over time (decades or more) the entire intertidal zone of the beach can become cemented as a shore-linear slab of beachrock (beach calcarenite). On micro-tidal beaches the slabs can range up to 4 m in thickness, while the most concentrated cementation is generally less than 1 m thick (Semeniuk and Searle, 1987). Thicker cementation can be expected on macro-tidal beaches.

The physical impact of the rock is to substantially reduce beach permeability, which in turn has dramatic effect on beach stability when the beachrock is close to or exposed at the surface. Lower permeability leads to less percolation of the swash, more backwash and a higher discharge or effluent zone. These will in turn impact on both beach morphology and stability. The raised effluent discharge line modifies the beach topography by raising the level of beach saturation, thereby producing a higher break in slope at the discharge line (see Section 8.4.2). Secondly, during erosive conditions beachrock will lead to accelerated beach erosion owing to the decreased percolation and increased effluent. Finally on a retreating shoreline, the unconsolidated beach may be completely eroded leaving behind a beachrock 'reef'. On some coasts these 'reefs' can extend for kilometres alongshore (Figure 10.9). As the shoreline retreats behind the 'reefs' they act like a fringing, then barrier, reef, modifying wave energy, which in turn impacts the shape and level of wave energy at the backing shoreline (e.g. Suhayda *et al.*, 1977). In some arid temperate coasts the same impact will occur with the formation of sub-aerial dunerock (aeolian calcarenite) which can impact many thousands of kilometres of coast, as in southern and western Australia (Figure 10.10; e.g. Short and Hesp, 1984).

### 10.4.3 Latitudinal impact

An indirect impact of low latitude is the low Coriolis effect. This is most manifest in the complete absence of cyclones from within 5° north and south of the equator, thereby precluding large locally produced waves. Even when tropical cyclones form they tend to migrate poleward, further reducing the potential impact of periodic high waves. The net result is low local wave generation (Section 2.6.3) and lower energy beaches.

**Figure 10.7** (a) Cape Tribulation, north Queensland, showing the low energy beach fronted by a fringing coral reef; and (b) fringing coral reefs with backing shoreline salients produced by the variable wave refraction across the reefs, Ningaloo Marine Park, Western Australia (A D Short).

**Figure 10.8** A seagrass 'berm' piled across the inter- to high tide beach at Nooneara, Great Australia Bight, Western Australia (A D Short).

## 10.5 MODIFICATION OF BEACH PARAMETERS

The degree to which beaches are modified by natural and/or human causes depends on the degree to which they in turn modify the primary parameters of wave height, wave period, sediment size and tide range. This section considers natural and human impact on these parameters.

### 10.5.1 Wave height

Wave height is the single most important process in controlling beach type and change. Wave sources and wave generation are naturally highly variable producing regular fluctuations in wave height at frequencies ranging from hours to seasonal to longer term (Section 2.6). Natural influences on wave height include all topographic controls that affect wave shoaling (attenuation), refraction, diffraction and breaking. These fluctuations in turn drive predictable beach changes as addressed in Chapters 7 and 9. This section considers three additional sources of wave modification, namely ice, tropical coasts and human interference.

#### *10.5.1.1 Ice effects*

As discussed in Section 10.3 coastal ice can have three effects on waves, namely eliminating waves at the shore, limiting wave generation, and reducing the time in which waves can be generated (see e.g. Short *et al.*, 1974). *Shorefast ice* will completely eliminate waves from reaching and reworking the shore. *Pack ice* lying offshore will reduce fetch and thereby the potential for wave generation, resulting in smaller, shorter waves. Finally *wave generation* is completely eliminated during the winter sea ice period. It is restricted to the limited open water season, and thereby the seasonal and long term duration of wave impacts is limited to this period, which may be as short as a few weeks in the high Arctic.

The net result is for cold temperatures to reduce the size of waves (height and period), restrict the occurrence of waves, and eliminate wave action at the shore completely when shorefast ice is present. All this restricts shoreline evolution to the open water period and encourages the predominance of lower energy reflective beaches, a fact also assisted by the dominance of coarse sediment on high latitude beaches (Section 2.4).

**Figure 10.9** (a) Three successive beachrock reefs (1, 2, 3) at Shoal Cape, south Western Australia. Each reef successively lowers the high breaker wave energy, resulting in a zero (low tide) to low energy (high tide) swash zone (A D Short). (b) Shore-parallel beachrock reef and backing wave shaped shoreline, Suape, Brazil (L D Wright)

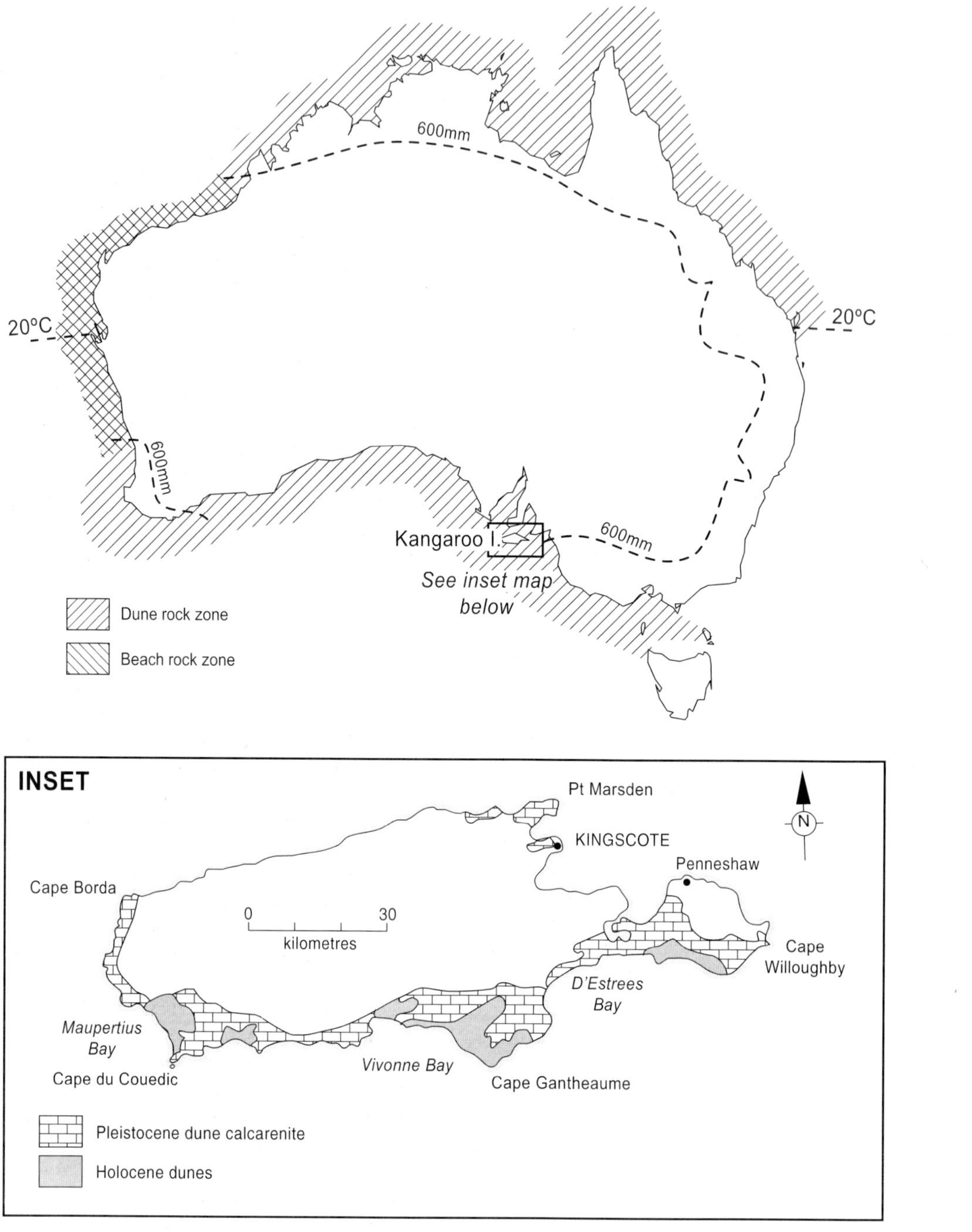


**Figure 10.10** Distribution of calcareous beachrock and dunerock around the Australian coast (source: Davies, 1980); and (inset) distribution of Pleistocene dunerock and Holocene dune along the coast of Kangaroo Island, South Australia (source: Short and Fotheringham, 1986).

#### *10.5.1.2 Tropical effects*

At the other extreme warm climates and coastal waters induce prolific coral and algal growth, which when manifest as coral and algal reefs, induces greater topographic influences, including wave breaking/dissipation over the reefs, and wave refraction around the reefs. The result is twofold. First, wave energy is lowered, and secondly, wave refraction and attenuation over and around reefs produce a more variable angle of wave approach and a longshore variation in wave height, both of which induce more crenulate beaches. The same impact occurs when beachrock in located at the base of, or off beaches (Figure 10.9).

#### *10.5.1.3 Human effects*

Human impact on wave height is normally related to structures that cause additional wave breaking, shoaling and refraction. Structures designed to lower or eliminate wave height, as in harbour construction, will substantially reduce wave height at the shore and change the consequent beach morphodynamics. In extreme cases the beach will be eliminated by construction and/or absence of waves. Where wave height is lessened and/or redirected the shoreline will readjust its shape to accommodate the change in height and direction. Other structures designed to interrupt or interfere with longshore sediment transport (training walls, groynes) will tend to produce a realignment of the up- and downdrift shoreline, as well as interrupting the longshore sediment transport (Figure 9.1). Finally, structures designed to prevent beach erosion, such as seawalls and revetments, will tend to reduce beach permeability and cause accelerated beach erosion and slower recovery (Figure 9.18; e.g. Griggs *et al.*, 1991; Fletcher *et al.*, 1997). The impact on wave characteristics of many types of structures is reviewed by Wiegel (1964) and examples are presented in Coastal Engineering Research Center (1984) and in Chapter 9.

### 10.5.2 Wave period

Wave period is independent of wave height and is not influenced by wave shoaling. Consequently there is little one can do to modify wave period, other than removing it altogether, as in the case of total wave breaking on a reef or against a structure.

### 10.5.3 Wave direction

In shallow water wave direction is modified through wave refraction, a common natural process. This process is intensified in lee of coral and beachrock reefs, resulting in more crenulate shores in their lee. It is also induced by any structures that project into the surf, resulting in either refraction or the creation of a wave shadow (Johnson, 1957; Dingler, 1982). These will have the same impact as presented in Figure 9.16.

In summary, all factors influencing wave characteristics lead to a lowering in wave height and/or redirection in wave approach, resulting in lower energy, more crenulate or discontinuous beach systems.

### 10.5.4 Sediment

A quantity of sediment is essential for beach formation. The available sediment will, with accommodation space, determine the overall dimensions of the beach and the sediment size and will interact with wave height and period to determine beach slope and beach type. Therefore changes in the quantity (sediment budget) will affect the size and occurrence of beaches, while changes in sediment size will impact beach morphodynamics.

#### *10.5.4.1 Sediment quantity*

The quantity for sediment available for a beach system is dependent on the sediment budget as defined by Davies (1974) and Komar (1976). A beach or barrier sediment budget is generally based on an accounting of the sediment sources or inputs, sediment losses or throughputs, combined with on-site attrition due to abrasion, solution, cementation or extraction. The net balance determines whether the system is being supplied with a surplus of sediment leading to storage in wave-, tide- and/or wind-deposited systems, or in deficit and loss of sediment leading to beach retreat and shoreline erosion.

The long term impact of sediment budgets is indirectly considered in Chapter 14 which deals with the nature and evolution of coastal barrier systems, all of which at some time require a positive sediment budget, and many of which subsequently lose sediment and erode. At contemporary time scales, in addition to natural changes to sediment budgets, human impact can and does change, modify and/or interrupt sediment budget, with consequent effects on shorelines. The major human impacts have already been considered in Chapter 9, and are listed in Table 10.2.

#### *10.5.4.2 Sediment size*

In natural systems, sediment size, while often segregated

**Table 10.2** Human impacts on sediment budgets.

| Decrease supply | Increase supply | Interrupt supply |
|---|---|---|
| Managed rivers<br>Dams<br>Extraction | Degraded river catchments<br>Deforestation<br>Cultivation, urban development | Groynes<br>Breakwaters<br>Training walls<br>Seawalls |
| Cliff and bluff protection<br>Seawalls<br>Beach mining | Beach nourishment | |

cross-shore (Section 12.2), and at times longshore (e.g. King, 1972), tends to remain relatively uniform through time at any particular location. Exceptions do occur however when major seasonal or longer term erosion events remove the usually finer beach veneer from an underlying coarser lag deposit. Such a phenomenon was first documented by Shepard (1963, p.175) who observed the winter storms stripping the beach sand to reveal a cobble/boulder lagged deposit, which was inherently reflective. Short (1984a) observed a similar phenomenon when a 1:100 year erosion event removed a fine surficial layer associated with a low tide terrace beach, to reveal a coarse sand lag, which produced a modally reflective beach. The fines returned over a period of six years rebuilding the beach to its modal low tide terrace state. Clearly this can happen in nature wherever coarser sediments underlie a finer layer, and storm, seasonal or aseasonal wave events remove the surface layers, temporarily changing sediment size, and thereby beach type.

Beach sediment size can be knowingly modified using two mechanisms. First, sand nourishment can be used to add finer to coarser sand. It is often argued that the nourishing sand must be the same as the native sand (Coastal Engineering Research Center, 1984). However this is only in order to reproduce the same beach characteristics. By knowingly adding coarser or finer sand, it is possible to produce a predictable change in beach type to more reflective, with coarser, and more dissipative, with finer. As many nourished beaches are important recreational and surfing beaches, this has further ramifications for modifying the suitability of the surf zone for swimming and surfing activities.

Second, beach mining, particularly where it involves the removal of a selected size fraction as in the case of coarse aggregate mining and heavy mineral sand mining, will have the impact of coarsening or fining the overall beach sand. This in turn will result in more reflective (coarsening) or dissipative (fining) conditions.

### 10.5.5 Surf zone circulation

On long natural beaches surf zone circulation is a product of the waves, tide and sediment. On beaches with topographic influences, either as boundary headlands, inlets or structures, or where rocks, reefs and structures occur along a beach, normal beach circulation will be modified to the extent that the boundary or intruding topography modifies waves and interferes with sediment movement.

Topographic impact on surf zone circulation can be divided into three categories: those that lower wave height, change wave direction or interrupt normal surf zone circulation. The major impacts, their sources and secondary impacts are listed in Table 10.3. As discussed in Section 9.3 this influence can range from local near the structure to dominating the entire surf zone circulation on shorter embayed beaches.

### 10.5.6 Wave breaking

A structure designed to predictably modify the pattern of wave breaking and surf zone circulation is surfing reefs. The design criteria for surfing reefs were pioneered by Walker (1974), and reefs have now been built in Italy (Fitzjones and Rainger, 1992) and Australia, while several are now in the planning stage (Jackson *et al.*, 1997; Pattiaratchi, 1997; Skelly, 1997), and it is likely that many more will be built in the coming decades. The intense scientific and engineering interest in surfing reefs was evident at the First International Surfing Reef Symposium held in 1997 (Pitt, 1997). Major advances have been made in both the design and materials used for reefs, as well as breaking wave dynamics over rock reefs (Black *et al.*, 1997).

While many surfing reefs will be constructed on existing headland and reefs well away from beaches, when constructed off a sandy beach the reefs will act in the same way as submerged breakwaters. They will induce wave breaking over and refraction around the reef, which will lead to a longshore variation in wave height and direction along the lee shore, which in turn will lead to a permanent foreland in lee of the reef and, if close to shore, possibly rip channels to either side. The proposed Narrowneck surfing reef on Queensland's sandy Gold Coast (Jackson *et al.*, 1997) is also designed to build a cuspate foreland in its lee, so as to purposely widen a section of the backing beach and barrier.

**Table 10.3** Topographic impacts on beach type and surf zone circulation.

| | | | |
|---|---|---|---|
| *Primary impact* | Lower wave height (wave attenuation) | Change wave direction (wave refraction) | Interrupt normal circulation (deflect longshore feeder currents) |
| *Cause of impact* | Breakwaters, reefs, inlet shoals | Headlands, breakwaters, training walls, groynes, ebb tide deltas | Headlands, training walls, groynes, inlet ramp, margin shoals |
| *Surf zone impact* | Lower energy beach type | Gradation in longshore wave energy | Induce (permanent) topographically controlled rips |
| *Other impacts* | Shoreline accretion | Realign shoreline | (Permanent) beach scour |

**Table 10.4** Summary of potential natural and human-induced changes in beach morphodynamics.

| Parameter/process | Modifying source | Impact on parameter | Impact on beach |
|---|---|---|---|
| Wave height | Ice | Lower $H_b$ | More reflective beach |
| | Coral reefs | Lower $H_b$ | |
| | Beachrock reefs | Lower $H_b$ | |
| | Breakwaters | Lower $H_b$ | |
| Wave period | None | | |
| Wave direction | Reefs | Wave refraction | Crenulate shoreline |
| | Surf zone and offshore structures | Wave refraction | Crenulate shoreline |
| | Surfing reefs | Lower $H_b$, wave refraction | Crenulate shoreline |
| Surf zone circulation | Surf zone structures, especially groynes | More cellular circulation | Crenulate beach, topographic rips, stronger rips |
| Sediment volume | Dams, mining | Less sediment | Beach erosion |
| | Land clearing, beach nourishment | More sediment | Beach accretion |
| Sediment size | Fine stripping | Coarser beach | More reflective |
| | Extraction of coarser sands | Finer beach | More dissipative |
| | Nourishment – finer | Finer beach | More dissipative |
| | Nourishment – coarser | Coarser beach | More reflective |
| Tide range | Entrance training walls | Modify tidal prism and range | Lower range – more wave-dominated<br>Higher range – more tide-dominated |

### 10.5.7 Tide range

Tide range has important influences on beach morphodynamics as discussed in Chapter 8. Tide range is controlled by a number of natural processes and can range from a few centimetres to more than 10 m. Human impact on tide range has largely been restricted to enclosed coastal bays and estuaries, where entrance training walls modify the tidal prism, and thereby tide range (e.g. Bird, 1978). There is no documented occurrence of human modification of tide range on the open coast, where it may impact beach morphodynamics. Where tide range is modified its impact through the *RTR* can be predicted using Table 10.1.

## 10.6 SUMMARY

Table 10.1 summarises the impact of waves, tides and sediment on beach morphodynamics, including beach type, bar number and impacts of embaymentisation. Using this table any beach can be located in relation to the dominant processes contributing to its maintenance and the likely beach type, bar number and circulation patterns. In Table 10.4 the impact of additional major natural and human modifications to sandy beach systems, through changes to wave height and direction, surf zone circulation and sediment size, are summarised. Each change will cause a predictable response in beach morphodynamics, as indicated in Table 10.1. Wave height is usually lowered by physical attenuation or barriers to movement, though in some cases wave reflection off highly reflective breakwaters can cause a wave interference pattern which will locally increase wave height (wave clapotis). Wave period alone among the wave parameters remains unchanged by interference. Wave direction will be modified by any structure causing additional wave shoaling. Structures located within the surf zone will interfere with normal surf zone circulation and on moderate to high energy beaches induce topographically controlled rips. Changes to sediment size and quantity will manifest by changed beach type and beach volume, the latter leading to beach accretion or erosion. Finally, tide range can be readily modified in lagoonal and estuarine environments through controls on tidal prism; however at present open coast tides are outside the scale of human interference.

# 11

# Beach Ecology

*Andrew D. Short and Patrick A. Hesp*

## 11.1 INTRODUCTION

Beaches and surf zones form the basis of ecological communities that consist of a potentially rich biotic population which inhabits the water column, sand surface and sub-surface. These organisms have direct and indirect impacts on the physical beach environment, as well as being part of larger coastal ecosystems. In this chapter the types of organisms that live in the beach environment are briefly reviewed, followed by the relation of beach ecology to other physical parameters, including beach type, and the overall role of beach ecology in beach systems. This review is based substantially on the work of McLachlan (1983) and Brown and McLachlan (1990).

## 11.2 BEACH HABITATS

Beaches offer a considerable range and diversity of biotic habitats. Horizontally these include the sub-aerial (upper dry beach and dunes), the intertidal (swash zone/shoreline), and the sub-tidal (surf zone and nearshore), while vertically there are pelagic (water column), benthic (surface) and sub-surface (interstitial) environments. All are relatively shallow and receive good penetration of sunlight. Most are also exposed to constant water movement which has the potential to circulate nutrients, and exchange water and nutrients with the coastal ocean, inlets and streams. Whilst the above favour a range of habitats and organisms there are disadvantages related to the soft, unstable nature of the benthic layer, including the potential for substrate burial and erosion, as well as the extreme turbulence generated in parts of the surf and swash zones. As a consequence of these conditions the flora is restricted to microscopic forms and the fauna to highly mobile and specially adapted species that permit the physical environment to play a major role in the type and number of species.

## 11.3 BEACH ORGANISMS

### 11.3.1 Interstitial

Interstitial organisms live both on and between the sand grains. The pore spaces between the grains can total 40% of beach volume, while the surface area of sand grains increases exponentially with decreasing grain size (McLachlan, 1983). The beach and surf zone is occupied by microscopic organisms ( < 200 mm in size) and meiofauna (200 mm to 1 mm) that live both in the pore spaces between the sand grains as well as on the surface of the grains. The surface of the sand grains hosts microscopic bacteria, fungi, diatoms and algae. Occupying the pore spaces are meiofauna including extremely small worms and shrimp; protozoa, namely small ciliates and foraminifera; and small metazoans particularly nematodes in finer sediment and copepods in coarser sediments (McLachlan, 1983). These may reach an abundance of one million individuals per square metre (McIntyre, 1969). They are commonly worm-shaped to assist movement through the sand grains (Jones and Short, 1995).

The vertical distribution of the meiofauna is related to the degree of drainage and oxygenation of the sediment, with abundance declining substantially in the reduced layers. Most bacteria and meiofauna tend to concentrate at and above mean sea level and below that part of the swash recharge surface where there is strong percolation and high water input.

*Handbook of Beach and Shoreface Morphodynamics*. Edited by A. D. Short.
© 1999 John Wiley & Sons Ltd.

### 11.3.2 Phytoplankton

The major form of plant life in the beach system, and the basis of much of the internal food chain, is phytoplankton; that is microscopic plant plankton that lives in the water column and depends on sunlight for energy. These are commonly called surf diatoms, and favour wide surf zones on gently sloping beaches (McLachlan, 1983). Under suitable conditions they can produce a diatom bloom in the surf zone. In all cases where surf diatom blooms have been recorded, the diatoms belong to one or two of four genera: *Chaetoceros*, *Aulacodiscus*, *Asterionella* and *Anaulus*. In most surf zones two species co-occur, but in certain cases only one is present (McLachlan and Hesp, 1984a). The blooms occur as accumulations of diatoms on the water surface and are visible as stains and/or semi-stable foam or scum. In most cases the accumulation disappears at night, reappearing in the morning owing to a pattern of diurnal periodicity in buoyancy (Lewin and Schaefer, 1983).

Diatom blooms are generally absent on reflective beaches although Gunter (1979) cites *Chaetoceros*-dominated strips along low energy Gulf Coast beaches. The blooms often occur as distinct accumulations associated with rip currents on high energy intermediate beaches. Here the blooms accumulate on the margins of rips in the areas of low velocity circulating flow (McLachlan and Lewin, 1981; Winter, 1983). Discrete blooms are also found in high energy, multi-barred, dissipative surf zones where rips are absent. McLachlan and Hesp (1984a) speculated that shore-normal standing waves may result in the discrete organisation of blooms at antinodes. In addition the vertically segregated onshore upper, and offshore lower flows characteristic of dissipative surf zones provide a mechanism for transporting the buoyant diatoms shorewards during the day and seawards at night when they sink to the bottom.

If washed onto the shore the diatoms can form a mass of dark brown organic matter, particularly in the upper swash limit. Phytoplankton is the basis of the beach food chain and as such its presence and abundance is critical to the entire beach ecology.

### 11.3.3 Zooplankton

Zooplankton is microscopic animal plankton that inhabits both the interstitial systems as well as the water column. They include small amphipods, isopods, mysids, shrimp (prawns) and crabs. These feed on the phyto- and zooplankton, as well as detritus in the water. At present little is known about their ecology other than that larger crustaceans dominate the biomass (McLachlan, 1983).

### 11.3.4 Macrofauna

Beach macrofauna ranges from 1 mm to 10 cm in size, and consists of those organisms too large to move between the sand grains. They require a high degree of mobility and are dominated by molluscs, crustaceans and polychaetes (Jones and Short, 1995). McLachlan (1983) suggests that unspecialised generalists are required to successfully inhabit the physically demanding beach and surf zone environment. For this reason beach macrofauna tend to be dominated by mobile filter feeders and scavengers, with deposit feeders in more sheltered locations, such as tidal flats.

*Molluscs* are shelled organisms, predominantly bivalves and gastropods. The most common beach bivalve is the genus *Donax*, known commonly as cockles, mussels and pippis, which have a broad global range (Ansell, 1993). McIntyre (1977) has also recorded the bivalve *Tellina* occurring in numbers of several hundred per square metre. Most bivalves are filter feeders, and it is common to find a strong association between high numbers of diatoms and large populations of filter feeding bivalves (e.g. *Siliqua*, *Donax*, *Amphidesma* and *Mesodesma*). Some bivalves such as *Bullia* are scavengers feeding on carrion. Mussels in general prefer temperate climates and moderate to high energy beach systems.

*Crustaceans* have external skeletons and in beach systems include crabs, amphipods, isopods, mysids and echinoderms. These occur as meiofauna in the interstitial systems and as benthic and pelagic macrofauna. Polychaetes (worms) tend to occur in the more sheltered locations and most are microscopic. However the giant beach worm (*Onuphis teres*) lives in energetic swash locations in southern Australia and can grow to a length of more than 2 m.

In a review of the literature, Bally (1981) found that macrofauna biomass increases with increasing wave energy, and is most abundant in high energy dissipative systems (Table 11.1) as well as preferring tropical beaches.

### 11.3.5 Fish

Fish are the most obvious inhabitant of the surf zone and the reason for the massive beach fishing industry. While the beach and surf zone is open to many deep water species, it is dominated by juvenile fish, which indicates that it acts as an important nursery area, a consequence of the relative abundance of food and shel-

**Table 11.1** Microfauna abundance and dry biomass (after Bally, 1981).

| | Low energy (reflective) | Medium energy | High energy (dissipative) |
|---|---|---|---|
| No. species | 11 | 17 | 30 |
| Abundance (no./$m^2$) | 400 | 75 | 1710 |
| Abundance (no./m) | 20 045 | 334 571 | 2 797 867 |
| Biomass (g/m) | 2.26 | 1.97 | 6.23 |
| Biomass (g/$m^2$) | 871 | 170 | 63 |
| Sand size ($\mu$m) | 310 | 257 | 238 |

ter from predators (Brown and McLachlan, 1990). Fish habitats include both pelagic and demersal, and benthic feeders include plantivores, picivores and omnivores. Fish communities tend to be diverse and highly dynamic. They may have only two to five resident species; however seasonal, strays and temporary associations can total 50 to 70 species (Brown and McLachlan, 1990). They are also opportunistic feeders with all types feeding on zooplankton and crustaceans.

Clark (1997) studied surf zone fish communities at four sites ranging from exposed to sheltered and found that the more exposed sites had the lowest density, but highest species richness and diversity, and lowest dominance, as compared to the sheltered sites. He attributed the trends to the availability and abundance of suitable food resources and to physical disturbance introduced by breaking waves.

### 11.3.6 Birds

Birds, both land and sea based, are a highly visible part of the beach ecosystem, and may rival the aquatic macrofauna in biomass. They also have a significant impact on the ecosystem due to predation and scavenging (Brown and McLachlan, 1990).

Land based birds feeding intertidally include Laridae (gulls), Scolopacidae (sandpipers, sanderlings, snipes), Chionididae (sheathbills), Haematopodidae (oystercatchers), Charadriidae (plovers) and Charadriiforms (all members). Phalacrocoridae (cormorants), the major sea based bird, feeds predominantly in the surf zone, as do terns.

Shorebirds may be partially or totally migratory in their habits, and avian densities vary according to season (generally higher in summer), overwintering, migratory and breeding patterns (Brown and McLachlan, 1990), and beach type.

Although there appears to have been very little research carried out, in view of the fact that the higher energy dissipative and intermediate beaches display significant numbers of macrofauna (and possibly fish also) compared to the lower and higher energy reflective beaches, numbers of avian fauna may also be positively related to beach types.

In summary, the beach ecosystem can contain a wide range of biota inhabiting the interstitial sand grains, the bed of the swash and surf zone, and the air and water above the surfaces. They range in size from microscopic plankton, to a wide range of molluscs, crustaceans and polychaetes, and birds and fish. Many live entirely within the beach and surf zone, while others move in from the land, air and ocean to periodically feed.

## 11.4 THE BEACH ENERGY SYSTEM

### 11.4.1 Food sources and biomass

The beach ecosystem comprises complex interactions between all the above organisms across a dynamic beach and surf zone system open to both land and sea influences and biota. To qualify and quantify the structure and energy flows within this system requires a knowledge of the resident and transient populations, their numbers, species and biomass, their food sources, and the energy flows within and between each.

Potential sources of food for sandy beaches are listed in Table 11.2. However, as flora are restricted to microscopic interstitial forms, most primary production is in the surf zone by phytoplankton (surf zone diatoms). Other food sources are imported, primarily from the sea.

### 11.4.2 Food chains

Three food chains operate within the beach and surf zone, with varying degrees of interaction. These are the macroscopic, the interstitial and the microbial loop. The macroscopic food chain consists of the benthos, including filter feeders, herbivore scavengers and carnivore scavengers/predators, zooplankton, fish and birds. The interstitial food chain within the sand consists of bacteria, protozoans and meiofauna, with no significant predation from macrofauna. Finally the microbial loop comprises bacteria, flagellates, ciliates and other microzooplankton in the water column, which subsists largely on phytoplankton and organic detritus.

### 11.4.3 Beach energy flows

Energy flow and faunal/environmental interactions oc-

**Table 11.2** Food sources for sandy beaches (source: Brown and McLachlan, 1990).

| Food types | Source | Remarks |
|---|---|---|
| Benthic microflora | Beach sand | On sheltered beaches |
| Surf diatoms, flagellates | Surf water | In well developed surf zones |
| Stranded macrophytes | The sea | Near kelp beds, estuaries, etc. |
| Carrion | The sea | |
| Particulates | The sea | |
| Dissolved organics | The sea | |
| Insects | The land | Particularly during strong offshore winds |
| Litter | The land | |

curring across a potentially wide system depend on a number of factors, the most important of which is the overall wave energy level and beach type, which in turn determines the width of the surf zone and physical characteristics. In order to summarise the wide range of potential systems Brown and McLachlan (1990), defined two main types of beach ecosystems: the beach interface system typical of lower energy reflective beaches with little or no surf zone; and the surf zone and beach ecosystem typical of higher energy dissipative beaches with a wider energetic surf zone. Intermediate beaches include elements of both types.

The beach interface system (Figure 11.1a) has just two food chains, the interstitial and macrofaunal, and is dependent on wave energy to drive water filtration, which delivers marine inputs. It exports macrofauna to larger predators, but generally has low faunal biomass. The surf zone and beach ecosystem (Figure 11.1b) is largely wave driven, with an additional food source, phytoplankton, and the microbial loop in the food chain. It also has higher primary production and generally higher biomass (Table 11.3).

Because of the complexity of beach ecosystems and the difficult in accurately sampling their structure and dynamics, only one system has been sufficiently studied to quantify its energy flow and the interactions of the four food chains. In this higher energy South African beach systems Brown and McLachlan (1990) found that the surf zone diatom *Anaulus* was the primary producer, together with flagellates, producing 200 000 $gC\,m^{-2}\,y^{-1}$. The interstitial fauna consumes 40 000 gC/m/a, 90% in the sub-tidal region. The macrofaunal food chain consists of mullet, zooplankton and filter feeding bivalves, which are eaten by crabs, fish and birds, with piscivorous fish on top of the food chain. This chain consumes 30 000 gC/m/a, mostly originating from phytoplankton, with 10 000 gC/m/a recycled as faeces. The microbial loop is well developed and subsists largely on phytoplankton, consuming approximately 100 000 gC/m/a. In total this system produces a surplus (20 000 gC/m/a) which is exported as living or dead particulate carbon across the boundary of the ecosystem.

## 11.5 BEACH ECOLOGY AND BEACH SYSTEMS

Because beach ecology is intimately tied to physical beach conditions including grain size, surf zone gradient, width and relative level of wave energy, it is also closely related to beach type. McLachlan and Lewin (1981) found that macrofaunal species diversity and abundance were related to physical parameters including slope and particle size. They found a substantial increase in each with decreasing beach slope and grain size. Likewise McLachlan (1983) and Bally (1981) found that as the level of breaker wave energy increased so too did the species number, abundance and biomass of major invertebrate groups occupying the intertidal zone, particularly crustaceans, molluscs and polychaetes.

### 11.5.1 Beach type

The beach types presented in Chapters 4 and 7 each represent a morphodynamic assemblage that presents a predictable range of environments (interstitial, benthic and water column) for beach organisms. As the number and geographical range of beach ecosystem studies have been increased it has become apparent that there is a strong relation between beach type and beach ecology. McLachlan (1990) was the first to clearly express this relationship in an examination of dissipative beaches and macrofaunal communities. McLachlan (1990) found three major trends across a gradient from reflective to dissipative beaches: (i) an increase in macrobenthic species richness; (ii) a logarithmic increase in total abundance; and (iii) a logarithmic increase in total biomass.

Subsequently, McLachlan *et al.* (1993), Jarmillo *et al.* (1996), and Hacking (1998) have extended the studies on a wide range of beach types, in a wide range of geographical locations.

An example of the relationship between beach type and biota is presented by Hacking (1998) for three NSW beaches all located within 25 km of each other. On the

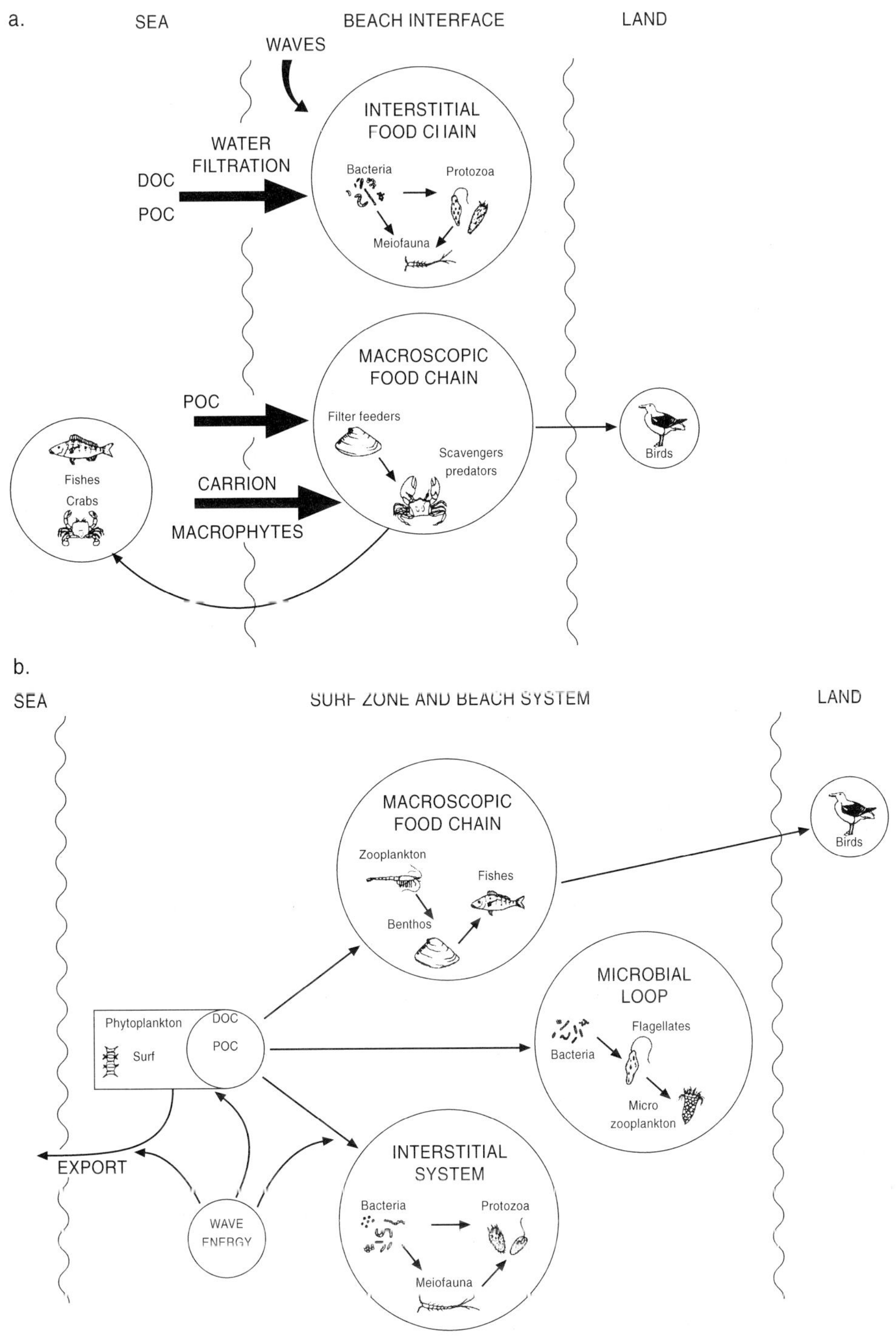


**Figure 11.1** The two main types of beach ecosystem based on carbon flux: (a) the beach interface system; and (b) the surf zone and beach ecosystem (source: Brown, A C and McLachlan, A, *Ecology of Sandy Shores*, 1990, p. 328, with kind permission from Elsevier Science – NL, Sara Burgerhartstraat 25, 1055 KV Amsterdam, The Netherlands).

**Table 11.3** Biomass in East Cape beach and surf zone (source: McLachlan, 1983).

| | Biomass (g/m) |
|---|---|
| Macrobenthos (molluscs, crustaceans, polychaetes) | 1500 |
| Zooplankton (higher turnover) | 1100 |
| Fish | 300 |
| Birds | 5 |

low energy reflective Korora beach (Figure 11.2) only limited numbers of five species are present. On moderate energy Ocean View, a transverse bar and rip beach, species number increases to 12, while on the most exposed Moonee beach, with rhythmic bar and beach topography, 19 species were found.

McLachlan *et al.* (1993) also found that species richness is controlled primarily by the physical environment, increasing from reflective to dissipative beach conditions. Species abundance and biomass had a similar though more variable relationship, as they were dependent on surf zone productivity which is in turn dependent on wave energy. They also found that while sand size and wave climate are important (Table 11.2) it is the swash climate that is most critical to controlling beach populations. They further found that the relationship to beach type was improved if tide range was included, and developed the beach state index (*BSI*)

$$BSI = \log(H_b M / W_s TE) + 1) \qquad (11.1)$$

where $M$ is the maximum tide range and $E$ the maximum theoretical equilibrium tide (for a theoretical earth covered by water, $E = 0.8$ m). This index improved on the use of the dimensionless fall velocity ($\Omega = H_b/W_s T$), as it included not only wave height and period, but also tide range. Using the *BSI* index beaches can be classified as follows:

- $BSI < 0.5$ reflective beaches
- $BSI = 0.5$–$1.0$ low to medium energy intermediate beaches
- $BSI = 1.0$–$1.5$ high energy intermediate/dissipative beaches
- $BSI = 1.5$–$2.0$ fully dissipative beaches
- $BSI > 2.0$ ultra-dissipative beaches

McLachlan *et al.* (1993) and Hacking (1998) found that beach macrofauna correlated well with beach morphodynamic states, with the best results used when biomass, abundance and species number are plotted against *BSI*.

These relationships indicate that macrofaunal richness, abundance and biomass are all predictably related to beach morphodynamic state, and that all increase toward more dissipative conditions. McLachlan *et al.* (1993) suggest that this relationship implies a dependence of certain taxonomic groups on swash climate. Dissipative beaches are characterised by swash with extended periods and lengths, variable speeds and with the predominant swash activity occurring below the effluent line. On reflective beaches swash speeds are high and waves break directly in the intertidal zone. Thus, dissipative swash conditions require minimal faunal adaptations, while as the more 'hostile' swash of the reflective beaches is approached, increasing specialisation is demanded, and consequently more species are excluded.

## 11.6 PHYSICAL IMPLICATION

While the ecology of sandy beaches is considerably influenced by beach morphodynamic state, the presence of a wide variety of species and the operation of various ecological interactions can in turn exert direct and indirect influences on the physical nature of the beach. The three most important ramifications are the supply of carbonate detritus to beaches, bioturbation of beach sediments, and the contribution to beach cementation.

### 11.6.1 Sediment supply

Most beaches have some proportion of carbonate material, indicating a biotic source. Some beaches have the majority of their sediments derived from such sources at which point the biotic environment becomes the major source of beach sediment. The occurrence of carbonate-rich beaches is well documented in the tropics (Russell, 1959; Russell and McIntire, 1965b; Clark, 1990), in arid temperate environments (Bird, 1982; Short *et al.*, 1986; Short and Fotheringham, 1986), with some occurrences in cooler temperate latitudes (Pilkey *et al.*, 1967).

Along the southern Australian coast, carbonate sediments dominate most of the beach and dune systems, including massive transgressive dunes extending several kilometres inland (Figure 10.10; e.g. Short and Hesp, 1982; Short and Fotheringham, 1986; Short *et al.*, 1986). Short (1987) calculated rates of carbonate sediment supply to high energy beaches peaking at $10\,m^3/m/a$. Most of this sediment is fine to very fine grained sand making it difficult to identify sources. While on some beaches local beach biotic production is so massive as to dominate the beach morphology (e.g. Sprigg, 1952), these are apparently the exception, and it appears that much of the material is of shelf origin and supplied during sea level transgressions. Along the cool temper-

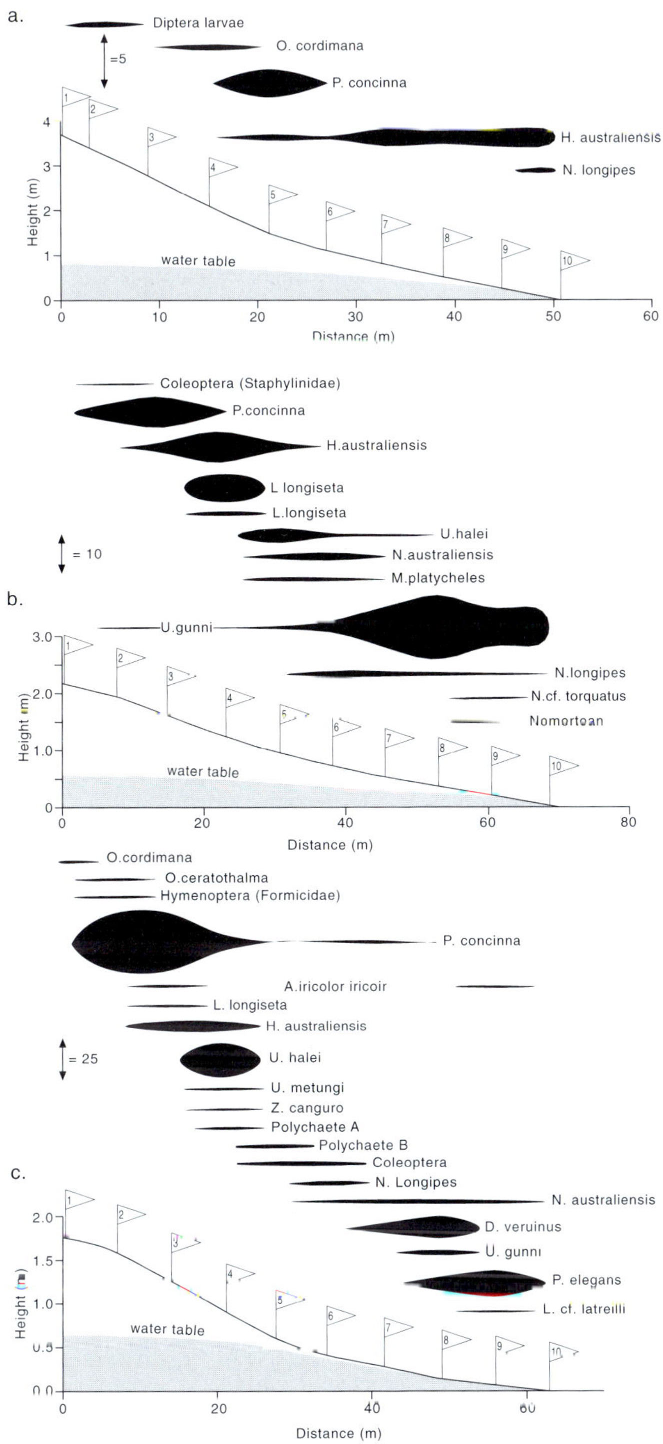


**Figure 11.2** Species presence and abundance for (a) Korora, (b) Ocean View and (c) Moonee beaches, New South Wales (source: Hacking, N, PhD thesis, 1998, with permission of the author).

ate carbonate-rich coast of southern Australia, Boreen *et al.* (1993) identified the inner high energy shelf between 30 and 70 m water depth as the major zone of carbonate production and transport, with sources dominated by bryozoans, but including molluscs, foraminifera, sponges, red coralline algae and echinoderms, as well as erosion of hard carbonate surfaces. Boreen and James (1993) found that rates of shelf carbonate production were of a similar magnitude in tropical and cool temperate locations.

Contemporary beach molluscs and other organisms do supply a limited amount of sediment directly to their beach systems, but they do not appear to contribute a majority. Unfortunately there are no data on the biotic composition of, and contribution to beach sand in southern Australia, nor on their abrasion and diminution rates. Until these issues are addressed the role of beach, nearshore and shelf carbonates, and their temporal and spatial supply remains unclear.

A more obvious impact of nearshore flora, including seagrass, seaweed and kelp, is the delivery of masses of eroded flora to the surf and swash zone. This occurs particularly during and following high wave events that induce the erosion. On many temperate, modally low energy southern Australian beaches fronted by seagrass meadows (*Zortra, Posidonia*), masses of seagrass detritus commonly litters the beaches, dominating the visual, and at times, physical beach environment (Figure 10.8).

### 11.6.2 Bioturbation

Mobile beach organisms bioturbate the surrounding sediments. There are numerous studies in the literature documenting the increase in bioturbation with depth (e.g. Hill and Hunter, 1976; Reineck and Singh, 1980; Frey and Pemberton, 1984). This does not imply that swash and surf zone benthic organisms do not bioturbate, but rather that their imprint has a lower preservation potential in the more energetic shallower beach environments. Therefore bioturbation is widespread across the beach–surf zone–nearshore, while preservation potential increases with depth and decreasing wave energy.

### 11.6.3 Cementation

The cementation of beaches and dunes to beach and dune calcarenite (commonly called beachrock and dunerock) is discussed in Section 10.4.2. The formation of beachrock occurs more frequently in tropical locations where warmer seawater assists the uptake of calcium carbonate; however it also occurs in some temperate locations such as south-western Australia (Semeniuk and Searle, 1987). Dunerock is more prevalent in drier tropical and temperate locations such as western and southern Australia (Figure 10.10), South Africa, North Africa and parts of the Mediterranean. Both inside and particularly outside the tropics beach cementation is encouraged locally by the presence of high carbonate content in the beach sediments (Krumbein, 1979), which may be of local or shelf origin.

## 11.7 SUMMARY

Beach systems at first glance often have the appearance of a barren ecology, particularly the bare dry sands, akin to desert environments. Research over the past two decades has dispelled this belief with the identification of potentially rich and diverse beach ecosystems, the more productive of which export nutrients to the coastal ocean, and act as nurseries for juvenile fauna. The recognition of the occurrence and importance of beach ecology, together with their interaction with adjacent land, sea and air environments, emphasises the importance of beach systems, and justifies the need to more fully understand and appreciate the role beaches play in both the physics and ecology of the coast.

Furthermore beach ecology does in places contribute substantially to the physical beach environment, through the supply of carbonate detritus as beach sediments, through the bioturbation of beach sediments, and indirectly through their contribution to the cementation of tropical beaches (beachrock) and some tropical and temperate dunes (dunerock).

# 12

# Beach and Dune Stratification

*Andrew D. Short and Patrick A. Hesp*

## 12.1 INTRODUCTION

Stratification refers to the primary sedimentary structures that are produced as a result of the interaction of sediment movement and deposition in various flow conditions (Chapters 4 and 5). As backshore, beach, surf zone and nearshore sediments are eroded, transported and deposited, their internal structures may be buried and preserved. These bedform structures provide evidence of flow conditions, water depth and velocity, landform or facies unit, and dynamic environment at the time of deposition. Certain sediments may retain evidence of the date of burial, and may be utilised to reconstruct beach evolution and chronology.

The study of beach stratification can therefore have a number of aims and associated techniques. Studies of contemporary sediment dynamics may only be concerned with boundary layer dynamics monitored using wave and current sensors together with some measure of suspended sediment concentration, bedform migration and depth of disturbance (e.g. Sherman *et al.*, 1993). Studies of the impact of specific wave, storm or seasonal events will require a record of the sediments and structures above the depth of disturbance, which can usually be recovered using a box corer or shallow vibrocorer (e.g. Aigner and Reineck, 1983). Studies of longer term sediment accumulation, and the chronostratigraphy of a particular beach barrier deposit, require deeper coring of sediments, which usually lie at both greater depth, and on prograded or regressive shores some distance inland from the shore (e.g. McCubbin, 1982).

There are a number of excellent texts covering boundary layer processes and sediment transport (Hardisty, 1994), sedimentary structures (Reineck and Singh, 1973; Allen, 1984; Collinson and Thompson, 1982; Harms *et al.*, 1982; Rubin, 1987), and beach stratification (McCubbin, 1982; Reinson, 1984). The following discussion draws on these, together with the results of a number of field investigations presented in Clifton (1969), Clifton *et al.* (1971), Hunter *et al.* (1979), Greenwood and Davis (1984) and Short (1986).

## 12.2 BOUNDARY LAYER PROCESSES

The distribution of beach sediments and the development of bedforms across the shoreface is a product of the sediment characteristics, water depth, $d$, wave height, $H$, and period. In shallow water the wave orbital velocity, $u$, is a product of wave height and water depth, and is given by

$$u = H/2\sqrt{(g/d)} \tag{12.1}$$

The current velocity together with the water depth are combined in the Froude number, $Fr$, which is the ratio between current velocity and water depth where

$$Fr = u/\sqrt{gd} \tag{12.2}$$

Using this parameter, flows can be classified into lower flow regime ($Fr < 1$), transitional flow regime ($Fr \cong 1$) and upper flow regime ($Fr > 1$). Finally, the relationship between the velocity and viscosity are combined in the Reynolds number

$$Re = \rho ud/\mu \tag{12.3}$$

where $\rho$ is density and $\mu$ is the viscosity. The Reynolds number indicates the type of flow, in particular whether it is laminar ($Re < 500$), transitional ($Re = 500$–$2000$) or turbulent ($Re > 2000$).

Simons *et al.* (1965) used these parameters, together with flume experiments, to examine flows over a sandy bed and developed a relationship between flow type and bedforms. Low flows produce a rippled sand bed and flat water surface. They also examined the impact of

*Handbook of Beach and Shoreface Morphodynamics*. Edited by A. D. Short.
© 1999 John Wiley & Sons Ltd.

grain size on both flow regime and bedform development. Harms and Fahnestock (1965) applied these parameters also to provide a conceptual overview of the relationship between flow regime and bedforms, and stream flow and sediment motion and concentration (Figure 12.1).

In moving from left to right across Figure 12.1 it is assumed that velocity is increasing and/or depth is decreasing. Once stream power exceeds 0.01 small current ripples are produced. As flow velocity increases the ripples are superimposed on larger dunes or megaripples, whose profile is out of phase with the 'boiled' water surface. The boils intensify as higher velocities reach the transitional flow regime, then as the highly turbulent upper flow regime is entered, the strong flow produces a plane bed and water surface. Increasing velocity (decreasing depth) will form antidunes, bedforms in phase with the rippled water surface. As the velocity increases the waves begin to break back on the flow producing in-phase antidunes and breaking standing waves, while in very high velocity flows a series of pools, chutes and breaking waves are produced. The increasing velocity also impacts the nature of sand and cobble sediment transport and concentration (Figure 12.1 centre), and the overall structure of the water surface and flow characteristics (Figure 12.1 lower).

While the full range of flow regimes can be produced in a laboratory flume, and commonly occur in swift, shallow streams, they are not all encountered across the shoreface, nor in the surf zone. Carter (1988) schematically summarised the impact of decreasing depth across the shoreface on both the Reynold and Froude numbers. The deeper waters seaward of the breaker zone are characterised by lower flow regimes, laminar flows and intermittent sediment transport. As the shallower breaker zone is approached flows become increasingly turbulent, with higher velocity. Within the turbulent surf zone water depth continues to decrease resulting in transitional to upper flow regimes in the swash zone.

Clifton *et al.* (1971) examined the variation in flow regimes across a shoaling beach zone, and found a predictable sequence of facies roughness and bedforms as waves shoal and break. The sequence shown in Figure 4.26 begins on the inner shelf where inactive symmetrical wave ripples are bioturbated during non-storm conditions. As depth gradually decreases across the shoreface, the increasing Froude Number produces essentially the same sequence of lower flow regime bedforms illustrated in Figure 12.1, with the transitional to upper flow regime only approached or reached under the breaker zone and in the shallow swash zone, both of which have flat 'planar' beds.

While there is general agreement on the occurrence of this sequence seaward of the breaker zone, within the surf zone, the potentially more complex morphodynamics, discussed in Chapter 4, produce a character-

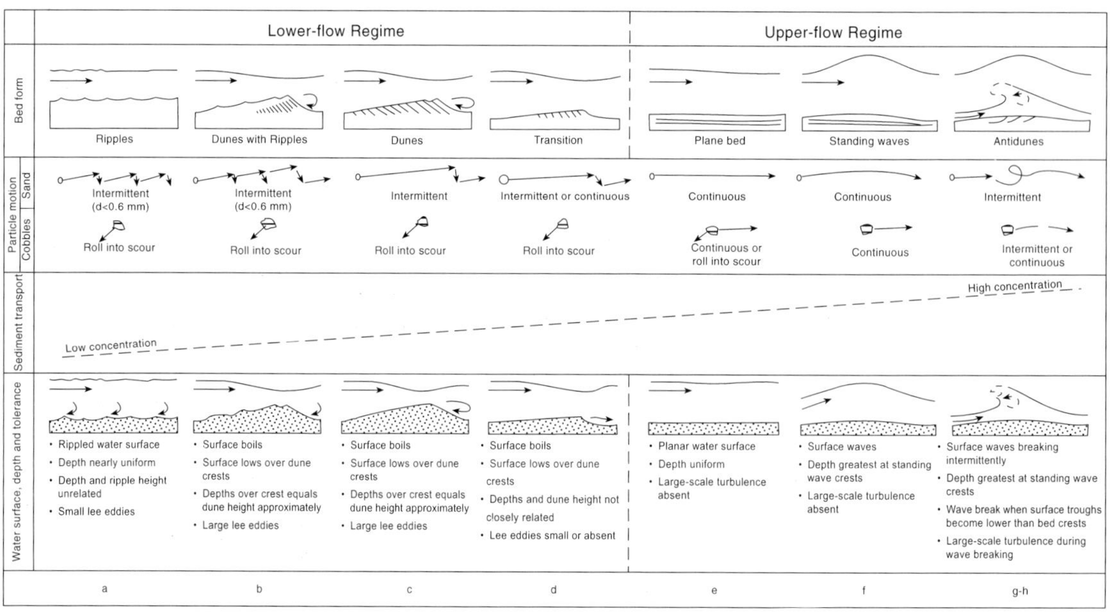


**Figure 12.1** Relationship between bedform, sediment motion and concentration, and water surface, depth and turbulence (modified from Harms and Fahnestock, 1965).

istically wider range of flow regimes associated with both wave orbital currents, on-, long- and offshore currents, as well as wave uprush and backwash, which in turn result in a more variable range of bedforms including planar bar crests, ripples, troughs and channels, and the planar swash zone.

The nature of barred surf zones has been described by Davidson-Arnott and Greenwood (1976), Hill and Hunter (1976), Shipp (1984) and Short (1984b, 1986). The inner rough zone consists of both wave and current ripples, while the shallow swash zone is dominated by transitional flows and a plane bed. Table 12.1 summarises the sequence of bedforms and structures (Figure 12.2) from the inner shelf to the swash zone.

## 12.3 BEACH TYPE AND STRATIFICATION

Clifton *et al.* (1971) recognised that the level of wave energy affected the boundary layer dynamics, and the nature and extent of the flow regimes and associated bedforms on beaches exposed to different levels of wave energy. Likewise, Howard and Reineck (1981) presented contrasting sections based on low (Georgia) and high (California) energy beach systems. While a number of stratigraphic studies have been undertaken in a variety of wave and tide environments (e.g. Greenwood and Davis, 1984), only Short (1983, 1984b, 1986) has presented a sequence of bedforms and structures from a range of representative wave-dominated beach types. The following is largely based on this work.

### 12.3.1 Reflective (non-barred) beaches

Modally reflective beaches occur on coasts with low waves and/or coarse sediments (Figure 7.5). Their morphodynamics are discussed in Section 7.3.4. They are characterised by a relatively steep beach face, possibly backed by a berm and/or cusps, with a coarser step at the toe of the beach face, then a lower gradient nearshore composed of finer sand. Because of the typically lower waves these systems are relatively narrow (tens to hundreds of metres) and shallow (1–10 m) (Figure 12.3).

**Table 12.1** Sequence of shoreface bedforms, structures and processes.

| Location | Zone | Bedforms | Structures | Processes |
|---|---|---|---|---|
| Inner shelf | Outer symmetrical (inactive) | Inactive, linear, symmetrical wave ripples | Relic, bioturbated ripple laminae (Figure 12.2a) | Storm waves |
| Outer shoreface | Outer (offshore) symmetrical (active) | Active, linear, symmetrical wave ripples | Ripple laminae (Figure 12.2b) | Modal wave orbital currents |
| Inner shoreface | Inner (offshore) asymmetrical | Irregular, sinuous, discontinuous, asymmetrical wave ripples | Ripple laminae (Figure 12.2c) | Increasing orbital currents |
| Immediately seaward of breaker zone | Outer rough | Cross-ripples to lunate megaripples | Cross-ripple laminae (Figure 12.2d) | Strong, wave orbital flows |
| Breaker zone | Outer planar | Plane bed | Horizontal to landward-dipping laminae (Figure 12.2e) | Sheet flow below breaking waves and waves of translation |
| Surf zone | Inner rough | Plane bed on bar; current ripples to megaripples in channels; wave ripples in troughs, and mixtures thereof | Bidirectional wave and current ripple laminae (Figure 12.2f, g) | Translatory waves, orbital currents, longshore currents, rip currents, infragravity bed return flow |
| Step | Inner planar | Steeply seaward-dipping plane bed | Coarse, poorly sorted, steeply seaward dipping, thick laminae (Figure 12.2h) | Wave surging, plus strong bidirectional uprush and backwash |
| Swash zone | Inner planar | Seaward-dipping plane bed | Parallel, seaward-dipping beach laminae (Figure 12.2i) | Uprush and backwash |

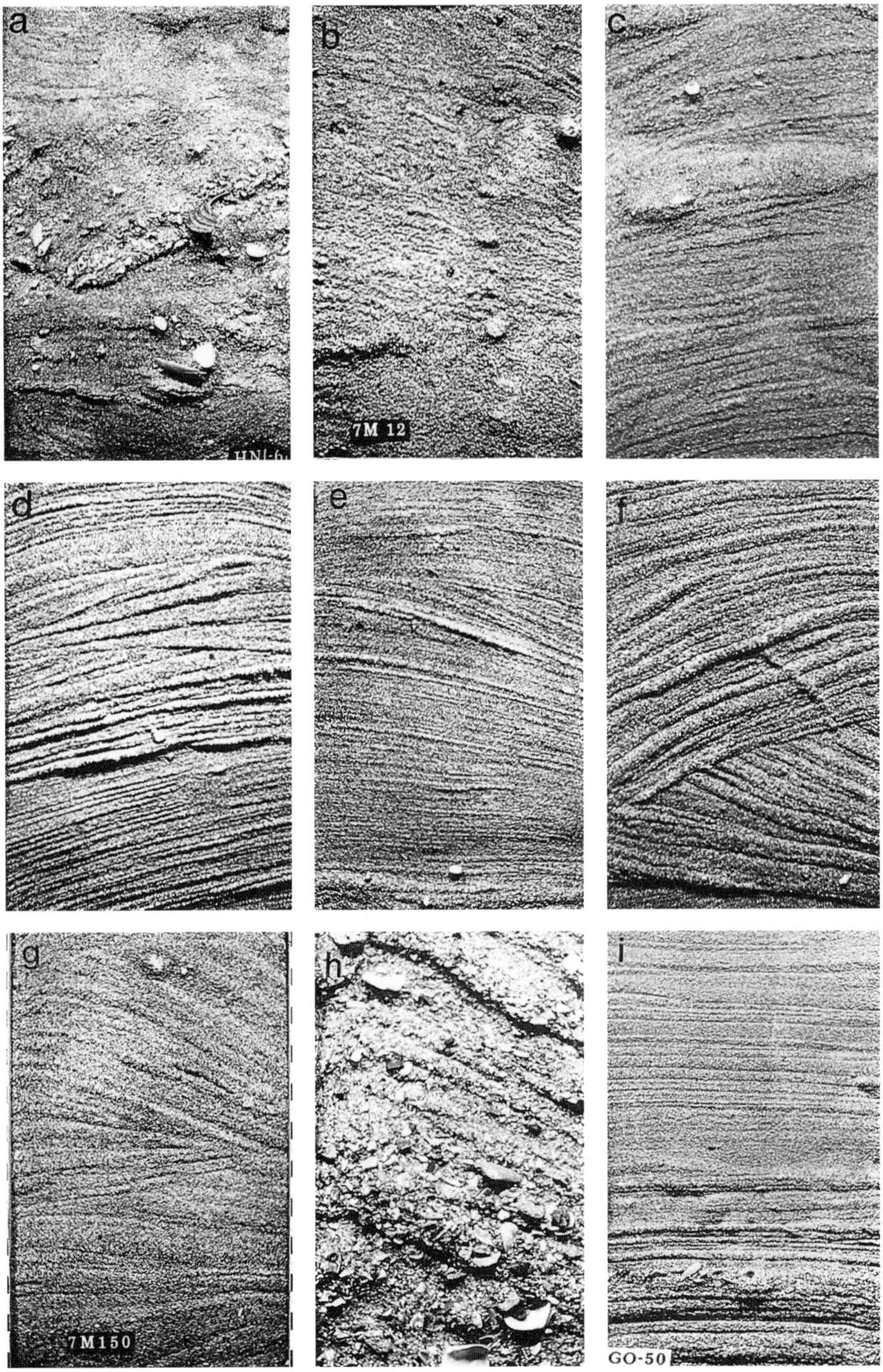


**Figure 12.2** Examples of primary sedimentary structures from a representative range of beach–shoreface environments. Refer to Table 12.1 for location. Scale: each image is 15 × 25 cm (A D Short).

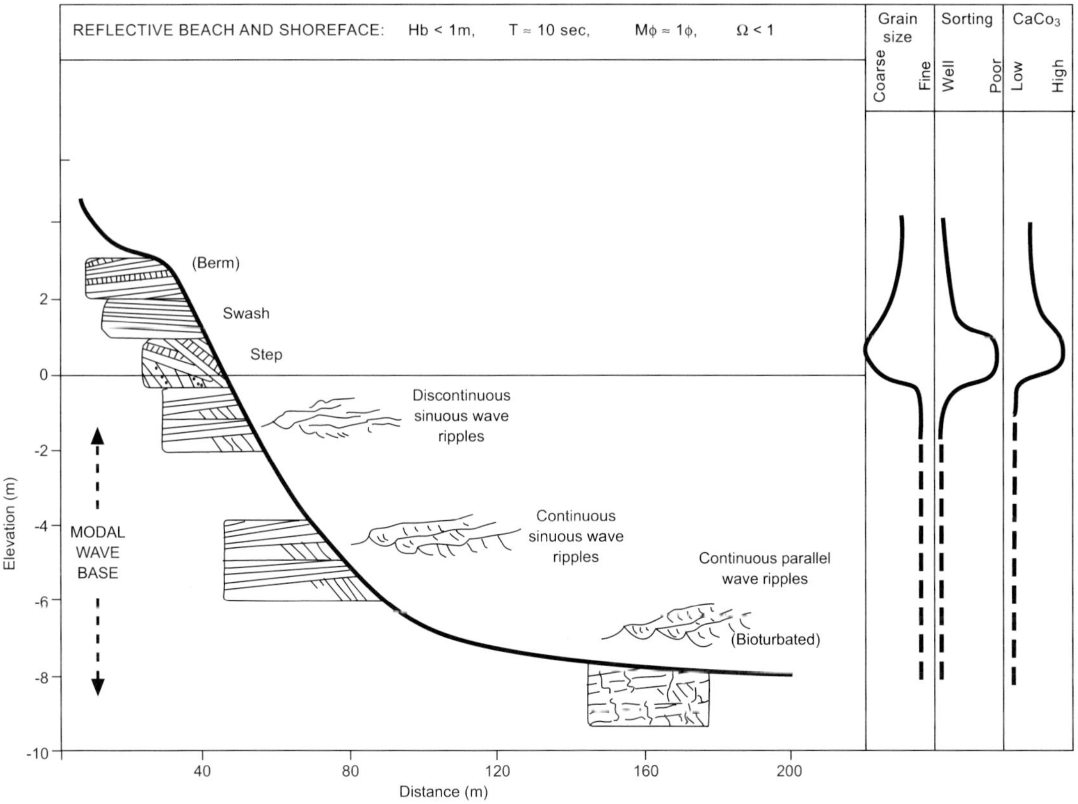


**Figure 12.3** Idealised morphology, bedforms and structures across a reflective beach–shoreface system (modified from Short, 1986).

The backshore (upper sub-aerial beach) stratification varies depending on beachface morphology. Relatively steeply seaward dipping, parallel, tabular laminae are typical of the lower energy reflective beaches which have simple, linear beach faces. A mixture of horizontal, seaward and landward dipping sets of laminae is typical of the moderate to high energy reflective beaches with berms and wider berm terraces (e.g. Thompson, 1937; Clifton *et al.*, 1971; Flemming and Fricke, 1983).

The swash zone is characterised by strong, high uprush, which overtops the berm/cusps at high tide. The intertidal beach consists of an upper berm/cusps facies consisting of horizontal to slightly landward dipping coarse (shelly) beach laminae, with alternating coarser and medium laminae. On lower energy, less active beaches, all swash laminae become increasingly discontinuous and sub-parallel in response to the lower frequency and energy of the swash.

The lower beachface contains slightly coarser sediments which dip increasingly (5–20°) seaward. Laminae become more discontinuous, together with truncations and some cross-laminae produced by wave ripples at high tide. At the base of the swash zone, around mean low water, a *step* will be encountered when there is a mixture of sediment sizes, with the coarser sediment composing the step. These sediments are arranged in steep (20°) seaward dipping foreset and cross-laminae, which may contain gravel and coarse shells (Figure 12.2h). The thickness of the step is a function of wave height and sediment sorting (see Section 5.5).

Immediately seaward of the lower beachface–step, sediments fine and asymmetrical wave ripples dominate. These become increasingly sinuous and parallel-crested with depth. Structures are predominantly gently landward dipping cross-laminae.

### 12.3.2 Intermediate (bar–tough) beach

Most beaches exposed to waves are characterised by at least an inner bar dominated by breaking waves and horizontal surf zone circulation, typified by rip currents (see Section 4.4). Such systems provide a range of water depths, current velocities and directions, and consequently have the most complex range of bedforms and structures. Intermediate beaches also range considerably in wave–sediment combinations (Figure 7.4) from lower energy low tide terraces, to the world's highest energy beach systems exposed to persistent high waves in a fine to medium sand environment, with surf zones dominated by massive rip systems. Consequently their dimensions also range considerably from widths of hundreds to thousands of metres, and depths from several metres to 30 m and more (Figure 12.4).

The *backshore* of intermediate beaches varies from relatively narrow to relatively wide, may comprise a simple, flat berm terrace, a berm crest and mid-backshore trough, a series of stacked berms, or very complex megacusp and horn-dominated berms. Thus, backshore stratification styles can vary considerably, and may comprise large scale (i.e. wide), tabular, horizontal to gently seaward dipping sets of laminae in the simple morphological cases, to a complex mixture of relatively large scale cross-bedding comprising many truncation surfaces separating sets of laminae dipping seawards and landwards (e.g. Hunter *et al.*, 1979; Davidson-Arnott and Greenwood, 1976). Erosion phases may produce very steep discontinuities between sets of accretional laminae.

Intermediate shorelines and *swash zones* are controlled by both the swash and the surf zone morphodynamics, with the characteristic longshore rhythmic morphology of transverse bar and rip and rhythmic bar and beach types. The morphodynamics of the swash zone can vary considerably alongshore between megacusp horns and swales (see Section 7.3.3). In general the structure of the swash zone is continuous longshore, tabular horizontal to seaward dipping beach laminae (Figure 12.2i). Horizontal width increases, gradient and grain size decrease within increasing beach type wave energy. Laminae usually dip seaward at 1–2°, and may display a transition from lower gradient, more continuous and parallel in the upper swash zone, to steeper gradient, more discontinuous laminae, with coarser, more poorly sorted and more carbonate-rich sediments

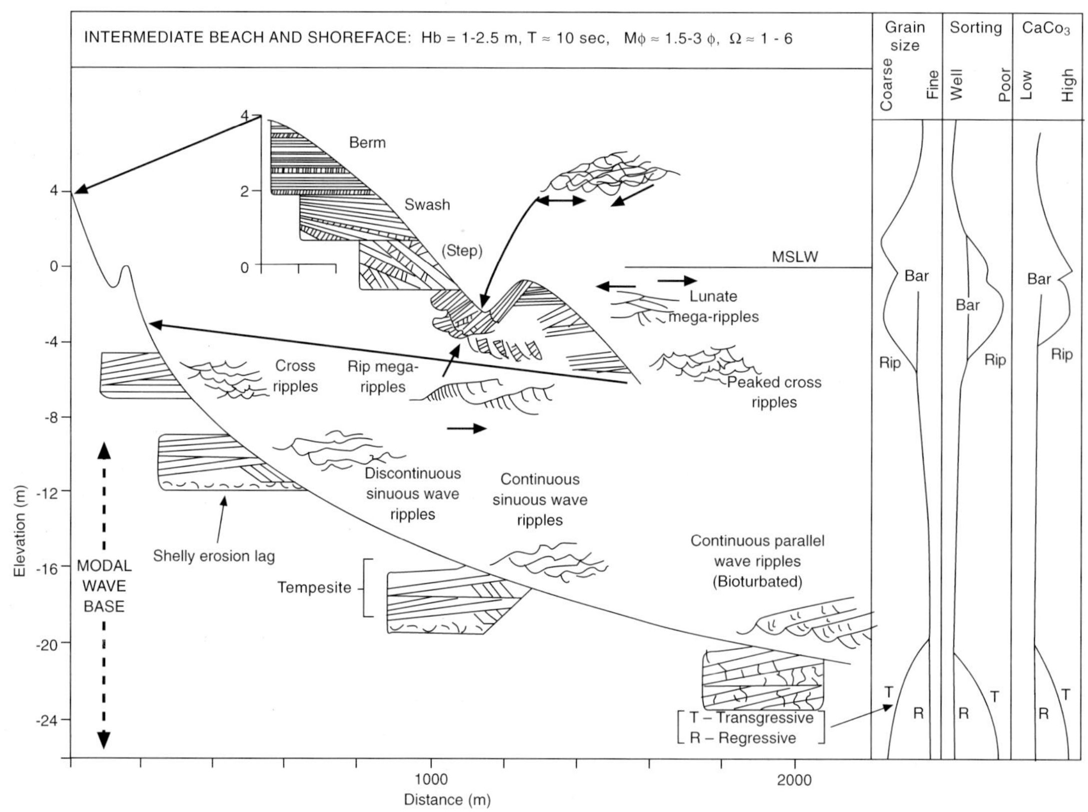


**Figure 12.4** Idealised morphology, bedforms and structures across an intermediate beach–shoreface system (modified from Short, 1986).

in the lower swash zone, peaking at the step, if present (Figure 12.2.h).

The intermediate *surf zones* are even more complex, as they contain morphologically distinct shallow bars, deeper rips channels, and shore-parallel rip feeder and longshore troughs. Flows include wave oscillations, especially in the deeper trough and channels, wave bores, especially on the bars, shore-parallel currents of variable speed and direction in the troughs, and strong seaward moving currents in the rip channels (see Section 4.4.2). Each of these sub-environments has a distinctive suite of structures and sediments and preservation potential.

The shallower *bars* are characterised by transitional flow regimes (shallow depth) and plane beds, with medium scale, horizontal and (shoreward) increasingly landward dipping laminae, composed of finer, well sorted sediments (Figure 12.2e). These may dip steeply (25°) into the trough on the inner bar edge, and exceed 1 m in thickness. Davidson-Arnott and Greenwood (1976) termed this the landward slope facies.

The longshore rip feeder and longshore *troughs* are dominated by both wave orbital and longshore currents, the former producing coarser asymmetrical wave ripples, the latter migrating current ripples, including megaripples. The superposition of these two crossing bedforms produces cross-trough stratification dominated by medium scale, shore-normal, megaripple foreset laminae (dipping 10–20°), and thinner, lower angle, landward dipping wave cross-laminae (Figure 12.2f). This is the trough facies of Davidson-Arnott and Greenwoord (1976).

The *rip* channels are the deepest part of the surf zone and are dominated by both wave orbital motions and seaward-moving rip currents. Rip currents tend to dominate, producing steeper, seaward-migrating, coarser, shell-rich megaripples, characterised by a medium scale basal trough unit, overlain by the coarse (shelly), seaward dipping (10–20°), foreset laminae of the ripples.

Overall, the surf zone from bar crest to rip trough is characterised by increasing sediment size and carbonate content (when present), and decreasing sorting, together with an increase in structure, size and dip, with direction varying from onshore (bar crest), mixed (troughs) to offshore (rip channel). Hill and Hunter (1976) identified three sets of depositional structures: small scale lenticular cross-bedding, usually without preferred orientation produced by sand ripples; planar lamination on higher energy bar crests; and medium scale cross-bedding produced by migrating megaripples, which dip in the direction of long or offshore (rip) current. These correspond to the seaward slope, bar crest and trough faces of Davidson-Arnott and Greenwood (1976), who also identified additional landward slope of the bar, composed of landward-dipping laminations.

Because of the mobility of the bar, the preservation potential of these structures increases with depth, with the rippled to laminar bar sequences having least potential, and the medium scale cross-bedding in the longshore troughs and rip channels having the best chance of preservation.

Seaward of the bar, across the *breaker zone* and beyond, the classic sequence identified by Clifton *et al.* (1971) (Figure 4.26) is encountered. With increasing wave depth and decreasing wave orbital velocity, the plane bed soon gives way to lunate megaripples ($\approx$ 50 m width), then peaked cross-ripples in the zone of immediate pre-breaking ($\approx$ 100–200 m width). Overall structures are composed of slightly finer sediments arranged in 10–20 cm beds which tend to dip more steeply under the lunate megaripple crests, with lower angles produced by the lunate troughs. The cross-ripples produce generally landward-dipping laminae in 10–15 cm thick beds which truncate the more gently landward-dipping laminae of the troughs (Figure 12.2d).

The mid- to outer *shoreface* is dominated by wave ripples. Initially they are discontinuous, sinuous wave ripples, then continuous sinuous wave ripples, and toward the modal wave base more continuous, parallel wave ripples. The typical structures are low angle truncations (5°), containing low to moderate angle (10–20°) landward-dipping cross-laminae (crests), with secondary, parallel and some truncated seaward-dipping laminae (troughs). These are contained in moderate scale beds (5–10 cm) (Figure 12.2b and c).

### 12.3.3 Dissipative (multi-barred) beaches

High energy dissipative multi-barred beaches represent the upper end of the beach spectrum. They typically require a combination of high waves and fine sand. As they are also favoured by short wave periods (Figure 7.5), they are relatively more common in fetch-limited sea environments than on open swell coasts, where very high waves are often required. While they have many of the characteristics of the high energy intermediate beaches, they also have some important differences. Firstly, on open coasts they are the largest of all beach systems, up to a few kilometres wide, and 30 m and more deep (Figure 12.5). Both their infragravity-dominated dynamics and fine sand ensure both overall low gradients (morphology and structures) and fine, well sorted laminae.

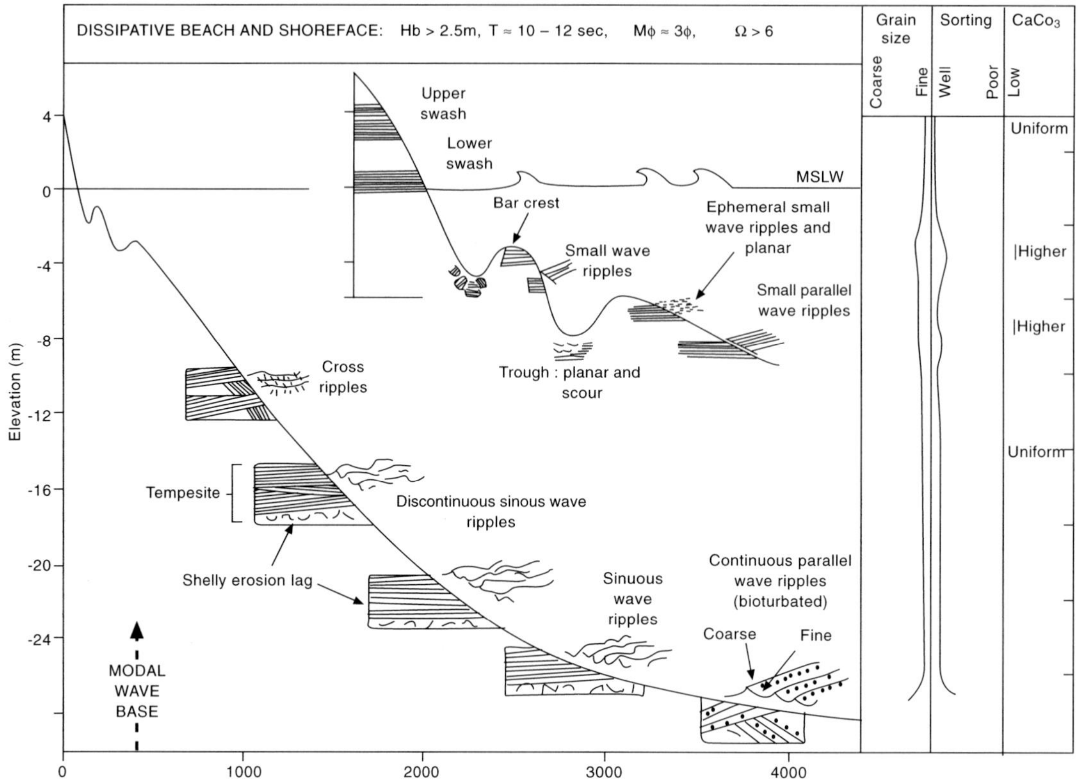


**Figure 12.5** Idealised morphology, bedforms and structures across a dissipative beach–shoreface system (modified from Short, 1986).

Dissipative beaches and surf zone are controlled by low frequency (> 30 s) infragravity standing waves acting across a wide (hundreds of metres), low gradient (0.01) beach–surf zone. The standing waves are manifest by substantial set-up and set-down across the swash zone, producing a wide (> 100 m) low gradient, uniform swash zone, usually devoid of berm/cusp features and rhythmic topography. The high tide swash commonly reaches to the toe of the dune.

Modal dissipative backshores are very wide and flat and devoid of berms. Backshore laminae are therefore very large scale, gently seaward-dipping, or almost horizontal. Truncation surfaces are relatively flat as storm waves tend to plane off the whole beach and backshore surface.

The *swash zone* is characterised by its extreme width, low gradient and fine sand. The surface is usually firm and featureless, apart from rhomboidal surface impressions. In the upper swash zone, near-horizontal, well developed, fine grained, continuous, parallel, thin laminae dominate (Figure 12.2i). Toward the lower beachface, increasing backwash velocity can shift the flow from lower to transition and even upper flow regimes, producing a hydraulic jump and associated standing waves and antidunes (Figure 12.1). When present these scour the laminae and leave landward-dipping truncations and laminae. Steps and accumulations of coarser sediments are usually absent.

The *surf zones* consist of alternating shore-parallel subdued bars and troughs. Wave breaking and transitional flow regimes dominate on the bar crests, and wave reformations and orbital wave flows in the troughs. Longshore flows and rip currents are usually weak or absent. The overall structures produced therefore grade from more horizontal, thin, parallel to slightly landward-dipping laminae on the bars, to thin, low angle, landward-dipping cross-laminae in the troughs.

Seaward of the wide breaker zone increasing depth in the inner *nearshore* leads to the formation of small wave ripples. These contain more steeply (5°) landward-dipping laminae, which truncate horizontal laminae. In the outer nearshore, wave ripples dominate; these are at first discontinuous, but become increasingly continuous, sinuous ripples in deeper water. Because of the fine sand they are moderate in size ($\lambda_r = 20$ cm, $\eta_r = 5$ cm). Such wave ripples produce predominantly landward-dipping cross-laminae.

### 12.3.4 Low to high energy beaches (reflective to dissipative)

As wave energy increases exponentially with a trend from low to high energy beach systems, there is a predictable change also in both the morphodynamics (Chapter 7), stratification and morphostratigraphy. Figure 12.6 illustrates the changes in morphostratification observed across a range of southern Australian beaches.

#### *12.3.4.1 Swash zone*

The lower energy *reflective* beaches are characterised by their shallowness and narrowness. Furthermore they contain the simplest suite of morphostratigraphic units, namely the sub-aerial berm/cusps, the intertidal beach-

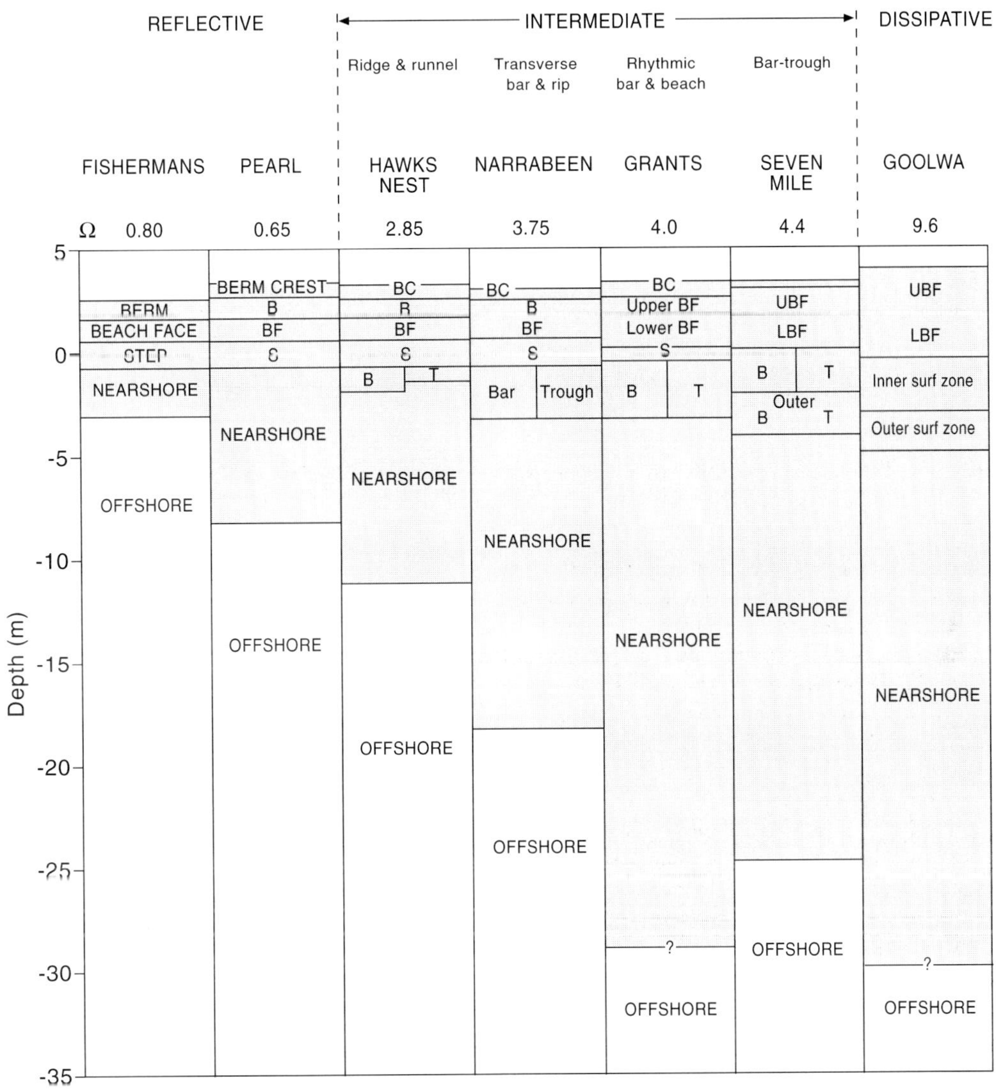


**Figure 12.6** Vertical sequences of beach environments based on field measurements. Note the increases in depth and diversity of environments with increasing wave energy (modified from Short, 1983).

face and basal step, and relatively shallow nearshore.

As wave energy increases across the range of *intermediate* beach types, changes in the morphodynamics are mirrored by predictable changes in the morphology. As incident wave swash is gradually replaced by infragravity set-up and set-down, the steep, narrow berms, are replaced by a wider, lower gradient swash zone. The step which characterises the lower energy beaches is usually absent in the higher energy systems. In all, there is greater mixing of the sediments across the swash and surf zone, resulting in their being finer and better sorted.

Finally in the *dissipative* beaches, berms and cusps are usually absent, as the wide low beachface is dominated by wave set-up and set-down that reaches the backing dunes.

#### *12.3.4.2 Surf zone*

The surf zone undergoes a greater transformation from reflective to dissipative, as does the underlying stratification. On *reflective* beaches wave breaking is confined to a narrow zone of wave collapse/surge at the base of the swash zone (Figure 5.2a), which produces the step and assists in maintaining the high, steep swash zone. Bars and troughs are absent.

On *intermediate* beaches, shore-parallel bars and troughs and shore-normal rip channels and their flows dominate. These initially produce a single, shallow bar unit on the low tide terrace, perhaps cut by small rip channel units. In the rip-dominated transverse bar and rip and rhythmic bar and beach, the deeper (coarser) rip channel and longshore trough units dominate at the expense of the shallower (finer) bar units. In higher energy and/or multi-bar systems this sequence deepens and thickens.

On *dissipative* beaches a broad, deep zone of wave breaking produces the thickest bar–tough sequences dominated by slightly coarser bar and rip sequences.

#### *12.3.4.3 Nearshore*

The nearshore units of all beaches are dominated by shoaling waves. The shift from lower energy reflective to higher energy dissipative is marked by a dramatic increase in the dimensions of the systems. Depth/thickness increases from 1 m to > 30 m, and width from a few tens of metres to a few kilometres, representing an increase in area and volume of two orders of magnitude. In the lower energy systems, the structures will tend to be coarser and steeper, with thicker, more discontinuous laminae arranged in thinner beds, while in the higher energy systems they tend toward finer, lower angle, thinner, more continuous laminae, but arranged in thicker beds.

The depth to the base of each major zone (beach, surf zone and nearshore) varies according to surf zone–beach type. Wave base on reflective beaches is in relatively shallow water depths, and may be limited to 1 to 3 m water depth below mean sea level on low energy beaches. As wave energy increases and with a trend from reflective to dissipative types, the wave base is at increasingly greater depths. Hesp and Hilton (1996) reviewed a wide range of evidence to show that the nearshore extends to approximately 30 to 35 m water depth below mean sea level on the higher energy intermediate and dissipative beaches.

## 12.4 DUNE STRATIFICATION

There have been several studies carried out on stratification and primary sedimentary structures in coastal dunes (e.g. Bigarella *et al.*, 1969; Bigarella, 1972; McKee 1979; Fryberger, 1991). However, far greater efforts have been concentrated on desert-type dunes (e.g. McKee, 1979; Brookfield, 1984). Much of this work has been concerned with delimiting the very detailed microstructures and bedding styles associated with different grain depositional patterns (e.g. Hunter, 1977a). The following briefly reviews some of the larger scale sedimentary structures which appear to typify the major coastal dune or dunefield types.

### 12.4.1 Foredunes

Foredunes may range from very sparsely vegetated to very well vegetated dunes. Depositional and erosional patterns and the resulting stratification strongly reflect the degree of vegetation cover and distribution, and wind velocity (Hesp, 1984a).

Shadow dunes, coppice dunes and nebkha are typified by discrete, semi-circular, small to medium scale, convex and triangular sets of low to high angle (up to 34°) laminae where groups of foresets dip in opposite directions (Goldsmith, 1973). Where foredunes are largely produced by closely spaced discrete vegetation clumps, and shadow and coppice dunes, the overall stratification is relatively chaotic, cross-bedding is pronounced, and the proportion of high angle beds can be significant (see McBride and Hayes, 1962; Goldsmith, 1973, 1978; Clemmensen, 1986).

As the degree of lateral and across-dune vegetation cover increases, large scale, tabular and gently convex cross-bedding prevails (Figure 12.7a; Hesp, 1988b, 1989; McCann and Byrne, 1989). Low angle beds domi-

**Figure 12.7** (a) A scarped upper beach and foredune at Muriwai Beach, New Zealand, showing the parallel beach laminations (lower) overlain by large scale dune cross bedding (upper). (b) Massive transgressive dune foreset beds dipping at up to 40°. The 20 m high multiple sequences are preserved in dunerock at Cape Dombey, South Australia (A D Short).

nate, reflecting gradual sand deposition on a relatively well vegetated surface (Bigarella, 1972; Hesp, 1988b; Pye and Tsoar, 1990). Traction and grainfall lamination predominate with some lamina disruption due to vegetation presence, although the latter is surprisingly limited on many vegetated foredunes. Backset (seaward) dip angles increase with increasing dune height, and foreset beds may be very steep where sand avalanches into dense vegetation (up to 46° recorded; Hesp, 1988b). Medium scale, more complex cross-bedding occurs as established foredunes become more erosional. Wave cut and aeolian eroded scarps may be preserved, and wave-deposited heavy mineral laminae, pumice and coarse sediments may be deposited on dune seaward slopes at quite high elevations (> 3 m above high tide lines observed).

### 12.4.2 Blowouts and parabolic dunes

Blowouts are characterised by a deflation basin, steep, erosional sidewalls and a depositional lobe. If sited within other dunes (e.g. a foredune), the sedimentary structures of the deflation basin will have resulted from depositional processes associated with the formation of the original dune or beach surface. There may therefore be backshore and beach laminae, or various kinds of dune cross-beds. Strata in the erosional walls (if preserved) will be dominated by avalanche and slump structures (see e.g. McKee and Bigarella, 1979; Fryberger, 1991), and wind ripple deposits. Strata forming the depositional lobe will be dominated by slipface (grainfall) and avalanche (grainflow) deposits.

Parabolic dunes have small to extensive deflation basins, short to very long trailing ridges, and small to extensive depositional lobes. The deflation basins and plains may be eroded down to a basement of calcrete, bedrock or non-aeolian sediments, or to the top of the lowest seasonal water table. Sedimentary structures may therefore be absent (as above for blowouts), comprise various dune strata, or be very low angle, wavy and planar strata formed by saltation and traction deposition on a bare or vegetating surface.

The depositional lobes may be simple, curved lobes, or they may have transverse and other dunes on the surface. Sedimentary structures may principally be high angle, grainfall and grainflow foreset strata formed on the downwind and outward-facing portion of the lobe slipface, and low angle to horizontal, large scale topsets (e.g. McKee, 1966). Since the depositional lobe has an arcuate form, slipface foreset strata will be formed right around the lobe encompassing around 180° of dip direction.

The trailing ridges are formed by vegetation trapping marginal lobe slipface sands. As the lobe moves downwind, the inside margin is eroded, the outside edge is retained by vegetation, and a ridge is formed. These ridges are therefore principally composed of slipface strata. Lower angle topset and bottomset strata are formed as the trailing ridges are vegetated (Figure 12.8).

### 12.4.3 Transgressive dunefields

Transgressive dunefields comprise a wide variety of dunes and landform units including deflation plains and basins, sand sheets, transverse, barchanoidal, linear, reversing and oblique dunes, interdune areas and swales, coppice mounds and nebkha, precipition ridges, erosional knobs, and sabhkas. A very wide range of sedimentary structures is therefore possible.

Brookfield (1984) provides an excellent review of the typical active (unvegetated) dune types and their stratification styles. Transgressive dunefields will generally look similar in overall structure to small scale draa – first-order bounding surfaces are gently inclined and enclose sets of climbing and migratory transverse, oblique, sometimes reversing dunes, with steep foresets inclined at around the angle of repose (Figure 12.7b; Hunter, 1977a; Hunter *et al.*, 1983; Hunter and Richmond, 1988; Kocurek, 1988; Fryberger, 1991; Figure 12.9). The classic and basic types of aeolian strata – ripple, grainfall and avalanche strata – occur in sets which, on occasions, reflect daily cycles of dune migration (Hunter, 1977b; Hunter and Richmond, 1988). Precipitation ridges and some dune foresets will be very steep ( > 40°) due to the presence of vegetation. As the number of different sized, superimposed bedforms increases, the structural complexity increases (e.g. figures 46 to 51 in Rubin, 1987; figure 7 in Brookfield, 1984).

Interdune areas, sand sheets and deflation plains and basins are common in transgressive dunefields. Interdune areas may be largely deflationary or principally accretional. Where erosion dominates, the sedimentary structures may have been previously formed by the migration of various types of dunes (Fryberger *et al.*, 1990). A thin veneer of ripple deposits may be present (Kocurek, 1981). Accretional interdune and sand sheet may have a predominance of wind ripple strata and/or very low angle, tabular, planar strata depending on the presence or absence of vegetation, degree of erosion versus accretion, and other factors. A wide range of other structures are possible depending on the range of depositional conditions prevailing (e.g. dry damp or wet interdunes; Kocurek, 1981).

Various other sedimentary structures may occur as a

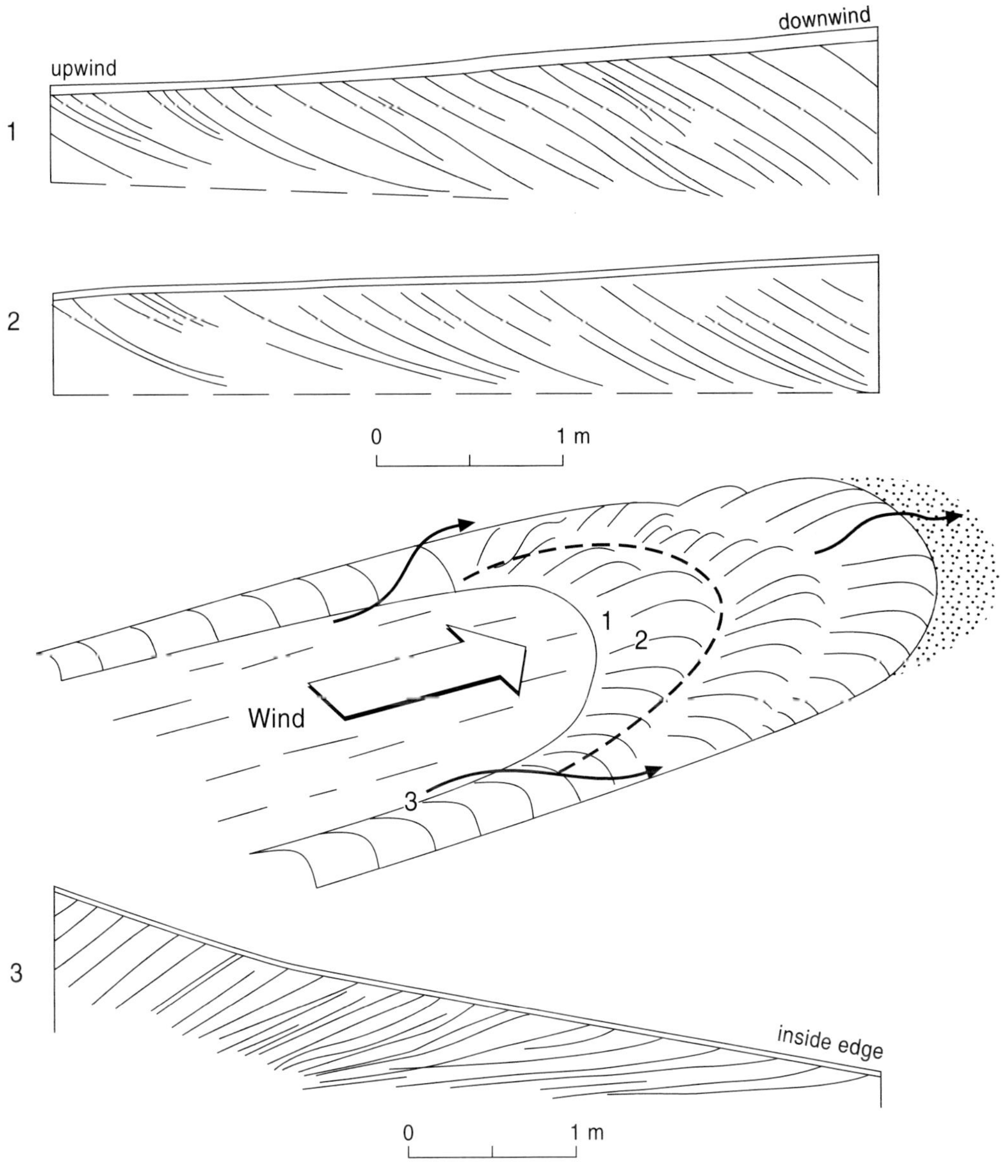


**Figure 12.8** Typical form and stratification through a parabolic dune.

result of the presence of vegetation. Coppice dunes, nebkha, foredunes and sparsely vegetated sand sheets may all be present, with various types of simple to complex cross-bedding produced.

## 12.5 SUMMARY

Beach and dune stratification is the internal expression and preservation of the sediments and processes that compiled the layers of sediment that comprise each beach–dune system, as well as potentially containing sediments that can provide a chronological history of the depositional sequence. Stratification is responsive and relevant to high frequency boundary layer processes, providing evidence of bedform migration and depth of disturbance, as well as potentially preserving records of similar events in past millennia.

As indicated in this chapter, while there is a generalis-

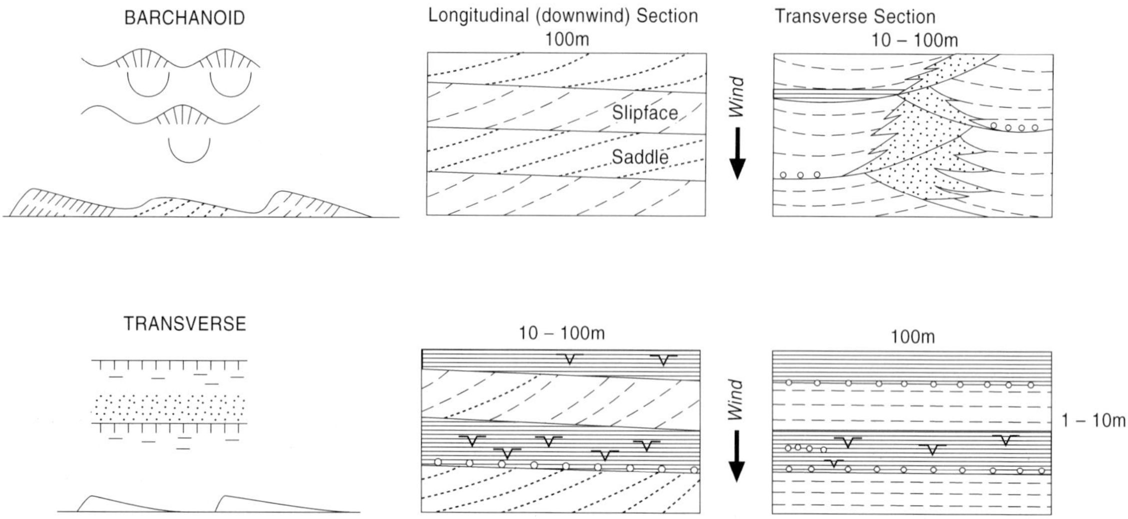


**Figure 12.9** Stratification models for the two main dune types that occur in transgressive dunefields. Arrow indicates wind direction (modified from Brookfield, 1984).

ed sequence of beach sediment characteristics and structures, both are also interdependent. Sediment size influences both beach type (Figure 7.8) and flow regime–bedform thresholds (Figure 12.1). Beach type in turn is associated with a characteristic sequence of meso-scale beach planforms and associated smaller scale bedforms and structures, such that the structures can be diagnostic of the beach type.

# 13

# Beach Hazards and Safety

*Andrew D. Short*

## 13.1 INTRODUCTION

Beaches are a major site of shoreline recreation and as such attract people to the marine environment. An estimated 233 million people visited United States beaches alone during 1996 (Gould, 1997). Beaches are, however, inherently hazardous. They form the boundary between dry land and deep water; they have a variable topography composed of bars and troughs; they contain breaking waves which can induce extreme turbulence, and wave bores which generate strong onshore currents; and they contain surf zone currents which can move onshore, longshore and offshore. Each of these factors can place people at risk on beaches. Additional hazards are also produced by regional factors such as tide range, tidal currents and strong winds, and local factors such as rocks, reefs and megarips. Beaches therefore contain an inherent range of physical hazards.

Beaches become a public risk when people enter the beach environment, the level of risk being a function of both the nature and level of hazards and the type and number of people. The presence of beach lifeguards and the large numbers of people who are rescued and those who drown at the beach attest to this risk (e.g. SLSA, 1995; Gould, 1997). This chapter is concerned with the nature and quantification of physical beach hazards and their mitigation, that is beach safety.

Biological hazards are not addressed in this book. For an excellent reference see *Venomous and Poisonous Marine Animals* (Williamson *et al.*, 1996).

## 13.2 BEACH HAZARDS

*Beach hazards* are elements of the beach–surf environment that expose the public to danger or harm. There are three hazards common to all beaches: water depth, breaking waves and surf zone currents.

### 13.2.1 Water depth

In the surf zone water depth can pose three types of hazards by being too deep, too shallow and spatially variable. *Deep water*, water greater than chest height, can drown the non-swimmer or inexperienced. It is also the relative, rather than the absolute, depth that is critical. Infants can drown in knee deep water, while adults will start experiencing problems in chest deep water. In the presence of a current, adults and children can only wade against the current when water depth is below chest height (Power, 1992). Therefore the age, size and experience of the swimmer, the relative water depth and the presence of currents all contribute to the water depth hazard.

*Shallow water* is also a hazard when people are propelled into shallow depths by diving or breaking waves. Both can result in injury and drowning if people hit the sand heavily, particularly head first. The latter results in scores of spinal injuries in Australia each year.

*Variable water depth* within the surf zone owing to the presence of bars, troughs, channels, steps and sloping swash zone is an additional hazard. Steeper slopes induce both less stable footing, while the juxtaposition of shallow bars and deep channels can result in people willingly or unwillingly moving quickly from the safety of a shallow bar or beachface to a deeper channel with possible currents.

### 13.2.2 Breaking waves

*Breaking waves* generate severe turbulence, wave bores and currents which generally move in a shoreward direction. The turbulence and wave bore can knock people over and hold them under water, while the wave bore and currents can propel them shoreward, and sideways if oblique waves.

*Handbook of Beach and Shoreface Morphodynamics*. Edited by A. D. Short.
© 1999 John Wiley & Sons Ltd.

**Box 13.1** Rip and surf zone current velocities.

| | |
|---|---|
| *Breaking waves* | travel at 3–4 m/s (10–15 km/h) |
| *Broken wave bores (white water)* | travel at 1–2 m/s (3–7 km/h) |
| *Rip feeder* and longshore currents | travel at 0.5–1.5 m/s (2–5 km/h) |
| *Rip currents* under average waves | ($<$ 1.5 m high) attain maximum velocities of 1.5 m/s = 5.4 km/h, and can reach 2 m/s |
| Note: Olympic swimmers swim at | 7 km/h (1.95 m/s) |

An average rip in a surf zone 50 m wide can transport a person outside the breakers in as little as 30 s.

Breaker height and type are equally important, as wave energy increases with the square of the *wave height*, and *breaker type* (spilling, plunging, surging) will determine over what distance the energy is expended. The most hazardous waves are high waves that break over a short distance, usually plunging and collapsing waves. The least hazardous are low spilling waves.

### 13.2.3 Surf zone currents

*Surf zone currents* and particularly *rip currents* pose the greatest threat to swimmers (Short and Hogan, 1994; Hoefel and Klein, 1998; Box 13.1). The problem with currents, particularly rip currents, is that they can move people willingly or unwillingly round the surf zone and ultimately seaward into deeper water and possibly the breaker zone (Figure 13.1). Short and Brander (in press) monitored a large rip current averaging 1.5 m/s and flowing up to 300 m out to sea. While currents are manageable in water below waist level, as water depth reaches chest height people become buoyant and cannot maintain their footing and will travel with the current. Currents are also the least obvious and usually the most difficult hazard for people to spot, even if they are aware of their presence (Short *et al.*, 1991).

### 13.2.4 Regional hazards

#### *13.2.4.1 Tide range and currents*

Tides in themselves are not a beach hazard; however they have two impacts that induce additional regional hazards, namely the impacts of tide range and tidal currents. As discussed in Chapter 8, *increasing tide range* induces predictable changes in beach morphology and surf zone dynamics. Basically with increasing tide range the surf zone and associated currents and topography are best developed at low tide causing the low tide beach to be more hazardous than the mid- to high tide beach. Even on micro-tidal beaches surf zone currents, in particular rip currents, are more intense at low tide (Short and Hogan 1994; Brander 1996). Consequently for the same wave conditions the low tide beach will be more hazardous than the mid- to high tide beach.

*Tidal currents* move large volumes of water to and from the shoreline. Whenever these currents are channelised such as in rip channels and tidal inlets, strong tidal currents, usually flowing in deeper water, present an additional hazard.

#### *13.2.4.2 Strong winds*

Strong onshore and particularly longshore winds induce two additional hazards. First they induce greater wind stress on the surf zone causing downwind currents to intensify, while rip currents may become highly skewed and even migrate downwind. Second, onshore winds induce an over-steepening of the waves and thereby premature and more erratic (seemingly confused) wave breaking.

### 13.2.5 Local hazards

Local hazards are any factors which intensify and/or obstruct wave breaking, water depth variation and currents (Figure 13.2). *Rocks, reef, headlands* and *structures* are all permanent localised hazards. Their presence may result in additional wave breaking and generate more and stronger rips (see Chapter 9), while their surfaces are usually hard, shallower and often dangerous. Their overall effect is to intensify the surf zone processes, as well as forming an often formidable physical presence.

The beach–surf zone environment will also possess a combination of the above hazards. What can make the surf zone particularly hazardous is that it always contains a combination, and in places all, of the above hazards. While some, such as breaking waves, are highly visible, others including current, tide effects and reefs

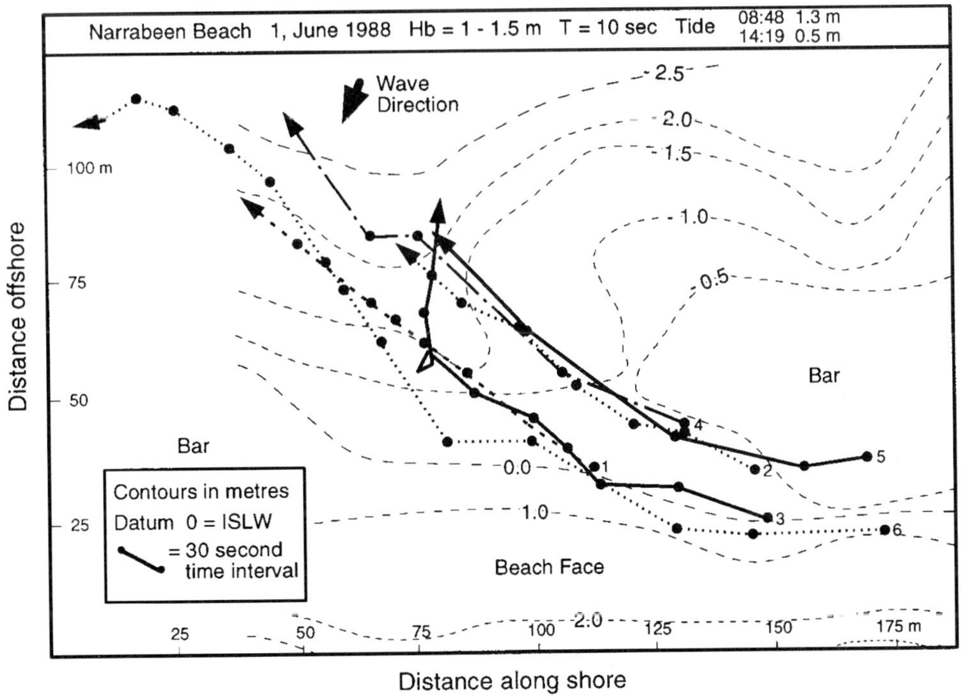


**Figure 13.1** Trajectory of volunteers drifting seaward in a small rip channel and current, under average wave and beach conditions at Narrabeen Beach, Australia. Velocities ranged from 0.4 to 0.73 m/s, moving all volunteers out of the surf zone within 5 min (from Short and Hogan, 1994; reproduced from the *Journal of Coastal Research* by permission of the Coastal Education and Research Foundation).

may be less obtrusive, but none the less extremely hazardous.

## 13.3 WAVE-DOMINATED BEACH HAZARDS

The wave- and tide-dominated beach models presented in Figures 7.4 and 8.15, indicate the inherent variation in water depth (gradient), breaker height and type, and surf zone currents within and between the beach types. In so far as the combination of these hazards varies between each type, so too does the nature and degree of beach hazards. This section briefly discusses the hazards associated with the wave- and tide-modified beaches.

### 13.3.1 Reflective beach hazards

Reflective sandy beaches are usually maintained by low waves ( $< 1$ m) which provide relatively safe swimming environments. However reflective beaches are also characterised by a relatively steep beachface, and possibly a step at the base of the beach, with an abrupt drop off into deeper water off the beach (Figure 13.3). In addition the waves, though usually low, expend much of their energy in a short distance, breaking over the base of the beach and running up and down the beach as a strong uprush and backwash. Above-average waves produce a more hazardous collapsing to plunging shore break over the step. Therefore while the least hazardous beach type, their steepness, deep inshore water and rapid wave breaking can all place beach users at risk.

### 13.3.2 Low tide terrace beach hazards

Low tide terrace beaches are the safest of the intermediate beaches, because of their characteristically lower waves ( $\approx 1$ m) and shallow terrace. However changing wave and tide conditions do produce a number of hazards. At high tide they behave like a reflective beach with a possibly steep beachface, deep water close to shore and surging breakers. At low tide waves plunge more heavily on the outer edge of the shallow bar with deep water beyond, while at mid-tide more general wave breaking and waist-deep water can induce weak to moderate rip currents. Because rip channels are often poorly developed or absent on these beaches, oblique and/or rising waves can induce spatially and temporally

**Figure 13.2** Nobby Beach, NSW, Australia, is a typical fine sand beach; however the addition of rocks and rock reefs in the surf zone induces additional permanent hazard to swimmers (A D Short).

variable pulsating rips, called 'flash' rips by Australian lifeguards (SLSA, 1995).

### 13.3.3 Transverse bar and rip beach hazards

Transverse bar and rip beaches have the most variable beach morphodynamics, with alternating shallow attached (transverse) bars adjacent to the deeper rip channels containing strong seaward-flowing rip currents. Furthermore as these two systems alternate alongshore the bars are always bordered by rip feeder currents close to shore and rip currents in the surf zone. They are also the lowest energy beach system dominated by rips and as such have waves low enough (1–1.5 m) to attract people into the surf, but high enough to induce strong rip currents. For this reason they are the beach system that contributes to most rescues along the New South Wales coast (Short *et al.*, 1991) and are the primary cause of rescues in the USA, with more than 20 000 swimmers rescued from rips each year (Gould, 1997).

### 13.3.4 Rhythmic bar and beach hazards

The rhythmic bar and beach system possesses alternating bars and rip channels, with a continuous trough both separating the bars from the beach and linking the rip channels via feeder channels and currents. This system therefore contains higher breaking waves, variable water depth, strong rip feeder currents against the shore, and strong rip currents. Because the bar is not attached to the shore, people are less likely to move into the deep trough. If they do, however, strong feeder and rip currents will move them long and/or offshore.

### 13.3.5 Longshore bar and trough hazards

The longshore bar and trough beach is the most hazardous of the intermediate beaches. It is produced by high breakers ( $> 2$ m), and has a relatively shallow outer bar separated from the shore by a wide, 2–3 m deep trough containing rip feeder currents and periodic rip currents. Once in the current-dominated surf zone there is essentially no shallow water, other than the bar some 50 to 100 m offshore. Swimmers in the trough will therefore be moved by the surf zone currents, and can only return to shore by swimming laterally across feeder currents, and/or by using wave bores, if present, to help propel them shoreward.

### 13.3.6 Dissipative beach hazards

Dissipative beaches are the most hazardous beach type owing to the high waves required for their formation,

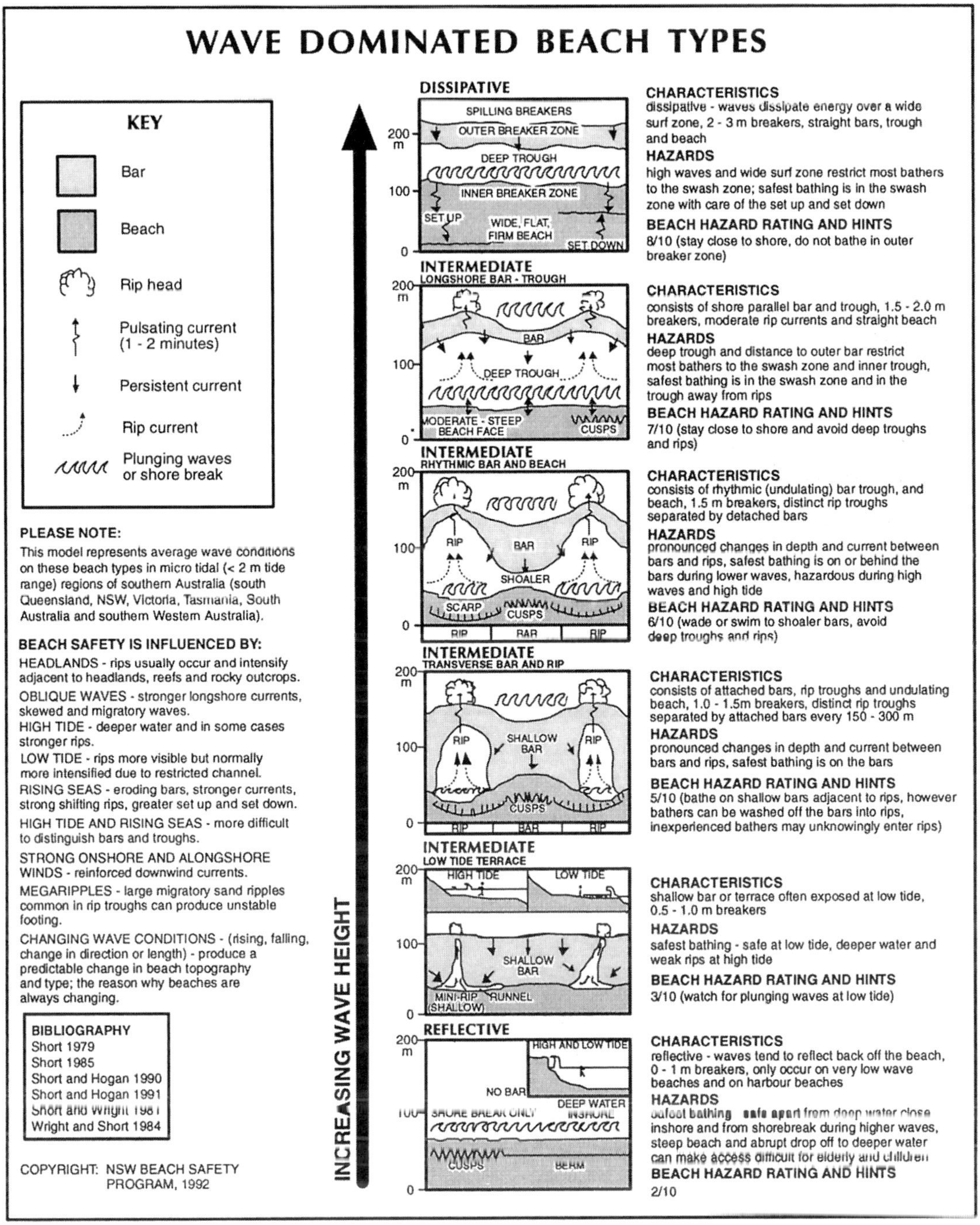


**Figure 13.3** A schematic illustration of the six micro-tidal, wave-dominated beach types, their general physical characteristics and patterns of wave breaking, bars, trough and currents. The characteristics of each and their hazards are noted, together with the influence of other factors that affect beach safety (modified from Short and Hogan, 1994).

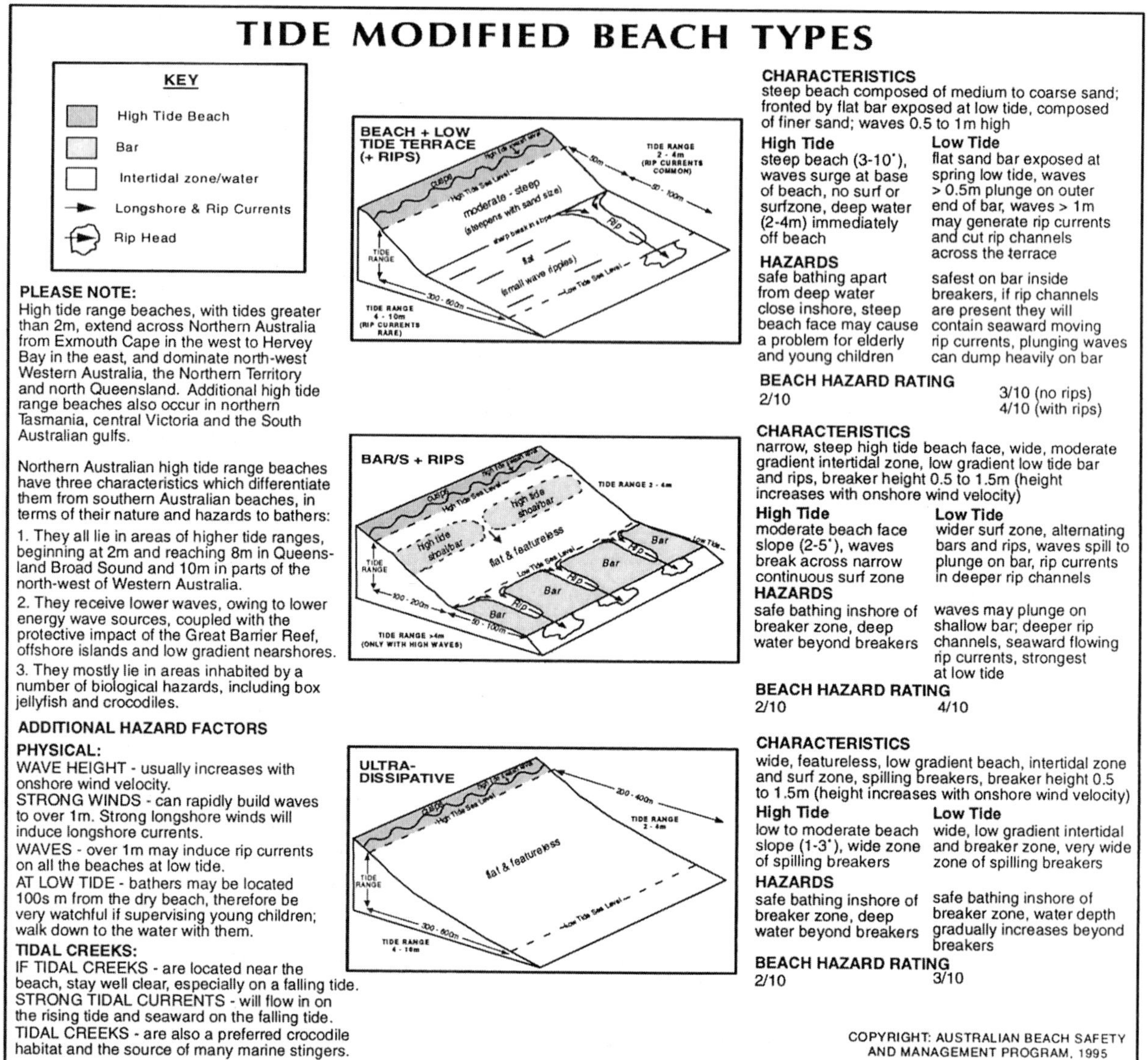


**Figure 13.4** A schematic illustration of the three tide-modified beach types, their general physical characteristics and hazards at high and low tide, together with the influence of other factors that affect beach safety (from Leahy *et al.*, 1996).

their wide surf zones and strong on–offshore currents, together with wave set-up at the shore. The waves break well offshore on the outer bar, with an outer trough, inner bar and inner trough occupying the surf zone. While currents are predominantly low frequency on–offshore, rip circulation is more likely in the inner surf zone. Fortunately the formidable high breakers ( > 2 m) and wide surf zone tend to keep most swimmers close inshore on the less hazardous lower beachface. Only if swimmers venture out into the breaker zone will the full range of hazards be encountered.

## 13.4 TIDE-MODIFIED BEACH HAZARDS

Increasing tide range relative to wave height produces a predictable shift in beach morphodynamics as presented in Chapter 8. The hazards associated with the three tide-modified beaches in Figure 13.4, are summarised below using three composite beach types.

### 13.4.1 Reflective and low tide bar (rip) hazards

This beach is an extension of the wave-dominated reflective and low tide terrace beaches, only with greater tide range, together with the possibility of well developed rip channels and currents on the low tide bar during higher wave conditions. At high tide and under low waves the high tide beach has the same hazards as the wave-dominated reflective beach (Section 13.3.1). At low tide the wide, low gradient terrace is exposed, and breaking waves and surf occupy the low tide bar, with rips forming when waves exceed 1 m. The rips are inactive at high tide.

### 13.4.2 Bar and rip hazards

Persistent moderate to high waves in an area of higher tide range produce the (multi-)bar and rip beach type. It has a high tide reflective beach with hazards similar to the wave-dominated reflective beach, a wide intertidal zone which may have a low shoal or bar, and a low tide beach characterised by intermediate surf zone topography and circulation, usually a transverse bar and rip or, under higher waves, rhythmic bar and beach. The hazards at high tide are relatively low compared to the more hazardous conditions associated with the low tide rips.

### 13.4.3 Ultra-dissipative hazards

The ultra-dissipative beach requires both high tide and relatively high waves, as well as fine sand. As with all tide-modified beaches the spring high tide beach is reflective with associated hazards. As the tide falls the beach becomes increasingly lower gradient with a wide surf zone containing spilling breakers and dissipative surf zone morphodynamics. However because of the mobility of the shoreline and surf zone dynamics, a planed down rather than a multi-bar beach results. The hazards at low tide are associated with the set-up and set-down across the wide flat surf zone, together with the often great distance (hundreds of metres) of the surf from the high tide shoreline.

## 13.5 BEACH HAZARD RATING

Short and Hogan (1994) introduced the *beach hazard rating* in order to empirically quantify the level of hazards associated with each wave-dominated beach type. Short *et al.* (1996) expanded this to include the tide-modified beaches. The hazard rating for each beach type together with comments on the associated morphodynamics and hazards are listed in Figure 13.5. The rating ranges from a low hazard of 1 to a maximum of 10. Each beach can be rated for its average or modal conditions, as well as day-to-day prevailing conditions, both of which also take into account additional local hazards, such as rocks, reefs and inlets.

### 13.5.1 Modal beach hazard rating

It is important to note that the rating in Figure 13.5 is *modal* or *average beach hazard rating* which is based on the hazards prevailing on a typical beach type under average wave conditions for that beach type. The modal hazard rating for each beach type and its associated modal wave height is indicated in bold, with the rating associated with higher and lower waves than average indicated. Therefore as the prevailing wave conditions change so too will the hazard rating.

### 13.5.2 Prevailing beach hazard rating

The *prevailing beach hazard rating* is the hazard level associated with a particular beach type under the prevailing wave conditions. Therefore for any beach type as wave conditions change so too will the hazard rating. As indicated in Figure 13.5, *all* beach types have low to moderate rating under low waves, and all have extreme hazards when waves exceed 3 m. Note that in nature changing wave conditions will ultimately lead to a change in beach type (Section 7.4).

Finally, the prevailing hazard rating must also include any additional regional and local hazards. Therefore

*modal hazard rating* = f (beach type × modal wave height)
*prevailing hazard rating* = f (beach type × prevailing wave height × local and regional hazards)

### 13.5.3 Spatial variation

Spatial variation in average breaker wave height produces changes in beach type and modal beach hazards Therefore as wave height changes along a beach owing to variable wave refraction and shoaling processes, so too will the beach type and the accompanying beach hazards. Figure 13.6 shows the Byron Bay region on the New South Wales coast, where the beaches south of the cape receive waves that average 1.6 m, while in lee of the cape they decrease to 0.5 m. The variable breaker wave height produces a range of beach types from an ener-

**BEACH HAZARD RATING GUIDE**
**Wave Dominated Beaches**
**Impact of changing breaker wave height on hazard rating for each beach type**

| BEACH TYPE \ WAVE HEIGHT | <0.5 (m) | 0.5 (m) | 1.0 (m) | 1.5 (m) | 2.0 (m) | 2.5 (m) | 3.0 (m) | >3.0 (m) |
|---|---|---|---|---|---|---|---|---|
| Dissipative | 4 | 5 | 6 | 7 | 8 | **9** | **10** | 10 |
| Long Shore Bar Trough | 4 | 5 | 6 | **7** | **7** | **8** | 9 | 10 |
| Rhythmic Bar Beach | 4 | 5 | 6 | **6** | **7** | 8 | 9 | 10 |
| Transverse Bar Rip | 4 | 4 | **5** | **6** | 7 | 8 | 9 | 10 |
| Low Tide Terrace | 3 | **3** | **4** | 5 | 6 | 7 | 8 | 10 |
| Reflective | **2** | **3** | 4 | 5 | 6 | 7 | 8 | 10 |

**BEACH HAZARD RATING**

Least hazardous: 1 - 3
Moderately hazardous: 4 - 6
Highly hazardous: 7 - 8
Extremely hazardous: 9 - 10

**KEY TO HAZARDS**

Water depth and/or weak currents
Shorebreak
Rips and surfzone currents
Rips, currents and large breakers

NOTE: All hazard level ratings are based on a bather being in the surf zone and will increase with increasing wave height or with the presence of features such as inlet, headland or reef induced rips and currents. Rips also become stronger with falling tide.

**BOLD** gradings indicate the average wave height usually required to produce the beach type and its average hazard rating.

**BEACH HAZARD RATING GUIDE**
**Tide Modified Beaches**
**(at high tide - at low tide add 1)**
**Impact of changing breaker wave height on hazard rating for each beach type**

| BEACH TYPE \ WAVE HEIGHT | <0.5 (m) | 0.5 (m) | 1.0 (m) | 1.5 (m) | 2.0 (m) | 2.5 (m) | >3.0 (m) |
|---|---|---|---|---|---|---|---|
| Ultradissipative | 1 | 2 | 4 | 6 | 8 | 10 | 10 |
| Bar & Rips | 1 | 2 | 3 | 5 | 7 | 9 | 10 |
| Beach & LTT | 1 | 1 | 2 | 4 | 6 | 8 | 10 |

**BEACH HAZARD RATING**

Least hazardous: 1 - 3
Moderately hazardous: 4 - 6
Highly hazardous: 7 - 8
Extremely hazardous: 9 - 10

**KEY TO HAZARDS**

Water depth and/or weak currents
Shorebreak (high tide only)
Rips and surf zone currents
Rips, currents and large breakers

**Figure 13.5** Beach hazard rating guide for wave- dominated and tide-modified beaches, based on the prevailing wave or breaker height and beach type. Note that the rating for individual beaches may also include additional local hazards such as rocks and reefs (from Leahy *et al.*, 1996).

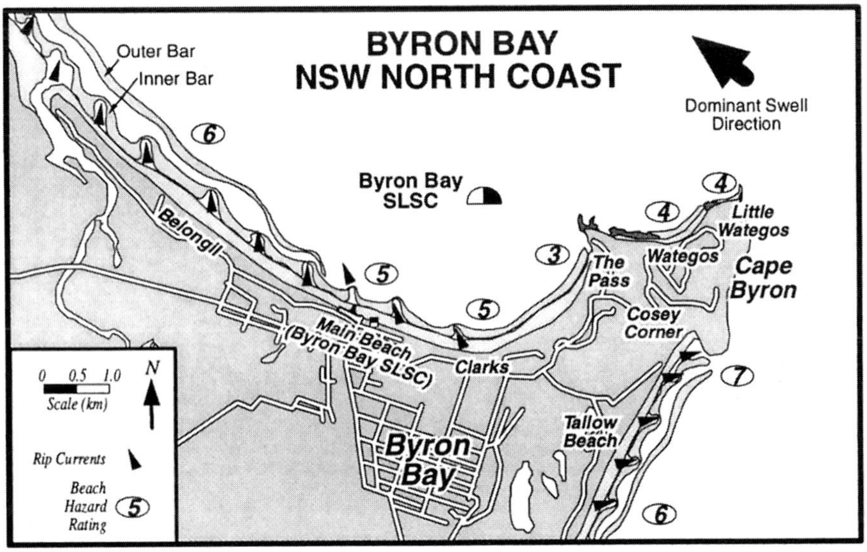


**Figure 13.6** A map showing the usual location of bars and rips around Cape Byron on the NSW north coast. Wave height averages 1.4 m at Belongil, from where it decreases toward Clarks Beach and The Pass, resulting in the rip-dominated double bar switching ultimately to a ripless attached low tide terrace. The beach types and corresponding beach safety ratings vary accordingly from a relatively high 6 to 7 on exposed Tallow Beach (average wave height 1.6 m) to a lower 3 to 4 at The Pass and the Wategos beaches (from Short and Hogan, 1994; reproduced from the *Journal of Coastal Research* by permission of the Coastal Education and Research Foundation).

getic double bar system north and south of the cape, to low tide terraces in lee of the cape, and modal beach hazard ratings from a high of 7 to a low of 3.

### 13.5.4 Temporal variation

Temporal changes in wave height and direction are a characteristic of all beaches, and these will cause a lagged adjustment in the beach type as well as the associated beach hazards. Figure 13.7 shows Mollymook Beach on the New South Wales coast under three different combinations of wave height and direction. Each set of wave conditions produces a different set of beach types along the beach, together with a different level and range in prevailing beach safety ratings. The result is a temporal change in both beach state and prevailing beach safety rating.

## 13.6 APPLICATIONS TO BEACH MANAGEMENT

### 13.6.1 Regional hazard assessment

The beach hazard rating provides a mechanism for quantifying the modal and prevailing hazard ratings on any beach or group of beaches. This permits a regional assessment of beach hazards. Table 13.1 lists the modal beach hazard rating for 2416 beaches in Queensland ($n = 961$), New South Wales ($n = 721$) and Victoria ($n = 588$) as well as Port Phillip Bay, a large bay on the Victoria coast ($n = 146$). On the NSW ocean coast the modal hazard rating is 6; however on the more energetic Victoria coast the modal rating is 7 and 40% by number rate between 7 and 10, indicating a high hazard level. In contrast, the largely enclosed Port Phillip Bay which receives no ocean swell and relies on wind waves, has substantially lower waves and hence ratings, with most beaches rating a low 1. The Queensland coast is largely protected from ocean waves by the Great Barrier Reef, resulting in lower wind waves and accompanying lower modal beach hazard rating of 1. All Queensland beaches rating greater than 5 are located on the exposed south-east coast, south of the Great Barrier Reef. Such figures permit an immediate assessment of the overall level of beach hazards along 7000 km of the Australian coast based on state regions, and by number of beaches (Table 13.1a) and length of beach coastline (Table 13.1b). By number of beaches the low rating of 1 dominates, primarily due to numerous usually short, low energy beaches in Queensland and Port Phillip Bay. However

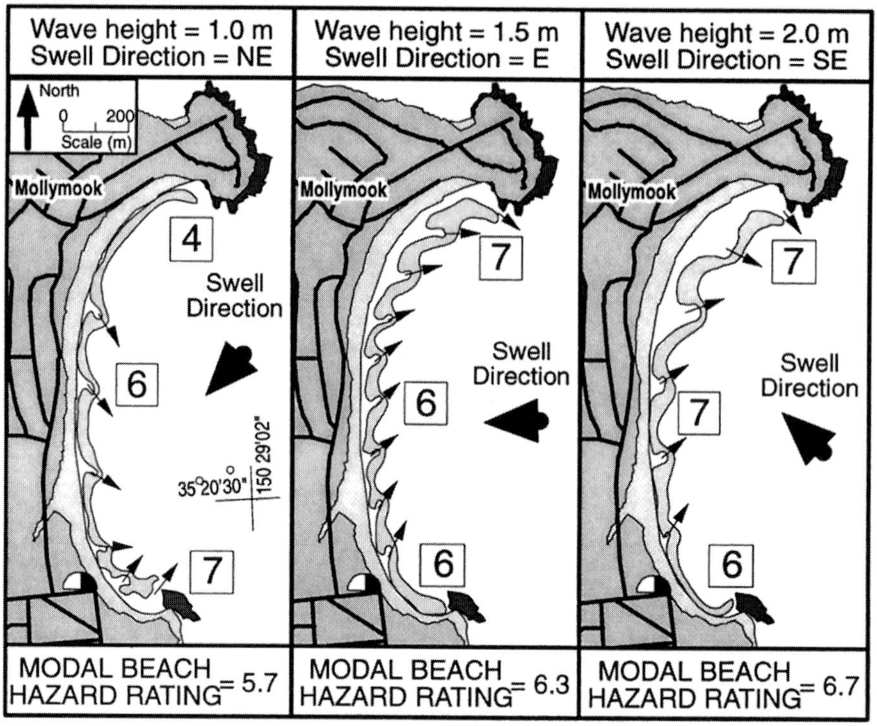


**Figure 13.7** Variation in both wave height and direction produce predictable changes in beach conditions and corresponding beach safety rating. At Mollymook Beach on the NSW south coast changing wave conditions result in alongshore variation in the occurrence, size and location of bars and rips, and beach safety rating (source: Short, 1993b).

by length of beach, the longer, more exposed beaches in New South Wales and Victoria dominate, resulting in a relatively hazardous rating of 6.

### 13.6.2 Beach risk parameter

In order to quantify the level of public risk, as opposed to hazard level, Short and Hogan (1990) introduced the beach risk parameter $R^*$

$$R^* = r/(p/1000) \quad (13.1)$$

where $r$ is the total number of rescues in a swimming season and $p$ the number of beachgoers for that season. On 12 popular Sydney beaches they found that $R^*$ ranged from 0.5 to 6 (rescues per 1000 people) indicating that the more hazardous beaches (rating 6) had up to 12 times more rescues per thousand users than adjoining less hazardous beaches (rating 2).

### 13.6.3 Beach hazard mitigation

Maintenance of public safety on beaches requires both hazard identification and mitigation of those hazards. The assessment of the modal beach hazard rating clearly identifies the general hazards, while the prevailing beach hazard rating identifies those present at any time. When combined with the level of beach usage then the level of public risk can be assessed, where public risk $P^*$ is

$$P^* = \text{f (beach hazard rating} \times \text{number and type of beach users)}$$

Leahy *et al.* (1996) applied this to beach management in Australia, developing a Beach Management Plan, which can be used in a planning and strategic mode. In the *planning mode* the modal beach hazard rating and average beach population are used to assess the average level of risk (Table 13.2). This is then used to determine the general level of beach safety resources required for any particular beach. In this way the beach safety resource requirement for a beach or series of beaches can be matched to the general level of risk.

In the *strategic mode* the lifeguards assess the prevailing beach hazard rating and beach population and thereby the prevailing level of risk. Based on the prevailing risk level, the actual beach safety resources required

**Table 13.1** Modal beach hazard ratings (1–10) by (a) percent number and (b) percent beach length, for all east Queensland, New South Wales and Victoria beaches. Modal rating is in bold (source: Short, 1993, 1996, in press).

(a) Percent number

| | | Beach hazard rating | | | | | | | | | |
|---|---|---|---|---|---|---|---|---|---|---|---|
| | *n* | 1 | 2 | 3 | 4 | 5 | 6 | 7 | 8 | 9 | 10 |
| Queensland | 961 | **68** | 27 | 1 | 1 | 1 | 2 | 0 | 0 | 0 | 0 |
| NSW | 721 | 0 | 6 | 14 | 19 | 12 | **32** | 16 | 1 | 0 | 0 |
| Victoria | 588 | 0 | 2 | 8 | 15 | 12 | 22 | **25** | 13 | 2 | 0 |
| Port Phillip | 146 | **44** | 19 | 32 | 5 | 0 | 0 | 0 | 0 | 0 | 0 |
| Total (*n*) | 2416 | **721** | 342 | 208 | 237 | 168 | 382 | 262 | 84 | 12 | 0 |

(b) Percent length

| | | Beach hazard rating | | | | | | | | | |
|---|---|---|---|---|---|---|---|---|---|---|---|
| | *n* | 1 | 2 | 3 | 4 | 5 | 6 | 7 | 8 | 9 | 10 |
| Queensland | 961 | 30 | **39** | 23 | 5 | 1 | 2 | 0 | 0 | 0 | 0 |
| NSW | 721 | 0 | 2 | 5 | 5 | 6 | **50** | 32 | 0 | 0 | 0 |
| Victoria | 588 | 0 | 1 | 5 | 5 | 15 | **38** | 23 | 12 | 1 | 0 |
| Port Phillip | 146 | **58** | 13 | 25 | 4 | 0 | 0 | 0 | 0 | 0 | 0 |
| Total (km) | 2416 | 377 | 412 | 323 | 114 | 143 | **604** | 366 | 71 | 6 | 0 |

**Table 13.2** Risk level (1–5) based on beach hazard rating (modal or prevailing) and beach usage (modal or prevailing) (from Leahy *et al.*, 1996).

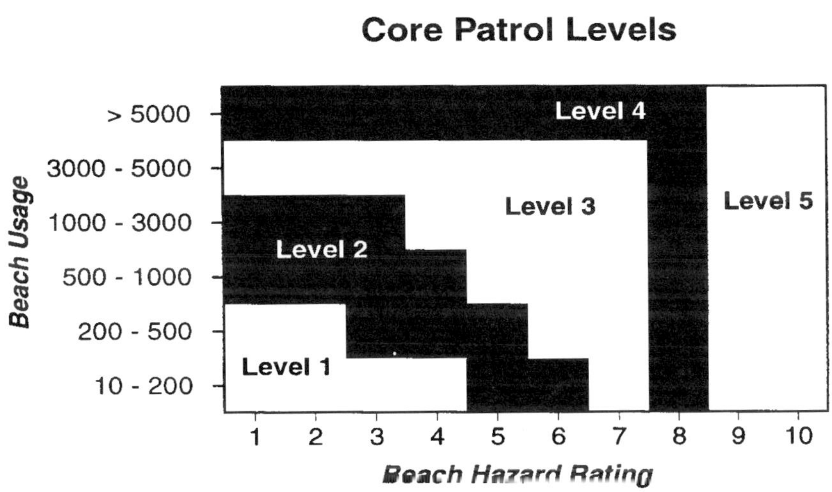


to maintain adequate public safety on the beach can be determined. This may require a temporary increase in safety resources when the risk level exceeds the average level for that beach.

Using this procedure the general level of safety resources is matched to the average level of risk, and adjustments can be made to the safety resources as conditions (hazards and usage) and risk level change. This plan ensures that beach hazards and risk level are clearly identified at all times and that safety resources are appropriate to maintain adequate public safety on the beach.

## 13.7 SUMMARY

While rips have long been recognised as a hazard to

people swimming in the surf (Shepard, 1949; McKenzie, 1955), hundreds of thousands of people have been rescued from rips this century (SLSA, 1995), and on the order of several thousand drownings have been caused by rips, the scientific community until recently paid little attention to the application of our improved knowledge of beach dynamics to the domain of public beach safety. Likewise, the lifeguards and lifesavers who for nearly a century have patrolled our beaches, whilst usually well trained in rescue techniques, received little training and often had little understanding of the cause of the problem, that is beach and surf zone dynamics.

This chapter summarises some of the advances made since the late 1980s in redressing this problem. The beach morphodynamic models developed in the late 1970s (Figure 7.4) have been used as a basis for the identification and quantification of beach hazards, while the addition of local beach hazards and conditions, together with level of beach usage, has led to the formulation of beach management plans that consider the beach hazard and the beach usage or population to assess the level of public risk. Based on the risk level then the beach safety resources required to maintain public safety can be more accurately determined. Furthermore there is growing demand for concerns for public beach safety to be incorporated in coastal management (Short and Hogan, 1994), so as to be more proactive in the mitigation of public risk from attractive and increasingly popular, but naturally hazardous beach systems.

This is particularly the case is countries with hazardous beaches but little public awareness of beach hazards and little or no lifeguarding facilities. J Buck (personal communication) reported that on the Pacific coast of Mexico rip currents are termed 'the claws of death', while Zauscher (1997) found that on some Pacific coast Colombian beaches 'it is common for four or more people to drown each month', many watched by fearful and helpless onlookers. Consequently while countries with well developed lifesaving procedures have a ready mechanism for implementing the beach safety management practices discussed above, in many countries a more basic approach to maintaining beach safety is still urgently required. The scientific as well as the lifesaving community should assist in this endeavour.

# PART E

# LARGE SCALE BEACH BEHAVIOUR



To many people a beach is simply the sand they sit on, together with perhaps the surf zone. In this book and in the coastal scientific community the spatial view of beaches is generally broader, encompassing the entire shoreface – the nearshore, surf zone, swash zone and backbeach. In this book the beach is also considered to have a broad temporal scale. Beaches have been forming for as long as wave and sediment have interacted at the shoreline, and many of these ancient systems are preserved in the rock record. The temporal perspective considered here however is essentially Holocene. Since sea level approached its present level beaches have formed a cross-shore and longshore conduit for sediment, which may be deposited around and above present sea level, by wave, tide and aeolian processes. In doing so beaches are an essential component of all barrier systems, and barrier evolution is intimately related to beach morphodynamics. Chapter 14 presents an attempt to examine the dynamics of barrier systems as well as integrating in a meaningful way the various barrier models into a more holistic approach to barriers, both their morphodynamics and classification.

# 14

# Barrier Morphodynamics

*Patrick A. Hesp and Andrew D. Short*

## 14.1 INTRODUCTION

Beaches form the seaward margin of coastal barriers. Whereas beaches represent the contemporary interface between shoaling–breaking waves and the shoreface sediments, barriers represent the time integral of the beach, and therefore incorporate the beach–shoreface, plus associated dune, washover, inlet and tidal delta deposits.

The barrier literature has traditionally been dominated by one form of barrier, the barrier island comprising six elements: the mainland, backbarrier lagoon, inlet and inlet deltas, barrier island, barrier platform and shoreface (Oretel, 1985). Barrier islands, however, represent one end member in a spectrum of barrier types. In one direction barrier islands grade, with increasing tide range and/or decreasing wave height, into mixed wave/tide-dominated barrier islands and inlets and finally into tide-dominated coastal flats (see Davis, 1994a for review). In the other direction, as the distance between the barrier island and mainland decreases due to increasing substrate gradient, the barrier may be partially or completely joined to the mainland as a spit (Dingler and Clifton, 1994), 'welded' barrier (Hayes, 1994), 'fringing' barrier (Carter and Orford, 1993), 'attached' barrier (Davis, 1994a), 'bay' barrier (Dingler and Clifton, 1994), 'pocket' barrier (Fitzgerald *et al.*, 1994), or 'embayed' barrier or 'mainland' beach (Roy *et al.*, 1994).

There is general consensus that barriers may range from barrier islands to some degree of mainland attachment, and there is generally universal agreement that they may be stable, transgressive or regressive (e.g. Davis, 1994a; Roy *et al.*, 1994). In order to incorporate these two elements, the approach used by Roy *et al.* (1994) is adopted, whereby sea level movement (transgressive/regressive/stable) is the prime control on barrier development, and substrate gradient the controlling factor on the location of the barrier relative to the mainland, as well as the regional coastal morphology. Other factors which will also influence barrier type and stability are the sediment supply, the contribution of wave/tide energy, (onshore) wind energy, geological inheritance and tectonism. As a consequence of the range of factors that can influence barrier evolution it is to be expected that barriers will range considerably in nature and location, as indicated by Figure 14.1.

In this book a *barrier* is defined as a shore-parallel, sub-aerial and sub-aqueous accumulation of detrital sediment (sand/boulders) formed by waves, tides and aeolian processes. It constitutes a definable coastal landform or sequence of landforms which is clearly separate in either age, lithology, and/or form from adjacent, underlying or landward landforms. The barrier may block off or impound drainage from the hinterland, but this is not a pre-requisite for definition as a barrier. Essentially a barrier is a coastal landform which acts as a 'barrier' between the sea and older coastal landforms or mainland bedrock.

## 14.2 THE ORIGIN OF BARRIERS

Discussion on the origin of barriers dates back to De Beaumont (1845) who favoured barrier emergence through upward shoaling of shallow sand bars. Later Gilbert (1885), working around Lake Bonneville, proposed that barriers evolve as spits from longshore drift McGee (1890) favoured the drowning of coastal ridges as a mechanism for barrier development. Subsequent to this early work there was little progress made on barriers until the 1960s when US workers began re-evaluating barrier models. First Hoyt (1967) favoured drowning, while Otvos (1970a,b) favoured an emergent bar,

*Handbook of Beach and Shoreface Morphodynamics.* Edited by A. D. Short.
© 1999 John Wiley & Sons Ltd.

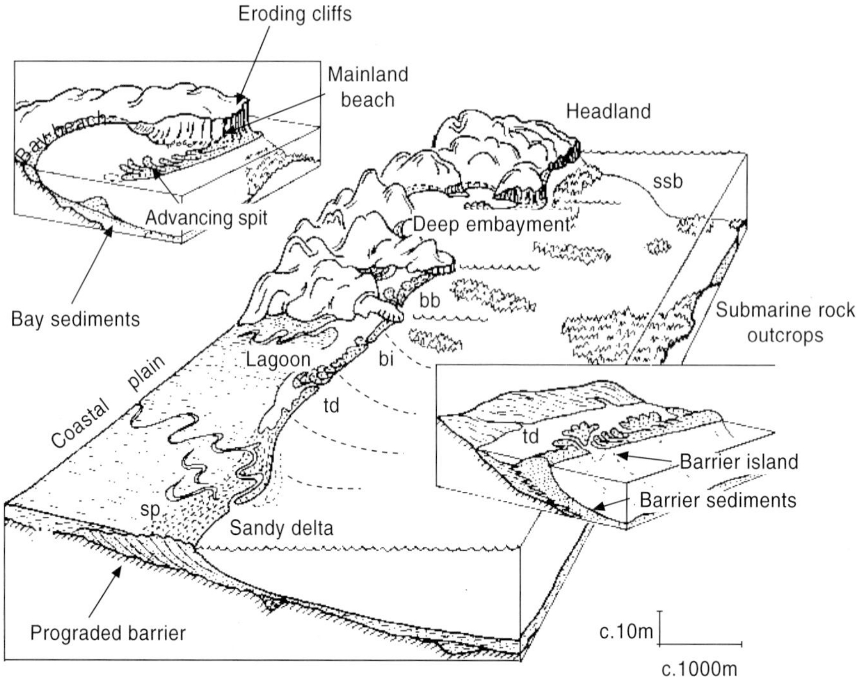


**Figure 14.1** Along a coast with variable substrate gradient and sediment supply, wave dominated barriers can range from fully attached bay and mainland beaches, to partially attached spits and bay barriers (bb), to detached barrier islands (bi) with tidal deltas (td), to progradational strandplains (sp) and inner shelf sand bodies (ssb) (source: Roy *et al.* in Carter, R W G and Woodroffe, C D (eds), *Coastal Evolution: Late Quaternary Shoreline Morphodynamics*, 1994, by permission of Cambridge University Press).

and Schwartz (1971) argued for multiple causality and supported all three modes under certain conditions.

However, it was not until the use of vibrocoring that more robust models emerged. Swift (1975) and Field and Duane (1976), utilising vibrocoring results, favoured onshore barrier migration during rising sea level. This model was further supported by Rampino and Sanders (1981) who found some barrier remnants were left behind on the shelf, and Niederoda *et al.* (1985b) who also found evidence for onshore barrier migration. Subsequently there has been a considerable body of field evidence (e.g. Leatherman, 1979b; Roy *et al.*, 1994) to support the conclusion reached by Davis (1994a) that 'most [barriers] appear to form as a result of some type of landward transport and upward accretion of sands'.

There is now common agreement that barriers form at sea level on suitable substrates by the action of waves (and tides). Most barrier models now contain these elements (e.g. Davis and Hayes, 1984; Otvos, 1985), while some also include sediment supply (e.g. Boyd *et al.*, 1992) and relative sea level (e.g. Penland *et al.*, 1985; Roy *et al.*, 1994). There is also agreement that during the last Postglacial marine transgression many low sea level barriers were eroded, truncated and reworked into new transgressive barriers, with some remnants left on the shelf (e.g. Roy *et al.*, 1994).

It now appears that most modern barriers were established during the final stages of the Postglacial marine transgression in the mid- to late Holocene and, depending on regional sea level, local sediment supply and other factors, may have since continued to retrograde (rising sea level, diminished sediment supply), remain stable, or prograde seaward (stable to falling sea level, positive sediment budget). In addition, factors including wave, tide and wind energy, tectonics and geological inheritance can all influence barrier behaviour.

## 14.3 FACTORS CONTRIBUTING TO BARRIER TYPE AND EVOLUTION

Barriers range considerably in size, type (cheniers to wave-dominated systems), position relative to the mainland (islands to attached), and stability (stable, progradational, retrogradational). As a consequence there are several factors that can contribute to the size, nature, position and stability of any barrier. While some factors are essentially universal, such as eustatic sea level rise, many, such as waves, tides, sediment supply and substrate, are regional to local in extent. Therefore, any general model of barrier evolution must consider all possible variables, so that the full range of barrier types can be predicted.

The only essential prerequisites for barrier formation are a suitable substrate, available sediment and waves. In other words, an accommodation space above the substrate, sediment to fill the space and waves to transport it. Such a combination will produce a shoreface–beach, with no dunes and no inlets. All other factors act to modify this basic model. Tides will modify the beach deposits and encourage the development of inlets and associated deposits; winds will cap the swash deposits and backbeach with aeolian sands; sea level transgression and regression will generally encourage barriers to erode or accrete; geological inheritance will modify the barrier through its impact on substrate, wave attenuation and sediment supply; while tectonism may strand (uplift) or reactivate (subsidence) barrier systems. This section briefly reviews the contribution of each of these factors to barrier size, type, position and stability.

### 14.3.1 Substrate gradient

Roy *et al.* (1994) have shown that beach deposits will only form when the substrate gradient is between 0.05° and 0.8° (Figure 1.5). On extremely shallow shelves (< 0.05°) frictional wave shoaling decreases wave energy to the point where there is insufficient energy to form a beach profile and sub-aqueous shoals and tidal deposits result. At the other extreme on steep gradients > 0.8°, sediment is moved off- rather than onshore and inner shelf sand deposits result. Beaches, and consequently barriers, are therefore developed on gradients between 0.05 and 0.8°, with Roy *et al.* (1994) predicting optimum beach development on a gradient of 0.1°.

Barrier formation is therefore dependent on an optimum window of substrate gradient and will not generally form on gradients either too low or too steep. In addition, substrate gradient will indirectly influence the level of wave and even tidal energy, and as Figure 1.5 indicates, it will also influence the width of the backbarrier depression. The depression will be wider on lower gradients, favouring the formation of barrier islands backed by wide lagoons, and non-existent on steeper gradients ( > 0.8°) producing mainland beaches and attached barriers. The position of the barrier, or distance from the mainland, is therefore a function of the difference between the substrate gradient (0.05–0.8°) and the barrier sand body gradient (usually between 0.1 and 1°).

An example of the role of gradient is provided by Penland *et al.* (1985) whose model of delta abandonment and barrier formation (Figure 14.2) goes through three gradient-dependent stages. In stage 1, erosional shoreface retreat associated with initial delta abandonment and steep gradients produces an attached overwashing barrier (erosional headland). As delta subsidence lowers the substrate gradient, a lagoon forms in the lee of the barrier forming a transgressive barrier island (stage 2, barrier island arc). Finally, as continued subsidence produces still lower gradients, wave energy is insufficient to maintain the barrier and it degrades into an inner shelf shoal (stage 3).

### 14.3.2 Sediment supply

Barrier sediments can range from mud to boulders. The size, mineralogy, quantity and rate of supply will depend on both local to regional geological and climatic factors (Section 2.3) together with the wave, tide and related processes which ultimately transport the sediment to the shoreface, and aeolian processes which build capping dune systems. Barriers are more common on sediment-rich, trailing edge coasts compared to sediment-deficient collision coasts (Section 2.1; Davies, 1980; Davis, 1994a), and are more likely to be capped by dunes on exposed windward shores (Short, 1985).

While many barriers consist of reworked shelf deposits (e.g. Davis, 1994a; Dingler and Clifton, 1994; Roy *et al.*, 1994), sediments can also originate from erosion of adjacent headlands, particularly in the formation of barrier spits (e.g. Hayes, 1994; Oretel and Kraft, 1994), erosion of drumlins (Ruz *et al.*, 1992; Carter and Orford, 1993), river sediments (Nummedal, 1983; Fitzgerald *et al.*, 1994; Hayes and Ruby, 1994), and in situ carbonate production (Short, 1987; Belperio *et al.*, 1988).

Sediments are not only essential for barrier formation, but their ongoing rate of supply will contribute to barrier type and stability (e.g. Psuty, 1988; Sherman and Bauer, 1993; Roy *et al.*, 1994). In areas with stable sea level the degree of barrier stability is largely dependent on the sediment budget, with a positive beach budget leading to shoreline progradation, a negative budget to retrogradation, while a balanced budget may tend towards shoreline stability.

### 14.3.3 Waves

Wave energy is an essential prerequisite for barrier for-

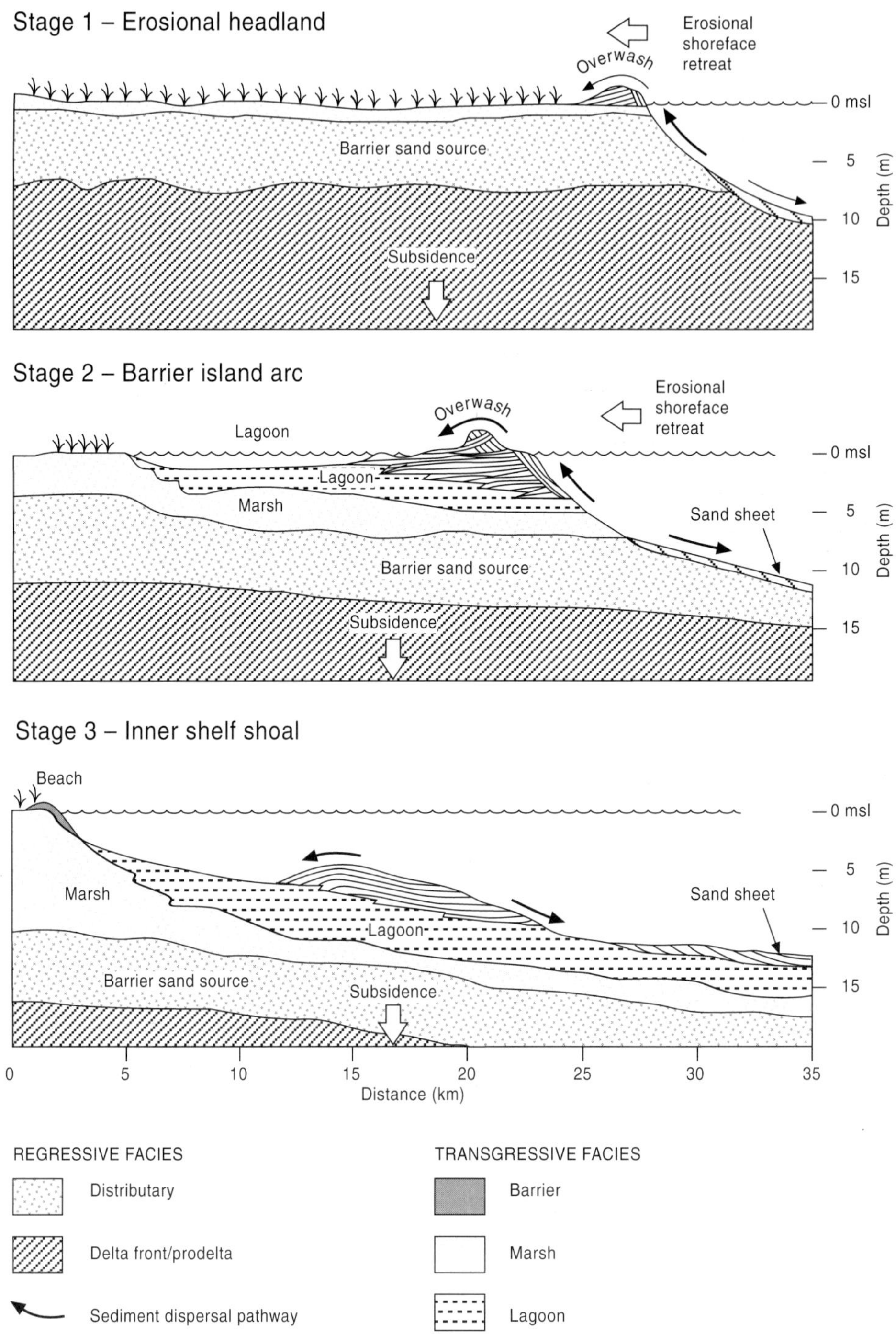


**Figure 14.2** Relative sea level rise and erosional shoreface retreat drive the barrier shoreline across a subsiding abandoned delta plain. Note that as the gradient decreases the attached barrier–headland (stage 1), becomes a barrier island (stage 2) and finally an inner shelf shoal (stage 3) (from: *Marine Geology*, Vol. 63, Penland *et al.*, 'Barrier island arcs along abandoned Mississippi River delta', pp. 197–233, 1985, with kind permission from Elsevier Science – NL, Sara Burgerhartstraat 25, 1055 KV Amsterdam, The Netherlands).

mation as it provides a major control on the rate and quantity of sediment supply and the barrier form. During and following sea level transgression, waves transport sediment from the inner shelf–shoreface to, and along the beach. The rate of onshore–longshore transport is directly related to the level of wave energy. In an analysis of Holocene onshore sediment transport on the southern Australian coast, Short (1987) found that low energy chenier deposits received sediment at a rate of $0.1\,m^3/m/a$, compared to moderate energy progradational beaches at around $3\,m^3/m/a$, and high energy retrogradational beaches averaging $10\,m^3/m/a$.

The relative contribution of wave energy also influences the shoreline configuration and barrier type, as indicated by Hayes (1979). He presented three depositional shoreline types based on the relative contribution of waves and tides, with wave-dominated shorelines consisting of long barrier islands. In contrast, mixed wave/tide shorelines consisted of shorter barriers separated by inlets (Figure 14.3). He applied this model to a wide range of barrier types in the Georgia Bight (Hayes, 1994). A more elaborate model was subsequently developed by Boyd *et al.* (1992).

### 14.3.4 Tides

Barriers do form in tideless environments. However all ocean barriers are influenced to some degree by tides and tidal currents. The role of tides in barrier form and evolution is indicated in the above models shown in Figure 14.3. Increasing tidal range has the effect of producing a greater tidal prism behind partially or fully detached barriers. The greater the tidal prism, the greater the number and/or size of tidal inlets required to exchange the prism with the ocean. Consequently, along barrier coasts, increasing tide range, relative to waves, usually leads to more inlets, and consequently shorter barriers (islands). The inlets produce substantial ebb and flood tide deltas that contain an increasing and ultimately major percentage of the barrier sediments. Furthermore, wave and tide flows through the inlets distort the barrier form, building recurved spits, while migratory inlets rework the barrier sediments to the depth of the inlet channel, leaving wave-capped tidal channel deposits as well as abandoned tidal deltas in their lee (Hayes, 1994).

### 14.3.5 Wind

Wind, like tides, is not an essential criterion for barrier formation; it is, however, essential for aeolian processes and the formation of coastal dunes which are an integral

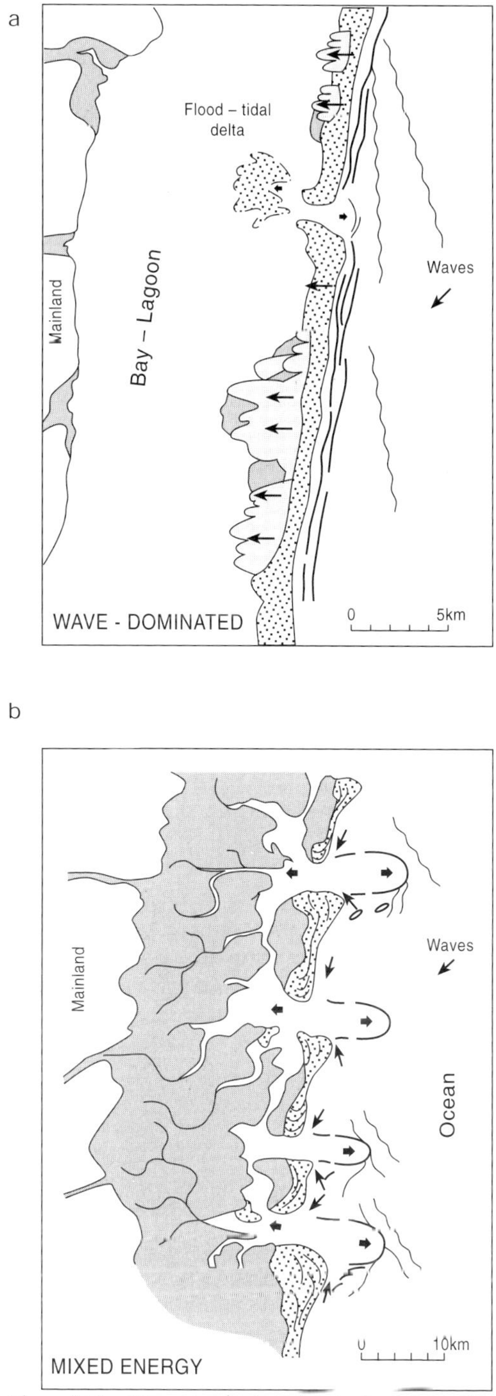


**Figure 14.3** Wave-dominated (a) and mixed wave–tide energy (b) barrier types (source: *Marine Geology*, Vol. 60, Davis and Hayes, 'What is a wave-dominated coast?', pp. 313–339, 1984, with kind permission from Elsevier Science – NL, Sara Burgerhartstraat 25, 1055 KV Amsterdam, The Netherlands).

part of many barrier systems. Whereas waves control the supply of sediment to the beach, it is wind which largely determines the quantity of the available supply of aeolian sediment to the backshore. Wind direction, velocity and frequency are therefore important secondary factors which can have major impacts on the nature and extent of many barrier systems, as indicated in Chapter 6.

Psuty (1988, 1992) proposes four dune (barrier) development scenarios which are dependent on wave supply of sediment to the beach and wind transport to the backing dunes. This model applies quite well to some barrier islands where there is longshore variation in sediment supply, erosion, accretion and stability. However, the model implies that erosional dune forms (e.g. blowouts and parabolics) will form in certain, predominantly negative sediment supply scenarios. Parabolic dunes, blowouts and transgressive dunefields are also found on many coasts experiencing a positive sediment supply. Foredunes may also be formed (rather than beach ridges) in positive sediment budget environments (Hesp, 1984b). Sherman and Bauer (1993) expanded this model to comprise nine sediment supply scenarios.

### 14.3.6 Sea level

The rate of Postglacial sea level rise between 18 000 and around 8000 years BP was relatively similar around the world (Clark *et al.*, 1978). However, since 8000 years BP there have been marked regional variations in sea level history (e.g. Finkl, 1995), which, in turn, have had a pronounced effect on gross barrier development.

Where sea level is still rising (hydro-isostatic loading, tectonic subsidence, etc.), receding barriers and eroding cliffs dominate, such as eastern Canada (Carter and Orford, 1993), eastern and Gulf coasts of the United States (Reinson, 1984; Fitzgerald *et al.*, 1994; Hayes, 1994; Morton, 1994), the Netherlands (Ehlers, 1988), western Poland (Kaszubowski, 1996) (Figure 14.4a).

In areas of falling sea level (isostatic rebound, tectonic uplift, etc.) a staircase sequence of beach ridge plains and foredune plains may exist where there is adequate sediment supply, as in Russia (Zenkovitch, 1967), Gulf of Carpentaria (Rhodes, 1982), and Norway (Fletcher *et al.*, 1993). A model of barrier progradation (or 'regression') under sea level regression is proposed by Roy, *et al.* (1994) (Figure 14.4b).

In areas of relative sea level stability for the past 6000–7000 years, a wide range of coastal barriers and landform types have developed, largely in response to variations in sediment supply, bedrock morphology and wave–beach–dune energies and interactions, as in New Zealand (Cowie, 1963; Muckersie and Shepherd, 1994), west coast USA (Hunter *et al.*, 1983; Orme, 1990) and Australian coasts (Short and Hesp, 1982; Thom *et al.*, 1992; Roy *et al.*, 1994).

The interaction of various sea level (rising/stable/falling) and substrate gradients (low to steep) scenarios has been examined by Roy *et al.* (1994; Figure 14.5). They proposed that rising sea level produces retrogradational (so-called 'transgressive') barrier islands and bay barriers on coastal plains and embayed coasts, with offshore transport to shelf sand bodies on steeper coasts. In contrast, during falling sea level, progradational ('regressive') barrier strandplains (beach, beach ridges and/or foredunes) are developed on coastal plains and embayed coasts, while the shelf deposits are reworked on steeper coasts. Finally, when sea level is stable, barrier islands persist on coastal plains; bay head and headland spit barriers and inner shelf sand bodies occur on embayed coasts; and on steeper, protruding cliffed coasts attached mainland beaches and shelf sand bodies occur.

Thus, while there is a finite range of barrier types, the predominance of a particularly strong forcing function such as rising, falling or stable sea level and/or substrate gradient can have an overriding influence on the nature and direction of barrier evolution.

### 14.3.7 Geological inheritance (bedrock and Pleistocene topography)

Coastal bedrock topography and rock lithology will influence barrier development through controls on sediment supply, substrate slope (occupation of accommodation space) and indirectly on wave–tide energy levels. Headland and bay topography in resistant rocks will favour the development of embayed bay-head and mid-bay beaches and barriers, as along much of the south-east Australian coast (Thom *et al.*, 1992). In contrast, highly erosional 'soft' rocks and sediments are likely to lead to the development of curvilinear (e.g. Nova Scotia – Forbes *et al.*, 1995) or straight (e.g. north Alaska – Short, 1979a; Delaware – Belknap and Kraft, 1985; south-east USA – Riggs *et al.*, 1995; western Poland – Kaszubowski, 1996) beaches and barriers. Submerged topography will also influence transgressive sedimentation and barrier form and evolution (e.g. Otvos, 1985; Delware – Belknap and Kraft, 1985; west Florida – Evans *et al.*, 1985).

Furthermore Pliocene and Pleistocene sea level oscillations have resulted in prior barrier formation at or close to present sea level approximately every 100 000 years. These deposits have varying rates of full or partial

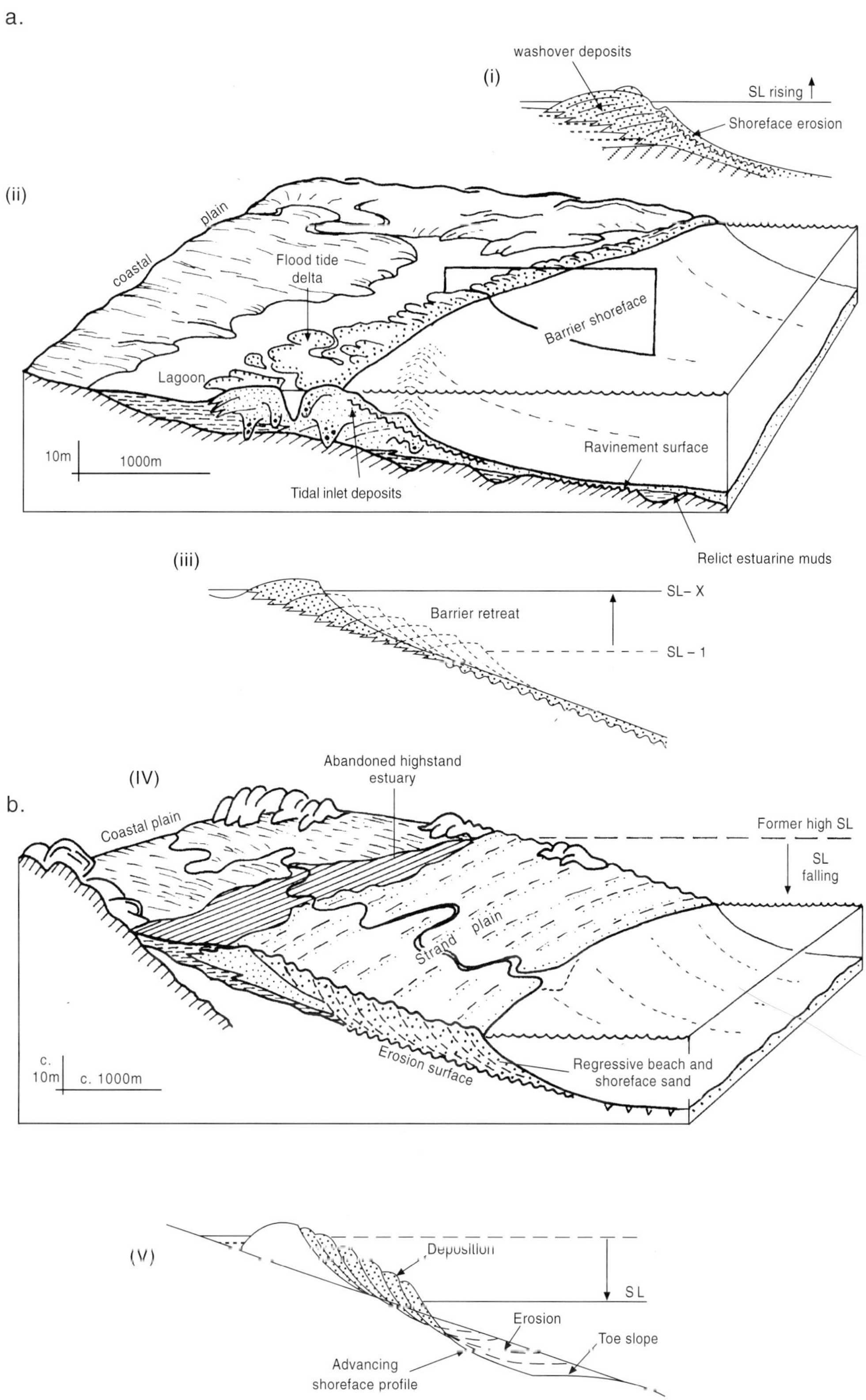


**Figure 14.4** Schematic models of (a) retrogradational 'transgressive' barrier during sea level rise; and (b) progradational 'regressive' barrier and strandplain formation accompanying sea level fall (source: Roy *et al.* in Carter, R W G and Woodroffe, C D (eds), *Coastal Evolution: Late Quaternary Shoreline Morphodynamics*, 1994, by permission of Cambridge University Press).

Local substrates steepening →

| Sea level | Coastal plain coasts | Embayed coasts | Protruding and cliffed coasts |
|---|---|---|---|
| rising | transgressive barrier islands and bay barriers | | transgressive sand sheets erosion surfaces |
| falling | regressive barrier strand plains | | erosional reworking |
| stable | barrier islands shelf sand ridges | bay barriers headland spits Inner shelf sand sheets | mainland beaches shelf sand bodies |

**Figure 14.5** Relationship between sea level stability, substrate gradient and the occurrence of various types of wave-formed coastal deposits (source: Roy *et al.* in Carter, R W G and Woodroffe, C D (eds), *Coastal Evolution: Late Quaternary Shoreline Morphodynamics*, 1994, by permission of Cambridge University Press).

**Figure 14.6** Section through a 40 m high Pleistocene barrier system, Beachport, South Australia. The entire section consists of aeolian calcarenite, with beach deposits (not visible) located at the base of the section. In the far distance is the interbarrier depression (Lake George) and Holocene barrier. (A D Short).

preservation and of vertical stability. Where they are preserved above present sea level they remain as relic barrier forms, as along the south-east coast of South Australia (Figure 14.6; Schwebel, 1984), while if preserved at or near to present sea level, they occupy potential accommodation space and will necessarily impact on the evolution and form of subsequent Holocene barrier deposits (e.g. Dingler and Clifton, 1994; Roy *et al.*, 1994).

In carbonate-dominated regions of the mid- to lower latitudes, cementation of former barriers may lead to substantial modification of Holocene coastal processes and subsequent Holocene beach and barrier formation. Cemented barriers can remain as reefs, rocks and cliffs (Figure 14.7), as well as occupying accommodation space. The former tend to modify wave processes leading to more irregular and more variable levels of wave energy along the shoreline, with, at times, barriers alternating with cliffs (lithified former barriers), while the latter act to disrupt sediment transport paths. Notable examples are the dunerock coasts of southern and western Australia (Short and Hesp, 1982; Short *et al.*, 1986) and South Africa (Tinley, 1985), and the beachrock coast of north-east Brazil (Suhayda *et al.*, 1977).

### 14.3.8 Tectonism

Tectonically active coasts may be constantly or periodically adjusting to new base levels as the land rises or falls (e.g. Japan, New Zealand, island arcs). Co-seismic uplift along parts of the New Zealand coast has resulted in the formation of a staircase of marine terraces extending back to the middle to early Pleistocene (e.g. Pillans, 1990, 1991, 1994). Holocene storm ridge plains, beach ridge plains, beach–foredune plains, and foredune plains also form staircase sequences in some very tectonically active regions (e.g. Zenkovitch, 1967; Stevens, 1973; Ota *et al.*, 1988, 1991; Ota and Kaizuka, 1991; Berryman *et al.*, 1992a,b; McSaveney *et al.*, 1993; Pillans and Huber, 1995). In these areas it can be virtually impossible for backbarrier and lagoonal landforms to develop, particularly if the uplift rate is rapid. Hence barriers may extend from the shore landward and up-

**Figure 14.7** View along the partially submerged Robe Range, South Australia. The lithified Pleistocene barrier produces a spatial variation in the deep water wave energy resulting in low energy bays, such as Nora Creina (foreground) backed by a single stable foredune, adjacent to higher energy beaches backed by active transgressive dunefields (background) (A D Short).

**Figure 14.8** (a) Reindeer Island, a low receding barrier island being overwashed during a storm surge, Beaufort Sea, north Alaska. (b) A stationary shingle barrier and infilled lagoon at Barmouth, Wales. (c) A prograding barrier capped by low foredunes, Aracaju, Brazil (A D Short).

wards, with little, sometimes no, ponding of drainage or formation of wetlands, swamps or lagoons.

In addition, the sediment supply from alongshore and offshore may be adjusted as base level changes, and formerly unavailable, deeper lying sediments may be tectonically lifted into the active shoreface zone.

## 14.4 BARRIER MORPHODYNAMICS AND TYPES

While barriers occupy a significant portion of the world's coasts (Davis, 1994a) there has, until recently, been little written about their evolution and types. Most texts have tended to treat beaches and dunes separately, with if anything only a passing reference to barriers, the integral of both beaches, dunes and more. The term barrier is not listed in the *Encyclopedia of Beaches and Coastal Environments* (Schwartz, 1982).

This situation partly reflects the poor state of knowledge of barriers prior to the commencement of barrier vibrocoring in the 1970s. It is further complicated by the wide range of variables that contributes to barrier formation, as discussed in Section 14.3, and to the wide range of environmental settings in which barriers occur (Figure 14.8). Furthermore the dominance of barrier islands in the literature, as opposed to barriers in general, has led to confusion and conflict in the definition of barriers and use of the term barrier. Some would argue that it should be applied only to barrier islands (e.g. Otvos, 1982a). However, while it is true that the term barrier was originally applied to barriers backed by a lagoon, the lagoon is a product of the substrate gradient (Figure 1.5) and not the formative barrier processes.

Subsequent work outside of east coast USA has shown that the same processes that form barrier islands form barriers on steeper gradients, often without lagoons (e.g. Roy *et al.*, 1994). A number of terms have been proposed to describe beaches in such situations, such as bayhead beaches and even bayhead barriers (Schwartz, 1982). However, on long straight or non-embayed coasts these are not appropriate and the term mainland beach (Roy *et al.*, 1994) and 'fringing' barrier (Carter and Orford, 1993) have been used.

Using the definition in Section 14.1, this book takes the approach that most beaches are part of a barrier system, with the dimensions of the barrier ranging from a narrow beach deposit to massive beach dune systems. The barrier may or may not be bordered by inlets and backed by lagoonal/swamp deposits. The major prerequisites are a beach and backbeach system. The latter may include, however, wave-deposited washover and backbarrier flats, associated tidal inlet deposits, and all beach-derived aeolian dune systems including clifftop dunes.

An overview of the barrier literature indicates that there are three basic marine settings within which most barriers have been formed. These are transgressive, regressive and stationary (stillstand) settings in which sea level is respectively rising, falling or approximately stationary. However while transgressive barriers may have relatively similar basal (sub-aqueous) stratigraphies, their sub-aerial landforms may include overwash terraces, foredunes, beach ridges, parabolic dunes or transgressive dunefields. Likewise the sub-aerial landforms of regressive barriers may also be composed of any of the landforms listed above. The same is true during stationary sea level (stillstand) conditions, where a wide range of barriers may be formed including so-called stationary barriers (i.e. typically comprising single, complex foredunes; e.g. Thom *et al.*, 1992), receding barriers (i.e. erosional barriers, in this case not receding due to sea level rise; e.g. Roy *et al.*, 1994), prograded and transgressive dune barriers. The sub-aerial barrier morphology is therefore not necessarily a good indication of the primary marine sea level setting.

Terminological problems have resulted because some barriers have been classified according to their (principally marine) stratigraphic nature (e.g. regressive or transgressive), while the stillstand or stationary barriers have been classified according to their marine stratigraphic framework and their sub-aerial morphology (e.g. figure 4.25 in Roy *et al.*, 1994).

In the following, barriers are first and more strictly classified according to their regressive, stationary or transgressive marine (sea level) setting, then their progradational, stationary or retrogradational nature, and finally according to their broadscale sub-aerial morphology.

### 14.4.1 Transgressive (rising) sea level and barrier types

A wide range of barrier types may be formed during transgressive sea level conditions. Depending on the rate of sea level rise, sediment supply and substrate slope, progradational and retrogradational barriers of various sizes and types may be formed. These barriers may display a range of surficial dune types and landforms.

#### *14.4.1.1 Transgressive sea level, prograded barriers and strandplains*

Much of the 'classic' early research on barriers was conducted on prograded foredune, chenier and beach ridge dominated barriers and strandplains formed under conditions of a rising (transgressive) sea level. It is something of an enigma that barriers have prograded during times in the last 7000 years in areas experiencing a sea level rise. However, most coasts behaving in this manner have developed and are developing in areas where marine sediments were presumably being swept landwards at a greater rate than could be removed by a rising sea, adjacent to a major sediment supply, or at the downdrift termini of significant (sometimes quite localised) littoral transport systems. Some of these barriers have been termed 'regressive barriers' even though they have formed under transgressive sea level conditions.

Examples of prograded barriers forming during sea level transgression include the Nayarit, Mexico coast (Curray *et al.*, 1969), the South Carolina coast (Nummedal, 1983), the Dutch coast (Jelgersma *et al.*, 1970; Klijn, 1990), and other coasts (Hayes and Ruby, 1994; Martinez *et al.*, 1995). Some of these barriers display complex stratigraphies comprising an initial progradational phase followed by an erosional phase.

Nummedal (1983) notes that the Holocene evolution of the Louisiana coast typifies the situation where coastal plain progradation has taken place at points where there is active delta lobe sedimentation, and coastal plain retrogradation has taken place adjacent to areas of abandoned delta lobes. The same rising (transgressive) sea level applies to each case. Penland *et al.* (1981) have modelled this situation and shown that barriers and barrier islands proceed through a sequence of flanking barriers and erosional headlands, through retrogradational ('transgressive') barrier island arcs to sub-aqueous shoals with time and progressive abandonment of the delta lobes (see Section 14.4.1.2).

Small to large scale transgressive dunefields (see Chapter 6) may be formed on barriers during sea level transgression. (Bakker *et al.*, 1990; Christiansen *et al.*, 1990; Thom *et al.*, 1992). Where such dunefields occur on barrier islands the distal and sometimes the middle parts of these barriers experience short to medium term progradation (tens to hundreds of years), but long term recession or retrogradation (e.g. Cape Cod; Nag's Head, Outer Banks, North Carolina – Goldsmith, 1978; Padre Island, Texas – Weise and White, 1991; Currituck Spit, Virginia – Goldsmith *et al.*, 1977).

#### *14.4.1.2 Transgressive sea level, retrogradational barriers/barrier islands*

Barrier retrogradation (also sometimes termed 'transgression') implies landward migration of the shoreline and barrier form. Barrier retrogradation in response to rising sea level is invoked as a major mechanism for barrier formation, with barriers originating on the shelf and moving shoreward with the rising sea level, stabilising at or near their present position as sea level stabilised (see Section 14.3.6). This mechanism requires a 'rollover' of beach and shoreface sediments. Sediments are continually reworked from the shoreface and eroded shoreline deposits are transported to and beyond the shoreline by waves to build berms, beaches, overwash and beach–barrier deposits; and by tides to build tidal inlets, recurved spits and flood tide deltas (Figure 14.4a).

Retrogradation of barrier islands is well documented and occurs particularly on coasts experiencing relative sea level rise. They are reported on the North Carolina coast (Monslow and Heron, 1994; Riggs *et al.*, 1995), Louisiana coast (McBride *et al.*, 1995), Florida (Davis, 1994b), New Jersey (Oretel and Kraft, 1994), Georgia Bight (Hayes, 1994), New England (Fitzgerald *et al.*, 1994) and Alaska (Short, 1979a; Hayes and Ruby, 1994). In most cases the retrogradation involves dune erosion and retreat and barrier washover, which causes the typically narrow barrier to migrate landward over existing backbarrier estuarine–lagoonal deposits. The eroded sequences may be exposed on the shoreline and across the shoreface (Figure 14.9).

On wave-dominated transgressive coasts the barrier islands are long, inlets few and overwash processes dominate. In mixed wave–tide environments, barriers shorten, inlets increase in number and size, and tidal processes increasingly dominate the coastal deposits as landward transfer of sediments via flood tide deltas and lateral spit development takes place (Hayes, 1994). Assemblages of transgressive wave–tide features have been described on the Nova Scotia coast (Forbes *et al.*, 1995), Louisiana coast (McBride *et al.*, 1995), New Jersey coast (Oretel and Kraft, 1994) and Georgia Bight (Hayes, 1994).

Barrier islands vary in morphology and stratigraphic framework as a function of local sediment supply and rate of sea level rise or land subsidence (Sherman and Bauer, 1993). The distal margins of spits and barrier islands commonly prograde via successive foredune development while the middle portions comprise eroding bluffs and/or various erosional dunes (and retrogradational shorelines), and the updrift proximal portions comprise overwash fans and terraces (e.g. Hayes, 1979, 1994; Leatherman, 1979a,b,c; Evans *et al.*, 1985; Oretel, 1985; Penland *et al.*, 1985; Ritchie and Penland, 1990; Psuty, 1992; Davis, 1994a; Fitzgerald *et al.*, 1994; Monslow and Heron, 1994; Oretel and Kraft, 1994). Hequette and Ruz (1991) found that barriers in the Canadian Beaufort Sea were retrograding in the long term, but sections of barriers and spits with the highest potential sediment supply retreated little over 30 to 40 years and foredunes were formed in these areas. The retreat rate increased as wave energy increased (cf. Short, 1979a; Forbes *et al.*, 1995). An overview of the Louisiana, Mississippi and Georgia coastlines allowed McBride *et al.* (1995) to identify eight megascale response types of the barrier shorelines. The drumstick

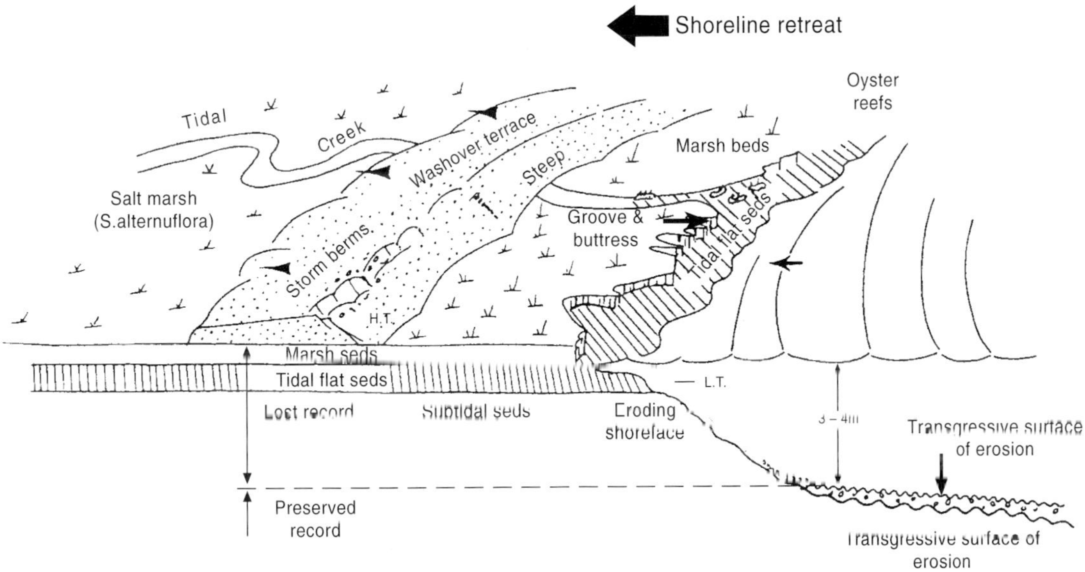


**Figure 14.9** Sketch of transgressive, retrogradational barrier island on Cape Romain, South Carolina (source: modified from Hayes, 1994).

and spit barriers can advance (prograde), retreat (retrograde) by overwash, rollover and breakup, maintain dynamic equilibrium (stable), or move laterally and rotate. Many islands have a combination of responses. Similar models have been proposed by Penland *et al.* (1985), Hayes (1994), Monslow and Heron (1994) and Oretel and Kraft (1994).

Ritchie and Penland (1990) demonstrated particularly well the longshore variation and spatial complexity of barrier islands forming in a relatively sediment-poor, transgressive sea level, retrogradational environment. Figure 14.10, shows how along a single retrograding barrier island variation in local–longshore sediment supply and rate of erosion can result in a range of barrier segments from eroding overwash sheets, to temporarily stable foredunes, to a prograding foredune plain with recurved spits.

#### *14.4.1.3 Transgressive sea level, retrogradational coastal plains*

On moderate gradient shelves, where transgressive barrier deposits have attached to the mainland, they may continue over the mainland deposits, as observed on the Texas coast (transgressive beaches; Morton, 1994), Georgia Bight (welded barriers and Pleistocene mainland-beach; Hayes, 1994) and western Poland (Kaszubowski, 1996). In south-east Australia, Thom *et al.* (1978) and Roy *et al.* (1980, 1994) report on a number of sites where mainland barriers are eroding into both Holocene and Pleistocene deposits, while across southern Australia there are many instances of Holocene barriers blanketing partially eroded Pleistocene barriers (Figure 14.10; Short *et al.*, 1986). Dune types are typically eroding foredunes and beach ridges, and scattered coppice (or nebkha) dune topography, in addition to various overwash built landforms. In extreme cases, on steep shorefaces the barriers have been completely eroded leaving as evidence either clifftop dunes and/or inner shelf regressive sand deposits (Figure 14.11).

Small scale to massive Pleistocene and Holocene dune transgression has occurred on retrogradational coasts exposed to both high waves and strong onshore winds, and an adequate to large sediment supply. On these coasts, waves are capable of transporting large quantities of sand and building substantial beach–barrier systems, even during rising sea level. Strong winds act to transport significant quantities of this sediment landward, building active dune systems extending up to several kilometres inland.

Large and moderate size parabolic and transgressive dunefields may therefore be found on coasts which have either been erosional due to relative sea level rise and transgressive for most, if not all, of their formation history, or have become erosional in more recent geological or historical times for other reasons (e.g. reduc-

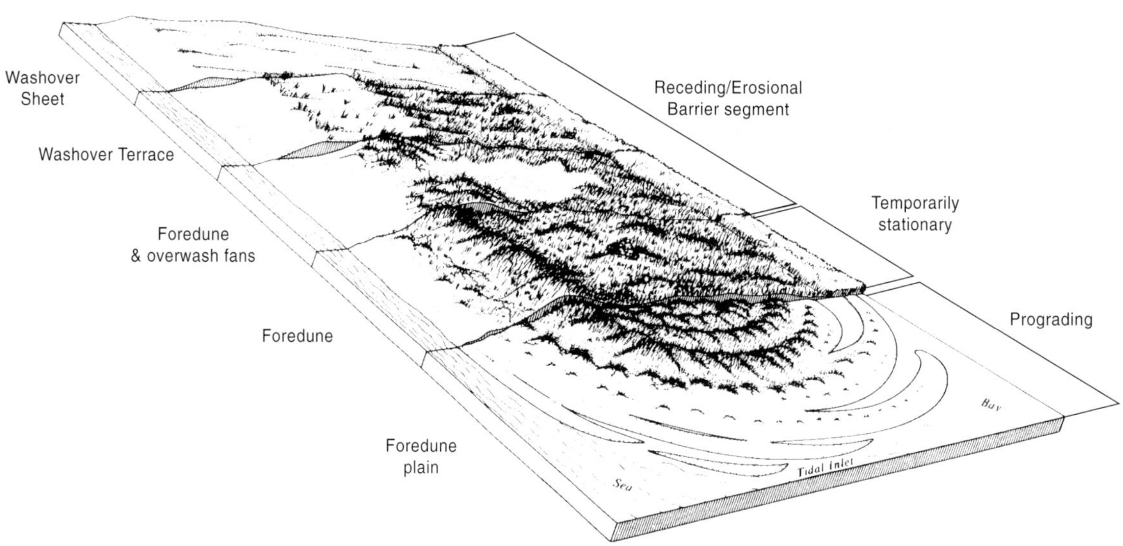


**Figure 14.10** Schematic sections along a south Louisiana barrier island, illustrating the variation in barrier type and dynamics (receding to prograding) based on local changes in sediment supply (deficient to surplus) and rate of erosion (modified from Ritchie and Penland, 1990).

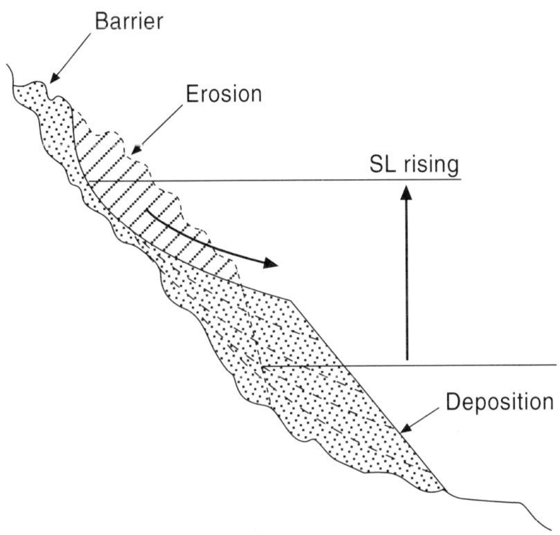


**Figure 14.11** Barrier erosion during rising sea level on a steep substrate gradient, with the eroded sediment deposited as inner shelf sand units (source: Roy *et al.* in Carter, R W G and Woodroffe, C D (eds), *Coastal Evolution: Late Quaternary Shoreline Morphodynamics*, 1994, by permission of Cambridge University Press).

tion in sediment supply during late stillstand conditions). Many of these barrier types are also receding barriers in the terminology of Thom *et al.* (1978) and Roy *et al.* (1994).

In some cases, long term shoreline erosion throughout the phase of reduced rate of sea level rise in the last 6000–7000 years has led to the continued landward migration of the shoreline, the development of dunefields and their migration over estuarine, lagoonal, backbarrier and other facies. Figure 14.12(iv) indicates a relatively large scale dunefield (Hesp and Hilton, 1996), but the aeolian units could be of various scales and comprise various dune types. Some of the larger, more slowly eroding USA Gulf and east coast barrier islands would also fit this model (e.g. Cape Cod, Massachusetts – Messinger, 1958; Goldsmith, 1975; Leatherman, 1979a; Padre Island – Weise and White, 1991; Currituck Spit, Virginia/North Carolina – Hennigar, 1979).

The USA Gulf and east coasts experience conditions of rising sea level, subsidence, in some cases a limited sediment supply, various levels of littoral drift and also occasional hurricanes. While foredune scarping and overwash development are common as transgression takes place (see above), small to moderate scale dunefields (blowouts, parabolic dunes and mobile dunes) have developed in some areas (Figure 14.12(vi)), and the stratigraphic models of Kraft (1971), Kraft *et al.* (1979) and others probably apply.

Some of the USA regional settings comprise Holocene lagoons and barriers lying seawards of Pleistocene barriers (e.g. Kraft *et al.*, 1979). The lagoon width may vary considerably. In some cases the Holocene dunes and barrier may be emplaced over Pleistocene barrier remnants or have gradually migrated to that position over the last 7000 years or so (Figure 14.12(v); Hesp and Hilton, 1996).

On the US west coast Dingler and Clifton (1994) classified all 125 west coast barriers. They found that many originated at a lower sea level when the coastline was more continuous, and were driven landward by rising sea level and a high energy wave climate to their present position on a rugged, cliffed coast, with most barriers occupying river mouths and embayments. They identified two basic barrier types: the barrier spit tied to a headland, with an inlet at the distal end, and the bay barrier. They also found that Pleistocene barriers occupied much of the accommodation space and were a reason for the limited Holocene barrier development.

### *14.4.1.4 Transgressive sea level, retrogradational attached barriers*

Cliffed coasts are common in many parts of the world. They may be composed of rocks of varying lithologies and ages. Active tectonic coasts may be characterised by geologically recent cliffs formed in Holocene and late Pleistocene marine and aeolian sediments (e.g. Japan and New Zealand; Pillans, 1990). Carbonate-dominated coasts often display Pleistocene sequences which have been lithified during low sea level events and uplifted to various degrees (e.g. South Australia, Western Australia; Cook *et al.*, 1977; Sprigg, 1979; Hesp and Chape, 1984; Short and Hesp, 1984; Short, 1988b). Some are cliffed, and in many cases beaches and dunes have been or are being formed at the base of and over such cliffs (Figure 14.13). Carter and Orford (1993) classify beaches and barriers at the base of cliffs as fringing barriers. Here we refer to such beaches and barriers as attached barriers.

Extensive coastal dune deposits, often sited well above present sea level, are documented along exposed sections of the north, east, west and south Australian coasts (Short and Hesp, 1982; Pye and Bowman, 1984; Schwebel, 1984; Short *et al.*, 1986; Hesp and Thom, 1990; Lees *et al.*, 1995; Cullen and Pemberton, 1997), South Africa (Tinley, 1985) and west coast USA (Orme, 1990). These deposits formed during rising sea level conditions between 5000 and 12 000 years BP. They indicate that, given sufficient shelf sediment sources and high wave energy, large quantities ( > 10 $m^3$/m/a, Short,

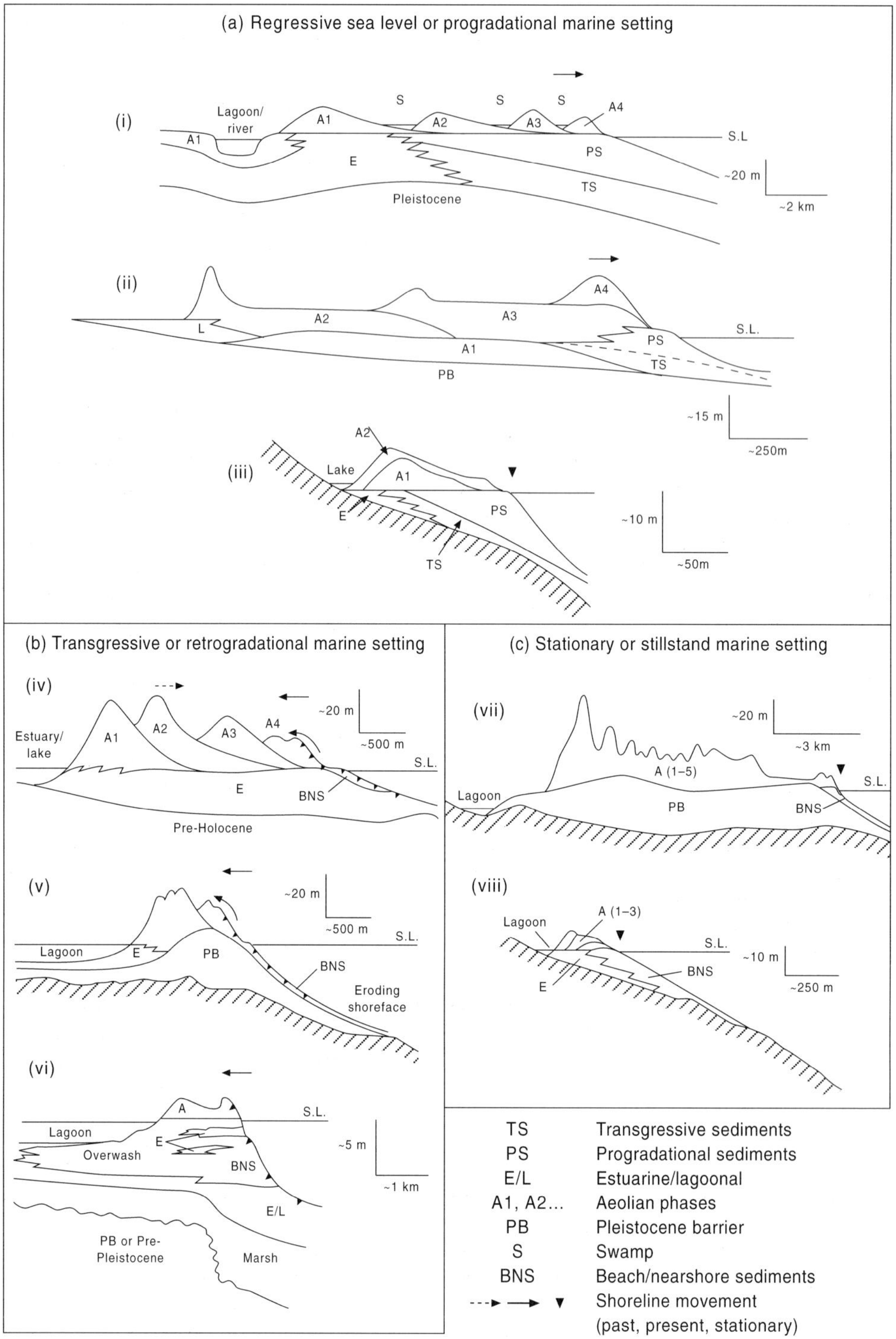


**Figure 14.12** Examples of dunefield-dominated barrier types. (a) Regressive sea level or progradational setting: (i) prograding, transgressive dunefield phases; (ii) limited–moderate progradation, pre-6000 BP (unit A1) and post-6000 BP transgressive dunefield phases; (iii) limited progradation, small aeolian dune phases or aggrading foredune. (b) Transgressive sea level or retrogradational setting: (iv) episodic dune emplacement; (v) small dunefield or large foredune; (vi) low overwashed beach and/or foredune. (c) Stationary (stillstand) sea level or stationary marine setting: (vii) large scale transgressive dunefield; (viii) small scale dunefield. Note that sea level may also be relatively stationary in both (a) and (b), and the coastline progrades or retrogrades for other reasons (e.g. sediment supply variations).

1987) of sand can be transported onshore during a marine transgression, and, together with strong onshore winds, can build extensive coastal dunes in advance of a retreating beach/barrier.

This process was termed *primary dune transgression* by Davies (1980) and is placed in the context of southern Australia barrier evolution by Short (1988a,b). In many instances, the beach/barriers were subsequently eroded, particularly where backed by cliffs and steep shoreface/shelf gradients. Proxy evidence for the former existence of eroded beaches remains in the form of massive late Pleistocene and early Holocene dune transgression (Cook *et al.*, 1977; Pye and Bowman, 1984; Hesp and Thom, 1990; Shulmeister and Lees, 1992), clifftop dunes on top of cliffs up to 150 m high, which occupy several hundred kilometres of southern Australia (Short, 1988a,b; Hesp, 1993; Thom *et al.*, 1994), and inner shelf, mid-Holocene sand bodies, which have been formed off some headland environments (often capped by clifftop dunes), along the south-east Australia coast (Ferland, 1993; Roy *et al.*, 1994).

In Figure 14.13a types (i) and (ii) comprise a type of transgressive dunefield barrier, in the first case, dunes that have been emplaced at the base or near a cliffed or steeply rising coast, and in the second case, one dominated by climbing dunes (Figure 14.14). The dunes may have been initiated and formed during the latter stages of the Postglacial marine transgression (PMT), or at times in the last 7000 years (when sea level may have stabilised, or fallen or risen a few metres; Jennings, 1967; Pye and Bowman, 1984; Hesp, 1993). Coasts with a high sediment supply, high wind energy and dissipative beaches will potentially have the largest dunefields. Climbing dune systems may be linked to falling dune systems where headland bypass dunefields occur (Tinley, 1985; Hellstrom and Lubke, 1993). Locally drainage may be impounded at the rear of the dunefield. Type (iii) indicates a barrier where the majority of sediment has been transferred to the clifftop, but a small beach/dunefield or foredune remains at the base.

Type (iv) attached barriers (Figure 14.13a) occur where the climbing dune ramp was removed once sea level attained its present position, or where occasional storms erode the beach and remove the dune base. The clifftop dunes may be vegetated and inactive, or periodically activated by cliff retreat, destabilisation and subsequent wind erosion. Foredunes, echo dunes or small climbing dunes may be a permanent feature of the cliff base/backshore area, or be ephemeral features formed between storm events. Gullies in the cliff may provide feeder points within which dune and beach sands are driven up to the clifftop (Tsoar and Blumberg, 1991; Jackson and Nevin, 1992).

Types (v) and (vi) barriers (Figure 14.13a) represent an evolutionary step from types (i), (ii) and (iii) where sea level rises sufficiently to erode all the climbing dune portion of the dune complex, or the sediment supply is diminished. A beach may subsequently be formed once sea level stabilises (iv), and sediment may be added to it by cliff and shore platform erosion. Alternatively, if there is little sediment available, or the height of sea level relative to the cliff morphology is such that it is impossible to construct a beach, then only a clifftop dune is left (v) (Tsoar and Blumberg, 1991). Several examples of these occur on the South and Western Australian coast. In some cases only a beach is constructed (vi) where either the sediment supply has been and is limited, or where the sediment emplaced during the rising PMT has largely been removed seawards.

Finally, when the clifftop dunes migrate across headlands and spill onto the downwind shore, aeolian headland bypassing occurs. Such systems many be active (Figure 14.13b), episodic or stable (Figures 14.14 and 14.15). Aeolian headland bypassing occurs on numerous suitable headlands in South Africa (Tinley, 1985) and parts of southern Australia (Short *et al.*, 1986; Short, 1996). Some authors would argue that some of these barrier types are not barriers at all, but dune and cliff complexes which fall outside the usual definition of barrier. By applying the term 'attached barrier' to these, it is possible to see them as an evolutionary continuum of types (e.g. Figure 14.13a types (i) to (vi)). While it may be difficult to envisage all beaches at the base of cliffs, or all stranded clifftop dunes as barriers, they may well have been initiated as barriers and/or evolved from barriers. In addition, there exist cliffed peninsulas which comprise multiple overlapping sequences of Pleistocene and Holocene dunes which enclose lagoons and estuaries (often with preserved sequences of Pleistocene Interglacial sediments) (Figure 14.13c). Such landforms satisfy most of the criteria (Section 14.1) to be classified as barriers with one addition – they are cliffed – and with one detraction – some have no marine facies present in their final evolutionary form (i.e. type (ivb)). There are many examples of these cliffed barriers on the south-western coast of Western Australia (Hesp and Chape, 1984), across high energy sections of the South Australian coast (Short *et al.*, 1986; Short and Fotheringham, 1986), in parts of western and eastern Victoria (Short, 1996), and on exposed headlands along the NSW coast (Thom *et al.*, 1994).

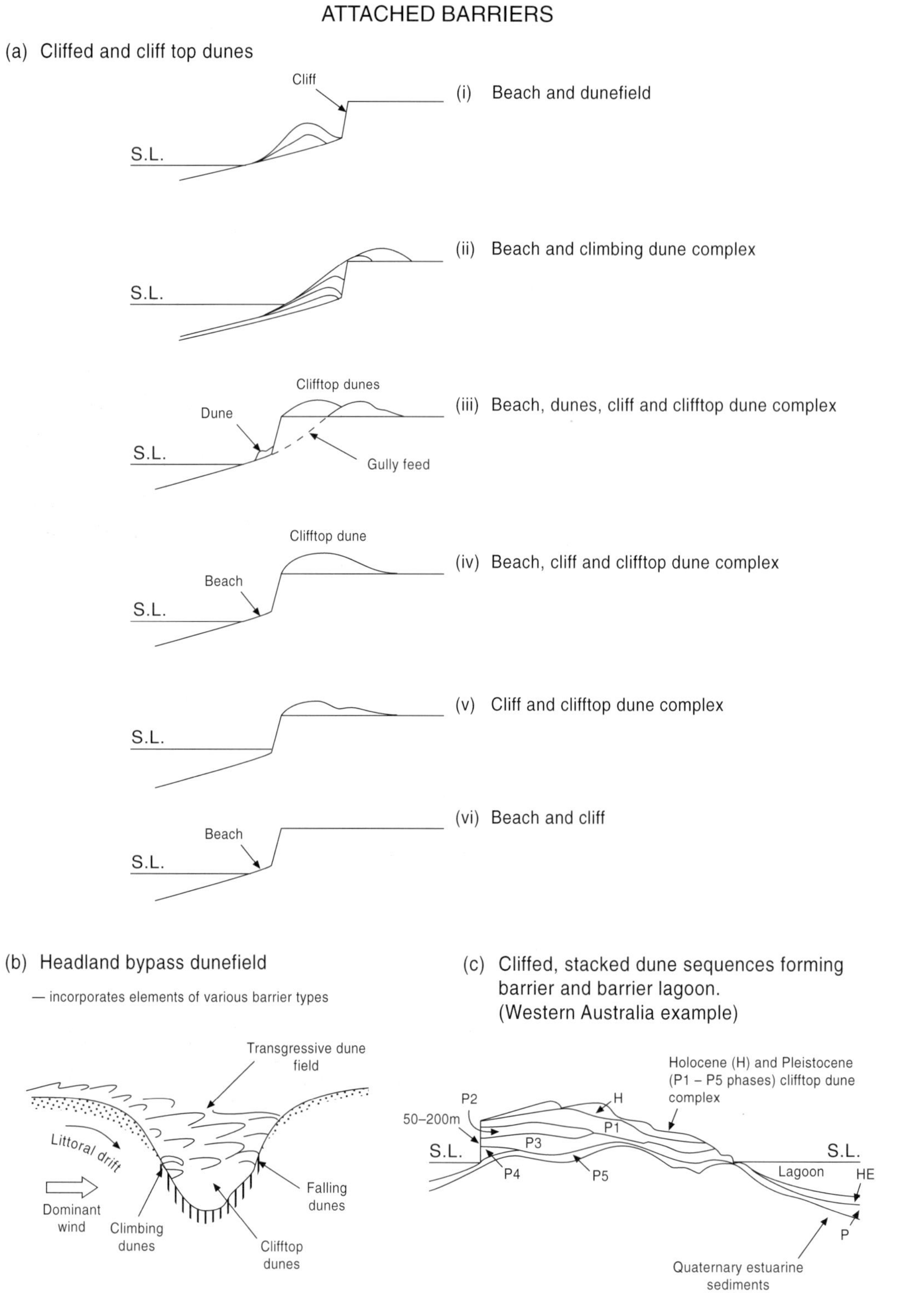


**Figure 14.13** Types of attached barriers. (a) On cliffed coasts: (i) barrier dunefield formed seaward of a cliff, (ii) with sand ramps and clifftop dunes; (iii) eroding barrier with active and inactive clifftop dunes; (iv) eroded barrier with remnant beach, cliffs and clifftop dunes; (v) eroded barrier with only cliff and clifftop dunes remaining; (vi) eroded barrier with only beach remaining. (b) Headland bypass dunefield fed by a bay to attached barrier. (c) Cross-section of a cliffed stacked Pleistocene and Holocene (lithified) dune sequence forming an attached barrier, common along the South and Western Australian coast. See text for full explanation.

**Figure 14.14** Attached barriers with active sand ramps and clifftop dunes moving over older vegetated clifftop dunes (foreground): (a) southern Western Australia, and (b) central Western Australia (A D Short).

**Figure 14.15** (a) Active aeolian headland bypassing at Injidup Bay, south Western Australia (A D Short); and (b) vegetated, clifftop parabolic dunes at Wilson Inlet, Western Australia. The calcarenite cliffs (foreground) consist of vertically stacked sequence of trangressive dunes, with a late Pleistocene–early Holocene dunefield on the surface (cf. Figure 14.13c) (S Chape).

### 14.4.2 Regressive (falling) sea level, progradational barriers

Shoreline and barrier progradation can be a product of falling sea level and/or a positive sand supply. Barriers composed of cheniers, beach ridges and foredune ridge plains are common expressions of prograding (also termed 'regressive') coasts. Falling sea level typically results in the stranding of barrier, beach and nearshore deposits, as the reverse Bruun Rule operates (Figure 14.16a). The most common cause of continuous sea level fall in the last several thousand years is glacio-isostatic rebound. However, in the long term eustatic changes associated with glacials will also result in the development of regressive coasts (Pirazzoli, 1991).

Glacio-isostatic and hydro-isostatic processes have been particularly operative on paraglacial and glacial coasts giving rise to barriers which range from extensive beach ridge complexes (Fletcher *et al.*, 1993), sandy beaches and aeolian dune systems, to single storm ridges (Carter and Orford, 1993) and overwash flats depending on the geological setting and sediment supply (Zenkovich, 1967; McCann, 1979; Nummedal *et al.*, 1987; Forbes *et al.*, 1990; Ruz *et al.*, 1992; Hill *et al.*, 1994; Forbes and Syvitski, 1994).

Relative sea level regression also occurs as a function of tectonic movements and hydro-isostatic adjustments. Staircase and more subtle, gently sloping barriers comprising cheniers (Rhodes, 1982), beach ridges (Belperio, 1983; Dominguez *et al.*, 1987; Short *et al.*, 1989; Fletcher *et al.*, 1993) and foredune sequences are found on such coasts (e.g. Pirazzoli, 1991; Berryman *et al.*, 1992b; Carter and Orford, 1993; Pillans and Huber, 1995; Angulo and Lessa, 1997). On the macro-tidal, tropical Queensland, Australian coast, Masselink and Lessa (1995) cored two barrier sequences. They found evidence of barrier retrogradation during the early Holocene sea level rise, followed by shoreline progradation, leaving the barrier capped by a shelf-supplied beach to foredune ridge sequence. They hypothesised that the shoreline progradation had been assisted by an isostatic induced 2 m relative fall in sea level.

Dunefields may be formed during shoreline/marine regression where there is a large supply of sediment usually coupled with high wave and wind energy and/or dissipative beaches (Figure 14.12a). A lack or paucity of vegetation to stabilise the sands, or disruption to the vegetation cover, is usually an added factor in producing active migrating dunes. Both transgressive dunefields and parabolic dunefields, or combinations of the two, may be formed. The Manawatu coast in New Zealand is the biggest Holocene parabolic dunefield in the country and one of the largest in the world. It extends up to 19 km inland from the coast, and has formed while the coast prograded around 4 km (Shepherd, 1987) possibly under conditions of slight sea level fall. Figure 14.12(i) illustrates this setting schematically. The extent of the transgressive and regressive sediment facies will vary considerably from place to place. In semi-arid environments, salt flats, saline lagoons and parent rock landscapes may back the dunefields. In arid environments, the dunefields may merge imperceptibly into desert dunescapes, overlie gravel stony plains or front extensive sabkhas.

### 14.4.3 Stationary (stillstand) marine setting and prograding barriers

While there is some debate about how 'still' or 'stationary' sea level actually is or was (it may have been higher by a metre or fluctuated a metre or two), there are several coasts in the world (e.g. a large part of Australia, South Africa, New Zealand) where the sea level has been roughly stable for the last 6000 to 7000 years. In such marine settings, the influence of sediment supply, wave energy and substrate slope are more clearly apparent.

Large prograded barriers comprising foredune plains (Figure 14.16b) are common on the east and west Australian micro-tidal coasts. On the east Australian coast and the west coast of South Island, New Zealand, they have formed in moderate wave energy (intermediate and dissipative beach type) environments with sediments principally derived from the shelf in the former area, and a mixture of shelf and fluvial supply in the latter case (Thom *et al.*, 1981, 1992). In Western Australia, such prograded foredune barriers principally occur in local or regional sediment traps and at the downdrift ends of littoral drift cells in low energy, seasonally variable, but principally reflective beach environments (Figures 14.16b and 14.17; Hesp, 1984a; Hesp and Chape, 1984; Sanderson and Eliot, 1996).

In the arid, tropical, lower wave energy, macro-tidal north-west of Western Australia, Seminiuk (1993) classified six coastal types, including prograded barriers associated with protruding active deltas and beach and foredune ridge sequences, and narrow retrograding barriers associated with inactive deltas.

Across humid, tropical northern Australia are a range of generally lower energy barrier types, including progradational beach ridges and cheniers (Smart, 1976; Rhodes 1982; Belperio, 1983; Chappell and Grindrod, 1984; Woodroffe *et al.*, 1985; Lees 1985, 1987; Lees and Clements, 1987; Short, 1989). However, in locations exposed to strong onshore trade winds, extensive active

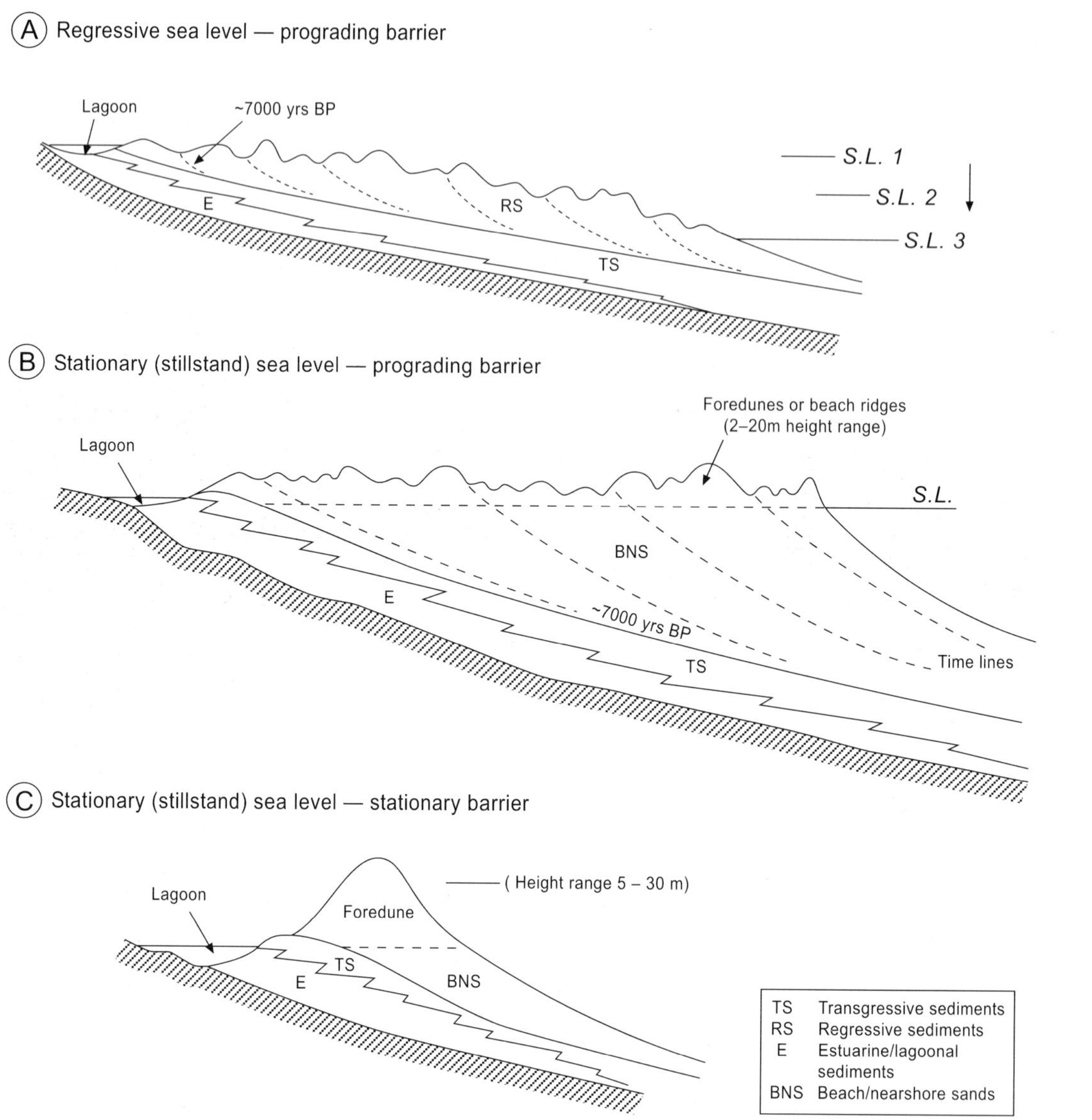


**Figure 14.16** Regressive and prograding barriers formed during (A) falling and (B) stationary sea level; and (C) a large aggrading foredune formed under stillstand conditions.

dune systems are found (Shulmeister and Lees, 1992).

Barrier islands, spits and barriers may comprise a single beach ridge, foredune, or coppice dunes and incipient foredune, a few foredunes, beach ridges or cheniers, or a suite of dune or beach ridges, foredunes or cheniers depending on gross sediment supply, relative shoreline stability (rate of erosion or accretion), and wind energy and aeolian sediment supply (e.g. Shulmeister and Kirk, 1993).

On the Texas coast, Morton (1994) identified three barrier island types. Regressive 'progradational' barriers, such as Galverston Island, occur with relatively stable sea level and abundant longshore and offshore sediment supply. Stable 'aggradational' barrier islands are also a response to both stable sea level and abundant sediment supply. However they represent a balance between the sediment supply and the accommodation space on the barrier platform as the sediment is

**Figure 14.17** Prograded barrier comprising a foredune plain formed under conditions of initially stable ($\approx$ 7000 to 4000 BP) then regressive (a rapid (instantaneous?) sea level fall around 4000 BP), then stable (4000 BP to present) sea level conditions: Warnboro Sound, Western Australia (S Chape).

deposited in foredune and transgressive dunes, while the shoreline remains stable. Transgressive, retrogradational barrier islands result when there is a minor rise in sea level or inadequate sediment supply. They are characterised by low, broken foredunes and washover deposits.

The number and size of foredunes are functions of the rate of shoreline progradation and the time a foredune remains in a stationary position. Rapid progradation leads to the development of multiple, low, small foredunes ( $<$ 5 m high toe-crest). Hesp (1984a) cites a Western Australian example where the beach prograded up to rates of $\geqslant$ 13 m per year. As a result very low (0.5 to 1 m high) foredunes were formed (cf. Shepherd, 1987).

However, the longer a foredune remains in a primary seawardmost position, the larger it potentially becomes, all other factors being equal (e.g. sediment supply). Thus the 'receded' and 'stationary' barriers of Thom *et al.* (1992), Roy *et al.* (1994) and others typically comprise a complex, often large established foredune up to 30 m high (Short and Hesp, 1982), usually erosional or scarped. Occasional scarping causes sand to be transported to the scarp face, crest and lee slope of foredunes and allows the dune to build upwards and landwards over time. The managed Dutch foredunes are a good (unnatural) example of this where some rise to $>$ 40 m in height as a function of continual wave scarping, maintaining the foredune position and artificially resupplying (nourishing) the foredune with new sediment.

### *14.4.3.1 Stationary sea level, stationary barrier*

Barriers comprising a single established foredune are also found on coasts where stillstand marine conditions predominate (Figure 14.18). Both the 'stationary' and 'receded' barriers of Thom *et al.* (1981), Roy *et al.* (1994) above have formed on stationary coasts (Figure 14.16c). The 'stationary' barrier tends to form in a sediment-deficient environment with minimal erosion or progradation and very gradual buildup of a foredune over several thousand years (Figure 14.16c). The 'receded' barrier in this context is typically an erosional foredune which may either have been part of a much more extensive barrier foredune system (as seemingly preferred by Roy *et al.*, 1994), or may be a single foredune which has very gradually been eroded on the seaward stoss face

**Figure 14.18** A stable (single established foredune) barrier and hypersaline lagoon, southern coast, Western Australia, with sea level conditions the same as Figure 14.17 (S. Chape).

and has migrated and still is slowly migrating landwards (Figure 14.12(vi)).

#### *14.4.3.2 Stationary sea level, transgressive dunefield*

Some transgressive dunefield barriers have been formed in a marine setting which is neither regressive nor transgressive but is, in fact, stationary (sea level at stillstand). In this setting, dunefield barriers are formed either during the latter stages of the PMT, or at stillstand, sometimes with significant progradation, sometimes with minimal shoreline recession or progradation (e.g. Semeniuk and Meagher, 1981; Morton, 1994).

Examples include the large, predominantly active transgressive dunefields of the south coasts of South Africa (Illenberger and Rust, 1988) and South Australia (Short, 1987, 1988a,b), west coast, North Island, New Zealand (Shepherd, 1987), Western Australia (Hesp and Chape, 1984), Brazil and the USA, especially California and Oregon (Cooper, 1958, 1967; Hunter *et al.*, 1983), and the arid coastal dunefields of Namibia (Lancaster, 1989), Peru, Chile (Araya-Vergara, 1986) and Baja California, Mexico (Fryberger *et al.*, 1990). Moderate size examples include some in New South Wales, Australia (Thom *et al.*, 1992; Roy *et al.*, 1994; Hesp and Hilton, 1996), and New Zealand (Brothers, 1954) and the Atlantic coast of France (Bressolier *et al.*, 1990).

In the case of Figure 14.12(vii), sediment may have been provided principally by aeolian means over the last 10 000 years with the shoreline at a stationary position. Alternatively, Thom *et al.* (1992) describe dune systems that have been emplaced in the last 7000–3000 years once sea level stabilised. Small dunefield barriers could potentially have been formed in the last 6000 to 7000 years with minimal shoreline recession or progradation (e.g. Figure 14.12(viii)).

#### *14.4.3.3 Stationary sea level, retrogradational barrier*

While all barriers will ultimately retrograde ('transgress') on suitable substrates during sea level rise, barriers can also erode and retrograde under stable sea level conditions. In these situations the controlling factor is usually a diminished sediment supply and/or negative sediment budget, leading to shoreline erosion and barrier retreat. Figure 14.19 illustrates a retrogradation barrier comprising active and relic dunes formed under largely stationary (stillstand) sea level condi-

**Figure 14.19** Leschenault Peninsula, Western Australia, is a retrograding barrier comprising active and inactive parabolic dunes formed under the same sea level conditions as Figure 14.17 (S Chape).

tions. Figure 14.12(v) and (vi) illustrate barriers which may be formed during stillstand conditions but continue to erode landward over Pleistocene or Holocene deposits for various reasons (typically limited sediment supply or removal of sediment off- or alongshore).

## 14.5 CONTROLS ON BARRIER EVOLUTION

As discussed in Section 14.3, barrier evolution is dependent on a range of environmental variables. Any modification to these parameters both spatial (longshore) or over time will therefore have a predictable impact on barrier evolution. Consequently changing environmental parameters will lead to spatial and temporally more variable and complex barrier systems.

### 14.5.1 Spatial controls

The role of spatial variation in wave, tide and sediment supply on barrier form and evolution has been discussed in Section 14.3. Short and Hesp (1982) cited regional and local variation in wave energy to explain both regional and local scale variation in the nature, extent and stability of southern Australian barrier systems. Their wave–beach–dune interaction model attributes prograded foredune ridge barriers to moderate rates of continuous sand supply on a low to moderate energy coast. Increasing wave energy supplies sediment more rapidly resulting in initial shoreline progradation, followed by an exhaustion of sediment and shoreline instability which triggers active (secondary) dune transgression. On high energy coasts active (primary) dune transgression accompanies the rising sea level and subsequent stillstand. While more sand is supplied to these barriers, it is also more rapidly exhausted, leading to barrier retrogradation, which on cliffed coasts can lead to complete removal of the beach and sand ramps, leaving stable clifftop dunes (Figure 14.15).

These models suggest that the barrier form, rates of evolution and stability will vary both locally and regionally as the input parameters change, particularly the level of wave, tide and wind energy, and substrate gradient and sediment supply (cf. Sherman and Bauer, 1993).

### 14.5.2 Temporal controls

Temporal variation in factors controlling barrier evol-

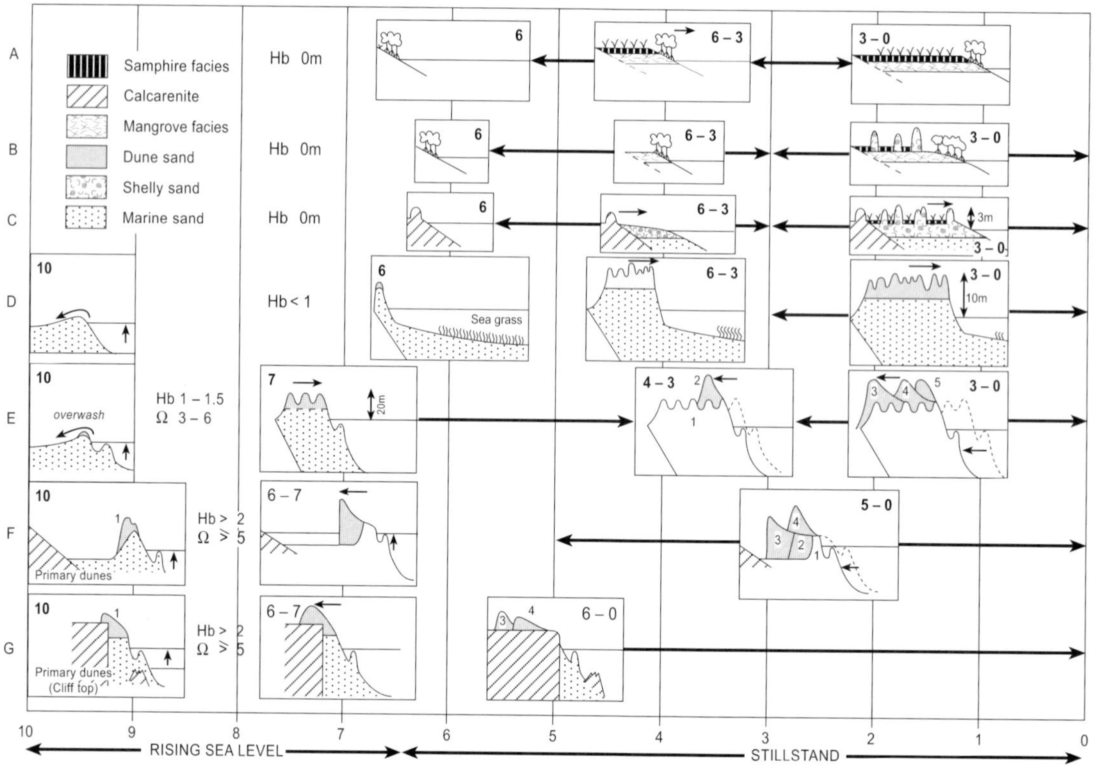


**Figure 14.20** Proposed sequence of seven southern Australian barrier types in relation to level of wave energy and rate, quantity and time of sediment supply (source: modified from Short, 1988a).

ution will induce predictable barrier responses. In the mid- to late Holocene sea level has regionally ranged from stable (e.g. Roy *et al.*, 1994), to slowly rising (e.g. Carter and Orford, 1993) to falling (e.g. Angulo and Lessa, 1997) depending on regional isostasy, subsidence, tectonism and loading.

One of the major controls on temporal variation in barrier stability and evolution is sediment supply. Short (1987, 1988a) used the wave–beach–dune model as a basis to combine the relative level of wave energy and indirectly its impact on sediment supply to propose a time–space matrix of barrier evolution for southern Australia (Figure 14.20). The barriers range with increasing wave energy from progradational chenier (low energy, low/slow sediment supply), through beach ridge and foredune ridge plains (moderate energy, continuing moderate sediment supply), to moderate energy stable and retrogradational to progradational (secondary dune transgression) barriers (moderate to high energy, cessation of moderate to abundant sediment supply), to high energy massive dune transgression, including primary barrier and clifftop dune transgression (high energy, abundant/fast sediment supply). In this model the wide range of barrier types, often located in close proximity, can be explained through the controlling influence of wave energy and in turn its impact on the rate, quantity and duration of sediment supply.

Likewise, in eastern Australia Roy *et al.* (1994) also indicated how a single barrier system might over time evolve from a progradational, to stationary, to receded, to mainland beach, in response to initial sediment surplus, followed by exhaustion and subsequent sediment deficit. It should be noted that neither of these models requires any change in sea level to induce progradation, stability or retrogradation.

On the glacially impacted Nova Scotia coast, Forbes *et al.* (1995) describe an evolutionary sequence of transgressing barriers attached to barrier islands in eroding drumlin fields, while on the New England coast, Fitzgerald *et al.* (1994) found a range of barrier types which they classified using sediment type (gravel or sand), sediment source (river, shelf or mixed), wave energy (sheltered, normal, exposed), and stability (transgressive, regressive or vertical building). They found that

**Table 14.1** Relationship between sea level stability, sediment supply (wave energy) and barrier type.

| Sea level | Barrier type | Sediment supply | Comments (low to high wave/wind energy) |
|---|---|---|---|
| Transgressive | Prograded barriers and strandplains | + | Chenier–beach foredune plains; parabolic–transgressive dunefields |
| | Retrograding barriers and barrier islands | – – | beaches, berms, overwash fans/terraces |
| | | – to + | Foredunes; blowouts |
| | Welded barriers | + | Blowouts; parabolic–transgressive dunefields |
| | | – | Beach and cliff |
| | | –/+ to + | Dunes and cliff |
| | | + | Clifftop dunes (at lower sea level) |
| Regressive | Prograded barriers and strandplains | + | Tidal flats–chenier–beach–foredune ridge plains |
| | | + + | Parabolic–transgressive dunefields |
| Stable | Prograded barriers and strandplains | + | Chenier–beach–foredune ridge plains |
| | | + + | Parabolic–transgressive dunefields |
| | Retrogradational barriers and barrier islands | – | 'Receded' barriers (foredune, parabolics) |
| | | – – | Overwash fans; erosional foredunes |
| | Welded barriers | + + | Beach, dunes, cliff, clifftop dunes |
| | | – | Cliff, clifftop dunes |

while the pocket gravel beaches tended to be transgressive, the other major types, namely river-derived sandy barriers, mixed sediment sandy barriers, sheltered sandy barriers and exposed sandy barriers, could be transgressive, regressive or vertical building depending on local conditions.

## 14.6 CONCLUSION

Davis (1994a) in his review of recent US barrier models concluded that most barriers 'appear to form as a result of some type of landward transport and upward accretion of sands'. The foregoing discussion indicates the wide range of barrier types that can occur, both in response to differing sea level scenarios, as well as local and regional factors such as substrate, sediment supply, wave and wind energy, and geological inheritance. Table 14.1 summarises the barrier types presented in this chapter, based on first, the relative sea level stability (rising/falling); second, the barrier stability (eroding/prograding); and third, the sediment supply (positive to negative). Because of the wide range of potential scenarios, it is not surprising that a wide range of barrier types have been identified to date, together with an equally wide range of models of barrier behaviour. Nummedal (1983) in a review of North American barrier islands stated that although barrier islands are of bewildering complexity at first glance '... they are controlled by relatively few factors' namely sea level history, sediment sources and relative contribution of waves and tides.

# References


Aagaard, T, 1985, Observations of beach cusps. *Geografisk Tidsskrift*, 85, 27–31.

Aagaard, T, 1988, Nearshore bar morphology on the low-energy coast of northern Zealand, Denmark. *Geografiska Annaler*, 70, 59–67.

Aagaard, T, 1990, Infragravity waves and nearshore bars in protected, storm-dominated coastal environments. *Marine Geology*, 94, 181–203.

Aagaard, T and Greenwood, B, 1994a, Sediment transport by wind waves, long waves and mean currents: an experiment on nearshore morphodynamics, Lake Huron, Canada. *Proceedings Coastal Dynamics '94*. ASCE, 14–28.

Aagaard, T and Greenwood, B, 1994b, Suspended sediment transport and the role of infragravity waves in a barred surf zone. *Marine Geology*, 118, 23–48.

Aagaard, T and Greenwood, B, 1995a, Suspended sediment transport and morphological response on a dissipative beach. *Continental Shelf Research*, 15, 1061–1086.

Aagaard, T and Greenwood, B, 1995b, Longshore and cross-shore suspended sediment transport at far infragravity frequencies in a barred environment. *Continental Shelf Research*, 15, 1235–1249.

Aagaard, T, Greenwood, B and Nielsen, J, 1997, Mean currents and sediment transport in a rip channel. *Marine Geology*, 140, 25–45.

Aagaard, T, Greenwood, B and Nielsen, J, 1998, Suspended sediment transport and nearshore bar evolution on a shallow intermediate-state beach. *Marine Geology*, 148, 203–225.

Ahlbrandt, T S, Swinehart, J B and Maroney, D G, 1983, The dynamic Holocene dunefields of the Great Plains and Rocky Mountain Basins, U.S.A. In Brookfield, M E and Ahlbrandt, T S (eds) *Eolian Sediments and Processes*. Developments in Sedimentology 38, Elsevier, Amsterdam, 379–406.

Ahmed, E, 1972, *Coastal Geomorphology of India*. Orient Longman, New Delhi, 222 pp.

Aigner, T and Reineck, H-E, 1983, Seasonal variation of wave-base on the shoreface of the barrier island Norderney, North Sea. *Senckenbergiana Marit*. 15, 87–92.

Allan, R, Lindesay, J and Parker, D, 1996, *The El Nino Southern Oscillation and Climatic Variability*. CSIRO, Collingwood, Victoria, 405 pp.

Allen, J R L, 1982, *Sedimentary Structures: Their Character and Physical Basis*. Volume 1. Elsevier, 593 pp.

Allen, J R L, 1984, *Sedimentary Structures*. Developments in Sedimentology 1, Elsevier, Amsterdam, 663 pp.

Allen, J R, 1985, Field evaluation of beach profile response to wave steepness as predicted by the Dean model. *Coastal Engineering*, 9, 71–80.

Allison, M A, Nittrouer, C A and Faria, L E C, Jr, 1995, Rates and mechanism of shoreface progradation and retreat down-drift of the Amazon river mouth. *Marine Geology*, 125, 373–392.

Andrews, E C, 1916, Shoreline studies at Botany Bay. *Journal Royal Society NSW*, 50, 165–176.

Angulo, R J and Lessa, G C, 1997, The Brazilian sea-level curves: a critical review with emphasis on the curves from the Paranagua and Cananeia regions. *Marine Geology*, 140, 141–166.

Angus, S and Elliott, M M, 1992, Erosion in Scottish machair with particular reference to the Outer Hebrides. In Carter, R W G, Curtis, T G F and Sheehy-Skeffington, M J (eds) *Coastal Dunes: Geomorphology, Ecology and Management for Conservation*. Proceedings of 3rd European Dune Congress, Galway, Ireland, 93–112.

Ansell, A D, 1993, The biology of the genus *Donax*. In McLachlan, A and Erasmus, T (eds) *Sandy Beaches as Ecosystems*. Junk, The Hague, 607–635.

Anthony, E J, 1995, Beach-ridge development and sediment supply: examples from West Africa. *Marine Geology*, 129, 175–186.

Araya-Vergara, J F, 1986, The evolution of modern coastal dune systems in central Chile. In Gardiner, V (ed.) *International Geomorphology Part II*. Proceedings 1st International Conference on Geomorphology, 1231–1243.

Arens, S M, 1994, *Aeolian processes in the Dutch foredunes*. PhD Thesis, University of Amsterdam, 150 pp.

Arens, S M and Wiersma, J, 1994, The Dutch foredunes; inventory and classification. *Journal of Coastal Research*, 10, 189–202.

Arens, S M, Van Kaam-Peters, H M E and Van Boxel, J H, 1995, Air flow over foredunes and implications for sand transport. *Earth Surface Processes and Landforms*, 20, 315–332.

Armon, J W, 1979, Landward sediment transfers in a transgressive barrier island system, Canada. In Leatherman, S P (ed.) *Barrier Islands from the Gulf of St. Lawrence to the Gulf of Mexico*. Academic Press, New York, 65–80.

Aseervatham, A M, Kang, H Y and Nielsen, P, 1993, Groundwater movement in beach watertables. *11th Australasian Conference on Coastal and Ocean Engineering*. 589–594.

Aubrey, D G, 1979, Seasonal patterns of onshore/offshore

sediment movement. *Journal of Geophysical Research*, 84, 6347–6354.

Aubrey, D G and Ross, R M, 1985, The quantitative description of beach cycles. *Marine Geology*, 69, 155–170.

Aubrey, D G, Inman, D L and Nordstrom, C E, 1976, Beach profiles at Torrey Pines. *Proceedings 15th International Conference on Coastal Engineering*. ASCE, 1297–1311.

Augustinus, P G E F (ed.) 1989, Cheniers and chenier plains. *Marine Geology*, 90, 219–351.

Avis, A M and Lubke, R A, 1985, The effect of wind-borne sand and salt spray on the growth of Scirpus nodosus in a mobile dune system. *South African Journal of Botany*, 51, 100–110.

Aylor, D E, Wang, Y and Miller, D, 1993, Intermittent wind close to the ground within a grass canopy. *Boundary-Layer Meteorology*, 66, 427–448.

Bagnold, R A, 1940, Beach formation by waves; some model experiments in a wave tank. *Journal Institute Civil Engineers*, 15, 27–52.

Bagnold, R A, 1941, *The Physics of Blown Sand and Desert Dunes*. Chapman and Hall, London, 265 pp.

Bagnold, R A, 1954, Experiments on a gravity free dispersion of large solid spheres in a Newtonian fluid under shear. *Proceedings Royal Society London* (Series A), 225, 46–63.

Bagnold, R A, 1956, Flow of cohesionless grains in fluids. *Philosophical Transactions Royal Society London*, 249, 235–297.

Bagnold, R A, 1963, Mechanics of marine sedimentation. In Hill, M N (ed.) *The Sea*, Vol. 3. Wiley, New York, 507–528.

Bagnold, R A, 1966, *An approach to the sediment transport problem from general physics*. Geological Survey Professional Paper 422–1.

Bailard, J A and Inman, D L, 1981, An energetics bedload model for a plane sloping beach: Local transport. *Journal of Geophysical Research*, 86, 10 938–10 954.

Baker Jr, B B, Deebel, W R, and Geisenderfer, R D (eds), 1966, *Glossary of Oceanographic Terms*. Special Publication, US Naval Oceanographic Research Office, Washington DC, 204 pp.

Bakker, Th W, Jungerius, P D and Klijn, J A (eds) 1990, *Dunes of the European Coasts*. *Catena*, Supplement 18.

Bakker, W T, 1968, A mathematical theory about sand waves and its application on the Dutch Wadden Sea Isle of Vlieland. *Shore and Beach*, 36, 4–14.

Bally, R, 1981, *The ecology of three sandy beaches on the west coast of South Africa*. PhD thesis, University of Cape Town.

Balsillie, J H, 1985, Redefinition of shore-breaker classification as a numerical continuum and a design shore-breaker. *Journal of Coastal Research*, 1, 247–254.

Barbour, M G, 1978, Salt spray as a microenvironmental factor in the distribution of beach plants at Point Reyes, California. *Oecologia*, 32, 213–224

Barbour, M.G., de Jong, T M and Pavlik, B M, 1985, Marine, beach and dune plant communities. In Chabot, B F and Mooney, H A (eds) *Physiological Ecology of North American Plant Communities*. Chapman and Hall, London, 296–322.

Barnes, P W, Kempema, E W, Reimnitz, E, McCormick, M, Weber, W S and Hayden, E C, 1993, Beach profile modification and sediment transport by ice: an overlooked process on Lake Michigan. *Journal of Coastal Research*, 9, 65–86.

Barrell, J, 1912, The criteria for the recognition of ancient delta deposits. *Geological Society of America Bulletin*, 23, 377–446.

Barthel, V and Funke, E R, 1984, Estimates of long waves in the Weser estuary. *Proceedings 19th International Conference on Coastal Engineering*. ASCE, 782–799.

Basco, D R, 1983, Surfzone currents. *Coastal Engineering*, 7, 331–355.

Bascom, W N, 1951, The relationship between sand size and beach face slope. *Transactions of the American Geophysical Union*, 32, 866–874.

Bascom, W N, 1954, Characteristics of natural beaches. *Proceedings 4th Conference on Coastal Engineering*. ASCE, 163–180.

Bascom, W N, 1964, *Waves and Beaches*. Doubleday, New York, 267 pp.

Battjes, J A, 1974, Surf similarity. *Proceedings 14th International Conference on Coastal Engineering*. ASCE, 466–480.

Battjes, J A, 1988, Surf-zone dynamics. *Annual Review of Fluid Mechanics*, 20, 257–293.

Battjes, J A and Janssen, J P F M, 1978, Energy loss and set-up due to breaking of random waves. *Proceedings 16th International Conference on Coastal Engineering*. ASCE, 569–587.

Battjes, J A and Stive, M J F, 1984, Calibration and verification of a dissipation model for random breaking waves. *Proceedings 19th International Conference on Coastal Engineering*. ASCE, 649–660.

Bauer, B O, 1990, Assessing the relative energetics of infragravity motions in lakes and seas. *Journal of Coastal Research*, 6, 853–865.

Bauer, B O and Allen, J R, 1995, Beach steps: An evolutionary perspective. *Marine Geology*, 123, 143–166.

Bauer, B O and Greenwood, B, 1990, Modification of a linear bar-trough system by a standing edgewave. *Marine Geology*, 92, 177–204.

Beach, R A and Sternberg, R W, 1988, Suspended sediment transport in the surf zone: Response to cross-shore infragravity motion. *Marine Geology*, 80, 61–79.

Beach, R A and Sternberg, R W, 1991, Infragravity-driven suspended sediment transport in the swash, inner and outer-surf zone. *Proceedings Coastal Sediments '91*. ASCE, 114–128.

Beach, R A and Sternberg, R W, 1992, Suspended sediment transport in the surf zone: response to incident wave and longshore current interaction. *Marine Geology*, 108, 275–294.

BEB, 1933, *Interim Report*. Beach Erosion Board, US Army Corps of Engineers.

Beets, D J, van der Valk, L and Stive, M J F, 1992. Holocene evolution of the coast of Holland. *Marine Geology*, 103, 423–443.

Belknap, D F and Kraft, J C, 1985, Influence of antecedent geology on stratigraphic preservation potential and evolution of Delaware's barrier systems. *Marine Geology*, 63, 235–262.

Belly, P-Y, 1964, *Sand Movement by Wind*. US Army Corps of Engineers CERC, Technical Memorandum 1, Washington, D.C., 38 pp.

Belperio, A J, 1983, Terrigenous sedimentation in the Great Barrier Reef lagoon: a model from the Burdekin region. *BMR Journal of Australian Geology and Geophysics*, 8, 1769–1790.

Belperio, A J, Gostin, V A, Cann, J H and Murray-Wallace, C V, 1988, Sediment–organism zonation and the evolution of Holocene tide sequences in southern Australia. In de Boer, P L, van Gelder, A and Nio, S D (eds) *Tide Influenced Sedimentary Environment and Facies*. Reidel, Dordrecht, 475–497.

Bernard, H A, Le Blanc, R J and Major, C F, 1962, Recent and Pleistocene geology of southeast Texas. In Rainwater, E H and Zingula, R P (eds) *Geology of the Gulf Coast and Central Texas: Guidebook of Excursion*. Houston Geological Society, 175–205.

Berryman, K R, Ota, Y and Hull, A G, 1992a, Holocene coastal evolution of an estuary on a tectonically rising coast: the Pakarae River locality, eastern North Island, New Zealand. *Journal of Sedimentary Geology*, 80, 151–165.

Berryman, K.R, Ota, Y and Hull, A G, 1992b, Holocene coastal evolution under the influence of episodic tectonic uplift: Examples from New Zealand and Japan. *Quaternary International*, 15/16, 31–45.

Bigarella, J J, 1972, Eolian environments: Their characteristics, recognition and importance. In Rigby, J K and Hamblin, W K (eds) *Recognition of Ancient Sedimentary Environments*. Society of Economic Paleontologists and Mineralogists, Special Publication, 16, 12–62.

Bigarella, J J, 1975, Lagoa dunefield, Santa Catarina, Brazil – A model of aeolian and pluvial activity. *Bulletin Parana Geoscience*, 33, 133–167.

Bigarella, J J, Becker, R N and Duarte, G M, 1969, Coastal dune structures from Parana (Brazil). *Marine Geology*, 7, 5–55.

Bird, E C F, 1965, *A Geomorphological study of the Gippsland Lakes*. Australian National University, Department of Geography Publication G/1, Canberra.

Bird, E C F, 1968, *Coasts*. Australian National University Press, Canberra, 246 pp.

Bird, E C F, 1974, Dune stability on Fraser Island. *Queensland Naturalist*, 21, 15–21.

Bird, E C F, 1976, *Coasts. An Introduction to Systematic Geomorphology*. Australian National University Press, Canberra, 320 pp.

Bird, E C F, 1978, *The Geomorphology of the Gippsland Lakes*. Environmental Studies Series, Publication No. 186, Ministry for Conservation, Victoria, 158 pp.

Bird, E C F, 1982, Australia, coastal morphology. In Schwartz, M L (ed.) *Encyclopedia of Beaches and Coastal Environments*. Hutchinson Ross, Stroudsburg, 110–120.

Bird, E C F, 1984, *Coasts, An Introduction to Coastal Geomorphology*, 3rd edition. Australian National University Press, Canberra, 320 pp.

Bird, E C F, 1985, *Coastline Changes*. Wiley Interscience, Chichester, 219 pp.

Bird, E C F, 1990, Classification of European dune coasts. In Bakker, Th W, Jungerius, P D and Klijn, J A (eds) *Dunes of the European Coasts. Catena* Supplement 18, 15–24.

Bird, E C F, 1993, *The Coast of Victoria*. Melbourne University Press, Melbourne, 324 pp.

Bird, E C F, 1996a, Lateral grading of beach sediments: a commentary. *Journal of Coastal Research*, 12, 774–785.

Bird, E C F, 1996b, *Beach Management*. John Wiley and Sons, Chichester, 281 pp.

Birkemeier, W A, 1985, Field data on the seaward limit of profile change. *Journal of Waterway, Port, Coastal and Ocean Engineering*, 111, 598–602.

Birkemeier, W A, Long, C E and Hathaway, K K, 1997, DELILAH, DUCK94 and SandyDuck: Three nearshore field experiments. *Proceedings of the International Coastal Engineering Conference*. ASCE, 4052–4065.

Bishop, C T, 1983, A shore protection alternative: artificial headlands. *Proceedings Canadian Coastal Conference*. National Research Council Canada, 305–319.

Bishop, P M and Cowell, P J, 1997. Lithological and drainage network determinants of the character of an embayed coast, New South Wales, Australia. *Journal of Geology*, 105, 685–689.

Black, K P, 1994, Suspended sediment load during an asymmetric wave cycle over a plane bed. *Coastal Engineering*, 23, 95–114.

Black, K P and Rosenberg, M A, 1991, Suspended sediment load at three timescales. *Proceedings Coastal Sediments '91*. ASCE, 313–327.

Black, K P, Andrews, C, Green, M, Gorman, R, Healy, T, Hume, T, Hutt, J, Mead, S and Sayce, A, 1997, Wave dynamics and shoreline response on and around surfing reefs. *Proceedings 1st International Surfing Reef Symposium*. Coastal Studies Unit, University of Sydney, 11 pp.

Bloom, A L, Pleistocene shorelines: a new test of isostasy. *Bulletin Geological Society of America*, 78, 1477–1494.

Bluck, B J, 1967, Sedimentation of beach gravels: examples from south Wales. *Journal of Sedimentary Petrology*, 37, 128–156.

Bodge, K R, 1989, A literature review of the distribution of longshore sediment transport across the surf zone. *Journal of Coastal Research*, 5, 307–328.

Bodge, K R, 1992, Representing equilibrium profiles with an exponential expression. *Journal of Coastal Research*, 8, 47–55.

Bodge, K R and Kraus, N C, 1991, Crtical examination of longshore transport magnitude. *Proceedings Coastal Sediments '91*. ASCE, 139–155.

Boon, J D and Green, M O, 1988, Caribbean beach-face slopes and beach equilibrium profiles. *Proceedings 21st International Conference on Coastal Engineering*. ASCE, 1618–1630.

Borden, P J S and Reed, R, 1996, *An Introduction to Coastal Ecology*. Blackie, London, 218 pp.

Boreen, T and James, N, 1993, Holocene sediment dynamics on a cool-water carbonate shelf: Otway, southeastern Australia. *Journal of Sedimentary Petrology*, 63, 574–588.

Boreen, T, James, N, Heggie, D and Wilson, C, 1993, Surficial cool-water carbonate sediments on the Otway continental margin, southeastern Australia. *Marine Geology*, 112, 35–56.

Borichansky, L S and Mikhailov, V N, 1966, Interaction of river and sea water in the absence of tides. In *Scientific Problems of the Humid Tropical Deltas and their Implications*. UNESCO, 175–180.

Borowka, R K, 1990, The Holocene development and present morphology of the Leba Dunes, Baltic coast of Poland. In Nordstrom, K F, Psuty, N P and Carter, R W G (eds) *Coastal Dunes: Form and Process*. John Wiley and Sons, Chichester, 289–313.

Bosman, J J, van der Valden, E T J M and Hulsbergen, C H, 1987, Sediment concentration measurement by transverse suction. *Coastal Engineering*, 11, 353–370.

Bourman, R P, 1986, Aeolian sand transport along beaches. *Australian Geographer*, 17, 30–34.

Bowen, A J, 1969a, Rip currents. 1. Theoretical investigations. *Journal of Geophysical Research*, 74, 5467–5478.

Bowen, A J, 1969b, The generation of longshore currents on a plane beach. *Journal of Marine Research*, 27, 206–215.

Bowen, A J, 1980, Simple models of nearshore sedimentation; beach profiles and longshore bars. In McCann, S B (ed.) *The Coastline of Canada*. Geological Survey of Canada, Paper 80–10, 1–11.

Bowen, A J and Guza, R T, 1978, Edge waves and surf beat. *Journal of Geophysical Research*, 83, 1913–1920.

Bowen, A J and Holman, R A, 1989, Shear instabilities of the mean longshore current: 1. Theory. *Journal of Geophysical Research*, 94, 18 023–18 030.

Bowen, A J and Inman, D L, 1969, Rip currents. 2. Laboratory and field observations. *Journal of Geophysical Research*, 74, 5479–5490.

Bowen, A J and Inman, D L, 1971, Edge waves and crescentic bars. *Journal of Geophysical Research*, 76, 8662–8670.

Bowen, A J, Inman, D L and Simmons, V P, 1968, Wave 'setdown' and setup. *Journal of Geophysical Research*, 73, 2569–2577.

Bowman, D, 1981, Efficiency of eigenfunctions for discriminant analysis of subaerial non-tidal beach profiles. *Marine Geology*, 39, 243–258.

Bowman, D, Arad, D, Rosen, D S, Kit, E, Goldberry, R and Slavicz, A, 1988, Flow characteristics along the rip current system under low-energy conditions. *Marine Geology*, 82, 148–167.

Boyd, R, Dalyrmple, R and Zaitlin, B A, 1992, Classification of clastic coastal depositional environments. *Sedimentary Geology*, 80, 139–150.

Bradley, E F, 1968, A micrometeorological study of velocity profiles and surface drag in the region modified by a change in surface roughness. *Quarterly Journal Royal Meteorological Society*, 94, 361–379.

Bradshaw, M P, 1980, Topographic control of runup variability. *Proceedings 17th Conference on Coastal Engineering*. ASCE, 1091–1105.

Brander, R W, 1991, *Bedform constraints on sediment re-suspension and transport under shoaling and breaking waves*. Unpublished MSc thesis, Department of Geography, University of Toronto, 182 pp.

Brander, R W, 1996, *Field observations on the morphodynamics of rip currents*. Unpublished PhD thesis, Department of Geography, University of Sydney, 240 pp.

Branner, J C, 1900, The origin of beach cusps. *Journal of Geology*, 8, 481–484.

Bressolier, C and Thomas, Y-F, 1977, Studies on wind and plant interactions on French Atlantic coastal dunes. *Journal of Sedimentary Petrology*, 47, 331–338.

Bressolier, C, Froidefond, J-M and Thomas, Y-F, 1990, Chronology of coastal dunes in the south-west of France. In Bakker, Th W, Jungerius, P D and Klijn, J A (eds) *Dunes of the European Coasts*. *Catena*, Supplement 18, 101–107.

Brodhead, J M B and Godfrey, P J, 1977, Off-road vehicle impact in Cape Cod National Seashore. *International Journal of Biometeorology*, 21, 299–306.

Brookfield, M E, 1984, Eolian sands. In Walker, R G (ed.) *Facies Models*, 2nd edition. GeoScience. Canada, Reprint Series 1, 91–103.

Broome, R and Komar, P.D, 1979, Undular hydraulic jumps and the formation of backwash ripples on beaches. *Sedimentology*, 26, 543–559.

Brothers, R N, 1954, A physiographical study of recent sand dunes on the Auckland west coast. *New Zealand Geographer* 10, 47–59.

Brown, A C and McLachlan, A, 1990, *Ecology of Sandy Shores*. Elsevier, Amsterdam, 328 pp.

Bruun, P, 1954a, Migrating sand waves and sand humps, with special reference to investigations carried out on the Dutch North Sea coast. *Proceedings 5th Conference Coastal Engineering*, 269–295.

Bruun, P, 1954b, *Coast Erosion and the Development of Beach Profiles*. Technical Memorandum No. 44, Beach Erosion Board, US Army Corps of Engineers, Washington, 79 pp.

Bruun, P, 1962, Sea-level rise as a cause of shore erosion. *Journal of the Waterways and Harbors Division*. ASCE, 88, 117–130.

Bruun, P, 1986. Sedimentary balances (land and sea) with special reference to the Icelandic south coast from Torlakshofn to Dyrholaey: River nourishment of shores – practical analogies on artificial nourishment. *Coastal Engineering*, 10, 193–210.

Bruun, P, 1988, The Bruun Rule of erosion by sea level rise: A discussion of large-scale two- and three-dimensional usages. *Journal of Coastal Research*, 4, 627–648.

Bryan, K R and Bowen, A J, 1996, Edge wave trapping and amplification on barred beaches. *Journal of Geophysical Research*, 101, 6543–6552.

Bryant, E A, 1979, Edge waves and sediment sorting relationships on beach foreshores, Broken Bay. *Search*, 10, 442–444.

Bryant, E A, 1982, Behaviour of grain size characteristics on reflective and dissipative foreshores, Broken Bay, Australia. *Journal of Sedimentary Petrology*, 52, 431–450.

Bryant, E A, Young, R W, Price, D M and Short, S A, 1994, Late Pleistocene dune chronology: Near-coastal NSW and eastern Australia. *Quaternary Science Reviews*, 13, 209–223.

Burkinshaw, J R, Illenberger, K W and Rust, I C, 1993, Wind-speed profiles over a reversing transverse dune. In Pye, K (ed) *The Dynamics and Environmental Context of Aeolian Sedimentary Systems.* Geological Society, Special Publication No. 72, 25–36.

Burnet, T, Gallagher, E, Okiro, M, Raubenheimer, B, Whitsel, R, Vanhoff, B, Elgar, S and Werner, B T, 1994, Field tests of theories for beach cusp formation. Abstracts AGU Fall meeting. *EOS, Transactions American Geophysical Union,* supplement, 75, 336.

Cacchione, D A and Drake, D E, 1982, Measurements of storm-generated bottom stresses on the continental shelf. *Journal of Geophysical Research,* 87, 1952–1960.

Carlson, C T, 1984, Field studies of run-up on dissipative beaches. *Proceedings 19th International Conference on Coastal Engineering,* ASCE, 399–414.

Carrier, G F and Greenspan, H P, 1958, Water waves of finite amplitude on a sloping beach. *Journal of Fluid Mechanics,* 4, 97–109.

Carson, M A and McLean, P A, 1986, Development of hybrid aeolian dunes: The William River dunefield, northwest Saskatchewan, Canada. *Canadian Journal of Earth Science,* 23, 1974–1990,

Carter, R W G, 1976, Formation, maintenance and geomorphological significance of an aeolian shell pavement. *Journal of Sedimentary Petrology,* 46, 418–429.

Carter, R W G, 1977, The rate and pattern of sediment interchange between beach and dune. In Tanner, W F (ed), *Coastal Sedimentology.* Geology Deptartment, Florida State University, Tallahassee, 3–34.

Carter, R W G, 1986, The morphodynamics of beach ridge formation: Magilligan, Northern Ireland. *Marine Geology,* 73, 192–214.

Carter, R W G, 1988, *Coastal Environments. An Introduction to the Physical, Ecological and Cultural Systems of Coastlines.* Academic Press, London, 617 pp.

Carter, R W G, 1990, The geomorphology of coastal dunes in Ireland. In Bakker, Th W, Jungerius, P D and Klijn, J A (eds) *Dunes of the European Coasts. Catena,* Supplement 18, 31–40.

Carter, R W G, and Orford, J, 1993, The morphodynamics of coarse clastic beaches and barriers: A short and long term perspective. *Journal of Coastal Research,* Special Issue, 15, 158–179.

Carter, R W G and Wilson P, 1988, Geomorphological, sedimentological and pedological influences on coastal dune development in Ireland. *Journal of Coastal Research,* Special Issue, 3, 27–31.

Carter, R W G and Wilson, P, 1990, The geomorphological, ecological and pedological development of coastal foredunes at Magilligan Point, Northern Ireland. In Nordstrom, K F, Psuty, N P and Carter, R W G (eds) *Coastal Dunes: Form and Process.* John Wiley and Sons, Chichester, 129–157.

Carter, R W G, Hesp, P A and Nordstrom, K F, 1990, Erosional landforms in coastal dunes. In Nordstrom, K F, Psuty, N P and Carter, R W G (eds) *Coastal Dunes: Form and Process.* John Wiley and Sons, Chichester, 217–249.

Carter, T G, Liu, P L-F and Mei, C C, 1973, Mass transport by waves and offshore sand bedforms. *Journal of the Waterways, Harbors and Coastal Engineering Division.* ASCE, 99, 165–184.

Cartica, R J and Quinn, J A, 1980, Responses to populations of Solidago sempervirens (Compositae) to salt spray across a barrier beach. *American Journal Botany,* 67, 1236–1242.

Chappell, J, 1974, Late Quaternary glacio- and hydro-isostasy on a layered earth. *Quaternary Research,* 4, 429–440.

Chappell, J and Grindrod, J, 1984, Chenier plain formation in northern Australia. In Thom, B G (ed.) *Coastal Geomorphology in Australia.* Academic Press, Sydney, 197–231.

Chappell, J and Wright, L D, 1978, Surf zone resonance and coupled morphology. *Proceedings 16th International Conference on Coastal Engineering.* ASCE, 1359–1377.

Chappell, J, Rhodes, E G, Thom, B G and Wallensky, E, 1982, Hydro-isostasy and the sea level isobase of 5500 BP in North Queensland, Australia. *Marine Geology,* 49, 81–90.

Chappell, T, Eliot, I G, Bradshaw, M P and Lonsdale, E, 1979, Experimental control of beach face dynamics by watertable pumping. *Engineering Geology,* 14, 29–41.

Cholnoky, E V, 1902, Die Bewegungsgesetze des Flugsandes. Foldtani Kozlony, 32, Budapest.

Christiansen, Ch, Dalsgaard, K, Moller, J T and Bowman, D, 1990, Coastal dunes in Denmark. Chronology in relation to sea level. In Bakker, Th W, Jungerius, P D and Klijn, J A (eds) *Dunes of the European Coasts. Catena,* Supplement, 18, 61–70.

Church, J C and Thornton, E B, 1993, Effects of breaking wave induced turbulence within a longshore current model. *Coastal Engineering,* 20, 1–28.

Church, J C, Thornton, E B and Oltman-Shay, J, 1994, Mixing by shear instabilities of the longshore current. *Proceedings Coastal Dynamics '94.* ASCE, 376–390.

Clark, B M, 1997, Variation in surf-zone fish community structure across a wave-exposure gradient. *Estuarine, Coastal and Shelf Science,* 4, 659–674.

Clark, J A, Farrell, W E and Peltier, W R, 1978, Global changes in Post-glacial sea level: a numerical calculation. *Quaternary Research,* 9, 265–287.

Clark, J R, 1996, *Coastal Zone Management.* CRC Lewis Publishers, Boca Raton, 694 pp.

Clark, R R, 1990, *The Carbonate Beaches of Florida: an Inventory of Monroe County Beaches.* Memorandum 90–1, Florida Department of Natural Resources, 63 pp.

Clarke, D J and Eliot, I G, 1982, Description of littoral, alongshore sediment movement from empirical eigen-function analysis. *Journal Geological Society Australia,* 29, 327–341.

Clarke, D J and Eliot, I G, 1983, Mean sea-level and beach-width variation at Scarborough, Western Australia. *Marine Geology,* 51, 251–267.

Clarke, D J and Eliot, I G, 1987, Groundwater-level changes in a coastal dune, sea-level fluctuations and shoreline movement on a sandy beach. *Marine Geology,* 77, 319–326.

Clarke, D J, Eliot, I G and Frew, J R, 1984, Variation in subaerial beach sediment volume on a small sandy beach over a monthly lunar tidal cycle. *Marine Geology,* 58,

319–344.

Clarke, T L, Lesht, B, Young, R A, Swift, D J P and Freeland, G L, 1982, Sediment resuspension by surface wave action: an examination of possible mechanisms. *Marine Geology*, 49, 43–59.

Cleary, W J and Hosier, P E, 1979, Geomorphology, washover history, and inlet zonation: Cape Lookout, N.C. to Bird Island, N.C. In Leatherman, S P (ed) *Barrier Islands from the Gulf of St. Lawrence to the Gulf of Mexico*. Academic Press, New York, 237–271.

Clemmensen, L B, 1986, Storm-generated eolian sand shadows and their sedimentary structures, Vejers Strand, Denmark. *Journal of Sedimentary Petrology*, 56, 520–527.

Clifton, H E, 1969, Beach lamination: nature and origin. *Marine Geology*, 7, 553–559.

Clifton, H E and Dingler, J R, 1984, Wave-formed structures and paleoenvironmental reconstruction. *Marine Geology*, 60, 165–198.

Clifton, H E, Hunter, R E and Phillips, R L, 1971, Depositional structures and processes in the non-barred high-energy nearshore. *Journal of Sedimentary Petrology*, 41, 651–670.

Cloud, P E, 1966, Beach cusps: response to Plateau's rule? *Science*, 154, 890–891.

Coaldrake, J E, 1962, The coastal dunes of southern Queensland. *Proceedings Royal Society of Queensland*, 72, 101–119.

Coastal Engineering Research Center, 1973, *Shore Protection Manual*. US Army Corps of Engineers, Washington, DC, Vols 1–3, 1160 pp.

Coastal Engineering Research Center, 1984, *Shore Protection Manual*. US Army Corps of Engineers, Vols 1 and 2.

Collins, J L, 1970, Probabilities of breaking wave characteristics. *Proceedings 12th International Conference on Coastal Engineering*. ASCE, 399–410.

Collinson, J D and Thompson, D B, 1982, *Sedimentary Structures*. George Allen and Unwin, London, 194 pp.

Conley, D C. and Inman, D L, 1992, Field observations of the fluid-granular boundary layer under near-breaking waves. *Journal of Geophysical Research*, 97, 9631–9643.

Cook, D O, 1970, The occurrence and geologic work of rip currents off southern California. *Marine Geology*, 9, 173–186.

Cook, D O, and Gorsline, D S, 1972, Field observations of sand transport by shoaling waves. *Marine Geology*, 13, 31–55.

Cook, P J, Colwell, J B, Firman, J B, Lindsay, J M, Schwebel, D A and Von der Bosch, C C, 1977, The late Cainozoic sequence of southeast South Australia and Pleistocene sea-level changes. *BMR Journal of Australian Geology and Geophysics*, 2, 81–88.

Cooper, W S, 1958, *Coastal Sand Dunes of Oregon and Washington*. Geological Society of America Memorandum 72, 169 pp.

Cooper, W S, 1967, *Coastal Sand Dunes of California*. Geological Society Amerorica Memorandum, 104, 131 pp.

Cornaglia, P, 1889. On beaches. In Fisher, J S and Dolan, R, (eds) *Beach Processes and Coastal Hydrodynamics*. Benchmark Papers in Geology, Vol. 39, Dowden, Hutchins and Ross, Stroudsberg (1977), 11–26.

Cowell, P J, 1986a, Wave-induced sand mobility and deposition on the south Sydney inner-continental shelf. In Frankel, E, Keene, J and Waltho, A E (eds) *Recent Sediments in Eastern Australia: Marine through Terrestrial*. Geology Society of Australia Special Publication, 2, 1–28.

Cowell, P J, 1986b, Australian 'Mega Rip' Study. *EOS*, 67, 449.

Cowell, P J and Kotvojs, F K, 1987, Design relationships for artificial beaches in rocky bays. *Proceedings 8th Australasian Conference on Coastal and Ocean Engineering*. 113–117.

Cowell, P J and Roy P S, 1988, *Shoreface Transgression Model: Programming Guide (Outline, Assumptions and Methodology)*. Unpublished Report, Coastal Studies Unit, Marine Studies Centre, University of Sydney, 23 pp.

Cowell, P J and Thom, B G, 1994, Morphodynamics of coastal evolution. In Carter, R W G and Woodroffe, C.D. (eds), *Coastal Evolution: Late Quaternary Shoreline Morphodynamics*, Cambridge University Press, Cambridge, 33–86.

Cowell, P J, Roy, P S, and Jones, R A, 1991, Shoreface translation model: application to management of coastal erosion. In Brierley, G and Chappell, J (eds) *Applied Quaternary Studies, Bio & Geo*. Australian National University, 57–73.

Cowell, P J, Roy, P S and Jones, R A, 1992, Shoreface translation model: computer simulation of coastal-sand-body response to sea level rise. *Mathematics and Computers in Simulation*, 33, 603–608.

Cowell, P J, Roy, P S and Jones, R A, 1995, Simulation of large-scale coastal change using a morphological behaviour model. *Marine Geology*, 126, 45–61.

Cowell, P J, Roy, P S, Cleveringa, J and de Boer, P L (in press), Simulating coastal systems tracts using the Shoreface-Translation Model. In Watney, L (ed.) *Numerical Experiments in Stratigraphy*. International Association of Sedimentologists Special Publication, Blackwell, Oxford.

Cowie, J D, 1963, Dune-building phases in the Manawatu district, New Zealand. *New Zealand Journal of Geology and Geophysics*, 6, 268–280.

Cowles, H C, 1899, The ecological relations of the vegetation on the sand dunes of Lake Michigan. *Botanical Gazette*, 27, 95–117, 167–202, 281–308, 361–391.

Crank, J, 1957, *The Mathematics of Diffusion*. Oxford University Press, London.

Csanady, G T, 1977, The Coastal Boundary Layer. *In Estuaries, Geophysics and the Environment*, National Research Council, Washington, DC, 57–68.

Csanady, G T, 1982, *Circulation in the Coastal Ocean*. Reidel, Dortrecht, 279 pp.

Cullen, P and Pemberton, D, 1997, Coastal dune development in south west Tasmania during the Holocene: implications for conservation and management. *Joint Australian Geographers and New Zealand Geographical Society Conference* (abstracts). Department Geography and Environmental Studies, University of Tasmania, Hobart, 34.

Curray, J R, 1964, Transgressions and regressions. In Miller, R C (ed) *Papers in Marine Geology*. Mcmillan, New York, 175–203.

Curray, J R, 1969, History of continental shelves. In Stanley, D

J (ed.) *New Concepts of Continental Margin Sedimentation.* American Geological Institute, Washington DC, JC-6-1 – JC-6-7.

Curray, J R, Emmel, F J and Crampton, P J S, 1969, Holocene history of a strand plain, lagoonal coast, Nayarit, Mexico. In Ayala-Castanares, A and Pheleger, F B (eds) *Coastal Lagoons, A Symposium.* Universidad Nacional Autonoma de Mexico, Cuidad Universitaria, 63–100.

Dally, W, 1980, *Numerical model for beach profile evolution.* MS Thesis, Department of Civil Engineering, University of Delaware, Newark.

Dally, W, 1987, Longshore bar formation – surf beat or undertow? *Proceedings Coastal Sediments '87.* ASCE, 71–86.

Dally, W, 1990, Random breaking waves: a closed-form solution for planar beaches. *Coastal Engineering*, 14, 233–263.

Dally, W and Dean, R G, 1984, Suspended sediment transport and beach profile evolution. *Journal Waterway, Port, Coastal and Ocean Engineering*, 110, 15–33.

Dally, W R and Dean, R G, 1986, Mass flux and undertow in a surf zone – discussion. *Coastal Engineering*, 10, 289–307.

Dally, W R, Dean, R G and Dalrymple, R A, 1985, Wave height variation across beaches of arbitrary profile. *Journal of Geophysical Research*, 90, 11 917–11 927.

Dalrymple, R A, 1975, A mechanism for rip current generation on an open coast. *Journal of Geophysical Research*, 80, 3485–3487.

Dalrymple, R A, 1992, Prediction of storm/normal beach profiles. *Journal Waterway, Port, Coastal, and Ocean Engineering*, 118, 193–200.

Dalrymple, R T and Lanan, G A, 1976, Beach cusps formed by intersecting waves. *Geological Society of America Bulletin*, 87, 57–60.

Davidson, M A, Russell, P E, Huntley, D A and Hardisty, J, 1993, Tidal asymmetry in suspended sand transport on a macrotidal intermediate beach. *Marine Geology*, 110, 333–353.

Davidson-Arnott, R G D, 1981, Computer simulation of nearshore bar formation. *Earth Surface Processes and Landforms*, 6, 23–34.

Davidson-Arnott, R G D, 1988a, Temporal and spatial controls on beach/dune interaction, Long Point, Lake Erie. *Journal of Coastal Research*, Special Issue 3, 131–136.

Davidson-Arnott, R G D, 1988b, Controls on formation of barred nearshore profiles. *The Geographical Review*, American Geographical Society, 78, 185–193.

Davidson-Arnott, R G D and Fisher, J D, 1992, Spatial and temporal controls on overwash occurrence on a Great Lakes barrier spit. *Canadian Journal of Earth Sciences* 29, 102–117.

Davidson-Arnott, R G D and Greenwood, B, 1974, Bedforms and structures associated with bar topography in the shallow-water wave environment, Kouchibouguac Bay, New Brunswick, Canada. *Journal of Sedimentary Petrology*, 44, 698–704.

Davidson-Arnott, R G D and Greenwood, B, 1976, Facies relationships on a barred coast, Kouchibouguac Bay, New Brunswick, Canada. *In Beach and Nearshore Sedimentation.* Society of Economic Paleontologists and Mineralogists, Special Publication 24, 149–168.

Davidson-Arnott, R G D and Law, M N, 1990, Seasonal patterns and control on sediment supply to coastal foredunes, Long Point, Lake Erie. In Nordstrom, K F, Psuty, N P and Carter, R W G (eds) *Coastal Dunes: Form and Process.* John Wiley, Chichester, 177–200.

Davidson-Arnott, R G D and Pyskir, N M, 1988, Morphology and formation of a Holocene coastal dune field, Bruce Peninsula, Ontario. *Geographie Physique et Quaternaire*, 42, 163–170.

Davies, J L, 1958, Wave refraction and the evolution of shoreline curves. *Geographical Studies*, 5, 1–14.

Davies, J L, 1964, A morphogenic approach to world shorelines. *Zeitschrift fur Geomorphologie*, 8, Mortensen Sonderheft, 127–142.

Davies, J L, 1972, *Geographical Variation in Coastal Development.* Oliver and Boyd, Edinburgh, 204 pp.

Davies, J L, 1974, The coastal sediment compartment. *Australian Geographical Studies*, 12, 139–151.

Davies, J L, 1980, *Geographical Variation in Coastal Development*, 2nd Edition. Longman, London, 212 pp.

Davies, J L, 1982a, High-latitude coasts. In Schwartz, M L (ed.) *Encyclopedia of Beaches and Coastal Environments.* Hutchinson Ross, Stroudsburg, 462–463.

Davies, J L, 1982b, Low-latitude coasts. In Schwartz, M L (ed.) *Encyclopedia of Beaches and Coastal Environments.* Hutchinson Ross, Stroudsburg, 522–524.

Davies, J L, 1982c, Mid-latitude coasts. In Schwartz, M L (ed.) *Encyclopedia of Beaches and Coastal Environments.* Hutchinson Ross, Stroudsburg, 543–544.

Davis, G A, Hanslow, D J, Hibbert, K and Nielsen, P, 1992, Gravity drainage: a new method of beach stabilisation through drainage of the watertable. *Proceeding 23rd Conference on Coastal Engineering.* ASCE, 1129–1141.

Davis, R A, 1982, Beach. In Schwartz, M L (ed.) *Encyclopedia of Beaches and Coastal Environments.* Hutchinson Ross, Stroudsburg, 140–141.

Davis, R A Jr, 1994a, Barrier island systems – a geologic overview. In Davis, R A Jr (ed.) *Geology of Holocene Barrier Island Systems.* Springer-Verlag, New York, 1–46.

Davis, R A Jr, 1994b, Other barrier systems of the world. In Davis, R A Jr (ed), *Geology of Holocene Barrier Island Systems.* Springer-Verlag, New York, 435–456.

Davis, R A and Hayes, M O, 1984, What is a wave-dominated coast? *Marine Geology*, 60, 313–329.

Davis, R A Jr and Schwartz, M L, 1982, Beach cycles. In Schwartz, M L (ed.) *Encyclopedia of Beaches and Coastal Environments.* Hutchinson Ross, Stroudsburg, 143–144.

Davis, R A, Fox, W T, Hayes, M O and Boothroyd, J C, 1972, Comparison of ridge and runnel systems in tidal and non-tidal environments. *Journal of Sedimentary Petrology*, 42, 413–421.

Dean, R G, 1973, Heuristic models of sand transport in the surf zone. *Proceedings of First Australian Coastal Engineering Conference*, Sydney. Institute Engineers Australia, 208–214.

Dean, R G, 1977, *Equilibrium beach profiles: US Atlantic and*

*Gulf Coast Ocean*. Engineering Report No. 12, University of Delaware, 45 pp.

Dean, R G, 1991, Equilibrium beach profiles: characteristics and applications. *Journal of Coastal Research*, 7, 53–84.

Dean, R G, 1997, Models for barrier island restoration. *Journal of Coastal Research*, 13, 694–703.

Dean, R G and Maurmeyer, E M, 1983, Models of beach profile response. In Komar, P D and Moore, J (eds) *CRC Handbook of Coastal Processes and Erosion*. CRC Press, Boca Raton, 151–165.

Dean-Rosati, J, Gravens, M B and Chasten, M A, 1992, Development of detached breakwater design criteria using a shoreline response model. *Coastal Engineering Practice '92*. 814–829.

De Beaumont, L E, 1845, Septieme lecon. In Bertand, P, *Lecons de Geologie Practique*. Paris, 221–252.

De Castro Lopo, L, 1979, Newcastle to Forster: A study of the coast. In Camm, J C R and Loughran, R J (eds) *Newcastle Studies in Geography*. University of Newcastle, NSW Australia, 47–59.

Deigaard, R, Fredsøe, J and Brøker Hedegaard, I, 1986, Suspended sediment in the surf zone. *Journal Waterway, Port, Coastal and Ocean Engineering*, 112, 115–128.

Deigaard, R, Christensen, E D, Damgaard, J S and Fredsoe, J, 1995, Numerical simulation of finite amplitude shear waves and sediment transport. *Proceedings 24th International Conference on Coastal Engineering*. ASCE, 1919–1933.

De Vriend, H J, 1991, Coastal morphodynamics. *Coastal Sediments '91*. ASCE, New York, 356–370.

De Vriend, H J, 1992a, Mathematical modelling and large-scale coastal behaviour: Part 1 – Physical Processes. *Maritime Hydraulics, Journal of Hydraulic Research*, Special Issue, 727–740.

De Vriend, H J, 1992b, Mathematical modelling and large-scale coastal behaviour: Part 2 – Predictive Models. *Maritime Hydraulics, Journal of Hydraulic Research*, Special Issue, 741–753.

De Vriend, H J, 1996, Mathematical modelling of meso-tidal barrier island coasts. In Liu. P L-F (ed.) *Advances in Coastal and Ocean Engineering*. World Scientific, Singapore.

De Vriend, H J, Capobianco, M, Chester, T, de Swart, H E, Latteux, B and Stive, M J F, 1993, Approaches to long-term modelling of coastal morphology: a review. *Coastal Engineering*, 21, 225–269.

Dibajnia, M, Shimizu, T and Watanabe, A, 1994, Profile change of a sheet flow dominated beach. *Proceedings 24th International Conference on Coastal Engineering*. ASCE, 1946–1960.

Dietz, R S, 1963, Wave base, marine profile of equilibrium, and wave-built terraces: a critical appraisal. *Bulletin Geological Society of America*, 74, 971–990.

Dietz, R S, 1964, Wave base, marine profile of equilibrium, and wave-built terraces: reply. *Bulletin Geological Society of America*, 75, 1275–1282.

Dingler, J R, 1982, Wave shadow. In Schwartz, M L (ed.) *Encyclopedia of Beaches and Coastal Environments*. Hutchinson-Ross, Stroudsberg, 870–871.

Dingler, J R and Clifton, H E, 1994, Barrier systems of California, Oregon, and Washington. In Davies, R A Jr (ed.) *Geology of Holocene Barrier Island Systems*. Springer-Verlag, New York, 115–165.

Dodd, N, Oltman-Shay, J and Thornton, E B, 1992, Shear instabilities in the longshore current: A comparison of observation and theory. *Journal of Physical Oceanography*, 22, 62–82.

Doering, J C and Bowen, A J, 1988, Wave-induced flow and nearshore suspended sediment. *Proceedings 21st International Conference on Coastal Engineering*. ASCE, 1452–1463.

Doing, H, 1985, Coastal foredune zonation and succession in various parts of the world. *Vegetation*, 61, 65–75.

Dolan, R, 1971, Coastal landforms: crescentic and rhythmic. *Geological Society of America Bulletin*, 82, 177–180.

Dolan, R and Davis, R E, 1992, An intensity scale for Atlantic northeast storms. *Journal of Coastal Research*, 8, 840–853.

Dolan, T J and Dean, R G, 1985, Multiple longshore bars in the upper Chesapeake Bay. *Estuarine Coastal and Shelf Science*, 21, 727–743.

Dolan, R and Ferm, J C, 1968, Concentric landforms along the Atlantic coast of the United States. *Science*, 159, 627–629.

Dominick, T F, Wilkins, B and Roberts, H, 1971, Mathematical model for beach groundwater fluctuations. *Water Resources Research*, 7, 1626–1635.

Dominguez, J M L, Martin, L and Bittencourt, A C S P, 1987, Sea-level history and Quaternary evolution of river mouth-associated beach-ridge plains along the east-southeast Brazilian coast: A summary. In Nummedal, D, Pilkey, O H and Howard, J D (eds) *Sea-level Fluctuations and Coastal Evolution*. Society Economic Paleontologists *and* Mineralogists, Special Publication No. 41, 115–127.

Douglas, S L, 1990, Influence of wind on breaking waves. *Journal Waterways, Port, Coastal and Ocean Engineering*, 116, 651–663.

Downing, J P, Sternberg, R W and Lister, C R B, 1981, New instrumentation for the investigation of sediment suspension processes in the shallow marine environment. *Marine Geology*, 42, 19–34.

Draper, L, 1980, Wave climatology of the U.K. continental shelf. In Branner, F T, Colins, M B and Massie, K S (eds) *The North-West European Shelf Seas: The Sea Bed and Sea in Motion: II Physical and Chemical Oceanography and Physical Resources*. Elsevier, Amsterdam, 353–368.

Dubois, R N, 1995, The transgressive barrier model: An alternative to two-dimensional volume balanced models. *Journal of Coastal Research*, 11, 1272–1286.

Dubois, R N, 1997, The influence of the shore slopes ratio on the nature of a transgressing shore. *Journal of Coastal Research*, 13, 1321–1327.

Duncan, J R, 1964, The effects of watertable and tide cycle on swash-backwash sediment distribution and beach profile development. *Marine Geology*, 2, 186–197.

Dury, G H, 1968, Streams- underfit. In Fairbridge, R W (ed.) *Encyclopedia of Geomorphology*. Reinhold, New York, 1070–1071.

Dyer, K R, 1986, *Coastal and Estuarine Sediment Dynamics*.

John Wiley and Sons, London, 342 pp.

Dyer, K R and Soulsby, R L, 1988, Sand transport on the continental shelf. *Annual Review of Fluid Mechanics*, 20, 295–324.

Dyhr-Nielsen, M and Sorensen, T, 1970, Sand transport phenomena on coasts with bars. *Proceedings 12th International Conference on Coastal Engineering*. ASCE, 855–866.

Eagleson, P S and Dean, R G, 1959, Wave-induced motion of bottom sediment particles. *Proceedings of the American Society of Civil Engineers*, (HY 10), 53–79.

Eagleson, P S and Dean, R G, 1961, Wave induced motion of bottom sediment particles. *Transactions of the American Society of Civil Engineers* 126, 1162–1189.

Eagleson, P S, Dean, R G and Peralta, L A, 1958, *The Mechanics of the Motion of Discrete Spherical Bottom Sediment Particles due to Shoaling Waves*. Beach Erosion Board, US Army Corps of Engineers, Technical Memorandum No. 104, Washington, DC.

Eagleson, P S, Glenne, B and Dracup, J A, 1963, Equilibrium characteristics of sand beaches. *Journal of the Hydraulics Division* ASCE, 89, 35–57.

Eckart, C, 1951, *Surface Waves in Water of Variable Depth*. Scripps Institute of Oceanography, Wave Report 100, ref. 51–12.

Ehlers, J, 1988, *The Morphodynamics of the Wadden Sea*. Balkema, Rotterdam, 397 pp.

Elgar, S and Guza, R T, 1985, Observations of bispectra of shoaling surface gravity waves. *Journal of Fluid Mechanics*, 161, 425–448.

Elgar, S, Herbers, T H C, Okihiro, M, Oltman-Shay, J and Guza, R T, 1992, Observations of infragravity waves. *Journal of Geophysical Research*, 97, 15 573–15 577.

Eliot, I G and Clarke, D J, 1986, Minor storm impact on the beachface of a sandy beach. *Marine Geology*, 73, 61–83.

Eliot, I G and Clarke, D J, 1988, Semi diurnal variation in beachface aggradation and degradation. *Marine Geology*, 79, 1–22.

Emery, K O and Foster, J F, 1948, Water tables in marine beaches. *Journal of Marine Research*, 7, 644–654.

Emery, K O and Gale, J F, 1951, Swash and swash marks. *EOS, Transactions American Geophysical Union*, 32, 31–36.

Emery, D and Myers, K, 1996, *Sequence Stratigraphy*. Blackwell Science, Oxford, 297 pp.

Enright, N J and Anderson, M J, 1988, Recent evolution of the Mangawhai Spit dunefield. *Journal Royal Society of New Zealand*, 18, 359–367.

Enright, N J, McLean, R F and Dodson, J R, 1988, Late Holocene development of two wetlands in the Te Paki region, far northern New Zealand. *Journal Royal Society of New Zealand*, 18, 369–382.

EOS, 1994, Dozer duck dominates the nearshore scene this summer. *Transactions, American Geophysical Union*, 75, 369–370.

Ericksen, N J, 1970, Measurement of tide induced changes to water table profiles in coarse and fine sand beaches along Pegasus Bay, Canterbury. *Earth Sciences Journal*, 4, 24–31.

Escher, B G, 1937, Experiments on the formation of beach cusps. *Leid. Geo. Meded*, 9, 79–104 (referenced in Komar, 1976).

Esler, A E, 1970, Manawatu sand dune vegetation. *Proceedings New Zealand Ecological Society*, 17, 41–46.

Evans, M W, Hine, A C, Belknap, D F and Dabis, R A Jr, 1985, Bedrock controls on barrier island development: west-central Florida coast. *Marine Geology*, 63, 263–283.

Evans, O F, 1938, The undertow. *Science*, 88, 279–281.

Evans, O F, 1940, The low and ball of the east shore of Lake Michigan. *Journal of Geology*, 48, 476–511.

Everts, C H, 1987, Continental shelf evolution in response to a rise in sea level. In Nummedal, D, Pilkey, O H and Howard, J D (eds) *Sea-level Fluctuations and Coastal Evolution*. Society of Economic Palaeontologists and Mineralogists, Special Publication No. 41, 49–57.

Fabbri, P (ed.) 1990, *Recreational Uses of Coastal Areas*. Kluwer Academic Publishers, Dordrecht, 285 pp.

Fairbridge, R W, 1968, Beach. In Fairbridge, R W (ed.) *The Encyclopedia of Geomorphology*. Reinhold, New York, 62–68.

Farrelly, W, 1972, *Form and sediments of an Antarctic beach*. Unpublished MA Thesis, Department of Geography, University of Canterbury, Christchurch, 109 pp.

Fenneman, N M, 1902, Development of the profile of equilibrium of the subaqueous shore terrace. *Geology*, 10, 1–32.

Ferland, M A, 1990, *Shelf Sand Bodies in Southeastern Australia*. Unpublished PhD Thesis, University of Sydney, Australia, 527 pp.

Ferland, M, 1993, Sedimentology and morphology of shelf sand bodies of southeastern Australia. In Thoms, M (ed.) *Catchments and Coasts in Eastern Australia*. Department of Geography, University of Sydney.

Fetter, C W, 1988, *Applied Hydrology*, 2nd edition. Merrill, Columbus, 487 pp.

Field, M E and Duane, D B, 1976, Post-Pleistocene history of the United States inner continental shelf: Significance of origin to barrier islands. *Geological Society of America Bulletin*, 87, 691–702.

Field, M E and Roy, P S, 1984. Offshore transport and sand body formation: evidence from a steep, high energy shoreface, southeastern Australia. *Journal of Sedimentary Petrology*, 54, 1292–1302.

Filion, L and Morisset, P, 1983, Eolian landforms along the eastern coast of Hudson Bay, Northern Quebec. *Nordicana*, 47, 73–94.

Finkel, J H, 1959, The barchans of southern Peru. *Journal of Geology*, 67, 614–647.

Finkl, C W Jr (ed.) 1995, Holocene cycles: climate, sea levels, and sedimentation. *Journal of Coastal Research*, Special Issue, 17, 402 pp.

Finnigan, J J, Neil, D, Lees, B G, Croome, R J and Woodgate, M, 1989, Modelling the wind flow pattern around a parabolic sand dune. *Proceedings 8th Biennial Conference of the Simulation Society of Australia and the International Association for Mathematics and Computers in Simulation*. 457–462.

Fisher, J S and Dolan, R (eds), 1977, *Beach Processes and Coastal Hydrodynamics*. Dowden, Hutchinson and Ross,

Stroudsburg, 382 pp.

Fitzgerald, D N, Rosen, P S and van Heteren, S, 1994, New England barriers. In Davies, R A Jr (ed.) *Geology of Holocene Barrier Island Systems*. Springer-Verlag, New York, 305–394.

Fitzjones, O and Rainger, T, 1992, *The Stormrider Guide: Europe*. Low Pressure Publication, London, 192 pp.

Fleming, N C, 1964, Tank experiments on the sorting of beach material during cusp formation. *Journal of Sedimentary Petrology*, 34, 112–122.

Flemming, B W and Fricke, A H, 1983, Beach and nearshore habitats as a function of internal geometry, primary sedimentary structures and grainsize. In McLachlan, A and Erasmus, T (eds) *Sandy Beaches as Ecosystems*. Junk, The Hague, 115–132.

Fletcher, C H, Fairbridge, R W, Moller, J J and Long, A J, 1993, Emergence of the Varanger Peninsula, Arctic Norway, and climate changes since deglaciation. *The Holocene*, 3, 116–127.

Fletcher, C H, Mullane, R A and Richmond, B M, 1997, Beach loss along armoured shorelines on Oahu, Hawaiian islands. *Journal of Coastal Research*, 13, 209–215.

Foote, Y L M and Huntley, D A, 1994, Velocity moments on a macro-tidal intermediate beach. *Proceedings Coastal Dynamics '94*. ASCE, 794–808.

Forbes, D L and Syvitski, J P M, 1994, Paraglacial coasts. In Carter, R W G and Woodroffe, C D (eds) *Coastal Evolution. Late Quaternary Shoreline Morphodynamics*. 373–424.

Forbes, D L, Taylor, R B, Shaw, J, Carter, R W G and Orford, J D, 1990, Development and stability of barrier beaches on the Atlantic coast of Nova Scotia. *Proceedings Canadian Coastal Conference*, Ottawa Natural Research Council, 83–98.

Forbes, D L, Orford, J D, Carter, R W G, Shaw, J and Jennings, S C, 1995, Morphodynamic evolution, self-organisation, and instability of coarse-clastic barriers on paraglacial coasts. *Marine Geology*, 16, 63–85.

Foster, D, Gordon, A D and Lawson, N Y, 1975, The storms of May–June 1974, Sydney, NSW. *Proceedings of the 2nd Australian Conference on Coastal and Ocean Engineering*. The Institution of Engineers, Australia, 1–11.

Fox, W T and Davis, R A, 1973, Simulation model for storm cycles and beach erosion on Lake Michigan. *Geological Society of America Bulletin*, 84, 1769–1790.

Fox, W T and Davis, R A, 1978, Seasonal variation in beach erosion and sedimentation on the Oregon coast. *Geological Society of America Bulletin*, 89, 1541–1549.

Fredsøe, J and Deigaard, R, 1992, *Mechanics of Coastal Sediment Transport*. World Scientific, Singapore, 369 pp.

Frey, R W and Pemberton, S G, 1984, Trace fossil facies models. In Walker, R G (ed.) *Facies Models*, 2nd edition. Geoscience Canada Reprint Series 1, 189–207.

Fryberger, S G, 1991, Unusual sedimentary structures in the Oregon coastal dunes. *Journal of Arid Environments*, 21, 131–150.

Fryberger, S G and Dean, G, 1979, Dune forms and wind regime. In McKee, E D (ed.) *A Study of Global Sand Seas*. Geological Survey Professional Paper 1052, US Government Printing Office, Washington, 137–169.

Fryberger, S G, Krystinik, L F and Schenk, C J, 1990, Tidally flooded back-barrier dunefield, Guerrero Negro area, Baja California, Mexico. *Sedimentology*, 37, 1–23.

Gallagher, B, 1971, Generation of surf beat by nonlinear wave interactions. *Journal of Fluid Mechanics*, 49, 1–20.

Galloway, J S, Collins, M B and Moran, A D, 1989, Onshore/offshore wind influence on breaking waves: an empirical study. *Coastal Engineering*, 13, 305–323.

Galvin, C J, 1967, Longshore current velocity: a review of theory and data. *Reviews of Geophysics*, 5, 287–304.

Galvin, C J, 1968, Breaker type classification on three laboratory beaches. *Journal of Geophysical Research*, 73, 3651–3659.

Gares P A and Nordstrom K F, 1987, Dynamics of a coastal foredune blowout at Island Beach State Park, N.J. *Proceedings Coastal Sediments '87*. ASCE, 213–221.

Gares, P A and Nordstrom, K F, 1995, A cyclic model of foredune blowout evolution for a leeward coast: Island Beach, New Jersey. *Annals Association of American Geographers*, 85, 1–20.

Gerritsen, F and van Heteren, J, 1984, Low frequency oscillations on the Dutch coast. *Proceedings 19th International Conference on Coastal Engineering*. ASCE, 625–641.

Gilbert, G K, 1877, *Report on the Geology of the Henry Mountains*. Government Printing Office, Washington, DC, 160 pp.

Gilbert, G K, 1885, The topographic features of lake shores. *US Geological Survey 5th Annual Report*. 69–123.

Gishler, E and Lomando, A J, 1997, Holocene cemented beach deposits in Belize. *Sedimentary Geology*, 100, 277–297.

Glenn, M, 1979, Glossary. In McKee, E D (ed.) *A Study of Global Sand Seas*. Geological Survey Professional Paper 1052, US Government Printing Office, Washington, 399–407.

Goda, Y, 1975, Irregular wave deformation in the surf zone. *Coastal Engineering in Japan*, 18, 13–27.

Godfrey, P J and Godfrey, M M, 1976, *Barrier Island Ecology of Cape Lookout National Seashore and Vicinity, North Carolina*. National Park Service, Scientific Monograph Series 9, 160 pp.

Godfrey, P J, Leatherman, S A and Buckley, P A, 1978, Impact of off-road vehicles on coastal ecosystems. *Proceedings Coastal Zone '78*. ASCE, 581–600.

Godfrey, P J, Leatherman, S P and Zaremba, R, 1979, A geobotanical approach to classification of barrier beach systems. In Leatherman, S P (ed.) *Barrier Islands from the Gulf of St Lawrence to the Gulf of Mexico*. Academic Press, New York, 99–126.

Goldsmith, V, 1973, Internal geometry and origin of vegetated coastal sand dunes. *Journal of Sedimentary Petrology*, 43, 1128–1142.

Goldsmith, V, 1975, *Eolian Sedimentation in Coastal Areas*. VIMS Contribution No. 533, Virginia Institute of Marine Science, Gloucester Point, 138 pp.

Goldsmith, V, 1978, Coastal Dunes. In Davis, R A Jr (ed.) *Coastal Sedimentary Environments*. Springer-Verlag, New

York, 171–230.

Goldsmith, V and Sofer, S, 1983, Wace climatology of the southeastern Mediterranean. *Israel Journal of Earth Sciences*, 32, 1–51.

Goldsmith, V, Hennigar, H F, Gutman, A L and Blake, N T, 1977, *Coastal processes and resulting forms of sediment accumulations, Currituck Spit, Virginia-North Carolina.* Field trip guide book, Society of Economic Paleontologists and Mineralogists and Department of Geological Oceanography, Virginia Institute of Marine Science, SRAMSOE 143.

Goldsmith, V, Bowman, D and Kiley, K, 1982a, Sequential stage development of crescentic bars: Hahoterim Beach, southeastern Mediterranean. *Journal of Sedimentary Petrology*, 52, 233–249.

Goldsmith, V, Bowman, D, Kiley, K, Burdick, B, Mart, Y and Sofer, S, 1982b, Morphology and dynamics of crescentic bar systems. *Proceedings 18th International Conference on Coastal Engineering.* ASCE, 941–953.

Gordon, A D, 1987, A tentative but tantalizing link between sea-level rise and coastal recession in New South Wales, Australia. In Pearman, I (ed.) *Greenhouse, Planning for Climate Change.* CSIRO, Melbourne, 121–133.

Gorycki, M A, 1973, Sheetflow structure: mechanism of beach cusp formation and related phenomena. *Journal of Geology*, 81, 109–117.

Gould, R, 1997, USLA lifesaving statistics, National Lifesaving Statistics Report 1996. *American Lifeguard Magazine*, 13, 8–9.

Gourlay, M R, 1968, *Beach and Dune Erosion Tests.* Delft Hydraulics Laboratory, Report No. M935/M936.

Gourlay, M R, 1992, Wave set-up, wave run-up and beach ground water table: Interaction between surf zone hydraulics and ground water hydraulics. *Coastal Engineering*, 17, 93–144.

Graf, J B, 1976, Comparison of measured and predicted nearshore grain size distribution patterns, southwestern Lake Michigan, U.S.A. *Marine Geology*, 22, 253–270.

Grant, U S, 1946, Effect of ground-water table on beach erosion (abstract). *Bulletin Geological Society of America*, 57, 1252.

Grant, U S, 1948, Influence of the water table on beach aggradation and degradation. *Journal Marine Research*, 7, 655–660.

Grant, W D and Madsen, O S, 1982, Movable bed roughness in unsteady oscillatory flow. *Journal of Geophysical Research*, 87, 469–481.

Grant, W D and Madsen, O S, 1986, The continental-shelf bottom boundary layer. *Annual Review of Fluid Mechanics*, 18, 265–305.

Greeley, R and Iversen, J D, 1985, *Wind as a Geological Process on Earth, Mars, Venus and Titan.* Cambridge University Press, London, 333 pp.

Green, M O and Vincent, C E, 1990, Wave entrainment of sand from a rippled bed. *Proceedings 22nd International Conference on Coastal Engineering.* ASCE, 2200–2212.

Greenspan, H P, 1958, On the breaking of water waves of finite amplitude on a sloping beach. *Journal of Fluid Mechanics*, 4, 330–334.

Greenwood, B G, 1987, Sediment balance and bar morphodynamics in a multiple bar system: Georgian Bay, Canada. In Gardiner, V (ed.) *International Geomorphology 1986* Part 1. John Wiley & Sons, Chichester, 1119–1143.

Greenwood, B G and Davidson-Arnott, R G D, 1979, Sedimentation and equilibrium in wave-formed bars: a review and case study. *Canadian Journal of Earth Sciences*, 16, 312–332.

Greenwood, B and Davis, R A, Jr (eds), 1984, *Hydrodynamics and sedimentation in wave-dominated coastal environments. Marine Geology*, 60, 473 pp.

Greenwood, B and Hale, P B, 1982, Lagrangian sediment motion in a crescentic nearshore bar under storm-induced waves and currents. *Canadian Journal of Earth Sciences*, 19, 424–433.

Greenwood, B and Osborne, P D, 1988, Vertical structure in shore-parallel flows and bed roughness. *Proceedings Workshop on Roughness and Friction.* National Research Council of Canada, 75–91.

Greenwood, B and Osborne, P D, 1990, Vertical and horizontal structure in cross-shore flows: An example of undertow and wave set-up on a barred beach. *Coastal Engineering*, 14, 543–580.

Greenwood, B and Osborne, P D, 1991, Equilibrium slopes and cross-shore velocity asymmetries in a storm-dominated, barred nearshore system. *Marine Geology*, 96, 211–235.

Greenwood, B and Sherman, D J, 1984, Waves, currents, sediment flux and morphological response in a barred nearshore system. *Marine Geology*, 60, 31–61.

Greenwood, B and Sherman, D J, 1986, Longshore current profiles and lateral mixing across the surf zone of a barred nearshore. *Coastal Engineering*, 10, 149–168.

Greenwood, B, Richards, R G and Brander, R W, 1993, Acoustic imaging of sea-bed geometry: a high resolution remote tracking sonar (HRRTSII). *Marine Geology*, 112, 207–218.

Griggs, G B, 1990, Littoral drift impoundment and beach nourishment in northern Monterey Bay, California. *Journal of Coastal Research*, Special Issue 6, 115–126.

Griggs, G B and Tait, J F, 1988, The effects of coastal protection structures on beaches along northern Monterey Bay, California. *Journal of Coastal Research*, Special Issue 4, 93–111.

Griggs, G B, Tait, J F, Scott, K and Plant, N, 1991, The interaction of seawalls and beaches: four years of field monitoring, Monterey Bay, California. *Proceedings Coastal Sediments '91*, ASCE, 1871–1885.

Guilcher, A, 1958, *Coastal and Submarine Morphology.* Methuen, London, 274 pp.

Guilcher, A, 1961, 'Le beachrock' ou gres dr plage. *Annales de Geographie*, 70, 113–125.

Guillen, J and Palanques, A, 1993, Longshore bar and trough systems in a microtidal, storm-wave dominated coast: the Ebro Delta (Northwestern Mediterranean). *Marine Geology*, 115, 239–252.

Gulliver, F, 1899, Shoreline topography. *American Academy of*

*Arts and Science Proceedings*, 34, 151–258.
Gunter, G, 1979, Notes on sea beach ecology: Food sources on sandy beaches and localised diatom blooms bordering Gulf beaches. Gulf Research Report 6, 305–307.
Guza, R T and Bowen, A J, 1975, The resonant instabilities of long waves obliquely incident on a beach. *Journal of Geophysical Research*, 80, 4529–4534.
Guza, R T and Bowen, A J, 1976, Resonant interactions from waves breaking on a beach. *Proceedings 15th International Coastal Engineering Conference*. ASCE, 560–579.
Guza, R T and Bowen, A J, 1981, On the amplitude of beach cusps. *Journal of Geophysical Research*, 86, 4125–4132.
Guza, R T and Davis, R E, 1974, Excitation of edge waves by waves incident on a beach. *Journal of Geophysical Research*, 79, 1285–1291.
Guza, R T and Inman, D L, 1975, Edge waves and beach cusps. *Journal of Geophysical Research*, 80, 2997–3012.
Guza, R T and Thornton, E B, 1981, Wave set-up on a natural beach. *Journal of Geophysical Research*, 86, 4133–4137.
Guza, R T and Thornton, E B, 1982, Swash oscillations on a natural beach. *Journal of Geophysical Research*, 87, 483–491.
Guza, R T and Thornton, E B, 1985a, Observations of surf beat. *Journal of Geophysical Research*, 90, 3161–3172.
Guza, R T and Thornton, E B, 1985b, Velocity moments in nearshore. *Journal Waterway, Port, Coastal and Ocean Engineering*, 111, 235–256.
Guza, R T, Thornton, E B and Holman, R A, 1984, Swash on steep and shallow beaches. *Proceedings 19th International Conference on Coastal Engineering*. ASCE, 708–723.
Hack, J T, 1960, Interpretation of erosion topography in humid temperate regions. *American Journal of Science*, 258–A, 80–97.
Hacking, N, 1998, Macrofaunal community structure and zonation of beaches in northern NSW, Australia. *Marine and Freshwater Research*, 49, 47–53.
Hails, J R and Bennett, J M, 1981, Wind and sediment movement in coastal dune areas. *Proceedings 17th International Coastal Engineering Conference*. ASCE, 1565–1575.
Haines, J W and Sallenger, A H, 1994, Vertical structure of mean cross-shore currents across a barred surf zone. *Journal of Geophysical Research*, 99, 14 223–14 242.
Hale, P B and McCann, S B, 1982, Rhythmic topography in a mesotidal, low-wave-energy environment. *Journal of Sedimentary Petrology*, 52, 415–429.
Hall, M J and Pilkey, O H, 1991, Effects of hard stabilization on dry beach width for New Jersey. *Journal of Coastal Research*, 7, 771–785.
Hallermeier, R J, 1981, A profile zonation for seasonal sand beaches from wave climate. *Coastal Engineering*, 4, 253–277.
Hands, E B, 1983, The Great Lakes as a test model for profile response to sea level changes. In Komar, P D (ed.) *Handbook of Coastal Processes and Erosion*. CRC Press, Boca Raton, Florida, 176–89.
Hanes, D M, 1988, Intermittent sediment suspension and its implications to sand tracer dispersal in wave-dominated environments. *Marine Geology*, 81, 175–183.
Hanes, D M and Huntley, D A, 1986, Continuous measurements of suspended sand concentration in a wave dominated nearshore environment. *Continental Shelf Research*, 6, 585–596.
Hanes, D M, Vincent, C E, Huntley, D A and Clarke, T L, 1988, Acoustic measurements of suspended sand concentration in the $C^2S^2$ experiment at Stanhope Lane, Prince Edward Island. *Marine Geology*, 81, 185–196.
Hanslow, D J, 1989, *Patterns of Grain-Size Distributions on the Shoreface*. MSc Thesis, University of Sydney, 182 pp.
Hanslow, D J and Nielsen, P, 1993, Shoreline setup on natural beaches. *Journal of Coastal Research*, Special Issue 15, 1–10.
Hanson, H, 1989, Genesis – A generalised shoreline change numerical model. *Journal of Coastal Research*, 5, 1–27.
Hardisty, J, 1986, A morphodynamic model for beach gradients. *Earth Surface Processes and Landforms*, 11, 327–333.
Hardisty, J, 1990, *Beaches, Form and Process*. Unwin Hyman, London, 324 pp.
Hardisty, J, 1994, Beach and nearshore sediment transport. In Pye, K (ed.) *Sediment Transport and Depositional Processes*. Blackwell, London, 216–255.
Hardisty, J and Whitehouse, R J S, 1988, Evidence for a new sand transport process from experiments on Saharan dunes. *Nature*, 322, 532–534.
Hardisty, J, Collier, J and Hamilton, D, 1984, A calibration of the Bagnold beach equation. *Marine Geology*, 61, 95–101.
Harms, J C and Fahnestock, R K, 1965, Stratification, bed forms, and flow phenomena (with an example from the Rio Grande). In *Sedimentary Processes*. Society of Economic Paleontologists and Mineralogists, Reprint Series 3, 84–115.
Harms, J C, Southard, J B and Walker, R G, 1982, *Structures and Sequences in Clastic Rocks*. Society of Economic Paleontologists and Mineralogists, Short Course No. 9.
Harris, C, 1974, Wind speed and sand movement in a coastal dune environment. *Area*, 6, 243–249.
Harrison, W, 1969, Empirical equations for foreshore changes over a tidal cycle. *Marine Geology*, 7, 529–551.
Harwood, P, 1993, *Modelling airflow and salt aerosol distribution over a storm shutter physiognomy on the north coast of NSW*. Unpublished BSc (Hons.) Thesis, University of Sydney, Australia.
Hashimoto, H and Uda, T, 1976, Shore processes at Ajigaura Beach. *Proceedings 23rd Japanese Conference Coastal Engineering*. JSCE, 245–249 (in Japanese).
Hashimoto, H and Uda, T, 1977, Shore processes at Ajigaura Beach. *Proceedings 24th Japanese Conference Coastal Engineering*. JSCE, 216–220 (in Japanese).
Hattori, M and Kawamata, R, 1980, Onshore–offshore transport and beach profile change. *Proceedings 17th International Conference on Coastal Engineering*. ASCE, 1175–1193.
Hawley, N, 1982, Intertidal sedimentary structures on macrotidal beaches. *Journal of Sedimentary Petrology*, 52, 785–795.
Hay, A E and Bowen, A J, 1993, Spatially-correlated depth changes in the nearshore zone during autumn storms. *Journal of Geophysical Research*, 98, 12 387–12 404.
Hay, A E and Bowen, A J, 1994, Coherence scales of wave-induced suspended sand concentration. *Journal of Geophysi-*

*cal Research*, 99, 12 749–12 765.

Hay, A E and Sheng, J, 1992, Vertical profiles of suspended sand concentration and size from multifrequency acoustic backscatter. *Journal of Geophysical Research*, 97, 15 661–15 677.

Hay, A E and Wilson, D J, 1994, Rotary sidescan images of nearshore bedform evolution during a storm. *Marine Geology*, 119, 57–65.

Hayes, M O, 1967a, Relationship between coastal climate and bottom sediment type on the inner continental shelf. *Marine Geology*, 5, 111–132.

Hayes, M O, 1967b, Hurricanes as geological agents, south Texas coast. *American Association of Petroleum Geologists Bulletin*, 51, 937–942.

Hayes, M O, 1979, Barrier island morphology as a function of tidal and wave regime. In Leatherman, S P (ed.) *Barrier Islands*. Academic Press, New York, 1–27.

Hayes, M O, 1994, The Georgia Bight barrier system. In Davis, R A Jr (ed.) *Geology of Holocene Barrier Island Systems*, Springer-Verlag, New York, 233–304.

Hayes, M O and Boothroyd, J C, 1969, Storms as modifying agents in the coastal environment. In Hayes, M O (ed.) *Coastal Environments: NE Massachusetts*. Department of Geology, University of Massachusetts, Amherst, 290–315.

Hayes, M O and Ruby, C H, 1994, Barriers of Pacific Alaska. In Davies, R A Jr (ed.) *Geology of Holocene Barrier Island Systems*. Springer-Verlag, New York, 395–433.

Haynes, D and Quinn, G P, 1995, Temporal and spatial variability in community structure of a sandy intertidal beach, Cape Paterson, Victoria, Australia. *Marine and Freshwater Research*, 46, 931–942.

Hedegaard, I B, Deigaard, R and Fredsoe, J, 1991, Onshore/offshore sediment transport and morphological modelling of coastal profiles. *Proceedings Coastal Sediments '91*. 643–654.

Hegge, B J and Masselink, G, 1991, Groundwater-table response to wave run-up: An experimental study from Western Australia. *Journal of Coastal Research*, 7, 623–634.

Hegge, B J and Masselink, G, 1996, Spectral analysis of geomorphic time series; auto-spectrum. *Earth Surface Processes and Landforms*, 21, 1021–1040.

Hellstrom, G B and Lubke, R A, 1993, Recent changes to a climbing–falling dune system on the Robberg Peninsula Southern Cape Coast, South Africa. *Journal of Coastal Research*, 9, 647–653.

Hennigar, H F Jr 1979, *Historical Evolution of Coastal Sand Dunes on Currituck Spit, Virginia/North Carolina*. MA Thesis, College of William and Mary, Virginia, 121 pp.

Hequette, A and Ruz, M-H, 1991, Spit and barrier island migration in the southeastern Canadian Beaufort Sea. *Journal of Coastal Research*, 7, 677–698.

Herbers, T H C, Elgar, S, Guza, R T and O'Reilly, W C, 1993, Infragravity frequency (0.005–0.05 Hz) motions on the shelf. *Proceedings 23rd International Conference on Coastal Engineering*. ASCE, 346–359.

Herbers, T H C, Elgar, S and Guza, R T, 1995a, Generation and propagation of infragravity waves. *Journal of Geophysical Research*, 100, 24 863–24 872.

Herbers, T H C, Elgar, S, Guza, R T and O'Reilly, W C, 1995b, Infragravity frequency (0.005–0.05 Hz) motions on the shelf. Part II: Free waves. *Journal of Physical Oceanography*, 25, 1063–1079.

Hesp, P A, 1981, The formation of shadow dunes. *Journal of Sedimentary Petrology*, 51, 101–112.

Hesp, P A, 1982, *Dynamics and Morphology of Foredunes in S.E. Australia*. Unpublished PhD Thesis, Deptartment of Geography, University of Sydney.

Hesp, P A, 1983, Morphodynamics of incipient foredunes in NSW, Australia. In Brookfield, M E and Ahlbrandt, T S (eds) *Eolian Sediments and Processes*. Elsevier, Amsterdam, 325–342.

Hesp, P A, 1984a, Foredune formation in southeast Australia. In Thom, B G (ed.) *Coastal Geomorphology in Australia*. Academic Press, London, 69–97.

Hesp, P A, 1984b, The formation of sand 'beach ridges' and foredunes. *Search*, 15, 289–291.

Hesp, P A, 1988a, Surfzone, beach and foredune interactions on the Australian southeast coast. *Journal of Coastal Research*, Special Issue 3, 15–25.

Hesp, P A, 1988b, Morphology, dynamics and internal stratification of some established foredunes in southeast Australia. *Journal of Sedimentary Geology*, 55, 17–41.

Hesp, P A, 1989, A review of biological and geomorphological processes involved in the initiation and development of incipient foredunes. In Gimmingham, C H, Ritchie, W, Willetts, BB and Willis, A J (eds) *Coastal Sand Dunes*. Proceedings of the Royal Society Edinburgh 96B, 181–202.

Hesp, P A, 1991, Ecological processes and plant adaptations on coastal dunes. *Journal of Arid Environments*, 21, 165–191.

Hesp, P A, 1993, Late Quaternary coastal geomorphology of the south Sydney – Illawarra Coast. In Thoms, M C (ed.) *Catchments and Coasts in Eastern Australia*. Department of Geography, University of Sydney, Research Monograph No. 5, 72–87.

Hesp, P A and Chape, S, 1984, *A 1:3 million map of the coastal environment of Western Australia*. Central Map Agency, WA Department of Lands and Survey, Perth, WA.

Hesp, P A and Hilton, M, 1996, Nearshore–surfzone system limits and the impacts of sand extraction. *Journal of Coastal Research*, 12, 726–747.

Hesp, P A and Hyde, R, 1996, Geomorphology and dynamics of a trough blowout. *Sedimentology*, 43, 505–525.

Hesp, P A and Pelham, A F, 1984, Vegetation succession in parabolic dune deflation basins, Dongara to Cliff Head, Western Australia. Abstract. *4th International Conference on Mediterranean Ecosystems*, Perth, WA.

Hesp, P A and Thom, B G, 1990, Geomorphology and evolution of active transgressive dunefields. In Nordstrom, K F, Psuty, N P and Carter, R W G (eds) *Coastal Dunes: Form and Process*. John Wiley, Chichester, 253–288.

Hesp, P A, Illenberger, W, Rust, I, McLachlan, A and Hyde, R, 1989, Some aspects of transgressive dunefield and transverse dune geomorphology and dynamics, south coast, South Africa. *Zietschrift fur Geomorphologie*, N.F. Suppl-Bd 73,

111–123.

Hibberd, S and Peregrine, D H, 1979, Surf and runup on a beach: A uniform bore. *Journal of Fluid Mechanics*, 95, 323–345.

Hill, G and Hunter, R E, 1976, Interaction of biological and geological processes in the beach and nearshore environments, northern Padre Island, Texas. In *Beach and Nearshore Sedimentation*. Society Economic Paleontologists Mineralogists, Special Publication 24, 169–187.

Hill, P R, Barnes, P W, Hequette, A and Ruz, M-H, 1994, Arctic coastal plain shorelines. In Carter, R W G and Woodroffe, C D (eds) *Coastal Evolution. Late Quaternary Shoreline Morphodynamics*, Cambridge University Press, New York, 341–372.

Hine, A C, 1979, Mechanisms of berm development and resulting beach growth along a barrier spit complex. *Sedimentology*, 26, 333–351.

Hino, M, 1974, Theory on formation of rip current and cuspidal coast. *Proceedings 14th International Conference on Coastal Engineering*. ASCE, 901–919.

Ho, D V and Meyer, R E, 1962, Climb of a bore on a beach: Part 1 Uniform beach slope. *Journal of Fluid Mechanics*, 14, 305–318.

Ho, D V, Meyer, R E and Shen, M C, 1963, Long surf. *Journal of Marine Research*, 21, 219–230.

Hoefel, F G and Klein, A H de F, 1998, Beach safety issues at oceanic beaches of central northern coast of Santa Catarina, Brazil: magnitude and nature. *Journal of Coastal Research*, Special Issue, 26, A13 (abstract).

Hogbom, I, 1923, Ancient inland dunes of northern and middle Europe. *Geografiska Annaler*, 5, 113–242.

Holland, G J, Lynch, A H and Leslie, L M, 1987, Australian east-coast cyclones, part 1. Synoptic overview and case study. *Monthly Weather Review*, 115, 304–326.

Holland, K T, Raubenheimer, B, Guza, R T and Holman, R A, 1995, Runup kinematics on a natural beach. *Journal of Geophysical Research*, 100, 4985–4993.

Holman, R A, 1981, Infragravity energy in the surf zone. *Journal of Geophysical Research*, 86, 6442–6450.

Holman, R A, 1983, Edge waves and the configuration of the shoreline. In Komar, P D (ed.) *CRC Handbook of Coastal Processes and Erosion*. CRC Press, Boca Raton, Florida, 21–33.

Holman, R A, 1986, Extreme value statistics for wave runup on a natural beach. *Coastal Engineering*, 9, 527–544.

Holman, R A, 1995, Nearshore processes. *Reviews of Geophysics*, Supplement 1995, 1237–1247.

Holman, R A and Bowen, A J, 1979, Edge waves on complex beach profile. *Journal of Geophysical Research*, 84, 6339–6346.

Holman, R A and Bowen, A J, 1982, Bars, bumps, and holes: Models for the generation of complex beach topography. *Journal of Geophysical Research*, 87, 457–468.

Holman, R A and Bowen, A J, 1984, Longshore structure of infragravity motions. *Journal of Geophysical Research*, 89, 6446–6452.

Holman, R A and Sallenger, A H, 1985, Setup and swash on a natural beach. *Journal of Geophysical Research*, 90, 945–953.

Holman, R A and Sallenger, A H, 1993, Sand bar generation: a discussion of the Duck experiment series. *Journal of Coastal Research*, Special Issue 15, 76–92.

Holman, R A, Huntley, D A and Bowen, A J, 1978, Infragravity waves in storm conditions. *Proceedings 16th International Conference on Coastal Engineering*. ASCE, 268–284.

Holman, R A, Howd, P, Oltman-Shay, J and Komar, P, 1990, Observations of the swash expression of far infragravity wave motions. *Proceedings 22nd International Conference on Coastal Engineering*. ASCE, 1242–1253.

Homma, M and Sonu, C J, 1962, Rhythmic pattern of longshore bars related to sediment characteristics. *Proceedings 8th Conference on Coastal Engineering*. Council on Wave Research, 248–278.

Homsi, A, 1978, Wave climate in some zones off the Brazilian coast. *Proceedings International Coastal Engineering Conference*. Hamburg, ASCE, 114–133.

Hopley, D, 1982, *The Geomorphology of the Great Barrier Reef*. John Wiley & Sons, Chichester, 453 pp.

Hopley, D, 1983, Evidence of 15,000 years of sea level change in tropical Queensland. In D. Hopley (ed.) *Australian Sea Levels in the last 15,000 Years: A Review*. Department of Geography, Monograph Series Occasional Paper No. 3, James Cook University North Queensland, 93–104.

Hopley, D, 1984, The Holocene 'high energy window' on the central Great Barrier reef. In Thom, B G (ed.) *Coastal Geomorphology in Australia*. Academic Press, Australia, 135–150.

Hopley, D, 1985, Geomorphological development of modern coastlines: A Review. In Pitty, A (ed.) *Themes in Geomorphology*, Croom Helm, London, 56–71.

Horikawa, K (ed.) 1988, *Nearshore Dynamics and Coastal Processes*. University of Tokyo Press, Tokyo, 522 pp.

Horikawa, K and Kuo, C T, 1966, A study on wave transformation inside the surf zone. *Proceedings 10th International Conference on Coastal Engineering*. ASCE, 217–233.

Horikawa, K., Hotta, S and Kraus, N, 1986, Literature review of sand transport on a dry sand surface. *Coastal Engineering*, 9, 503–526.

Horn, D P, 1993, Sediment dynamics on a macrotidal beach: Isle of Man, U.K. *Journal of Coastal Research*, 9, 189–208.

Hosier, P E and Cleary, W J, 1977, Cyclic geomorphic patterns of washover on a barrier island in southeastern North Carolina. *Environmental Geology*, 2, 23–31.

Hotta, S, 1988, Sand transport by wind. In Horikawa, K (ed.) *Nearshore Dynamics and Coastal Processes. Theory, Measurement and Predictive Models*. University of Tokyo Press, Tokyo, 218–238.

Howard, A D, 1977, Effect of slope on the threshold of motion and its application to orientation of wind ripples. *Geological Society of America Bulletin*, 88, 853–856.

Howard, A D, Morton, J B, Gad-El-Hak, M and Pierce, D B, 1978, Sand transport model of barchan dune equilibrium. *Sedimentology*, 25, 307–338.

Howard, J D and Reineck, H-E, 1981, Depositional facies of high energy beach to offshore sequence: comparison with low-energy sequence. *American Association of Petroleum Geologists*, 65, 807–830.

Howd, P A and Holman, R A, 1987, A simple model of beach foreshore response to long-period waves. *Marine Geology*, 78, 11–22.

Howd, P A, Oltman-Shay, J, Holman, R A and Komar, P D, 1990, Evolution of infragravity variance during a storm. *Proceedings 22nd International Conference on Coastal Engineering*. ASCE, 1118–1130.

Howd, P A, Oltman-Shay, J and Holman, R A, 1991, Wave variance partitioning in the trough of a barred beach. *Journal of Geophysical Research*, 96, 12 781–12 795.

Howd, P A, Bowen, A J and Holman, R A, 1992, Edge waves in the presence of strong longshore currents. *Journal of Geophysical Research*, 97, 11 357–11 371.

Hoyt, J H, 1967, Barrier island formation. *Geological Society of America Bulletin*, 78, 1125–1136.

Hsu, J R C and Evans, C, 1989, Parabolic bay shapes and applications. *Proceedings Institution Civil Engineers, London*, Part 2, 87, 557–570.

Hsu, J R C, Silvester, R and Xia, Y M, 1987, New characteristics of equilibrium shaped bays. *Proceedings 8th International Conference on Coastal Engineering*. ASCE, 140–144.

Hsu, J R C, Silvester, R and Xia, Y M, 1989a, Static equilibrium bays: new relationships. *Journal of Waterway, Port, Coastal and Ocean Engineering*, 115, 285–298.

Hsu, J R C, Silvester, R and Xia, Y M, 1989b, Applications of headland control. *Journal of Waterway, Port, Coastal and Ocean Engineering*, 115, 299–310.

Hsu, S A, 1977, Boundary-layer meteorological research in the coastal zone. In Walker, H J (ed.) *Geoscience and Man*. Vol. XVIII, *Research Techniques in Coastal Environments*. School of Geoscience, Louisiana State University, 99–111.

Hughes, M G, 1992, Application of a non-linear shallow water theory to swash following bore collapse on a sandy beach. *Journal of Coastal Research*, 8, 562–578.

Hughes, M G, 1995, Friction factors for wave runup. *Journal of Coastal Research*, 11, 1089–1098.

Hughes, M G and Cowell, P J, 1987, Adjustment of reflective beaches to waves. *Journal of Coastal Research*, 3, 153–167.

Hughes, M G, Masselink, G and Brander, R W, 1997. Flow velocity and sediment transport in the swash zone of a steep beach. *Marine Geology*, 138 91–103.

Hunt, L A, 1959, Design of seawalls and breakwaters. *Proceedings American Society of Civil Engineers*, 85, 123–152.

Hunter, R E, 1977a, Terminology of cross-stratified sedimentary layers and climbing ripple structures. *Journal of Sedimentary Petrology*, 47, 697–706.

Hunter, R E, 1977b, Basic types of stratification in small eolian dunes. *Sedimentology*, 24, 361–387.

Hunter, R E and Richmond, B M, 1988, Daily cycles in coastal dunes. *Sedimentary Geology*, 55, 43–67.

Hunter, R E, Clifton, H E and Phillips, L R, 1979, Depositional processes, sedimentary structures, and predicted vertical sequences in barred nearshore systems, southern Oregon coast. *Journal of Sedimentary Petrology*, 49, 711–726.

Hunter, R E, Richmond, B R and Alpha, T R, 1983, Storm-controlled oblique dunes of the Oregon coast. *Bulletin Geological Society of America*, 94, 1450–1465.

Hunter R E, Dingler, J R, Anima, R J and Richmond, B M, 1988, Coarse-sediment bands on the inner shelf of southern Monterey Bay, California. *Marine Geology*, 80, 81–98.

Huntley, D A, 1976, Long-period waves on a natural beach. *Journal of Geophysical Research*, 81, 6441–6449.

Huntley, D A, 1980, Edge waves in a crescentic bar system. In McCann, S B (ed.) *The Coastline of Canada*. Geological Survey of Canada, Paper 80–10, 111–121.

Huntley, D A, 1988, Evidence of phase coupling of edge wave modes. *Journal of Geophysical Research*, 93, 12 393–12 408.

Huntley, D A and Bowen, A J, 1975, Comparison of the hydrodynamics of step and shallow beaches. In Hails, J R and Carr, A (eds) *Nearshore Sediment Dynamics and Sedimentation*. John Wiley, New York, 69–109.

Huntley, D A and Bowen, A J, 1978, Beach cusps and edge waves. *Proceedings 16th International Conference on Coastal Engineering*. ASCE, 1378–1393.

Huntley, D A and Bowen, A J, 1990, Modeling sand transport on continental shelves. In Davies, A M (ed.) *Modelling Marine Systems*. CRC Press, Florida, 221–254.

Huntley, D A and Hanes, D M, 1987, Direct measurement of suspended sand transport. *Proceedings Coastal Sediments '87*. ASCE, 723–737.

Huntley, D A and Kim, C S, 1984, Is surf beat forced or free? *Proceedings 19th International Conference on Coastal Engineering*. ASCE, 871–885.

Huntley, D A and Short, A D, 1992, On the spacing between observed rip currents. *Coastal Engineering*, 17, 211–225.

Huntley, D A, Guza, R T and Thornton, E B, 1981, Field observations of surf beat. 1. Progressive edge waves. *Journal of Geophysical Research*, 86, 6451–6466.

Huntley, D A, Davidson, M, Russell, P, Foote, E Y and Hardisty, J, 1993, Long waves and sediment movement on beaches: Recent observations and implications for modelling. *Journal of Coastal Research*, Special Issue 15, 215–229.

IAHR, 1989, List of sea-state parameters. *Journal of Waterway, Port, Coastal and Ocean Engineering*, 115, 793–808.

Illenberger, W K and Rust, I C, 1988, A sand budget for the Alexandria coastal dunefield, South Africa. *Sedimentology*, 35, 513–521.

Inden, R F and Moore, C H, 1983, Beach environment. In Scholle, P A, Bebout, D G and Moore, C G (eds), *Carbonate Depositional Environments*. American Association of Petroleum Geologists, Tulsa, 211–265.

Ingle, J C, 1966, *The Movement of Beach Sand*. Elsevier, Amsterdam, 221 pp.

Inman, D L, 1953, *Areal and seasonal variations in beach and nearshore sediments at La Jolla, California*. Beach Erosion Board, Technical Memorandum No. 39.

Inman, D L, 1960, Shore processes. In *McGraw-Hill Encyclopedia of Science and Technology*. McGraw-Hill, 299–306.

Inman, D L and Bagnold, R A, 1963, Beach and nearshore processes, II: Littoral processes. In Hill, M N (ed.) *The Sea*, Vol. 3. Interscience, New York, 529–533.

Inman, D L and Bowen, A J, 1963, Flume experiments on sand

transport by waves and currents. *Proceedings 8th International Conference on Coastal Engineering*. ASCE, 137–150.

Inman, D L and Brush, B M, 1973, The coastal challenge. *Science*, 181, 20–32.

Inman, D L and Dolan, R, 1989, The Outer Banks of North Carolina: Budget of sediment and inlet dynamics along a migrating barrier system. *Journal of Coastal Research*, 5, 193–237.

Inman, D L and Filloux, J, 1960, Beach cycles related to tide and local wind regime. *Journal of Geology*, 68, 225–231.

Inman, D L and Guza, R T, 1982, The origin of swash cusps on beaches. *Marine Geology*, 49, 133–148.

Inman, D L and Quinn, Q H, 1952, Currents in the surf zone. *Proceedings 2nd Conference on Coastal Engineering*. Council on Wave Research, 24–36.

Inman, D L and Nordstrom, C E, 1971, On the tectonic and morphologic classification of coasts. *Journal of Geology*, 79, 1–21.

Inman, D L, Ewing, G C and Corliss, J B, 1966, Coastal sand dunes of Guerrero negro, Baja California, Mexico. *Bulletin Geological Society of America*, 77, 787–802.

Inman, D L, Elwany, M H and Jenkins, S A, 1993, Shore rise and bar-berm on ocean beaches. *Journal of Geophysical Research*, 98, 18 181–18 199.

Ippen, A T (ed.) 1966, *Estuary and Coastline Hydrodynamics*. McGraw-Hill, New York, 744 pp.

Ippen, A T and Eagleson, P S, 1955, *A Study of Sediment Sorting by Waves Shoaling on a Plane Beach*. Beach Erosion Board, Technical Memorandum No. 63, Army Corps of Engineers, Washington, DC, 83 pp.

Iwagaki, Y and Noda, H, 1962, Laboratory study of scale effects in two-dimensional beach processes. *Proceedings 8th International Conference on Coastal Engineering*. ASCE, 194–210.

Jackson, A, Tomlinson, R, Black, K and Couriel, E, 1997, Proposed surfing reef for northern Gold Coast. *Proceedings 1st International Surfing Reef Symposium*. Coastal Studies Unit, University of Sydney, 7 pp.

Jackson, D W T and Nevin, G H, 1992, Sand transport in a cliff top dune system at Fonte de Telha, Portugal. In Carter, R W G, Curtis, T G and Sheehy-Skeffington, M J (eds) *Coastal Dunes*. Balkema, Rotterdam, 81–92.

Jaffe, B E, Sternberg, R W and Sallenger, A H, 1984, The role of suspended sediment in shore-normal beach profile change. *Proceedings 19th International Conference on Coastal Engineering*. ASCE, 1983–1996.

Jago, C F and Barusseau, J P, 1981, Sediment entrainment on a wave graded shelf, Roussillon, France. *Marine Geology*, 42, 279–299.

Jago, C F and Hardisty, J, 1984, Sedimentology and morphodynamics of a macrotidal beach, Pendine Sands, SW Wales. *Marine Geology*, 60, 123–154.

James, C P and Brenninkmeyer, B M, 1977, Sediment entrainment within bores and backwash. *Geoscience and Man*, 18, 61–68.

James, N P, 1984, Facies Models 10. Shallowing-upward sequences in carbonates. In Walker, R G (ed.) *Facies Models*. Geoscience Canada Reprint Series 1, 109–119.

James, W R, 1974, Beach fill stability and borrow material texture. *Proceedings 14th Conference on Coastal Engineering*. ASCE, 1334–1344.

Jarmillo, E, 1996, Seasonal variability of the sand beach macroinfauna in south-central Chile. *Revista Chilena de Historia Natural*, 69, 641–653.

Jelgersma, S and van Regteren Altena, J F, 1969, An outline of the geological history of the coastal dunes in the western Netherlands. *Geologie en Mijnbouw*, 48, 335–342.

Jelgersma, S, De Jong, J D, Zagwijn,W H and van Regteren Altena, J F, 1970, The coastal dunes of the western Netherlands; geology, vegetational history and archaeology. *Med. Rijks. Geol. Dienst*, N.S. nr. 21.

Jennings, J N, 1957, On the orientation of parabolic or U-dunes. *Geographical Journal*, 123, 474–480.

Jennings, J N, 1967, Cliff-top dunes. *Australian Geographical Studies*, 5, 40–49.

Johnson, D W, 1919, *Shore Processes and Shoreline Development*. Hafner, New York, 584 pp (facsimile edition published in 1972).

Johnson, D W, 1925, *The New England–Arcadian Shoreline*. John Wiley & Sons, New York, 608 pp.

Johnson, J W, 1949, Scale effects in hydraulic model involving wave motion. *Transactions American Geophysical Union*, 30, 517–525.

Johnson, J W (ed.) 1951, *Proceedings First Conference on Coastal Engineering*. Council on Wave Research, The Engineering Foundation, University of California, Berkeley.

Johnson, J W, 1956, Dynamics of nearshore sediment transport. *Bulletin American Association of Petroleum Geologist*, 40, 2211–2232.

Johnson, J W, 1957, The littoral drift problem at shoreline harbours. *Journal of Waterways and Harbour Division*. ASCE, 85 (WW3), 227–251 Paper 2177.

Johnson, J W and Eagleson, P S, 1966, Coastal processes. In Ippen, A J (ed.) *Estuary and Coastline Hydrodynamics*. MGraw-Hill, New York, 404–492.

Jones, A R and Short, A D, 1995, Sandy beaches. In Underwood, A J and Chapman, M G (eds) *Coastal Marine Ecology of Temperate Australia*. University of New South Wales Press, 136–151.

Jungerius, P D, 1984, A simulation model of blowout development. *Earth Surface Processes and Landforms*, 9, 509–512.

Jungerius, P D and van der Meulen, F, 1988, Erosion processes in a dune landscape along the Dutch coast. *Catena*, 15, 217–228.

Jungerius, P D and van der Meulen, F, 1989, The development of dune blowouts, as measured with erosion pins and sequential air photos. *Catena*, 16, 369–376.

Jungerius, P D, Verheggen, A J and Wiggers, A J, 1981, The development of blowouts in 'De Blink', a coastal dune area near Noordwijkerhout, The Netherlands. *Earth Surface Processes and Landforms*, 6, 375–396.

Kaimal, J C and Finnigan, J J, 1994, *Atmospheric Boundary Layer Flows. Their Structure and Measurement*. Oxford University Press, Oxford, 289 pp.

Kana, T W, 1978, Surf zone measurements of suspended sediment. *Proceedings 16th International Conference on Coastal Engineering*. ASCE, 1725–1743.

Kang, H Y and Nielsen, P, 1994, Wave runup and the coastal watertable. *Conference on Hydraulics in Civil Engineering*. 4 pp.

Kang, H Y, Nielsen, P and Hanslow, D J, 1994, Overheight due to wave runup on a sandy beach. *Proceedings 24th International Conference on Coastal Engineering*. ASCE, 330–331.

Kaszubowski, L J, 1996, Geomorphology and contemporary evolution of the western Polish coast. *Journal of Coastal Research*, 12, 484–495.

Katoh, K., 1981, Analysis of edge waves by means of empirical eigenfunctions. *Report Port Harbour Research Institute*, 20, 3–51.

Katoh, K and Yanagishima, S, 1992, Berm formation and berm erosion. *Proceedings 23rd International Conference on Coastal Engineering*, Vol. 2, Venice, Italy, ASCE. 2136–2149.

Kaufman, W and Pilkey, O H, 1979, *The Beaches are Moving: the Drowning of America's Shoreline*. Anchor Press, Garden City, 326 pp.

Kawamura, R, 1951, *Study of sand movement by wind*. Research Report HEL-2–9, Hydraulic Engineering Laboratory, University of California, Berkeley (English translation 1964).

Keller, H B, Levine, D A and Whitham, G B, 1960, Motion of a bore over a sloping beach. *Journal of Fluid Mechanics*, 7, 302–316.

Kemp, P H and Plinston, D T, 1968, Beaches produced by waves of low phase difference. *Journal of the Hydraulics Division*. ASCE, 94, 1183–1195.

Keulegan, G H, 1948, *An experimental study of submarine bars*. Beach Erosion Board, US Army Corps of Engineers, Technical Report No. 3, 40 pp.

Keulegan, G H and Krumbein, W C, 1949, Stable configuration of bottom slope in a shallow sea and its bearing on geological processes. *EOS, Transactions of the American Geophysical Union*, 30, 855–861.

Kimura, R, 1976, *Changes of beach profiles – a case study at Tatado beach, Shimoda, Shizuoka Prefecture, Japan*. Unpublished BS Thesis, Geography Department, Tokoyo Kyoiku University, 53 pp (in Japanese).

King, B A, Blackley, M W L, Carr, A P and Hardcastle, P J, 1990, Observations of wave-induced set-up on a natural beach. *Journal of Geophysical Research*, 95, 22 289–22 297.

King, C A M, 1972, *Beaches and Coasts*. Edward Arnold, London, 570 pp.

King, C A M and Williams, W W, 1949. The formation and movement of sand bars by wave action. *Journal of Geography*, 113, 70–85.

Kirby, J T, Dalrymple, R A and Liu, P L-F, 1981, Modification of edge waves by barred beach topography. *Coastal Engineering*, 5, 35–49.

Kirk, R M, 1966, *Depth of disturbance by waves on a pebble beach over shingle*. Coastal Studies Research Report, Department of Geography, University of Canterbury.

Kirk, R M, 1980, Mixed sand and gravel beaches: morphology, processes and sediments. *Progress in Physical Geography*, 4, 189–210.

Kirkgoz, M S, 1981, A theoretical study of plunging breakers and their runup. *Coastal Engineering*, 5, 353–370.

Klijn, J A, 1990, The younger dunes of the Netherlands; chronology and causation. In Bakker, Th W, Jungerius, P D and Klijn, J A (eds) *Dunes of the European Coasts. Catena*, Supplement 18, 89–100.

Knight, J H, 1981, Steady periodic flow through a rectangular dam. *Water Resources Research*, 17, 1222–1224.

Kocurek, G, 1981, Significance of interdune deposits and bounding surfaces in aeolian dune sands. *Sedimentology*, 28, 753–780.

Kocurek, G, 1988, First order and super bounding surfaces in aeolian sequences – bounding surfaces revisited. *Sedimentary Geology*, 56, 193–206.

Kocurek, G, Townsley, M, Yeh, E, Havholm, K and Sweet, M L, 1992, Dune and dune-field development on Padre Island, Texas, with implications for interdune deposition and water-table-controlled accumulation. *Journal of Sedimentary Petrology*, 62, 622–635.

Komar, P D, 1971a, The mechanics of sand transport on beaches. *Journal of Geophysical Research*, 76, 713–721.

Komar, P D, 1971b, Nearshore cell circulation and the formation of giant cusps. *Geological Society of American Bulletin*, 82, 2643–2650.

Komar, P D, 1973a, Computer models for delta growth due to sediment input from rivers and longshore transport. *Geological Society of America Bulletin*, 84, 2217–2226.

Komar, P D, 1973b, Observations of beach cusps at Mono Lake, California. *Geological Society of America Bulletin*, 84, 3593–3600.

Komar, P D, 1975, Nearshore currents: Generation by oblique incident waves and longshore variations in breaker height. In Hails, J and Carr, A (eds) *Nearshore Sediment Dynamics and Sedimentation*. Wiley Science, London, 17–45.

Komar, P D, 1976, *Beach Processes and Sedimentation*, Prentice-Hall, Inc., New Jersey, 429 pp.

Komar, P D, 1978, Relative quantities of suspension versus bed-load transport on beaches. *Journal of Sedimentary Petrology*, 48, 921–932.

Komar, P D, 1983a, The erosion of Siletz Spit, Oregon. In Komar, P D (ed.) *CRC Handbook of Coastal Processes and Erosion*, CRC Press, Boca Raton, 65–76.

Komar, P D, 1983b, Nearshore currents and sand transport on beaches. In Johns, B (ed.) *Physical Oceanography Coastal and Shelf Seas*. Elsevier, Amsterdam, 67–109.

Komar, P D, 1984, Coastal erosion in response to the construction of jetties and breakwaters. In Komar, P D (ed.) *Handbook of Coastal Processes and Erosion*. CRC Press, Boca Raton, 191–204.

Komar, P D and Gaughan, M K, 1972, Airy wave theory and breaker height prediction. *Proceedings 13th International Conference on Coastal Engineering*. ASCE, 405–418.

Komar, P D and Holman, R A, 1986, Coastal processes and the development of shoreline erosion. *Annual Review Earth and Planetary Sciences*, 14, 237–265.

Komar, P D and Inman, D L, 1970, Longshore sand transport

on beaches. *Journal of Geophysical Research*, 75, 5914–5927.

Komar, P D and McDougal, W G, 1994, The analysis of exponential beach profiles. *Journal of Coastal Research*, 10, 59–69.

Kotvojs, F J and Cowell, P J, 1991, Refinement of the Dean profile model for beach design. *Australian Civil Engineering Transactions*, Institute Engineers Australia, CE33, 9–15.

Kraft, J C, 1971, Sedimentary facies pattern and geologic history of a Holocene marine transgression. *Bulletin Geological Society of America*, 82, 2131–2138.

Kraft, J C, Allen, E A, Bellknap, D F, John, C J and Maurmeyer, E M, 1979, Processes and morphologic evolution of an estuarine and coastal barrier system. In Leatherman S P (ed.) *Barrier Islands from the Gulf of St. Lawrence to the Gulf of Mexico*. Academic Press, New York, 149–183.

Kraus, N C, 1987, Application of portable traps for obtaining point measurements of sediment transport rates in the surf zone. *Journal of Coastal Research*, 3, 139–152.

Kraus, N C and Dean, J L, 1987, Longshore sediment transport rate distributions measured by trap. *Proceedings Coastal Sediments '87*. ASCE, 881–896.

Kraus, N C and Horikawa, K, 1990, Nearshore sediment transport. In LeMehaute, B and Hanes D (eds) *The Sea: Ideas and Observations on Progress in the Study of the Seas*, Vol. 9B. Wiley, New York, 775–813.

Kraus, N C and Larson, M, 1988, *Beach profile changes measured in the tank for large waves, 1956–1957 and 1962*. Technical Report CERC-88-6, US Army Coastal Engineering Research Center, Vicksburg.

Kraus, N C and McDougall, W G, 1996, The effects of seawalls on beaches, part I, an updated literature review. *Journal of Coastal Research*, 12, 691–701.

Kraus, N and Pilkey, OH (eds), 1988, *The Effects of Seawalls on Beaches, Journal of Coastal Research*, Special Issue 4, 146 pp.

Kraus, N C, Farinato, R S and Horikawa, K, 1981, Field experiments on longshore sand transport in the surf zone. *Coastal Engineering in Japan*, 24, 171–194.

Kraus, N C, Gingerich, K J and Rosati, J D, 1988, Toward an improved formula for longshore sand transport. *Proceedings 21st International Conference on Coastal Engineering*. ASCE, 1182–1196.

Kraus, N C, Larson, M and Kriebel, D L, 1991, Evaluation of beach erosion and accretion predictors. *Proceedings Coastal Sediments '91*. ASCE, 572–587.

Kriebel, D L and Dean, R G, 1985, Numerical simulation of time-dependent beach and dune erosion. *Coastal Engineering*, 9, 221–245.

Kriebel, D L, Dally, W R and Dean, R G, 1986, Undistorted Froude scale model for surf zone sediment transport. *Proceedings 20th International Conference on Coastal Engineering*. ASCE, 1296–1310.

Kriebel, D L, Kraus, N C and Larson, M, 1991, Engineering methods for predicting beach profile response. *Coastal Sediments '91*. ASCE, New York, 557–571.

Kroon, A, 1994, *Sediment transport and morphodynamics of the beach and nearshore zone near Egmond, the Netherlands*. Faculteit Ruimtelijke Wetenschappen, University of Utrecht, The Netherlands, 275 pp.

Krumbein, W C and Monk, G D, 1942, *Permeability as a function of the size parameters of unconsolidated sand*. American Institute of Mining and Metallurgical Engineering, Technical Publication No. 1492, 11 pp.

Krumbein, W E, 1979, Photolithotropic and chemoorganotropic activity of bacteria and algae as related to beachrock formation and degradation (Gulf of Aquaba, Sinai). *Geomicrobiological Journal*, 1, 139–203.

Kuenen, P H, 1948, The formation of beach cusps. *Journal of Geology*, 56, 34–40.

Kunte, P D, 1995, On some aspects of barrier islands of the west coast, India. *Journal of Coastal Research*, 11, 508–515.

Laing, C C, 1954, *The ecological life-history of the Ammophila breviligulata community on the Lake Michigan dunes*. Unpublished PhD Thesis, University of Chicago, 108 pp.

Lancaster, N, 1986, Dynamics of deflation hollows in the Elands Bay area, Cape Province, South Africa. *Catena*, 13, 139–153.

Lancaster, N, 1989, *The Namib Sand Sea – Dune Forms, Processes and Sediments*. Balkema, Rotterdam, 180 pp.

Lancaster, N and Nickling, W G, 1994, Aeolian sediment transport. In Abrahams, A D and Parsons, A J (eds) *Geomorphology of Desert Environments*. Chapman and Hall, London, 447–473.

Landsberg, H and Riley, N A, 1943, Wind influences on the transportation of sand over a Michigan sand dune. *Proceedings 2nd Hydraulics Conference*. Bulletin 27, University Iowa Studies in Engineering.

Landsberg, S Y, 1956, The orientation of dunes in Britain and Denmark in relation to wind. *Geographical Journal*, 122, 176–189.

Langbein, W B, 1963, The hydraulic geometry of a shallow estuary. *Bulletin International Association of Scientific Hydrology*, 8, 84–93.

Lanyon, J A, Eliot, I G and Clarke, D J, 1982, Groundwater variation during semi diurnal spring tidal cycles on a sandy beach. *Australian Journal of Marine and Freshwater Research*, 33, 377–400.

Larson, M, 1988, *Quantification of Beach Prile Change*. Department of Water Resources Engineering, Institute of Science and Technology, University of Lund, Report No. 1008, 293 pp.

Larson, M and Kraus, N C, 1989, *SBEACH, numerical model for simulating storm induced beach change*. Technical Report CERC-89-9, US Army Corps of Engineers, Coastal Engineering Research Center, Vicksburg.

Larson, M and Kraus, N C, 1995, Prediction of cross-shore sediment transport at different spatial and temporal scale. *Marine Geology*, 126, 111–127.

Larson, M. and Sunamura, T., 1993, Laboratory experiment on flow characteristics at a beach step. *Journal of Sedimentary Petrology*, 63, 495–500.

Leahy, S, Short, A D and McLeod, K, 1996, *Beach Management Plan*. Surf Life Saving Australia, Sydney, 27 pp.

Leatherman, S P, 1976, Barrier island dynamics, Overwash processes and aeolian transport. *Proceeding 15th Coastal Engineering Conference*.. ASCE, 1958–1974.

Leatherman, S P (ed.), 1979a, *Environmental Geologic Guide to Cape Cod National Seashore*. National Park Service, Cooperative Research Unit, University of Massachusetts Field Trip Guide Book, 249 pp.

Leatherman, S P, 1979b, *Barrier Island Handbook*. National Park Service, Cooperative Research Unit, University of Massachusetts, 101 pp.

Leatherman, S P (ed.), 1979c, *Barrier Islands from the Gulf of St. Lawrence to the Gulf of Mexico*. Academic Press, New York, 325 pp.

Leatherman, S P, 1979d, Barrier dune systems, a reassessment. *Sedimentary Geology*, 24, 1–16.

LeBlond, P H, 1979, An explanation of the logarithmic spiral plan shape of headland bay beaches. *Journal of Sedimentary Petrology*, 49, 1093–1100.

Lee, G-H, Nicholls, R J and Birkemeier, W A, 1995, A conceptual fairweather-storm model of beach rearshore profile evolution at Duck, North Carloina, U.S.A. *Journal of Coastal Research*, 11, 1157–1166.

Lee, G-h, Nicholls, R J and Birkemeier, W A, 1998, Storm-driven variability of the beach-nearshore profile at Duck, North Carolina, USA, 1981–1991. *Marine Geology*, 148, 163–177.

Lees, B G, 1985, Geomorphology and development of the Victoria delta. In Bardsley, K N, Davie, J O S and Woodroffe, C D (eds) *Coasts and Tidal Wetlands of the Australian Monsoon Region*. Mangrove Monograph 1, Australian National University, Northern Australian Research Unit, Darwin, 31–42.

Lees, B G, 1987, Age structure of the Point Stuart chenier plain: a reassessment. *Search*, 18, 257–259.

Lees, B G and Clements, A M, 1987, Climatic implications of chenier dates in northern Australia. *Radiocarbon*, 29, 311–317.

Lees, B G, Yanchou, Lu and Head, J, 1990, Reconnaissance thermoluminescence dating of northern Australian coastal dune systems. *Quaternary Research*, 34, 169–185.

Lees, B G, Stanner, J, Price, D M and Yanchou, Lu, 1995, Thermoluminescence dating of dune podzols at Cape Arnhem, northern Australia. *Marine Geology*, 129, 63–75.

LeMehaute, B, 1962, On non-saturated breakers and then wave run-up. *Proceedings 8th International Conference on Coastal Engineering*. ASCE, 77–92.

Lemm, A, 1996, *Offshore wave climate, Perth, Western Australia*. Honours Thesis, Department of Environmental Engineering, University of Western Australia, Perth, 179 pp.

Leopold, L B and Langbein, W B, 1962, *The Concept of Entropy in Landscape Evolution*. US Geological Survey Professional Paper, 500–A, 20 pp.

Lewin, J and Schaefer, C T, 1983, The role of phytoplankton in surf ecosystems. In McLachlan, A and Erasmus, T (eds) *Sandy Beaches as Ecosystems*. Junk, The Hague, 381–389.

Lippman, T C and Holman, R A, 1990, The spatial and temporal variability of sand bar morphology. *Journal of Geophysical Research*, 95, 1575–1590.

Lippman, T C, Thornton, E B and Reniers, A J H M, 1995, Wave stress and longshore current on barred profiles. *Proceedings Coastal Dynamics '95*, Gdansk, Poland. ASCE, New York, 401–412.

List, J H, 1986, Wave groupiness as a source of nearshore long waves. *Proceedings 20th International Conference on Coastal Engineering*. ASCE, 497–511.

List, J H, 1992, A model for the generation of two-dimensional surf beat. *Journal of Geophysical Research*, 97, 5623–5635.

List, J H and Terwindt, J H J (eds), 1995, Large-scale coastal behaviour. *Marine Geology*, Special Issue, 126, 331 pp.

Lofquist, K E B, 1975, *An Effect of Permeability on Sand Transport by Waves*. US Army Corps of Engineers, Engineering Research Center, Technical Memorandum No. 62, 74 pp.

Logan, B W, 1970, Carbonate sedimentation and environments, Shark Bay, Western Australia. *American Society of Petroleum Geologists, Memoir* 13, 1–38.

Longuet-Higgins, M S, 1952, On the statistical distribution of the heights of sea waves. *Journal of Marine Research*, 11, 245–266.

Longuet-Higgins, M S, 1953, Mass transport in water waves. *Transactions of the Royal Society of London*, 245, 525–581.

Longuet-Higgins, M S, 1970a, Longshore currents generated by obliquely incident sea waves 1. *Journal of Geophysical Research*, 75, 6778–6789.

Longuet-Higgins, M S, 1970b, Longshore currents generated by obliquely incident sea waves 2. *Journal of Geophysical Research*, 75, 6790–6801.

Longuet-Higgins, M S, 1983, Wave set-up, percolation and undertow in the surf zone. *Proceedings of the Royal Society of London*, A 390, 283–291.

Longuet-Higgins, M S and Parkin, D W, 1962, Sea waves and beach cusps. *Geographical Journal*, 128, 194–201.

Longuet-Higgins, M S and Stewart, R W, 1962, Radiation stresses and mass transport in gravity waves with applications to surf beat. *Journal of Fluid Mechanics*, 13, 481–504.

Longuet-Higgins, M S and Stewart, R W, 1964, Radiation stresses in water waves – a physical discussion with applications, *Deep-Sea Research*, 11, 529–562.

Lowe, D R, 1976, Subaqueous liquefied and fluidised sediment flows and their deposits. *Sedimentology*, 23, 285–308.

Lowe, D R, 1982, Sediment gravity flows II: depositional model with special reference to deposits of high density turbidity currents. *Journal of Sedimentary Petrology*, 52, 279–297.

Luck, G, 1978, Islands in front of the southern North Sea coast. *Die Kuste*, 32, 85.

Lushine, J B, 1991, Rip currents, human impact and forcastibility. *Coastal Zone 91*, Long Beach. ASCE, 3558–3569.

Madsen, O S and Grant, W D, 1976, *Sediment Transport in the Coastal Environment*. MIT Technical Report 209, 105 pp.

Mansard, E P D and Barthel, V, 1984, Shoaling properties of bounded long waves. *Proceedings 19th International Confer-*

*ence on Coastal Engineering*. ASCE, 798–814.

Marques, M A and Julia, R, 1988, St. Pere Pescador beach-dune interaction. In Psuty, N P (ed.), *Dune/Beach Interaction, Journal of Coastal Research*, Special Issue 3, 57–61.

Marsh, W M, Marsh, B D and Dozier, J, 1973, Formation, structure, and geomorphic influence of Lake Superior ice-foots. *American Journal of Science*, 273, 38–64.

Marston, R A, 1986, Maneuver-caused wind erosion impacts, south central New Mexico. In Nickling, W G (ed.), *Aeolian Geomorphology*. The Binghampton International Series in Geomorphology, No. 17, Allen and Unwin, London, 273–290.

Marta, M D, 1958, Coastal dunes, a study of the dunes at Vera Cruz. *Proceedings Coastal Engineering Conference*, 6, 520–530.

Martens, D, Williams, T and Cowell, P J, in press, Mega-rip dimensional analysis on the Sydney coast, Australia, and implications for beach-state recognition and prediction. *Journal of Coastal Research*.

Martin, C S, 1970, Effect of a porous sand bed on incipient sediment motion. *Water Resources Research*, 6, 1162–1174.

Martin, C S and Aral, M M, 1971, Seepage force on interfacial bed particles. *Journal of the Hydraulics Division*. ASCE, 7, 1081–1100.

Martinez, J O, Gonzalez, J L, Pilkey, O H and Neal, W J, 1995, Tropical barrier islands of Columbia's Pacific coast. *Journal of Coastal Research*, 11, 432–453.

Martinez, P A, 1987, Wave: programs for simulating onshore–offshore transport in two dimensions using a Macintosh computer. *Computers and Geoscience*, 13, 513–540.

Martini, I P, 1981, Coastal dunes of Ontario, distribution and geomorphology. *Geogr. Phys. Quat.*, 35, 219–229.

Mase, H, 1995, Frequency down-shift of swash oscillations compared to incident waves. *Journal of Hydraulic Research*, 33, 397–411.

Massel, S R, 1989, *Hydrodynamics of Coastal Zones*. Elsevier, Amsterdam, 336 pp.

Masselink, G, 1993, Simulating the effects of tides on beach morphodynamics. *Journal of Coastal Research*, Special Issue No. 15, 180–197.

Masselink, G, 1995a, *Morphodynamics of Macrotidal Beaches, Examples from Central Queensland, Australia*. Department of Geography, University of Sydney, Coastal Studies Unit Technical Report 95/1.

Masselink, G, 1995b, Group bound long waves as a source of infragravity energy in the surf zone. *Continental Shelf Research*, 15, 25–47.

Masselink, G, 1996, Sea breeze activity and its effect on coastal processes near Perth, Western Australia. *Journal of the Royal Society of Western Australia*, 79, 199–205.

Masselink, G and Black, K P, 1995, Magnitude and cross-shore distribution of bed return flow measured on natural beaches. *Coastal Engineering*, 25, 165–190.

Masselink, G and Hegge, B J, 1995, Morphodynamics of meso- and macrotidal beaches. Examples from central Queensland. *Marine Geology*, 129, 1–23.

Masselink, G and Lessa, G, 1995, Barrier stratigraphy on the macrotidal central Queensland coastline, Australia. *Journal of Coastal Research*, 11, 454–477.

Masselink, G and Pattiaratchi, C, 1998, The effect of sea breeze on beach morphology, surf zone hydrodynamics and sediment resuspension. *Marine Geology*, 146, 115–135.

Masselink, G and Short, A D, 1993, The effect of tide range on beach morphodynamics, A conceptual beach model. *Journal of Coastal Research*, 9, 785–800.

Mather A S and Ritchie W, 1978, *The Beaches of the Highlands and Islands of Scotland*. Countryside Commission for Scotland, Redgorton, Perth.

Matsunaga, N and Honji, H, 1980, The backwash vortex. *Journal of Fluid Mechanics*, 99, 813–815.

Matsunaga, N and Honji, H, 1983, The steady and unsteady backwash vortices. *Journal of Fluid Mechanics*, 135, 189–197.

McBride, E F and Hayes, M O, 1962, Dune cross-bedding on Mustang Island,Texas. *Bulletin American Association of Petroleum Geologists*, 46, 546–551.

McBride, R A, Byrnes, M R and Hiland, M W, 1995, Geomorphic response-type for barrier coastlines, a regional prespective. *Marine Geology*, 126, 143–159.

McCann, S B, 1979, Barrier islands in the southern Gulf of St. Lawrence, Canada. In Leatherman, S P (ed.), *Barrier Islands from the Gulf of St. Lawrence to the Gulf of Mexico*. Academic Press, New York, 29–63.

McCann, S B (ed.), 1980, *The Coastline of Canada*. Geological Survey Paper 80–10, 439 pp.

McCann, S.B. and Byrne, M.-L., 1989, Stratification models for vegetated coastal dunes in Atlantic Canada. In Gimmingham, C H, Ritchie, W, Willetts, B B and Willis, A J (eds), *Proceedings of the Royal Society Edinburgh*, 96B, 203–215.

McCann, S B and Carlisle, R J, 1972, The nature of the ice-foot on the beaches of Radstock Bay, south-west of Devon Island, N.W.T., Canada. In *Polar Geomorphology. Institute of British Geographers*, Special Publication No. 4, 175–186.

McCave, I N, 1978, Grain size trends and transport along beaches, Examples from eastern England. *Marine Geology*, 23, M43–M52.

McCubbin, D G, 1982, Barrier-island and strand plain facies. In Scholle, P A and Spearing, D (eds), *Sandstone Depositional Environments*, American Association of Petroleum Geologists, Tulsa, 247–279.

McGee, W J, 1890, Encroachment of the sea. *Forum*, 9, 437–449.

McGowan, J, 1894, On the highest wave of permanent type. *Philosophical Magazine*, Series 5, 32, 351–358.

McIntyre, A D, 1969, Ecology of marine meiobenthos. *Biological Review*, 44, 245–290.

McIntyre, A D, 1977, Sandy foreshores. In Barnes, R S K (ed.) *The Coastline*. John Wiley & Sons, London, 31–47.

McKee, E D, 1966, Structures of dunes at White Sands National Monument, New Mexico (and a comparison with structures of dunes from other selected areas). *Sedimentology*, 7, 1–69.

McKee, E D, 1979, *A Study of Global Sand Seas*. US Geological Survey Professional Paper 1052, 429 pp.

McKee, E.D. and Bigarella J.J., 1979, Sedimentary structures in dunes. In McKee, E D (ed.) *A Study of Global Seas*. U.S. Geological Survey Professional Paper 1052, 83–134.

McKenna Neuman, C, 1993, A review of aeolian transport processes in cold environments. *Progress in Physical Geography*, 17, 137–155.

McKenzie, P, 1958, Rip-current systems. *Journal of Geology*, 66, 103–113.

McLachlan, A and Erasmus, T (eds), 1983, *Sandy Beaches as Ecosystems*. Junk, The Hague.

McLachlan, A, 1983, Sandy beach ecology – a review. In McLachlan, A and Erasmus, T (eds) *Sandy Beaches as Ecosystems*, Junk, The Hague, 321–380.

McLachlan, A, 1990, Dissipative beaches and macrofaunal communities on exposed intertidal sands. *Journal of Coastal Research*, 6, 57–71.

McLachlan, A and Hesp, P A, 1984a, Surf zone diatom accumulations on the Australian coast. *Search* 15, 230–231.

McLachlan, A and Hesp, P A, 1984b, Faunal response to morphology and water circulation of a sandy beach with cusps. *Marine Ecology – Progress Series*, 19, 133–144.

McLachlan, A and Lewin, J, 1981, Observations of surf phytoplankton blooms along the coasts of South Africa. *Botanical Mar*. 24, 553–557.

McLachlan, A., Ascaray, C and du Toit, P, 1987, Sand movement, vegetation succession and biomass spectrum in a coastal dune slack in Algoa Bay, South Africa. *Journal of Arid Environments*, 12, 9–25.

McLachlan, A, Jaramillo, E, Donn, T E and Wessels, F, 1993, Sandy beach macrofauna communities and their control by the physical environment, A geographical comparison. *Journal of Coastal Research*, Special Issue 15, 27–38.

McLean, R F and Kirk, R M, 1969, Relationship between grain size, size-sorting, and foreshore slope on mixed sand-shingle beaches. *New Zealand Journal of Geology and Geophysics*, 12, 138–155.

McSaveney, M J, Begg, J and Hull, A G, 1993, Turakirae raises its head again (abstract). *Geological Society of New Zealand*, Miscellaneous Publication 79A, 115.

Meade, R H, 1996, River-sediment inputs to major deltas. In Milliman, J D and Haq, B U (eds), *Sea-level Rise and Coastal Subsidence*, Kluwer, Dordrecht, 63–85.

Medina, R, Losada, M A, Dalrymple, R A and Roland, A, 1991, Cross-shore sediment transport determined by EOF method. In Kraus, N C, Gingerich, K J and Kriebel, D L (eds), *Coastal Sediments '91*. ASCE, 2160–2174.

Meleo, J F, 1994, *Shoreface variability in Southeastern Australia*. unpub. Hons thesis, Department of Geography, University of Sydney, 142 pp.

Melton, F A, 1940, A tentative classification of sand dunes, its application to dune history in the southern High Plains. *Journal of Geology*, 48, 113–174.

Messinger, C, 1958, *A Geomorphic Study of the Dunes of the Provincetown Peninsula, Cape Cod, Massachusetts*. MSc Thesis, University of Massachusetts, 144 pp.

Meyer, R D, 1933, *A model of wave action on beaches*. Unpublished MSc Thesis, University of California. (reported in King, 1972).

Meyer, R E, (ed.) 1972, *Waves on Beaches*. Academic Press, New York, 462 pp.

Meyer, R E and Taylor, A D, 1972, Runup on beaches. In Meyer, R E (ed.) *Waves on Beaches and Resulting Sediment Transport*, Academic Press, New York, 357–412.

Miche, A, 1951, Exposes a l'action de la houle. *Ann. Ponts. Chaussees*, 121, 285–319.

Middleton, R A, 1966, Experiments on density and turbidity currents I, II, III. *Canadian Journal of Earth Science*, 3, 523–546, 627–637, 475–505.

Miller, M C and Komar, P D, 1980, Oscillation sand ripples generated by laboratory apparatus. *Journal of Sedimentary Petrology*, 50, 173–182.

Miller, R L, 1968, Experimental determination of runup of undular and fully developed bores. *Journal of Geophysical Research*, 73, 4497–4510.

Miller, R L, 1976, Role of vortices in surf zone prediction, sedimentation and wave forces. In Davis, R A and Ethington, R L (eds) *Beach and Nearshore Sedimentation*. SEPM Special Publication 24, 92–114.

Miller, R L and Ziegler, J M, 1958, A model relating dynamics and sediment pattern in equilibrium in the region of shoaling waves, breaker zone and foreshore. *Journal of Geology*, 66, 417–441.

Miller, W R and Mason, T R, 1994, Erosional features of coastal beachrock and aeolianite outcrops in Natal and Zululand, South Africa. *Journal of Coastal Research*, 10, 374–394.

Milliman, J D and Meade, R H, 1983, World-wide delivery of river sediment to the oceans. *Journal of Geology*, 91, 1–21.

Mogi, A, 1963, On the shore types of the coasts of Japanese islands. *Geographical Review of Japan*, 36, 245–266 (in Japanese with English abstract).

Monslow, T F and Heron, S D, 1994, Texas barriers. In Davis, R A Jr (ed.) *Geology of Holocene Barrier Island Systems*. Springer Verlag, New York, 75–114.

Moore, D G and Curray, J R, 1964, Wave-base, marine profile of equilibrium and wave built terraces: discussion. *Geological Society of America Bulletin*, 75, 1267–1274.

Morton, R A, 1981, Formation of storm deposits by wind-forced currents in the Gulf of Mexico an the North Sea. *International Association of Sedimentologists Special Publication*, 5, 303–396.

Morton, R A, 1994, Texas barriers. In Davis, R A Jr (ed.), *Geology of Holocene Barrier Island Systems*. Springer-Verlag, New York, 75–114.

Muckersie, C and Shepherd, M J, 1994, Dune phases as time-transgressive phenomena, Manawatu. *New Zealand Quaternary International*, 26, 61–67.

Mulhearn, P.J., 1978, A wind-tunnel boundary-layer study of the effects of a surface roughness change, rough to smooth. *Boundary-Layer Meteorology*, 15, 3–30.

Mulrennan, M E, 1992, Ridge and runnel beach morphodynamics. An example from the central east coast of Ireland. *Journal of Coastal Research*, 8, 906–918.

Munk, W H, 1949, Surf beats. *Transactions, American Geo-*

*physical Union*, 30, 849–854.

Murray, M, 1967, Control of grain dispersion by particle size and wave state. *Journal of Geology*, 75, 612–634.

Murray, R J, Robinson, D A and Soward, C L, 1993, Monitoring of the southern Gold Coast beach nourishment projects. *Proceedings 11th Australasia Conference Coastal Ocean Engineering*, Townsville. ASCE, 119–125.

Naffa, M G, 1995, Wave climate along the Nile delta coast. *Journal of Coastal Research*, 11, 219–229.

Nakato, T, Locher, F A, Glover, J R and Kennedy, J F, 1977, Wave entrainment of sand from rippled beds. *Journal of the Waterway, Port, Coastal and Ocean Division*, ASCE, 103, 83–99.

Namikas, S.L. and Sherman, D J, 1995, A review of the effects of surface moisture content on aeolian sand transport. In Tchakerian, V (ed.) *Desert Aeolian Processes*. Chapman and Hall, London, 269–292.

Neal, W J, Blakeney, W C, Pilkey, O H, Jr and Pilkey, O H, Sr. 1984, *Living with the South Carolina Shore*. Duke University Press, Durham.

Nelson, C L and Miller, R L, 1974, *The Interaction of Fluid and Sediment on the Foreshore*. Fluid Dynamics and Sediment Transport Laboratory, Department of Geophysical Sciences, University of Chicago, Technical Report 15, 175 pp.

Nelson, R C, 1994, Depth limited design wave heights in very flat regions. *Coastal Engineering*, 23, 43–59.

Neumann, C J, 1993, Global overview. In Holland, G J (ed.) *The Global Guide to Tropical Cyclone Forecasting*. World Meteorological Organisation, Geneva, 1–143.

Nicholls, R J, Birkemeier, W A and Lee, G-h, 1998, Evaluation of depth of closure using data from Duck, North Carolina, USA. *Marine Geology*, 148, 179–201.

Nichols, R L, 1961, Characteristics of beaches formed in polar climates. *Journal of Glaciology*, 7, 499–478.

Nickling, W G and Davidson-Arnott, R G D, 1990, Aeolian sediment transport on beaches and coastal dunes. In Davidson-Arnott, R G D (ed.) *Proceedings of the Symposium on Coastal Sand Dunes*. National Research Council, Ottawa, 1–35.

Niedoroda, A W and Swift, D J P, 1991, Shoreface processes. In Herbich, J B (ed.) *Handbook of Coastal and Ocean Engineering*, Vol. 2. Gulf Publishing, Houston, 736–769.

Niedoroda, A W, Swift, D J P and Hopkins, T S, 1985a, The shoreface. In Davis, R.A. (ed.) *Coastal Sedimentary Environments*, 2nd edition. Springer-Verlag, New York, 533–624.

Niedoroda A W, Swift, D J P, Figeiredo, A G and Freeland, G L, 1985b, Barrier island evolution, middle Atlantic shelf, USA, part II, evidence from the shelf floor. *Marine Geology*, 63, 363–396.

Niedoroda, A W, Reed, C W, Swift, D J P, Arato, H and Hoyanagi, K, 1995, Modeling shore-normal large-scale coastal evolution. *Marine Geology*, 126, 181–199.

Nielsen, A F, 1995, Subaqueous beach fluctuations on the Australian south-eastern seaboard. *Australian Civil Engineering Transactions*. Institute of Engineers Australia, CE36 (1), 57–67.

Nielsen, P, 1981, Dynamics and geometry of wave-generated ripples. *Journal of Geophysical Research*, 86, 6467–6472.

Nielsen, P, 1982, Explicit formulae for practical wave calculations. *Coastal Engineering*, 6, 389–398.

Nielsen, P, 1983, Entrainment and distribution of different sand sizes under water waves. *Journal of Sedimentary Petrology*, 53, 423–428.

Nielsen, P, 1984, Field measurements of time-averaged suspended sediment concentrations under waves. *Coastal Engineering*, 8, 51–72.

Nielsen, P, 1986, Suspended sediment concentration under waves. *Coastal Engineering*, 10, 23–31.

Nielsen, P, 1988a, Three simple models of wave sediment transport. *Coastal Engineering*, 12, 43–62.

Nielsen, P, 1988b, Wave setup. A field study. *Journal of Geophysical Research*, 93, 15 643–15 652.

Nielsen, P, 1989, Wave setup and runup. An integrated approach. *Coastal Engineering*, 13, 1–9.

Nielsen, P, 1990, Tidal dynamics of the water table in beaches. *Water Resources Research*, 26, 2127–2134.

Nielsen, P, 1992, *Coastal Bottom Boundary Layers and Sediment Transport*. World Scientific, Singapore, 324 pp.

Nielsen, P and Hanslow, D J, 1991, Wave runup distributions on natural beaches. *Journal of Coastal Research*, 7, 1139–1152.

Nielsen, P, Davis, G A, Winterbourne, J M and Elais, G, 1988, *Wave Setup, and the Watertable in Sandy Beaches*. Public Works Department NSW Coastal Branch, Technical Memorandum. 88/1, 132 pp.

Noda, E K, 1974, Wave-induced nearshore circulation. *Journal of Geophysical Research*, 79, 4097–4106.

Nordstrom, K F, 1980, Cyclic and seasonal beach response, a comparison of oceanside and bayside beaches. *Physical Geography*, 1, 177–196.

Nordstrom, K F, 1992, *Estuarine Beaches*. Elsevier, Barking, 225 pp.

Nordstrom, K F and Jackson, N L, 1990, Migration of swash zone, step and microtopographic features during tidal cycles on an estuarine beach, Delaware Bay, New Jersey, U.S.A. *Marine Geology*, 92, 147–154.

Nordstrom, K F and Jackson, N L, 1993, The role of wind direction in eolian transport on a narrow sandy beach. *Earth Surface Processes and Landforms*, 18, 675–685.

Nordstrom, K F and McCluskey, J M, 1984, Considerations for control of house construction in coastal dunes. *Journal of Coastal Zone Management*, 12, 385–402.

Nordstorm, K F, Psuty, N P and Carter, R W G, 1990, *Coastal Dunes: Form and Process*. John Wiley, Chichester, 392 pp.

Nossin, J J, 1965, Analysis of younger beach ridge deposits in eastern Malaya. *Zeitschrift fur Geomorphologie*, 9, 186–208.

Nummedal, D, 1983, Barrier Islands. In Komar, P D (ed.) *Handbook of Coastal Processes and Erosion*. CRC Press, Florida, 77–121.

Nummedal, D and Finley, R J, 1978, Wind-generated longshore currents. *Proceedings 16th International Conference on Coastal Engineering*. ASCE, 1428–1438.

Nummedal, D, Sonnenfield, D L and Taylor, K, 1984, Sediment transport and morphology at the surf zone of Presque Isle, Lake Erie, Pennsylvania. *Marine Geology*, 60, 99–122.

Nummedal, D, Hine, A C and Boothroyd, J C, 1987, Holocene evolution of the south-central coast of Iceland. In Fitzgerald, D M and Rosen, P S (eds) *Glaciated Coasts*. Academic Press, San Diego, 115–150.

Oertel, G F, 1985, The barrier island system. *Marine Geology*, 63, 1–18.

Oertel, G F and Kraft, J C, 1994, New Jersey and Delmarva barrier islands. In Davies, R A Jr (ed.) *Geology of Holocene Barrier Island Systems*. Springer-Verlag, New York, 207–232.

O'Hare, T J and Huntley, D A, 1994, Bar formation due to wave groups and associated long waves. *Marine Geology*, 116, 313–325.

Okazaki, S and Sunamura, T, 1991, Re-examination of breaker-type classification on uniformly inclined laboratory beaches. *Journal of Coastal Research*, 7, 559–564.

Oldenziel, D M and Brink, W E, 1974, Influence of suction and blowing on entrainment of sand particles. *Journal of the Hydraulics Division*. ASCE, 100, 935–950.

Olson, J S, 1958, Lake Michigan dune development. I. Wind velocity profiles. *Journal of Geology*, 66, 254–263.

Oltman-Shay, J and Guza, R T, 1987, Infragravity edge wave observations on two California beaches. *Journal of Physical Oceanography*, 17, 644–663.

Oltman-Shay, J, Howd, P A and Birkemeier, W A, 1989, Shear instabilities of the mean longshore current, 2. Field observations. *Journal of Geophysical Research*, 94, 18 031–18 042.

O'Neill, R V, de Angelis, D L, Waide, J B and Allen, T F H, 1986, *A hierarchical concept of ecosystems*. Monographs in Population Biology, 23, 253 pp.

Onno, M, 1933, Die Strandformationen an der mittleren Lubecker Bucht. *Bericht. Deuts. Bot. Gesells.* 51, 232–266.

Oost, A P, 1995, Dynamics and sedimentary development of the Dutch Wadden Sea with emphasis on the Frisian inlet. *Geologica Iltraiectina*, University of Utrecht, 126, 455 pp.

Oosting, H J and Billings, W D, 1942, Factors affecting vegetational zonation on coastal dunes. *Ecology*, 23, 131–142.

Oretel, G F, 1985, The barrier island system. *Marine Geology*, 63, 1–18.

Oretel, G F and Kraft, C, 1994, New Jersey and Delmarva barrier islands. In Davis, R A Jr (ed.) *Geology of Holocene Barrier Island Systems*. Springer Verlag, New York, 207–232.

Orford, J D and Wright, P, 1978, What's in a name? – Descriptive or genetic implications of 'ridge and runnel' topography. *Marine Geology*, 28, M1–M8.

Orme, A R, 1985, The behaviour and migration of longshore bars. *Physical Geography*, 5, 142–164.

Orme, A R, 1990, The instability of Holocene coastal dunes, the case of the Morro dunes, California. In Nordstrom, K F, Psuty, N P and Carter, R W G (eds) *Coastal Dunes: Form and Process*. Wiley, Chichester, 315–338.

Orme, A R and Orme, A J, 1988, Ridge-and-runnel enigma. *Geographical Review*, 78, 169–184.

Orme, A R and Tchakerian, V P, 1986, Quaternary dunes of the Pacific Coast of the Californias. In Nickling, W (ed.) *Aeolian Geomorphology*. Allen and Unwin, Boston, 149–175, 461.

O'Rourke, J G and LeBlond, P H, 1972, Longshore currents in a semicircular bay. *Journal of Geophysical Research*, 77, 444–452.

Osborne, P D and Greenwood, B, 1990, Set-up driven undertows on a barred beach. *Proceedings 22nd International Conference on Coastal Engineering*. ASCE, 227–240.

Osborne, P D and Greenwood, B, 1992, Frequency dependent cross-shore suspended sediment transport. 1. A non-barred shoreface. *Marine Geology*, 106, 1–24.

Osborne, P D and Greenwood, B, 1993, Sediment suspension under waves and currents, time scales and vertical structure. *Sedimentology*, 40, 599–622.

Osborne, P D and Vincent, C E, 1993, Dynamics of large and small scale bedforms on a macro-tidal shoreface under shoaling and breaking waves. *Marine Geology*, 115, 207–226.

Ota, Y and Kaizuka, S, 1991, Tectonic geomorphology at active plate boundaries – Examples from the Pacific rim. *Zietschrift fur Geomorphologie* N.F. Supplementband 82, 119–146.

Ota, Y, Berryman, K R, Hull, A G, Miyauchi, T and Iso, N, 1988, Age and height distribution of Holocene transgressive deposits in eastern North Island, New Zealand. *Palaeogeography, Palaeoclimatology, Palaeoecology*, 68, 135–151.

Ota, Y, Hull, A G and Berryman, K R, 1991, Coseismic uplift of Holocene marine terraces in the Pakarae River area, eastern North Island. *New Zealand Quaternary Research*, 35, 331–346.

Otvos, E G, 1964, Observations of beach cusps and pebble ridge formation on the Long Island Sound. *Journal of Sedimentary Petrology*, 34, 554–560.

Otvos, E G, 1965, Sedimentation–erosion cycles of single tidal periods on Long Island Sound beaches. *Journal of Sedimentary Petrology*, 35, 604–609.

Otvos, E G, 1970a, Development and migration of barrier islands, northern Gulf of Mexico. *Geological Society of America Bulletin*, 81, 241–246.

Otvos, E G, 1970b, Post-Pleistocene history of the United States inner continental shelf, significance to origin of barrier islands, Discussion. *Geological Society of America Bulletin*, 88, 69–702.

Otvos, E G, 1982a, Barrier beaches. In Schwartz, M L (ed.) *Encyclopedia of Beaches and Coastal Environments*. Hutchinson Ross, Stroudsburg, 123.

Otvos, E G, 1982b, High-energy coasts. In Schwartz, M L (ed.) *Encyclopedia of Beaches and Coastal Environments*. Hutchinson Ross, Stroudsburg, 461.

Otvos, E G, 1982c, Low-energy coasts. In Schwartz, M L (ed.) *Encyclopedia of Beaches and Coastal Environments*. Hutchinson Ross, Stroudsburg, 522.

Otvos, E G, 1985, Barrier platforms; northern Gulf of Mexico. *Marine Geology*, 63, 285–305.

Otvos, E G, 1995, Multiple Pliocene–Quaternary marine highstands, NE Gulf coastal plain – fallacies and facts. *Journal of*

*Coastal Research*, 11(4), 984–1002.

Owens, E H, 1976, The effects of ice on the littoral zone at Richibucto Head, eastern New Brunswick. *Revue de Geographie Montreal*, 30, 95–104.

Owens, E H, 1977, Temporal variations in beach and nearshore dynamics. *Journal of Sedimentary Petrology*, 47, 168–190.

Owens, E H and Harper, J R, 1977, Frost-table and thaw depths in the littoral zone near Peard Bay, Alaska. *Arctic*, 30, 155–168.

Owens, E H and McCann, S B, 1970, The role of ice in the Arctic beach environment with special reference to Cape Ricketts, southwest Devon Island, Northwest Territories, Canada. *American Journal of Science*, 269, 397–414.

Owens, E H and Roberts, H H, 1978, Variations in wave-energy levels and coastal sedimentation, eastern Nicaragua. *Proceedings 16th International Coastal Engineering Conference*, Hamburg. ASCE, 1195–1214.

Packwood, A R, 1983, The influence of beach porosity on wave uprush and backwash. *Coastal Engineering*, 7, 29–40.

Parker, W R, 1975, Sediment mobility and erosion on a multi-barred foreshore (southwest Lancashire, UK). In Hails, J and Carr, A (eds) *Nearshore Sediment Dynamics and Sedimentation*. Wiley and Sons, London, 151–179.

Pattiaratchi, C, 1997, Design studies for an artificial surfing reef at Cable Station, Western Australia. *Proceedings 1st International Surfing Reef Symposium*, Sydney Coastal Studies Unit, University of Sydney, 4 pp.

Penland, P S, Boyd, R, Nummedal, D and Roberts, H H, 1981, Deltaic barrier development on the Louisiana coast. *Gulf Coast Association Geological Society Transactions*, 31, 471.

Penland, S, Suter, J R and Boyd, R, 1985, Barrier island arcs along abandoned Mississippi River delta. *Marine Geology*, 63, 197–233.

Peregrine, D H, 1972, Equations for water waves and the approximations behind them. In Meyer, R E (ed.) *Waves on Beaches and Resulting Sediment Transport*. Academic Press, New York, 95–122.

Peregrine, D H, 1974, Water–wave interaction in the surf zone. *Proceedings 14th Coastal Engineering Conference*. ASCE, 500–517.

Pethick, J, 1984, *An Introduction to Coastal Geomorphology*. Edward Arnold, London, 260 pp.

Philip, J R, 1973, Periodic non-linear diffusion, an integral relation and its physical consequences. *Australian Journal of Physics*, 26, 513–519.

Phillips, D A, 1985, Headland-bay beaches revisited, an example from Sandy Hook, New Jersey. *Marine Geology*, 65, 21–31.

Phillips, J D, 1992, Nonlinear dynamical systems in geomorphology, revolution or evolution. *Geomorphology*, 5, 219–229.

Pierce, J W, 1970, Tidal inlets and washover fans. *Journal of Geology*, 78, 230–234.

Pilarczyk, K W and Zeidler, R B, 1996, *Offshore Breakwaters and Shore Evolution Control*. Balkema, Rotterdam, 560 pp.

Pilkey, O H and Dixon, K L, 1996, *The Corps and the Shore*. Island Press, Covelo, 272 pp.

Pilkey, O H and Neal, W J, 1984, *Living with the Shore*. Duke University Press, Durham.

Pilkey, O H and Wright, H L, 1988, Seawalls versus beaches. *Journal of Coastal Research*, SI4, 41–64.

Pilkey, O H, *et al.*, 1967, The carbonate fraction of beach and dune sands. *Sedimentology*, 8, 311–327.

Pilkey, O H, Pilkey, O H Sr and Turner, R, 1975, *How to Live with an Island*. North Carolina Department of Natural and Economic Resources, Raleigh, 191 pp.

Pilkey, O H, Young, R S, Riggs, S R, Smith, A W S, Wu, H and Pilkey, W D, 1993, The concept of shoreface profile of equilibrium: a critical review: *Journal of Coastal Research*, 9, 255–278.

Pillans, B, 1990, Pleistocene marine terraces in New Zealand, a review. *New Zealand Journal of Geology and Geophysics*, 33, 219–231.

Pillans, B, 1991, New Zealand Quaternary stratigraphy, an overview. *Quaternary Science Reviews* 10, 405–418.

Pillans, B, 1994, Direct marine–terrestrial correlations, Wanganui basin, New Zealand, The last 1 million years. *Quaternary International*, 13, 189–200.

Pillans, B and Huber, P, 1995, Interpreting coseismic deformation using Holocene coastal deposits, Wellington, New Zealand. *Quaternary International* 26, 87–95.

Pirazzoli, P A, 1991, *World Atlas of Holocene Sea Level Changes*. Elsevier Oceanography Series 58, Elsevier, Amsterdam.

Pitt, A B (ed.), 1997, *Proceedings First International Surfing Reef Symposium*. Coastal Studies Unit, University of Sydney.

Pluis, J L A, 1992, Relationships between deflation and near surface wind velocity in a coastal dune blowout. *Earth Surface Processes and Land Forms*, 17, 663–673.

Power, J, 1992, *Beach hazard analysis and bather safety – an investigation using a geographical information system*. BSc Honours Thesis, Department of Geography, University of Sydney, 178 pp.

Press, W W, Flannery, B P, Teukolsky, S A and Vetterling, W T, 1989, *Numerical Recipes (Fortran Version)*. Cambridge University Press, 702 pp.

Price, W A, 1954, Correlation of shoreface type with offshore conditions in the Gulf of Mexico. *Second National Coastal Geography Conference*. Academy of Science, 11–30.

Price, W A, 1982, Beach ridge plain. In Schwartz, M L (ed.) *Encyclopedia of Beaches and Coastal Environments*. Hutchinson Ross, Stroudsburg, 159–160.

Psuty, N P, 1965, Beach ridge development in Tabasco, Mexico. *Annals of the Association of American Geographers*, 55, 112–124.

Psuty, N P, 1988, Sediment budget and dune/beach interaction. *Journal of Coastal Research*, Special Issue 3, 1–4.

Psuty, N P, 1989, An application of science to the management of coastal dunes along the Atlantic coast of the USA. In Gimmingham, C H, Ritchie, W, Willetts, B B and Willis, A J (eds) *Proceedings of the Royal Society of Edinburgh*, 96B, 289–307.

Psuty, N P, 1992, Spatial variation in coastal foredune development. In Carter, R W G, Curtis, T G F and Sheehy-Skeffington, M J (eds) *Coastal Dunes*. Balkema, Rotterdam, 3–13.

Psuty, N P, 1993, Foredune morphology and sediment budget, Perdido Key, Florida, USA. In Pye, K (ed.) *The Dynamics and Environmental Context of Aeolian Sedimentary Systems*. Geological Society, Special Publication No. 72, 145–157.

Pugh, D T, 1987, *Tides, Surges and Mean Sea-Level*. John Wiley & Sons, Chichester, 472 pp.

Putrevu, U and Svendsen, I A, 1993, A mixing mechanism in the nearshore region. *Proceedings 23rd International Conference on Coastal Engineering*. ASCE, 2758–2771.

PWD, 1992, *New South Wales wave climate annual summary 1991/92*. Report MHL627, Manly Hydraulics Laboratory, Public Works Department, Sydney, 7pp.

Pye, K, 1980, Beach salcrete and aeolian sand transport, evidence from north Queensland. *Journal of Sedimentary Petrology*, 50, 257–261.

Pye, K, 1982, Morphological development of coastal dunes in a humid tropical environment, Cape Bedford and Cape Flattery, north Queensland. *Geografiska Annaler*, 64A, 213–227.

Pye, K, 1983, Formation and history of Queensland coastal dunes. *Zietschrift fur Geomorphologie*. N.F. Supplementband 45, 175–204.

Pye, K, 1990, Physical and human influences on coastal dune development between the Ribble and Mersey estuaries, northwest England. In Nordstrom, K F, Psuty, N P and Carter, R W G (eds), *Coastal Dunes: Form and Process*. John Wiley, Chichester, 339–359.

Pye, K and Bowman, G, 1984, The Holocene marine transgression as a forcing function in episodic dune activity on the eastern Australian coast. In Thom, B G (ed.) *Coastal Geomorphology in Australia*. Academic Press, Sydney, 179–196.

Pye, K and Tsoar, H, 1990, *Aeolian Sand and Sand Dunes*. Unwin Hyman, London, 396 pp.

Quatro Rodas, 1995, *Guia de Praias*. [Guide to the Beaches (of Brazil)]. Quatro Rodas, Sao Paulo, 210 pp.

Quinn, C M, 1977, *Sand Dunes. Formation, Erosion and Management*. An Foras Forbartha, Dublin.

Rampino, M R and Sanders, J E, 1981, Holocene transgression in south-central Long Island, New York. *Journal of Sedimentary Petrology*, 50, 1063–1080.

Randall, R E, 1983, Management for survival – A review of the plant ecology and protection of the machair beaches of north-west Scotland. In McLachlan, A and Erasmus T (eds) *Sandy Beaches as Ecosystems*. Junk, The Hague, 733–740.

Ranwell, D S, 1972, *Ecology of Salt Marshes and Sand Dunes*. Chapman and Hall, London.

Ranwell, D S (ed.), 1980, *Sand Dune Machair 3*. Institute of Terrestrial Ecology, Norwich.

Rasmussen, K R, 1989, Some aspects of flow over coastal dunes. In Gimmingham, C H, Ritchie, W, Willetts, B B and Willis, A J (eds) *Proceedings of the Royal Society of Edinburgh*, 96B, 129–147.

Raubenheimer, B, Guza, R T, Elgar, S and Kobayashi, N, 1995, Swash on a gently sloping beach. *Journal of Geophysical Research*, 100, 8751–8760.

Raudkivi, A J, 1976, *Loose Boundary Hydraulics*, Pergamon Press, Oxford, 397 pp.

Raymond, P E and Stetson, H C, 1932, A calcareous beach on the coast of Maine. *Journal of Sedimentary Petrology*, 2, 51–62.

Rea, C C and Komar, P D, 1975, Computer simulation models of a hooked beaches shoreline configuration. *Journal of Sedimentary Petrology*, 45, 866–872.

Rector, R L, 1954, *Laboratory study of equilibrium profiles of beaches*. US Army Corps of Engineers, Beach Erosion Board, Technical Memorandum 41, 38 pp.

Reimnitz, E, Toimol, L J, Shepard, F P and Guiterrez-Estrada, M, 1976, Possible rip current origin for bottom ripple zones to 30ft depth. *Geology*, 4, 395–400.

Reimnitz, E, Barnes, P W and Harper, J R, 1990, A review of beach nourishment from ice transport of shoreface materials, Beaufort Sea. *Journal of Coastal Research*, 6, 439–469.

Reineck, H-E and Singh, I B, 1973, *Depositional Sedimentary Environments*. Springer-Verlag, Berlin, 439 pp.

Reineck, H-E and Singh, I B, 1980, *Depositional Sedimentary Environments, with Reference to Terrigenous Clastics*, second revised edition. Springer-Verlag, Berlin, 549 pp.

Reinson, G E, 1984, Barrier island and associated strand-plain systems. In Walker, R G (ed.) *Facies Models*. Geoscience Canada Reprint Series 1, 119–140.

Rhodes, E G, 1982, Depositional model for a chenier plain, Gulf of Carpentaria, Australia. *Sedimentology*, 29, 201–222.

Ribberink, J S and Al-Salem, A A, 1994, Sediment transport in oscillatory boundary layers in cases of rippled beds and sheet flow. *Journal of Geophysical Research*, 99, 12 707–12 727.

Richmond, B M and Sallenger, A H, 1984, Cross-shore transport of bi-modal sands. *Proceedings 19th Coastal Engineering Conference*. ASCE, 1997–2008.

Riggs, S R, Cleary, W J and Snyder, S W, 1995, Influence of inherited geologic framework on barrier shoreface morphology and dynamics. *Marine Geology*, 126, 213–234.

Ritchie, W, 1972, The evolution of coastal sand dunes. *Scottish Geographical Magazine*, 88, 19–35.

Ritchie, W and Penland, S, 1988, Rapid dune changes associated with overwash processes on the deltaic coast of south Louisiana. *Marine Geology*, 81, 97–112.

Ritchie, W and Penland, S, 1990, Aeolian sand bodies of the south Louisiana coast. In Nordstrom, K F, Psuty, N P and Carter, R W G (eds) *Coastal Dunes, Form and Process*. John Wiley and Sons, London, 105–127.

Robertson-Rintoul, M J, 1990, A quantitative analysis of the near-surface wind flow pattern over coastal parabolic dunes. In Nordstrom, K F, Psuty, N P and Carter, R W G (eds), *Coastal Dunes, Form and Process*. John Wiley and Sons, London, 57–78.

Robertson-Rintoul, M and Ritchie, W, 1990, The geomorphology of coastal sand dunes in Scotland. A review. In Bakker,

Th W, Jungerius, P D and Klijn, J A (eds). *Dunes of the European Coasts, Catena*. Supplement 18, 41–49.

Roelvink, J A and Brøker, I, 1993, Cross-shore profile models. *Coastal Engineering*, 21, 163–191.

Roelvink, J A and Stive, M J F, 1991, Sand transport in the shoreface of the Holland coast. *Proceedings 22nd International Conference on Coastal Engineering*, New York. ASCE, 1909–1921.

Rosen, P S, 1979, Aeolian dynamics of a barrier island system. In Leatherman, S P (ed.) *Barrier Islands from the Gulf of St. Lawrence to the Gulf of Mexico*. Academic Press, New York, 81–98.

Rosengren, N, 1981, Dune systems on cuspate forelands, east Gippsland, Victoria. *Proceedings of the Royal Society of Victoria*, 92, 137–147.

Roy, P S and Cowell, P J, 1991, Geological-inheritance effects in the response of the shoreface to coastal storms. *Geography Society of New South Wales Conference Papers*, 9, 74–82.

Roy, P S and Crawford, E A, 1980, Quaternary geology of Newcastle Bight inner continental shelf, New South Wales, Australia, *New South Wales Geological Survey Records*, 19, 145–188.

Roy, P S and Stephens, A W, 1981, Geological controls on process-response, S.E. Australia. *Proceedings 17th International Coastal Engineering Conference*. ASCE, 23–28.

Roy, P S and Thom, B G, 1981, Late Quaternary marine deposition in New South Wales and southern Queensland – an evolutionary model. *Journal of the Geology Society of Australia*, 28, 471–489.

Roy, P S, Thom, B G and Wright, L D, 1980, Holocene sequences on an embayed high-energy coast, an evolutionary model. *Sedimentary Geology*, 16, 1–9.

Roy, P S, Zhuang, W-Y, Birch, G F and Cowell, P J, 1992, *Quaternary geology and placer mineral potential of the Forster-Tuncurry shelf, SE Australia*. Geological Survey Report GS 1992/201, Department of Mineral Resources, NSW, 164 pp.

Roy, P S, Cowell, P J, Ferland, M A and Thom, B G, 1994, Wave dominated coasts. In Carter, R W G and Woodroffe, C.D. (eds) *Coastal Evolution, Late Quaternary Shoreline Morphodynamics*. Cambridge University Press, Cambridge, 121–186.

Rozema, J, Biljwaard, P, Prast, B and Broekman, R, 1985, Ecophysiological adaptations of coastal halophytes from foredunes and salt marshes. *Vegetatio*, 62, 499–521.

Rubin, D M, 1987, *Cross-bedding, bedforms and palaeocurrents*. SEPM Concepts in Sedimentology and Paleontology, 1, 187 pp.

Russell, P E, 1993, Mechanisms for beach erosion during storms. *Continental Shelf Research*, 13, 1243–1265.

Russell, P, Davidson, M, Huntley, D, Cramp, A, Hardisty, J and Lloyd, G, 1991, The British Beach And Nearshore Dynamics (B-BAND) programme. *Proceedings Coastal Sediments '91*. 371–384.

Russell, R J, 1959, Caribbean beach rock observations. *Zietschrift fur Geomorphologie*, 3, 227–236.

Russell, R J, 1962, Origin of beach rock. *Zietschrift fur Geomorphologie*, 6, 1–16.

Russell, R J, 1967a, Aspects of coastal morphology. *Geografiska Annaler*, A, 2–4, 299–309.

Russell, R J, 1967b, *River Plains and Sea Coasts*. University of California Press, Berkeley, 173 pp.

Russell, R J, 1968, Foreword, in Kaiser, K (ed.) *Kustengeomorphologie, Zietschrift fur Geomorphologie*, Supplement 7, v–vii.

Russell, R J and McIntire, W G, 1965a, Beach cusps. *Geological Society of America Bulletin*, 76, 307–320.

Russell, R J and McIntire, W G, 1965b, Southern hemisphere beach rock. *Geographical Review*, 55, 17–45.

Rutin, J, 1983, *Erosional processes on a coastal sand dune, De Blink, Noordwijkerhiut, The Netherlands*. PhD Dissertation, University of Amsterdam, Publicaties van het Fysisch Geografisch en Bodemkundig Lab., 35, 144 pp.

Ruz, M-H and Allard, M, 1994, Foredune development along a subarctic emerging coastline, eastern Hudson bay, Canada. *Marine Geology*, 117, 57–74.

Ruz, M-H, Hequette, A and Hill, P R, 1992, A model of coastal evolution in a transgressed thermokarst topography, Canadian Beaufort Sea. *Marine Geology* 106, 251–278.

Salisbury, E, 1952, *Downs and Dunes. Their Plant Life and Environment*. G. Bell and Sons, London.

Sallenger, A H, 1979a, Beach-cusp formation. *Marine Geology*, 56, 23–37.

Sallenger, A H, 1979b, Inverse grading and hydraulic equivalence in grain-flow deposits. *Journal of Sedimentary Petrology*, 49, 553–562.

Sallenger, A H, 1981, Swash mark and grain flow. *Journal of Sedimentary Petrology*, 51, 261–264.

Sallenger, A H and Holman, R T, 1985, Wave energy saturation on a natural beach with variable slope. *Journal of Geophysical Research*, 90, 11 939–11 944.

Sallenger, A H and Howd, P A, 1989, Nearshore bars and the break-point hypothesis. *Coastal Engineering*, 12, 301–313.

Sallenger, A H and Richmond, B M, 1984, High frequency sediment-level oscillations in the swash zone. *Marine Geology*, 60, 155–164.

Sallenger, A H, Howard, P C, Fletcher, C H and Howd, P A, 1983, A system for measuring bottom profile, waves and currents in the high-energy nearshore environments. *Marine Geology*, 51, 63–76.

Sallenger, A H, Holman, R A and Birkemeier, W A, 1985, Storm-induced response of a nearshore-bar system. *Marine Geology*, 64, 237–257.

Sand, S E, 1982, Wave grouping described by bounded long waves. *Ocean Engineering*, 9, 567–580.

Sanderson, P G and Eliot, I, 1996, Shoreline salients, cuspate forelands and tombolos on the coast of Western Australia. *Journal of Coastal Research*, 12, 761–773.

Sarre, R, 1988, Evaluation of aeolian sand transport equations using intertidal zone measurements, Saunton Sands, England. *Sedimentology* 35, 671–679.

Sasaki, T, 1977, *Field investigations of nearshore currents on a gently sloping bottom*. Nearshore Environment Research

Center TR-77-2, Tokyo, 209 pp.

Sasaki, T. 1982, *Three-dimensional topographic changes of the foreshore zone of sandy beaches.* Doctoral Thesis, University of Tsukuba, 172 pp.

Sasaki, T and Horikawa, K, 1975, Nearshore current systems on a gently sloping bottom. *Coastal Engineering in Japan*, 18, 123–142.

Sasaki, T and Horikawa, K, 1978, Observation of nearshore current and edge waves. *Proceedings 16th Coastal Engineering Conference, Hamburg, American Society Civil Engineers, New York, 791–809.*

*Sasaki, T, Horikawa, K and Hotta, S, 1976, Nearshore current on a gently sloping beach. 15th Coastal Engineering Conference*, Honolulu. ASCE, 626–644.

Sato, S and Kishi, T, 1954, Shearing force on sea bed and movement of bed material due to wave motion. *Journal of Research, Public Works Research Institute*, 1, 1–11.

Sato, M, Kuroki, K and Shinohara, T, 1992, A field experiment on the formation of beach cusps. *Proceedings, 23rd International Conference on Coastal Engineering*, Venice, Italy, Vol. 2. ASCE, 2205–2218.

Saunders, K E and Davidson-Arnott, R G D, 1991, Coastal dune response to natural disturbances. In Davidson-Arnott, R G D (ed.) *Proceedings Symposium on Coastal Sand Dunes.* Natural Resources Council of Canada, Ottawa, 321–346.

Savage, R P, 1957, *Model tests for hurricane protection project.* Beach Erosion Board, US Corps of Engineers, Washington, DC.

Saville, T, 1957, Scale effects in two-dimensional beach studies. *Proceedings 7th General Meeting of the International Association of Hydraulic Research*, A3, 1–8.

Saville, T, 1961, Experimental determination of wave set-up. *Proceedings 2nd Conference on Hurricanes.*

Schalk, M, 1938. A textural study of the Outer Beach of Cape Cod, Massachusetts. *Journal of Sedimentary Petrology*, 8, 41–54.

Scheffer, H J, Fernando, K R M D and Fittschen, T, 1994, *Directional wave climate study, south-west coast of Sri Lanka.* Report, Sri Lankan-German Cooperation CCD-GTZ Coast Conservation Project, Colombo, 91 pp.

Schepers, J D, 1978, *Zandtrtansport onder van Golven en een Eenparige Stroom bij Varierende Korreldiameter*. Masters Thesis, Delft University of Technology.

Schoonees, J S and Theron, A K, 1995, Evaluation of ten cross-shore sediment transport/morphological models. *Coastal Engineering*, 25, 1–41.

Schumm, S A, 1991, *To Interpret the Earth: Ten Ways to be Wrong.* Cambridge University Press, 133 pp.

Schumm, S A and Lichty, RW, 1965, Time, space and causality in geomorphology. *American Journal of Science*, 263, 110–119.

Schwartz, M L, 1967, The Bruun theory of sea-level rise as a cause of shore erosion. *Journal of Geology*, 75, 76–92.

Schwartz, M L, 1971, The multiple causality of barrier islands. *Journal of Geology*, 79, 91–94.

Schwartz, M L, 1972a, Theoretical approach to the origin of beach cusps. *Geological Society of America Bulletin*, 83, 1115–1116.

Schwartz, M L (ed.), 1972b, *Spits and Bars.* Dowden, Hutchinson & Ross, Stroudsburg, 542 pp.

Schwartz, M L (ed.), 1973, *Barrier Islands.* Benchmark Papers in Geology, Dowden, Hutchinson and Ross, Stroudsburg.

Schwartz, M L (ed.), 1982, *The Encyclopedia of Beaches and Coastal Environments.* Hutchinson Ross, Stroudsburg, 940 pp.

Schwartz, M L and Bird, E C F (eds), 1990, *Artificial Beaches, Journal of Coastal Research*, Special Issue, 6, 140 pp.

Schwebel, D A, 1984, Quaternary stratigraphy and sea-level variation in the southeast of South Australia. In Thom, B G (ed.) *Coastal Geomorphology in Australia.* Academic Press, Sydney, 291–311.

SCOR, 1991, The response of beaches to sea-level changes, a review of predictive models. *Journal of Coastal Research*, 7, 895–921.

Seely, M (ed.), 1994, *Deserts. The Illustrated Library of the Earth.* Weldon Owen, Sydney, 160 pp.

Semeniuk, V, 1993, The Pilbara coast, a riverine coastal plain in a tropical arid setting, northwestern Australia. *Sedimentary Geology*, 83, 235–256.

Semeniuk, V and Meagher, T D, 1981, The geomorphology and surface processes of the Australind-Leschenault Inlet coastal area. *Journal of the Royal Society of Western Australia*, 64, 33–51.

Semeniuk, V and Searle, D J, 1987, Beach rock ridges/bands along a high-energy coast in southwestern Australia – their significance and use in coastal history. *Journal of Coastal Research*, 3, 331–342.

Sexton, W J, 1995, The post-storm Hurrican Hugo recovery of the undeveloped beaches along the south Carolina coast, 'Capers Island to the Santee Delta'. *Journal of Coastal Research*, 11, 1020–1025.

Seymour, R J, 1986, Nearshore autosuspending turbidity currents. *Ocean Engineering*, 13, 435–447.

Seymour, R J (ed.), 1989, *Nearshore Sediment Transport.* Plenum, New York, 418 pp.

Seymour, R J and King, D B, 1982, Field comparisons of cross-shore transport models. *Journal of the Waterway, Port, Coastal and Ocean Division.* ASCE, 108, 163–179.

Shen, M C and Meyer, R E, 1963, Climb of a bore on a beach, Part 3 Runup. *Journal of Fluid Mechanics*, 16, 113–125.

Shepard, F P, 1948, *Submarine Geology.* Harper and Row, New York, 557 pp.

Shepard, F P, 1949, Dangerous currents in the surf. *Physics Today*, 2, 20–29.

Shepard, F P, 1950, Beach cycles in Southern California. US Army Corps of Engineers, Beach Erosion Board, Technical Memorandum No.15, 31 pp.

Shepard, F P, 1952, Revised nomenclature for depositional coastal features. *Bulletin American Association of Petroleum Geologist*, 36, 1902–1912.

Shepard, F P, 1963, *Submarine Geology*, 2nd Edition, Harper and Row, New York, 557 pp.

Shepard, F P and Inman, D L, 1950, Nearshore water circulation related to bottom topography and wave refraction.

*Transactions of the American Geophysical Union*, 31, 196–212.

Shepard, F P and Inman, D L, 1951, Nearshore circulation. *Proceedings 1st Conference Coastal Engineering*, Long Beach. Council on Wave Research, 50–59.

Shepard, F P and LaFonda, E C, 1940, Sand movement along the Scripps Institution Pier. *American Journal of Science*, 238, 272–284.

Shepard, F P, Emery, M O and Lafond, E C, 1941. Rip currents. A process of geological importance. *Journal of Geology*, 49, 337–369.

Shepherd, M J, 1987, Holocene alluviation and transgressive dune activity in the lower Manawatu Valley, NZ. *New Zealand Journal of Geology and Geophysics*, 30, 175–187.

Shepherd, M J, 1990, Relict and contemporary foredunes as indicators of coastal processes. In Brierley, G and Chappell, J (eds) *Proceedings Applied Quaternary Studies Symposium*, Canberra. 17–24.

Sherman, D J and Bauer, B O, 1993, Dynamics of beach-dune systems. *Progress in Physical Geography*, 17, 413–447.

Sherman, D J and Greenwood, B, 1984, Boundary roughness and bedforms in the surf zone. *Marine Geology*, 60, 199–218.

Sherman, D J and Hotta, S, 1990, Aeolian sediment transport, Theory and measurement. In Nordstrom, K F, Psuty, N P and Carter, R W G (eds) *Coastal Dunes, Form and Process*. John Wiley, Chichester, 17–37.

Sherman, D J and Lyons, W, 1994, Beach-state controls on aeolian sand delivery to coastal dunes. *Physical Geography*, 15, 381–395.

Sherman, D J, Short, A D and Takeda, I, 1993, Sediment mixing depth and bedform migration in rip channels. *Journal of Coastal Research*, Special Issue 15, 39–46.

Shi, N C and Larsen, L H, 1984, Reverse sediment transport induced by amplitude-modulated waves. *Marine Geology*, 54, 181–200.

Shields, A, 1936. Anwendung der Ähnlichkeitsmechanik und Turbulensforschung auf die Geschiebebewegung. Mitteilungen Preussisches Versuchsanstalt für Wasserbau and Schiffbau, 26.

Shipp, C, 1984, Bedforms and depositional sedimentary structures of a barred nearshore system, eastern Long Island, New York. *Marine Geology*, 60, 235–259.

Short, A D, 1975a, Offshore bars along the Alaskan Arctic coast. *Journal of Geology*, 83, 209–221.

Short, A D, 1975b, Multiple offshore bars and standing waves. *Journal of Geophysical Research*, 80, 3838–3840.

Short, A D, 1976, Observations of ice deposited by waves on Alaskan Arctic beaches. *Revue de Geographie Montreal*, 30, 115–122.

Short, A D, 1978, Wave power and beach-stages, A global model. *Proceedings 16th International Conference on Coastal Engineering*. ASCE, 1145–1162.

Short, A D, 1979a, Barrier island development on the north Alaskan Coast. *Bulletin Geological Society of America*, Part II, 90, 77–103.

Short, A D, 1979b, Three dimensional beach-stage model. *Journal of Geology*, 87, 553–571.

Short, A D, 1981, Beach response to variations in breaker height. *Proceedings 17th International Coastal Engineering Conference*, Sydney. ASCE, 1016–1035.

Short, A D, 1983, Sediments and structures in beach-nearshore environments, southeast Australia. In McLachlan, A and Erasmus, T (eds) *Sandy Beaches as Ecosystems*. Junk, The Hague, 145–155.

Short, A D, 1984a, Temporal change in beach type resulting from a change in grain size. *Search*, 15, 228–230.

Short, A D, 1984b, Beach and nearshore facies, southeast Australia. *Marine Geology*, 60, 261–282.

Short, A D, 1985, Rip current type, spacing and persistence, Narrabeen Beach, Australia. *Marine Geology*, 65, 47–71.

Short, A D, 1986, Sandy shore facies in southeast Australia. In Frankel, E, Keene, J B and Waltho, A E (eds) *Recent Sediments in Eastern Australia*. NSW Division, Geological Society of Australia, 87–100.

Short, A D, 1987, Modes, timing and volume of Holocene cross-shore and aeolian sediment transport, southern Australia. *Coastal Sediments '87*, New Orleans. ASCE, 1925–1937.

Short, A D, 1988a, Response of a high through low energy sandy coast to sea level transgression and stillstand, South Australia. *Geographical Review*, 78, 119–136.

Short, A D, 1988b, Holocene coastal dune formation in southern Australia – A case study. *Sedimentary Geology*, 55, 121–142.

Short, A D, 1989, Chenier research on the Australian coast. *Marine Geology*, 90, 345–351.

Short, A D, 1991a, *Beach morphodynamic systems of the central Netherlands coast, Den Helder to Hoek van Holland*. Rapport GEOPRO 1991,01, Institute of Geographical Research, University of Utrecht, 106 pp.

Short, A D, 1991b, Macro-meso tidal beach morphodynamics – An overview. *Journal of Coastal Research*, 7, 417–436.

Short, A D, 1992, Beach systems of the central Netherlands coast; processes, morphology and structural impacts in a storm driven, multi-bar system. *Marine Geology*, 107, 103–137.

Short, A D (ed.), 1993a, *Beach and Surf Zone Morphodynamics*, *Journal of Coastal Research*, Special Issue 15, 231 pp.

Short, A D, 1993b, *Beaches of the New South Wales Coast*. Australian Beach Safety and Management Program, Sydney, 358 pp.

Short, A D, 1996, *Beaches of the Victorian Coast and Port Phillip Bay*. Australian Beach Safety and Management Program, Sydney, 298 pp.

Short, A D, in press, *Beaches of the Queensland Coast, Cooktown to Coolongatta*. Australian Beach Safety and Management Program, Sydney.

Short, A D and Aagaard, T, 1993, Single and multi-bar beach change models. *Journal of Coastal Research*, Special Issue 15, 141–157.

Short, A D and Brander, R W, in press, Regional variation in rip density. *Journal of Coastal Research*.

Short, A D and Fotheringham, D G, 1986, *Coastal mor-*

*phodynamics and Holocene evolution of the Kangaroo Island coast, South Australia.* Technical Report No. 86/1, Coastal Studies Unit, University of Sydney, 112 pp.

Short, A D and Hesp, P A, 1982, Wave, beach and dune interactions in southeast Australia. *Marine Geology*, 48, 259–284.

Short, A D and Hesp, P A, 1984, *Beach and dune morphodynamics of the south east coast of South Australia.* Technical Report 84/1, Coastal Studies Unit, University of Sydney, 142 pp.

Short, A D and Hogan, C L, 1990, Sydney's southern surfing beaches, characteristics and hazards. In Fabbri, P (ed.) *Recreational Uses of Coastal Areas.* Kluwer Academic, Dordrecht, 199–210.

Short, A D and Hogan, C L, 1994, Rip currents and beach hazards, their impact on public safety and implications for coastal management. In Finkl, C W (ed.) *Coastal Hazards, Journal of Coastal Research*, Special Issue 12, 197–209.

Short, A D and Ni, M, 1997, Regional beach characteristics of the New South Wales coast. *Institute of Australian Geographers and New Zealand Geographical Society Joint Conference*, Hobart. 143.

Short, A D and Wright, L D, 1983, Physical variability of sandy beaches. In McLachlan, A and Erasmus, T (eds), *Sandy Beaches as Ecosystems.* Junk, The Hague, 133–144.

Short, A D and Trenaman, N L, 1992, Wave climate of the Sydney region, an energetic and highly variable ocean wave regime. *Australian Journal of Marine and Freshwater Research*, 43, 765–791.

Short, A D and Wiseman, W J Jr, 1974, Freezeup processes on Arctic beaches. *Arctic*, 27, 215–224.

Short, A D and Wiseman, W J Jr, 1975, Coastal breakup in the Alaskan Arctic. *Geological Society of America Bulletin*, 86, 199–202.

Short, A D, Coleman, J M and Wright, L D, 1974, Beach dynamics and nearshore morphology of the Beaufort Sea coast, Alaska. In Reed, J C and Sater, J E (eds) *The Coast and Shelf of the Beaufort Sea.* The Arctic Institute of North America, Arlington, VA, 477–488.

Short, A D, Buckley, R and Fotheringham, D G, 1986, *Coastal morphodynamics and coastal evolution of the Eyre Peninsula coast, South Australia.* Technical Report 86/2, Coastal Studies Unit, University of Sydney, 178 pp.

Short, A D, Buckley, R and Fotheringham, D G, 1989, Preliminary investigations of beach ridge progradation on Eyre Peninsula and Kangaroo island. *Transactions of the Royal Society of South Australia*, 113, 145–161.

Short, A D, May, A and Hogan, C L, 1991, *A three year study into the circumstances behind surf based rescues.* Report 91–1, NSW Beach Safety Program, Surf Life Saving NSW, Sydney, 19 pp.

Short, A D, Cowell, P J, Cadee, M, Hall W and van Dijk, B, 1995, Beach rotation and possible relation to the Southern Oscillation. In Aung T H (ed.) *Ocean Atmosphere Pacific Conference.* National Tidal Facility, Adelaide, 329–334.

Shulmeister, J and Kirk, R M, 1993, Evolution of a mixed sand and gravel barrier system in North Canterbury, New Zealand, during Holocene sea-level rise and still-stand. *Sedimentary Geology*, 87, 215–235.

Shulmeister, J and Kirk, R M, 1996, Holocene history and a thermoluminescence based chronology of coastal dune ridges near Leithfield, North Canterbury, New Zealand. *New Zealand Journal of Geology and Geophysics*, 39, 25–32.

Shulmeister, J and Lees, B G, 1992, Morphology and chronostratigraphy of a coastal dune field, Goote Eylandt, Northern Australia. *Geomorphology*, 5, 521–534.

Shulmeister, J, Short, S A, Price, D M and Murray, A S, 1993, Pedogenic uranium/thorium and thermoluminescence chronologies and evolutionary history of a coastal dunefield, Groote Eylandt, northern Australia. *Geomorphology*, 8, 47–64.

Shyuer-Ming, S and Komar, P D, 1994, Sediments, beach morphology and sea cliff erosion within an Oregon coast littoral cell. *Journal of Coastal Research*, 10, 144–157.

Silvester, R, 1960, Stabilization of sedimentary coastlines. *Nature*, 188, 467–469.

Silvester, R, 1974, *Coastal Engineering*, Elsevier, Amsterdam, 2 vols.

Silvester, R and Ho, S K, 1970, Use of crenulate shaped bays to stabilise coasts. *Proceedings of 13th International Conference on Coastal Engineering.* ASCE, 1347–1365.

Silvester, R and Hsu, J R C, 1993, *Coastal Stabilization, Innovative Concepts.* Prentice Hall, Englewood Cliffs, New Jersey, 578 pp.

Silvester, R, Tsuchiya, Y and Shibano, Y, 1980, Zeta bays, pocket beaches and headland control. *Proceedings of 17th International Conference on Coastal Engineering.* ASCE, 1306–1319.

Simons, D B, Richardson, E V and Nordin, C F Jr, 1965, Sedimentary structures generated by flow in alluvial channels. In *Primary Sedimentary Structures and their Hydrodynamic Interpretation.* Society of Economic Paleontologists and Mineralogists, Special Publication 12, 34–52.

Skelly, D, 1997, Surfrider Foundation El Segundo surf enhancement project. *Proceedings 1st International Surfing Reef Symposium*, Sydney. Coastal Studies Unit, University of Sydney, 8 pp.

Sleath, J F A, 1984, *Sea Bed Mechanics.* John Wiley and Sons, New York, 335 pp.

Sleath, J F A, 1990, Seabed boundary layers. In LeMehaute, B and Hanes, D M (eds) *The Sea*, Vol. 9B. Wiley Interscience, New York, 693–727.

SLSA, 1995, *Surf Lifesaving Training Manual.* Surf Life Saving Australia, Sydney, 152 pp.

Smart, J, 1976, The nature and origin of beach ridges, western Cape York Peninsula. *Journal of Australian Geology and Geophysics*, 1, 211–218.

Smiles, D E and Stokes, A N, 1976, Periodic solutions of a non-linear diffusion equation used in groundwater flow theory, using a Hele-Shaw model. *Journal of Hydrology*, 31, 27–35.

Smith, H T U, 1960, *Physiography and photo interpretation of coastal sand dunes.* Final Report Contract NONR – 2242(00), Office of Naval Research, Geography Branch.

Smith, H T U, 1964, Dune morphology and chronology in central and western Nebraska. *Journal of Geology*, 73, 557–578.

Smith, J M, 1994, Undertow at Supertank. *Proceedings Coastal Dynamics '94*. ASCE, 220–232.

Smith, J M, Svendsen, I A and Putrevu, U, 1992, Vertical structure of the nearshore current at Delilah, measured and modeled. *Proceedings 23rd International Conference on Coastal Engineering*. ASCE, 2825–2838.

Smith, J M, Larson, M and Kraus, N C, 1993, Longshore current on a barred beach. *Journal of Geophysical Research*, 98, 22 717–22 731.

Sonu, C J, 1969, Collective movement of sediment in littoral environment. *Proceedings 11th Conference Coastal Engineering*, London. ASCE, 373–400.

Sonu, C J, 1972, Field observation of nearshore circulation and meandering currents. *Journal of Geophysical Research*, 77, 3232–3247.

Sonu, C J, 1973, Three dimensional beach changes. *Journal of Geology*, 81, 42–64.

Sonu, C J and James, W R, 1973, A Markov model for beach profile changes. *Journal of Geophysical Research*, 78, 1462–1471.

Sonu, C J and Russell, R J, 1967, Topographic changes in the surf zone profile. *Proceedings 10th Conference Coastal Engineering*. ASCE, 502–522.

Sonu, C J and van Beek, J L, 1971, Systematic beach changes on the Outer Banks, North Carolina. *Journal of Geology*, 79, 416–425.

Sonu, C J, Murray, S P, Hsu, S A, Suhayda, J N and Waddell, E, 1973, Sea breeze and coastal processes. *Transactions American Geophysical Union*, 54, 820–833.

Soulsby, R L, 1983, The bottom boundary layer of shelf seas. In Johns, B (ed.) *Physical Oceanography of Coastal and Shelf Seas*. Elsevier, Amsterdam, 189–266.

Sprigg, R C, 1952, *The geology of the South East Province, South Australia, with special reference to Quaternary coastline migration and modern beach development*. South Australian Department of Mines, Geological Survey of South Australia Bulletin 29, 120 pp.

Sprigg, R C, 1979, Stranded and submerged sea-beach systems of southeast South Australia and the aeolian desert cycle. *Sedimentary Geology*, 22, 53–96.

Stapor, F W Jr, 1975, Holocene beach ridge plain development, northwest Florida. *Zeitschrift fur Geomorphologie*, Supplement 22, 116–144.

Stapor, F W Jr, 1982, Beach ridges and beach ridge coasts. In Schwartz, M L (ed.) *Encyclopedia of Beaches and Coastal Environments*. Hutchinson Ross, Stroudsburg, 160–161.

Stapor, F W Jr, Mathews, T D and Lindfors-Kearns, F E, 1991, Barrier-island progradation and Holocene sea-level history in southwest Florida. *Journal of Coastal Research*, 7, 815–838.

Steers, J A, 1953, *The Sea Coast*. Collins, London, 276 pp.

Steers, J A, 1964, *The Coastline of England and Wales*. Cambridge University Press, 750 pp.

Steers, J A, 1973, *Coastline of Scotland*. Cambridge University Press, Cambridge, 335 pp.

Sternberg, R W, Shi, N C and Downing, J P, 1984, Field investigations of suspended sediment transport in the nearshore zone. *Proceedings 19th International Conference on Coastal Engineering*. ASCE, 1782–1798.

Stevens, G R, 1973, Late Holocene marine features adjacent to Port Nicholson, Wellington, New Zealand. *New Zealand Journal of Geology and Geophysics*, 16, 455–484.

Stive, M J F and Battjes, J A, 1984, A model for offshore sediment transport. *Proceedings 19th International Conference on Coastal Engineering*. ASCE, 1420–1436.

Stive, M J F and De Vriend, H J, 1995, Modelling shoreface profile evolution. *Marine Geology*, 126, 235–248.

Stive, M J F and Wind, H G, 1986, Cross-shore mean flow in the surf zone. *Coastal Engineering*, 10, 325–340.

Stive, M J F, Roelvink, D J A and de Vriend, H J, 1991, Large-scale coastal evolution concept. *Proceedings 22nd International Conference on Coastal Engineering*, New York. ASCE, 1962–1974.

Stive M J F, De Vriend, H J, Cowell, P J and Niedoroda, A W, 1995. Behaviour-oriented models of shoreface evolution. *Proceedings Coastal Dynamics '95*. ASCE, 998–1005.

Stoker, J J, 1957, *Water Waves, The Mathematical Theory with Applications*. Wiley Interscience, New York, 567 pp.

Stokes, G G, 1880. On the theory of oscillatory waves. Mathematical and Physical Papers 1, Cambridge University Press, 225–228.

Strahler, A N, 1952. Dynamic basis of geomorphology. *Bulletin Geological Society of America*, 63, 923–938.

Strahler, A N, 1966, Tidal cycle of changes in an equilibrium beach, Sandy Hook, New Jersey. *Journal of Geology*, 74, 247–268.

Sturman, A and Trapper, N, 1996, *The Weather and Climate of Australia and New Zealand*. Oxford University Press, Melbourne, 476 pp.

Suhayda, J N, 1974, Standing waves on beaches. *Journal of Geophysical Research*, 79, 3065–3071.

Suhayda, J N, Hsu, S A, Roberts, H H and Short, A D, 1975, *Documentation and analysis of coastal processes, northeast coast of Brazil*. Coastal Studies Institute, Louisiana State University, Baton Rouge, Louisiana, Technical Report 238, 98 pp.

Sunamura, T, 1980, A laboratory study of offshore transport of sediment and a model for eroding beaches. *Proceedings 17th International Conference on Coastal Engineering*. ASCE, 1051–1070.

Sunamura, T, 1984a, Onshore–offshore sediment transport rate in the swash zone of laboratory beaches. *Coastal Engineering in Japan*, 27, 205–212.

Sunamura, T, 1984b, Quantitative prediction of beach-face slopes. *Geological Society of America Bulletin*, 95, 242–245.

Sunamura, T, 1986, A parameter for cross shore sediment transport direction and its application to beach erosion/accretion problem. *Annual Report of the Institute of Geoscience*. University of Tsukuba, No.12, 52–54.

Sunamura, T, 1988, Beach morphologies and their change. In Horikawa, K. (ed.) *Nearshore Dynamics and Coastal Pro-*

*cesses*. University of Tokyo Press, Tokyo, 136–166.

Sunamura, T, 1989, Sandy beach geomorphology elucidated by laboratory modelling. In Lakhan, V C and Trenhaile, A S (eds) *Applications in Coastal Modelling*. Elsevier, Amsterdam, 159–213.

Sunamura, T and Mizuno, O, 1987, *A Study on Depositional Shoreline Forms behind an Island*. Annual Report Institute Geosciences, University of Tsukuba, Tsukuba, No. 13, 71–73.

Sunamura, T and Takeda, I, 1984, Landward migration of inner bars. *Marine Geology*, 60, 63–78.

Svasek, J N and Terwindt, J H J, 1974, Measurements of sand transport by wind on a natural beach. *Sedimentology*, 21, 311–322.

Svendsen, I A, 1984a, Wave heights and set-up in a surf zone. *Coastal Engineering*, 8, 303–329.

Svendsen, I A, 1984b, Mass flux and undertow in a surf zone. *Coastal Engineering*, 8, 347–365.

Svendsen, I A and Buhr-Hansen, J, 1988, Cross-shore currents in surf-zone modelling. *Coastal Engineering*, 12, 23–42.

Svendsen, I A and Putrevu, U, 1996, Surf zone hydrodynamics. In Liu, P L F (ed.) *Advances in Coastal and Ocean Engineering, Vol 2*. World Scientific, Singapore, 1–78.

Svendsen, I A, Madsen, P A and Buhr Hansen, J, 1978, Wave characteristics in the surf zone. *Proceedings 16th International Conference on Coastal Engineering*. ASCE, 520–539.

Svendsen, I A, Schäffer, H A and Buhr-Hansen, J, 1987, The interaction between the undertow and the boundary layer flow on a beach. *Journal of Geophysical Research*, 92, 11 845–11 856.

Swain, A, 1989, Beach profile development. In Lakhan, V C and Trenhaile, A S (eds) *Applications in Coastal Modeling*. Elsevier, Amsterdam, 215–232.

Swart, D H, 1974, *Offshore sediment transport and equilibrium beach profiles*. Delft Hydraulics Laboratory, Publication 131, 247 pp.

Swift, D J P, 1968, Coastal erosion and transgressive stratigraphy. *Journal of Geology*, 76, 444–456.

Swift, D J P, 1975, Barrier island genesis, Evidence from the central Alantic shelf, eastern USA. *Sedimentary Geology*, 14, 1–43.

Swift, D J P, 1976, Continental shelf sedimentation. In Stanley, D J and Swift, D J P (eds) *Marine Sediment Transport and Environmental Management*. John Wiley and Sons, New York, 311–350.

Swift, D J P, 1984, Response of the shelf floor to flow. In Tilman, R W, Swift, D J P and Walker, R G, (eds) *Sand and Sandstone Reservoirs*. Society of Economic Palaeontologists and Mineralogists, Tulsa, Short Course Notes No. 13.

Swift, D J P and Thorne, J A, 1991, Sedimentation on continental margins, I, a general model for shelf sedimentation. In Swift, D J P, Oertel, G F, Tillman, R W and Thorne, J A (eds) *Shelf Sand and Sandstone Bodies – Geometry, Facies and Sequence Stratigraphy*. Special Publication No. 14, International Association of Sedimentologists, Blackwell Scientific Publications, Oxford, 3–31.

Swift D J P, Sanford, R., Dill, C E and Avignone, N, 1971, Textural differentiation on the shoreface during erosion retreat of an unconsolidated coast, Cape Henry to Cape Hatteras, western North Atlantic Shelf. *Sedimentology*, 16, 221–250.

Swift, D J P, Figueiredo, A G Jr, Freeland, G L and Oertel, G F, 1983, Hummocky cross stratification and megaripples: a geological double standard? *Journal of Sedimentary Petrology*, 53, 1295–1317.

Swift, D J P, Phillips, S and Thorne, J A, 1991, Sedimentation on continental margins, IV, lithofacies and depositional systems. In Swift, D J P, Oertel, G F, Tillman, R W and Thorne, J A (eds) *Shelf Sand and Sandstone Bodies – Geometry, Facies and Sequence Stratigraphy*. Special Publication No. 14, International Association of Sedimentologists, Blackwell Scientific Publications, Oxford, 89–152.

Symonds, G and Bowen, A J, 1984, Interactions of nearshore bars with incoming wave groups. *Journal of Geophysical Research*, 89, 1953–1959.

Symonds, G, Huntley, D A and Bowen, A J, 1982, Two-dimensional surf beat, Long wave generation by a time-varying breakpoint. *Journal of Geophysical Research*, 87, 492–498.

Synolakis, C E, 1987, The runup of solitary waves. *Journal of Fluid Mechanics*, 185, 523–545.

Takeda, I and Sunamura, T, 1982, Formation and height of berms. *Transactions, Japanese Geomorphological Union*, 3, 145–157.

Takeda, I and Sunamura, T, 1983, A wave flume experiment of beach steps. *Annual Report the Institute Geoscience*, University of Tsukuba, 9, 45–48.

Tanner, W F, 1988, Beach ridge data and sea level history from the Americas. *Journal of Coastal Research*, 4, 81–91.

Taylor, M and Stone, G W, 1996, Beach ridges, a review. *Journal of Coastal Research*, 12, 612–621.

Tedesco, L P, Wanless, H R, Scusa, L A, Risi, J A and Gelsanliter, S, 1995, Impact of Hurricane Andrew on south Florida's sandy coastlines. *Journal of Coastal Research*, Special Issue 21, 59–82.

Terwindt, J H J and Battjes, J A, 1991, Research on large-scale coastal behaviour. *Proceedings 22nd International Conference on Coastal Engineering*, New York. ASCE, pp. 1975–1983.

Thevenot, M M and Kraus, N C, 1995, Longshore sand waves at Southampton Beach, New York, observation and numerical simulation of their movement. *Marine Geology*, 126, 249–269.

Thieler, E R, Brill, A L, Cleary, W J, Hobbs, C H and Gammisch, R A, 1995, Geology of the Wrightsville Beach, North Carolina shoreface: implications for the concept of shoreface profile of equilibrium. *Marine Geology*, 126, 271–287.

Thom, B G, 1984, Transgressive and regressive stratigraphies of coastal sand barriers in eastern Australia. *Marine Geology*, 56, 137–158.

Thom, G B (ed.), 1984a, *Coastal Geomorphology in Australia*. Academic Press, Sydney.

Thom, B G and Hall, W, 1991, Behaviour of beach profiles during accretion and erosion dominated periods. *Earth Surface Processes and Landforms*, 16, 113–127.

Thom, B G and Roy, P S, 1985, Relative sea levels and coastal sedimentation in southeast Australia in the Holocene. *Journal of Sedimentary Petrology*, 55, 257–264.

Thom, B G, Polach, H A and Bowman, G M, 1978, *Holocene Age Structure of Coastal Sand Barriers in NSW, Australia*. Geography Department Report, RMC Duntroon, Canberra, Australia, 86 pp.

Thom, B G, Bowman, G, Gillespie, R, Temple, R and Barbetti, M, 1981, *Radiocarbon dating of Holocene beach ridge sequences in south east Australia*. Monograph 11, Department of Geography, Royal Military College, Canberra, 36 pp.

Thom, B G, Shepherd, M J, Ly, C, Roy, P S, Bowman, G M and Hesp, P A, 1992, *Coastal Geomorphology and Quaternary Geology of the Port Stephens–Myall Lakes Area*. Department of Biogeography and Geomorphology, ANU Monograph No. 6, Canberra, 407 pp.

Thom, B G, Hesp, P A and Bryant, T, 1994, Last glacial 'coastal dunes' in Eastern Australia and implications for landscape stability during the Last Glacial Maximum. *Palaeogeography, Palaeoclimatology, Palaeoecology*, 111, 229–248.

Thompson, C H, 1983, Development and weathering of large parabolic dune systems along the subtropical coast of eastern Australia. *Zietschrift fur Geomorphologie* NF, Supplementband 45, 205–225.

Thompson, C H and Bowman, G M, 1984, Subaerial denudation and weathering of vegetated coastal dunes in Eastern Australia. In Thom, B G (ed.) *Coastal Geomorphology in Australia*. Academic Press, Australia, 263–290.

Thompson, W O, 1937. Original structures of beaches, bars and dunes. *Geological Society of American Bulletin*, 48, 723–752.

Thorne, J A and Swift, D J P, 1991, Sedimentation on continental margins, VI, a regime model for depositional sequences, their component system tracts, and bounding surfaces. In Swift, D J P, Oertel, G F, Tillman, R W and Thorne, J A (eds) *Shelf Sand and Sandstone Bodies – Geometry, Facies and Sequence Stratigraphy*. Special Publication 14, International Association of Sedimentologists, Blackwell Scientific Publications, Oxford, 189–255.

Thornton, E B and Abdelrahman, S, 1991, Sediment transport in the swash due to obliquely incident wind-waves modulated by infragravity waves. *Proceedings Coastal Sediments '91*. ASCE, 100–113.

Thornton, E B and Guza, R T, 1982, Energy saturation and phase speeds measured on a natural beach. *Journal of Geophysical Research*, 87, 9499–9508.

Thornton, E B and Guza, R T, 1983, Transformation of wave height distribution. *Journal of Geophysical Research*, 88, 5925–5938.

Thornton, E B and Guza, R T, 1986, Surf zone longshore currents and random waves, field data and models. *Journal of Physical Oceanography*, 16, 1165–1178.

Thornton, E B and Guza, R T, 1989, Models for surf zone dynamics. In Seymour, R J (ed.) *Nearshore Sediment Transport*. Plenum Press, New York, 337–369.

Thornton, E B and Kim, C S, 1993, Longshore current and wave height modulation at tidal frequency inside the surf zone. *Journal of Geophysical Research*, 98, 16 509–16 519.

Thornton, E B and Lippman, T C, 1994, Longshore currents measured in the field. Abstracts, *Coastal Dynamics '94*. ASCE, 121–122.

Tilmans, W, 1979, *Zandtransport in de Golfrichting in Relatief Ondiep Water*. Masters Thesis, Delft University of Technology.

Tinley, K L, 1985, *Coastal Dunes of South Africa*. South African National Scientific Programme, Report No. 109, FRD CSIR, Pretoria, South Africa, 300 pp.

Todd, D K, 1980, *Groundwater Hydrology*, second edition. John Wiley, Chichester, 530 pp.

Trask, P D, 1955, *Movement of Sand around Southern California Promontories*. Beach Erosion Board, Technical Memorandum No. 76, 60 pp.

Trowbridge, J and Madsen, O S, 1984, Turbulent wave boundary layers. 2. Second-order theory and mass transport. *Journal of Geophysical Research*, 89(C5), 7999–8007.

Tsoar, H and Blumberg, D, 1991, The effect of sea cliffs on inland encroachment of aeolian sand. *Acta Mechanica*, Supplement, 2, 131–146.

Tucker, MH, 1950, Surf beat-sea waves of 1 to 5 min period. *Proceedings Royal Society of London*, 202, 565–573.

Tucker, M J, Carr, A P and Pitt, E G, 1983, The effect of an offshore bank in attenuating waves. *Coastal Engineering*, 7, 133–144.

Tunstall, E B and Inman, D L, 1975, Vortex generation by oscillatory flow over rippled surfaces. *Journal of Geophysical Research*, 80, 3475–3484.

Turner, I L, 1993a, Water table outcropping on macro-tidal beaches, a simulation model. *Marine Geology*, 115, 227–238.

Turner, I L, 1993b, The total water content of sandy beaches. *Journal of Coastal Research*, Special Issue 15, 11–26.

Turner, I L, 1994, *Watertable outcropping on macro-tidal beaches and its influence on the morphology of the intertidal prile*. Unpublished PhD Thesis, Coastal Studies Unit, University of Sydney, 288 pp.

Turner, I L, 1995a, Simulating the influence of groundwater seepage on sediment transported by the sweep of the swash zone across macro-tidal beaches. *Marine Geology*, 125, 153–174.

Turner, I L, 1995b, Simulating the time-varying extent of groundwater seepage on tidal beaches. *Software and Technical Bulletin, Earth Surface Processes and Landforms*, 20, 833–843.

Turner, I L and Masselink, G, 1998, Swash infiltration and exfiltration and sediment transport. *Journal of Geophysical Research*, 103, 30 813–30 824.

Turner, I L and Nielsen, P, 1997, Rapid water table fluctuations within the beachface: implications for swash zone sediment mobility? *Coastal Engineering*, 32, 45–60.

Turner, I L, Coates, B P and Acworth, R I, 1995, Tides, waves and the super-elevation of groundwater at the coast. *Journal of Coastal Research*, 13, 46–60.

Ursell, F, 1952, Edge waves on a sloping beach. *Proceedings of the Royal Society of London*, Series A, 214, 79–97.

Ursell, F, 1953, The long wave paradox in the theory of gravity waves. *Proceedings of Cambridge Philosophical Society*, 49, 685–692.

US Army, 1954, *Shore Protection Planning and Design*. Beach Erosion Board, Technical Report 4, US Army Corps of Engineers, Vicksburg.

van Bendegom, L, 1949, *Consideration of the fundamentals of coastal protection*. PhD Thesis, Technical University, Delft, The Netherlands (in Dutch).

van de Meene, J W H, 1994, *The Shoreface-Connected Ridges Along the Central Dutch Coast*. Netherlands Geographical Studies 174, 1222 pp.

van der Merwe, D, 1988, *The effects of off-road vehicles (ORV's) on coastal ecosystems – A review*. Institute for Coastal Research Report No. 17, University of Port Elizabeth, South Africa.

van der Meulen, F and Jungerius, P D, 1989, Landscape development in Dutch coastal dunes, the breakdown and restoration of geomorphological and geohydrological processes. *Proceedings of the Royal Society of Edinburgh*, 96B, 219–229.

van Dieren, J W, 1934, *Organogene Dunenbildung*. Haag.

van Dijk, D and Law, J, 1995, Sublimation and aeolian sand movement from a frozen surface, experimental results from Presqu'ile Beach, Ontario. *Geomorphology*, 11, 177–187.

van Rijn, L C, 1993, *Principles of Sediment Transport in Rivers, Estuaries and Coastal Seas*. Aqua Publications, Amsterdam.

van Rijn, L C and Kroon, A, 1993, Sediment transport by currents and waves. *Proceedings 23rd International Conference on Coastal Engineering*. ASCE, 2613–2628.

van Straaten, LM J, 1965, Coastal barrier deposits in south and north Holland – in particular in the area around Scheveningen and Ijmuiden. *Meded. Geol. Sticht.*, NS, 17, 41–75.

Vellinga, P, 1986, *Beach and dune erosion during storm surges*. Delft Hydraulics Communications No. 372, 169 pp.

Verstappen, H Th, 1968, On the origin of longitudinal (seif) dunes. *Zeitschrift fur Geomorphologie*, NF, 12, 200–220.

Viles, H and Spencer, T, 1995, *Coastal Problems*. Edward Arnold, London, 350 pp.

Vincent, C E and Green, M O, 1990, Field measurements of the suspended sand concentration profiles and fluxes and of the resuspension coefficient $g_0$ over a rippled bed. *Journal of Geophysical Research*, 95, 11 591–11 601.

Vincent, C E and Osborne, P D, 1993, Bedform dimensions and migration rates under shoaling and breaking waves. *Continental Shelf Research*, 13, 1267–1280.

Waddell, E, 1976, Swash–groundwater–beach profile interactions. In Davis, R A and Ethington, R L (eds) *Beach and Nearshore Sedimentation*. SEPM 24, 115–125.

Wal, A and McManus, J, 1993, Wind regime and sand transport on a coastal beach-dune complex, Tentsmuir, eastern Scotland. In Pye, K (ed.) *The Dynamics and Environmental Context of Aeolian Sedimentary Systems*. Geological Society, Special Publication 72, 159–171.

Walcott, R I, 1972, Past sea levels, eustasy and deformation of the earth. *Quaternary Research*, 2, 1–14.

Walker, Jr 1974, *Recreational Surf Parameters*. Look Laboratory Report TR-30, Department of Ocean Engineering, University of Hawaii, Honolulu.

Walker, R G, 1979, Facies models 7: Shallow marine sands. In Walker, R G (ed.) *Facies Models*. Geoscience Canada Reprint Series 1, Geology Association Canada, 75–89.

Walker, H J and Matsukura, Y, 1979, Barchans and barchan-like dunes as developed in two contrasting areas with restricted source regions. *Annual Report, Institute of Geoscience, University of Tsukuba*, 5, 43–46.

Ward, E M, 1922, *English Coastal Evolution*. Methuen, London, 262 pp.

Ward, J D, 1987, *The Cainozoic succession in the Kuiseb valley central Namib desert*. Memoir 9, Geological Survey, Department of Economic Affairs, South West Africa/Namibia, 124 pp.

Wasson, R J and Nannings, P M, 1986, Estimating wind transport of sand on vegetated surfaces. *Earth Surface Processes and Landforms*, 11, 505–514.

Watters, G Z, Manam, V P and Rao, A M, 1971, Hydrodynamic effects of seepage on bed particles. *Journal of the Hydraulics Division*. ASCE, 97(HY3), 421–439.

Watts, G M and Dearduff, R F, 1954, *Laboratory Study of the Effect of Tidal Action on Wave-Formed Beach Profiles*. US Army Corps of Engineers, Beach Erosion Board, Technical Memorandum 52, 21 pp.

Weggel, J R, 1972, Maximum breaker height. *Journal of the Waterways and Harbours Division*, ASCE, 98, 529–548.

Weise, B R and White, W A, 1991, *Padre Island National Seashore. A Guide to the Geology, Natural Environments, and History of a Texas Barrier Island*. Bureau of Economic Geology, Guidebook 17, 94 pp.

Weisser, P and Backer, A P 1983, Monitoring beach and dune advancement and vegetation changes 1937–1977 at the farm Twinstreams, Mtunzini, Natal, RSA. In McLachlan, A and Erasmus, T (eds) *Sandy Beaches as Ecosystems*. Junk, The Hague, 727–732.

Wells, D R, 1967, Beach equilibrium and second order wave theory. *Journal of Geophysical Research*, 72, 497–504.

Werner, B T and Fink, T M, 1993, Beach cusps as self-organised patterns. *Science*, 260, 968–970.

White, B R, 1979, Soil transport by winds on Mars. *Journal of Geophysical Research* 84, 4643–4651.

White, T E and Inman, D L, 1989, Measuring longshore transport with tracers. In Seymour, R J (ed.) *Nearshore Sediment Transport*. Plenum Press, New York, 287–312.

Whitford, D J and Thornton, E B, 1993, Comparison of wind and wave forcing of longshore currents. *Continental Shelf Research*, 13, 1205–1218.

Whitham, G B, 1958, On the propagation of shock waves through regions of non-uniform area or flow. *Journal of Fluid Mechanics*, 4, 337–360.

Whittow, J, 1984, *The Penguin Dictionary of Physical Geography*. Penguin Books, Harmondsworth, 591 pp.

Wiberg, P L and Harris, C K, 1994, Ripple geometry in wave-dominated environments. *Journal of Geophysical Research*, 99, 775–789.

Wiegel, R L, 1953, *Waves, Tides, Currents and Beaches, Glossary of Terms and List of Standard Symbols*. US Council on

Wave Research, The Engineering Foundation, 113 pp.

Wiegel, R L, 1964, *Oceanographical Engineering*. Prentice-Hall, Englewood Cliffs, 532 pp.

Willetts, B B and Drossos, M E 1975, Local erosion caused by rapid forced infiltration. *Journal of the Hydraulics Division*, ASCE, 13, 1478–1488.

Willetts, B B and Rice, M A, 1986, Collision in aeolian transport, the saltation/creep link. In Nickling, W G (ed.) *Aeolian Geomorphology*, Proceedings 17th Binghampton Symposium, Guelph. Allen and Unwin, Boston, 1–17.

Williams, W W, 1960, *Coastal Change*. Routledge and Kegan Paul, London, 220 pp.

Williamson, J A, Fenner, P J, Burnett, J W and Rifkin, J F, 1996, *Venomous and Poisonous Marine Animals*. University of New South Wales Press, Sydney, 504 pp.

Willis, D H, 1987, The Canadian Coastal Sediment Study, an overview. *Coastal Sediments '87*, New Orleans. ASCE, 682–693.

Wilson, J D, Clarkson, T S and Hadfield, M G, 1984, Observations of wind flow and tracer gas dispersion over sand dunes. *New Zealand Journal of Science*, 27, 237–242.

Winant, C D, Inman, D L and Nordstrom, C E, 1975, Description of seasonal beach changes using empirical eigenfunctions. *Journal of Geophysical Research*, 80, 1979–1986.

Winter, D F, 1983, A theoretical model of surf zone circulation and diatom growth. In McLachlan, A and Erasmus, T (eds) *Sandy Beaches as Ecosystems*. Junk, The Hague, 157–167.

Wiseman, W J, Coleman, J M, Gregory, A, Hsu, S A, Short, A D, Suhayda, J N, Walters, C D and Wright, L D, 1973, *Alaskan Arctic Coastal Processes and Morphology*. Technical Report No. 149, Coastal Studies Institute, Louisiana State University, Baton Rouge, 171 pp.

Wolfe, S A and Nickling, W G, 1993, The protective role of sparse vegetation in wind erosion. *Progress in Physical Geography*, 17, 50–68.

Wolman, M G, 1955, *The natural channel of Brandywine Creek, Pennsylvania*. US Geological Survey Professional Paper 271, 128 pp.

Wolman, M G and Gerson, R, 1978, Relative scales of time and effectiveness of climate in watershed geomorphology. *Earth Surface Processes*, 3, 198–208.

Woodroffe, C D, Chappell, J, Thom, B G and Wallensky, E, 1985, Geomorphology of the South Alligator tidal river and plains, Northern Territory. In Bardsley, K N, Davie, J O S and Woodroffe, C D (eds) *Coasts and Tidal Wetlands of the Australian Monsoon Region*. Mangrove Monograph 1, Australian National University, Northern Ausralian Research Unit, Darwin, 3–15.

Wright, L D, 1970, The influence of sediment availability on patterns of beach ridge development in the vicinity of the Shoalhaven River delta, NSW. *Australian Geographer*, 11, 336–348.

Wright, L D, 1976, Nearshore wave-power dissipation and the coastal energy regime of the Sydney-Jervis Bay region, New South Wales, a comparison. *Australian Journal of Marine and Freshwater Research*, 27, 633–640.

Wright, L D, 1980, Beach cut in relation to surf zone morphodynamics. *Proceedings 17th International Conference on Coastal Engineering*. ASCE, 978–996.

Wright, L D, 1982, Field observations of long-period, surf-zone standing waves in relation to contrasting beach morphologies. *Australian Journal of Marine and Freshwater Research*, 33, 181–201.

Wright, L D, 1987, Shelf-surfzone coupling: diabathic shoreface transport. *Coastal Sediments '87, Proceedings Special Conference*, ASCE, 25–40.

Wright, L D, 1995, *Morphodynamics of Inner Continental Shelves*. CRC Press, Boca Raton, 241 pp.

Wright, L D and Short, A D, 1983, Morphodynamics of beaches and surf zones in Australia. In Komar, P D (ed.) *Handbook of Coastal Processes and Erosion*. CRC Press, Boca Raton, 35–64.

Wright, L D and Short, A D, 1984, Morphodynamic variability of beaches and surf zones, a synthesis. *Marine Geology*, 56, 92–118.

Wright, L D and Thom, B G, 1977, Coastal depositional landforms, a morphodynamic approach, *Progress in Physical Geography*, 1, 412–59.

Wright, L D, Coleman, J M and Thom, B G, 1973, Processes of channel development of a high-tide range environment, Cambridge Gulf–Ord River Delta, Western Australia. *Journal of Geology*, 81, 15–41.

Wright, L D, Chappell, J, Thom, B G, Bradshaw, M P and Cowell, P J, 1979, Morphodynamics of reflective and dissipative beach and inshore systems, Southeastern Australia. *Marine Geology*, 32, 105–140.

Wright, L D, Guza, R T and Short, A D, 1982a, Dynamics of a high-energy dissipative surf zone. *Marine Geology*, 45, 41–62.

Wright, L D, Nielsen, P, Short, A D and Green, M O, 1982b, Morphodynamics of a macrotidal beach. *Marine Geology*, 50, 97–128.

Wright, L D, Nielsen, P, Short, A D, Coffey, F C and Green, M O, 1982c, *Nearshore and surf zone morphodynamics of a storm wave environment: eastern Bass Strait, Australia*. Coastal Studies Unit Technical Report 82/3, Department of Geography, University of Sydney, 154 pp.

Wright, L D, May, S K, Short, A D and Green, M O, 1984, Beach and surf zone equilibria and response time. *Coastal Engineering Conference*, Houston. ASCE, 2150–2164.

Wright, L D, Boon, J D, Green, M O and List, J H, 1986a, Response to the mid-shoreface of the southern Mid-Atlantic Bight to a 'northeaster'. *Geo-Marine Letters*, 6, 153–160.

Wright, L D, Nielsen, P, Shi, N C and List, J H, 1986b, Morphodynamics of a bar-trough surf zone. *Marine Geology*, 70, 251–285.

Wright, L D, Short, A D, Boon, J D, Kimball, S and List, J H, 1987, The morphodynamic effects of incident wave groupiness and tide range on an energetic beach. *Marine Geology*, 74, 1–20.

Wright, L D, Boon, J D, Kim, S C and List, J H, 1991, Modes of cross-shore sediment transport on the shoreface of the Middle Atlantic Bight. *Marine Geology*, 96, 19–51.

Wright, L D, Xu, J P and Madsen, O S, 1994a, Across-shelf

benthic transports on the inner shelf of the Middle Atlantic Bight during the Halloween storm of 1991. *Marine Geology*, 118, 61–77.

Wright, L D, Madsen O S and Chisholm T A, 1994b, Inner continental shelf bottom boundary layer processes and cross shelf sediment transport: Field experiments in the Middle Atlantic Bight. *Proceedings of Coastal Dynamics '94*. ASCE.

Wright, L D, Short, A D and Green, M, 1995, Short-term changes in the morphodynamic state of beaches and surf zones, an empirical predicative model. *Marine Geology*, 62, 339–364.

Wright, P, 1984, Facies development on a barred (ridge and runnel) coastline. The case of south-west Lancashire (Merseyside), U.K. In Clark, M W (ed.) *Coastal Research, U.K. Perspectives*. GeoBooks, Norwich, 105–118.

Yanagishima, S and Katoh, K, 1990, Field observation on wave set-up near the shoreline. *Proceedings 22nd International Conference on Coastal Engineering*. ASCE, 95–108.

Yasso, W E, 1965, Plan geometry of headland-bay beaches. *Marine Geology*, 73, 702–714.

Yeh, H H, Ghazali, A and Marton, I, 1989, Experimental study of bore run-up. *Journal of Fluid Mechanics*, 206, 563–578.

Young, D R, Shao, G and Brinson, M, 1995, The impact of the October 1991 northeaster storm on barrier island shrub thickets (*Myrica cerifera*). *Journal of Coastal Research*, 11, 1322–1328.

Young, I R and Holland, G J, 1996, *Atlas of the Oceans, Wind and Wave Climate*. Elsevier, Oxford, 241 pp.

Young, R W, Bryant, E A, Price, D M, Wirth, L M and Pease, M, 1993, Theoretical constraints and chronological evidence of Holocene coastal development in central and southern NSW, Australia. *Geomorphology*, 7, 317–329.

Yu, Y, Sternberg, R W and Beach, R A, 1993, Kinematics of breaking waves and associated suspended sediment in the nearshore zone. *Continental Shelf Research*, 13, 1219–1242.

Zauscher, B, 1997, The beaches of Puerto Colombia. *American Lifeguard Magazine*, 13, 23–25.

Zenkovich, V P, 1946. On the study of shore dynamics. *Akademiya nauk U.S.S.R., Institut. Okeandogi, Trudy*, 1, 99–112

Zenkovich, V P, 1967, *Processes of Coastal Development*. Oliver and Boyd, London, 738 pp.

Zenkovich, V P, 1969, Origin of barrier beaches and lagoon coast. *Lagunas Costeras, Un Simposio, Mem. Simp. Intern. lagunas Costeras*. UNAM-UNESCO, 27–37.

Zhang, D P, Sunamura, T, Tanaka, S and Yamamoto, K, 1994, Laboratory experiments of longshore bars produced by breaker-induced vortex action. *Proceedings Coastal Dynamics '94*. ASCE, 29–43.

Zi-Fang Lee, P, 1994, The submarine equilibrium profile, a physical model. *Journal of Coastal Research*, 10, 1–17.

# Author index

# Location index

# Subject index